Quantum Gravity

Quantum gravity is perhaps the most important open problem in fundamental physics. It is the problem of merging quantum mechanics and general relativity, the two great conceptual revolutions in the physics of the twentieth century.

This book discusses the many aspects of the problem, and presents technical and conceptual advances towards a background-independent quantum theory of gravity, obtained in the last two decades. The first part of the book is an exploration on how to re-think basic physics from scratch in the light of the general-relativistic conceptual revolution. The second part is a detailed introduction to *loop quantum gravity* and the *spinfoam* formalism. It provides an overview of the current state of the field, including results on area and volume spectra, dynamics, extension of the theory to matter, applications to early cosmology and black-hole physics, and the perspectives for computing scattering amplitudes. The book is completed by a historical appendix which overviews the evolution of the research in quantum gravity, from the 1930s to the present day.

CARLO ROVELLI was born in Verona, Italy, in 1956 and obtained his Ph.D. in Physics in Padua in 1986. In 1996, he was awarded the Xanthopoulos International Prize, for the development of the loop approach to quantum gravity and for research on the foundation of the physics of space and time. Over the years he has taught and worked in the University of Pittsburgh, Université de la Méditerranée, Marseille, and Università La Sapienza, Rome. Professor Rovelli's main research interests lie in general relativity, gravitational physics, and the philosophy of space and time. He has had over 100 publications in international journals in physics and has written contributions for major encyclopedias. He is senior member of the Institut Universitaire de France.

CAMBRIDGE MONOGRAPHS ON
MATHEMATICAL PHYSICS

General editors: P. V. Landshoff, D. R. Nelson, S. Weinberg

[†] Issued as a paperback

Quantum Gravity

CARLO ROVELLI

Centre de Physique Théorique de Luminy
Université de la Méditerranée, Marseille

CAMBRIDGE
UNIVERSITY PRESS

CAMBRIDGE UNIVERSITY PRESS
Cambridge, New York, Melbourne, Madrid, Cape Town,
Singapore, São Paulo, Delhi, Tokyo, Mexico City

Cambridge University Press
The Edinburgh Building, Cambridge CB2 8RU, UK

Published in the United States of America by
Cambridge University Press, New York

www.cambridge.org
Information on this title: www.cambridge.org/9780521715966

First published 2004
Reprinted 2005
First paperback edition published with correction 2008
Reprinted 2010

A catalogue record for this publication is available from the British Library

Library of Congress Cataloguing in Publication data

ISBN 978-0-521-83733-0 Hardback
ISBN 978-0-521-71596-6 Paperback

Contents

Foreword

The problem of what happens to classical general relativity at the extreme short-distance Planck scale of 10^{-33} cm is clearly one of the most pressing in all of physics. It seems abundantly clear that profound modifications of existing theoretical structures will be mandatory by the time one reaches that distance scale. There exist several serious responses to this challenge. These include effective field theory, string theory, loop quantum gravity, thermogravity, holography, and emergent gravity. Effective field theory is to gravitation as chiral perturbation theory is to quantum chromodynamics – appropriate at large distances, and impotent at short. Its primary contribution is the recognition that the Einstein–Hilbert action is no doubt only the first term in an infinite series constructed out of higher powers of the curvature tensor. String theory emphasizes the possible roles of supersymmetry, extra dimensions, and the standard-model internal symmetries in shaping the form of the microscopic theory. Loop gravity most directly attacks the fundamental quantum issues, and features the construction of candidate wave-functionals which are background independent. Thermogravity explores the apparent deep connection of semiclassical gravity to thermodynamic concepts such as temperature and entropy. The closely related holographic ideas connect theories defined in bulk spacetimes to complementary descriptions residing on the boundaries. Finally, emergent gravity suggests that the time-tested symbiotic relationship between condensed matter theory and elementary particle theory should be extended to the gravitational and cosmological contexts as well, with more lessons yet to be learned.

In each of the approaches, difficult problems stand in the way of attaining a fully satisfactory solution to the basic issues. Each has its band of enthusiasts, the largest by far being the string community. Most of the approaches come with rather strong ideologies, especially apparent when they are popularized. The presence of these ideologies tends to isolate the communities from each other. In my opinion, this is extremely unfortunate, because it is probable that all these ideologies, including my own (which is distinct from the above listing), are dead wrong. The evidence is history: from the Greeks to Kepler to Newton to Einstein there has been no shortage of grand ideas regarding the Basic Questions. In the presence of new data available to us and not them, only fragments of those grand visions remain viable. The clutter of thirty-odd standard model parameters and the descriptive nature of modern cosmology suggests that we too have quite a way to go before ultimate simplicity is attained. This does not mean abandoning ideologies – they are absolutely essential in driving us all to work hard on the problems. But it does mean that an attitude of humility and of high sensitivity toward alternative approaches is essential.

This book is about only one approach to the subject – loop quantum gravity. It is a subject of considerable technical difficulty, and the literature devoted to it is a formidable one. This feature alone has hindered the cross-fertilization which is, as delineated above, so essential for progress. However, within these pages one will find a much more accessible description of the subject, put forward by one of its leading architects and deepest thinkers. The existence of such a fine book will allow this important subject, quite likely to contribute significantly to the unknown ultimate theory, to be assimilated by a much larger community of theorists. If this does indeed come to pass, its publication will become one of the most important developments in this very active subfield since its onset.

James Bjorken

Preface

A dream I have long held was to write a "treatise" on quantum gravity once the theory had been finally found and experimentally confirmed. We are not yet there. There is neither experimental support nor sufficient theoretical consensus. Still, a large amount of work has been developed over the last twenty years towards a quantum theory of spacetime. Many issues have been clarified, and a definite approach has crystallized. The approach, variously denoted,[1] is mostly known as "loop quantum gravity."

The problem of quantum gravity has many aspects. Ideas and results are scattered in the literature. In this book I have attempted to collect the main results and to present an overall perspective on quantum gravity, as developed during this twenty-year period. The point of view is personal and the choice of subjects is determined by my own interests. I apologize to friends and colleagues for what is missing; the reason so much is missing is due to my own limitations, for which I am the first to be sorry.

It is difficult to over-estimate the vastitude of the problem of quantum gravity. The physics of the early twentieth century has modified the very foundation our understanding of the physical world, changing the meaning of the basic concepts we use to grasp it: matter, causality, space and time. We are not yet able to paint a consistent picture of the world in which these modifications taken together make sense. The problem of quantum gravity is nothing less than the problem of finding this novel consistent picture, finally bringing the twentieth century scientific revolution to an end.

Solving a problem of this sort is not just a matter of mathematical skill. As was the case with the birth of quantum mechanics, relativity, electromagnetism, and newtonian mechanics, there are conceptual and foundational problems to be addressed. We have to understand which (possibly

---

[1] See the notation section.

new) notions make sense and which old notions must be discarded, in
order to describe spacetime in the quantum relativistic regime. What we
need is not just a technique for computing, say, graviton–graviton scattering amplitudes (although we certainly want to be able to do so, eventually). We need to re-think the world in the light of what we have learned
about it with quantum theory and general relativity.

General relativity, in particular, has modified our understanding of the
spatio-temporal structure of reality in a way whose consequences have
not yet been fully explored. A significant part of the research in quantum
gravity explores foundational issues, and Part I of this book ("Relativistic
foundations") is devoted to these basic issues. It is an exploration of how
to rethink basic physics from scratch, after the general-relativistic conceptual revolution. Without this, we risk asking any tentative quantum
theory of gravity the wrong kind of questions.

Part II of the book ("Loop quantum gravity") focuses on the loop
approach. The loop theory, described in Part II, can be studied by itself,
but its reason and interpretation are only clear in the light of the general
framework studied in Part I. Although several aspects of this theory are
still incomplete, the subject is mature enough to justify a book. A theory
begins to be credible only when its original predictions are reasonably
unique and are confirmed by new experiments. Loop quantum gravity is
not yet credible in this sense. Nor is any other current tentative theory of
quantum gravity. The interest of the loop theory, in my opinion, is that
at present it is the only approach to quantum gravity leading to well-
defined physical predictions (falsifiable, at least in principle) and, more
importantly, it is the most determined effort for a genuine merging of
quantum field theory with the world view that we have discovered with
general relativity. The future will tell us more.

There are several other introductions to loop quantum gravity. Classic reports on the subject [1–10, in chronological order] illustrate various
stages of the development of the theory. For a rapid orientation, and to
appreciate different points of view, see the review papers [11–15]. Much
useful material can be found in [16]. Good introductions to spinfoam theory are to be found in [11, 17–19]. This book is self-contained, but I have
tried to avoid excessive duplications, referring to other books and review
papers for nonessential topics well developed elsewhere. This book focuses
on physical and conceptual aspects of loop quantum gravity. Thomas
Thiemann's book [20], which is going to be completed soon, focuses on
the mathematical foundation of the same theory. The two books are complementary: this book can almost be read as Volume 1 ("Introduction and
conceptual framework") and Thiemann's book as Volume 2 ("Complete
mathematical framework") of a general presentation of loop quantum
gravity.

The book assumes that the reader has a basic knowledge of general relativity, quantum mechanics and quantum field theory. In particular, the aim of the chapters on general relativity (Chapter 2), classical mechanics (Chapter 3), hamiltonian general relativity (Chapter 4), and quantum theory (Chapter 5) is to offer the fresh perspective on these topics which is needed for quantum gravity to a reader already familiar with the conventional formulation of these theories.

Sections with comments and examples are printed in smaller fonts (see Section 1.3.1 for first such example). Sections that contain side or more complex topics and that can be skipped in a first reading without compromising the understanding of what follows are marked with a star (∗) in the title. References in the text appear only when strictly needed for comprehension. Each chapter ends with a short bibliographical section, pointing out essential references for the reader who wants to go into more detail or to trace original works on a subject. I have given up the immense task of collecting a full bibliography on loop quantum gravity. On many topics I refer to specific review articles where ample bibliographic information can be found. An extensive bibliography on loop quantum gravity is given in [9] and [20].

I have written this book thinking of a researcher interested in working in quantum gravity, but also of a good Ph.D. student or an open-minded scholar, curious about this extraordinary open problem. I have found the journey towards general relativistic quantum physics, towards quantum spacetime, a fascinating adventure. I hope the reader will see the beauty I see, and that he or she will be capable of completing the journey. The landscape is magic, the trip is far from being over.

Preface to the paperback edition

Three years have lapsed since the first edition of this book. During these three years, the research in loop gravity has been developing briskly and in several directions. Remarkable new results are for instance: the proof that spinfoam and hamiltonian loop theory are equivalent in 3d; the proof of the unicity of the loop representation (the "LOST" theorem); the resolution of the $r = 0$ black hole singularity; major advances in loop cosmology; the result that in 3d loop quantum gravity plus matter yields an effective non-commutative quantum field theory; the "master constraint" program for the definition of the quantum dynamics; the idea of deriving particles from linking; the recalculation of the Immirzi parameter from black hole thermodynamics; and, last but not least, the first steps toward calculating scattering amplitudes from the background independent quantum theory. I am certainly neglecting something that will soon turn out to be important...

I have added notes and pointers to recent literature or recent review papers, where the interested reader can find updates on specific topics. In spite of these rapid developments, however, it is too early for a full-fledged second edition of this book: it seems to me that the book, as it is, still provides a comprehensive introduction to the field. In fact, several of these developments reinforce the point of view of this book, namely that the lines of research considered form a coherent picture, and define a common language in which a consistent quantum field theory without (background) space and time can be defined.

When I feel pessimistic, I see the *divergence* between research lines and the impressive number of problems that are still open. When I feel optimistic, I see their remarkable *coherence*, and I dream we might be, with respect to quantum gravity, as Einstein was in 1914: with all the machinery ready, trying a number of similar field equations... Then, it seems to me that a quantum theory of gravity (certainly not the final theory of everything) is truly at hand; maybe we have it, maybe what we need is just the right combination of techniques, a few more details, or one last missing key idea.

Once again, my wish is that among the readers of this paperback edition there is she or he, who will give us this last missing idea.

Acknowledgements

I am indebted to the many people that have sent suggestions and corrections to the draft of this book posted online and to its first edition. Among them are M Carling, Alexandru Mustatea, Daniele Oriti, John Baez, Rafael Kaufmann, Nedal, Colin Hayhurst, Jürgen Ehlers, Chris Gauthier, Gianluca Calcagni, Tomas Liko, Chang Chi-Ming, Youngsub Yoon, Martin Bojowald and Gen Zhang. Special thanks in particular to Justin Malecki, Jacob Bourjaily and Leonard Cottrell.

My great gratitude goes to the friends with whom I have had the privilege of sharing this adventure:

To Lee Smolin, companion of adventures and friend. His unique creativity and intelligence, intellectual freedom and total honesty, are among the very best things I have found in life.

To Abhay Ashtekar whose tireless analytical rigor, synthesis capacity and leadership have been a most precious guide. Abhay has solidified our ideas and transformed our intuitions into theorems. This book is a result of Lee's and Abhay's ideas and work as much as my own.

To Laura Scodellari and Chris Isham, my first teachers who guided me into mathematics and quantum gravity.

To Ted Newman, who, with Sally, parented the little boy just arrived from the Empire's far provinces. I have shared with Ted a decade of intellectual joy. His humanity, generosity, honesty, passion and love for thinking, are the example against which I judge myself.

I would like to thank one by one all the friends working in this field, who have developed the ideas and results described in this book, but they are too many. I can only mention my direct collaborators, and a few friends outside this field: Luisa Doplicher, Simone Speziale, Thomas Schucker, Florian Conrady, Daniele Colosi, Etera Livine, Daniele Oriti, Florian Girelli, Roberto DePietri, Robert Oeckl, Merced Montesinos, Kirill Krasnov, Carlos Kozameh, Michael Reisenberger, Don Marolf,

Berndt Brügmann, Junichi Iwasaki, Gianni Landi, Mauro Carfora, Jorma Louko, Marcus Gaul, Hugo Morales-Tecotl, Laurent Freidel, Renate Loll, Alejandro Perez, Giorgio Immirzi, Philippe Roche, Federico Laudisa, Jorge Pullin, Thomas Thiemann, Louis Crane, Jerzy Lewandowski, John Baez, Ted Jacobson, Marco Toller, Jeremy Butterfield, John Norton, John Barrett, Jonathan Halliwell, Massimo Testa, David Finkelstein, Gary Horowitz, John Earman, Julian Barbour, John Stachel, Massimo Pauri, Jim Hartle, Roger Penrose, John Wheeler and Alain Connes.

With all these friends I have had the joy of talking about physics in a way far from problem-solving, from outsmarting each other, or from making weapons to make "us" stronger than "them." I think that physics is about escaping the prison of the received thoughts and searching for novel ways of thinking the world, about trying to clear a bit the misty lake of our insubstantial dreams, which reflect reality like the lake reflects the mountains.

Foremost, thanks to Bonnie – she knows why.

Terminology and notation

- In this book, "relativistic" means "*general* relativistic," unless otherwise specified. When referring to *special* relativity, I say so explicitly. Similarly, "nonrelativistic" and "prerelativistic" mean "non-*general*-relativistic" and "pre-*general*-relativistic." The choice is a bit unusual (special relativity, in this language, is "nonrelativistic"). One reason for it is simply to make language smoother: the book is about *general* relativistic physics, and repeating "*general*" every other line sounds too much like a Frenchman talking about de Gaulle. But there is a more substantial reason: the complete revolution in spacetime physics which truly deserves the name of relativity is general relativity, not special relativity. This opinion is not always shared today, but it was Einstein's opinion. Einstein has been criticized on this; but in my opinion the criticisms miss the full reach of Einstein's discovery about spacetime. One of the aims of this book is to defend in modern language Einstein's intuition that his gravitational theory is the full implementation of relativity in physics. This point is discussed at length in Chapter 2.

- I often indulge in the physicists' bad habit of mixing up function f and function values $f(x)$. Care is used when relevant. Similarly, I follow standard physicists' abuse of language in denoting a field such as the Maxwell potential as $A_\mu(x)$, $A(x)$, or A, where the three notations are treated as equivalent manners of denoting the field. Again, care is used where relevant.

- All fields are assumed to be smooth, unless otherwise specified. All statements about manifolds and functions are local unless otherwise specified; that is, they hold within a single coordinate patch. In general I do not specify the domain of definition of functions; clearly equations hold where functions are defined.

- Index notation follows the most common choice in the field: Greek indices from the middle of the alphabet $\mu, \nu, \ldots = 0, 1, 2, 3$ are 4d spacetime tangent indices. Upper case Latin indices from the middle of the alphabet $I, J, \ldots = 0, 1, 2, 3$ are 4d Lorentz tangent indices. (In the special relativistic context the two are used without distinction.) Lower case Latin indices from the beginning of the alphabet $a, b, \ldots = 1, 2, 3$ are 3d tangent indices. Lower case Latin indices from the middle of the alphabet $i, j, \ldots = 1, 2, 3$ are 3d indices in R^3. Coordinates of a 4d manifold are usually indicated as $x, y, \ldots$, while 3d manifold coordinates are usually indicated as $\vec{x}, \vec{y}$ (also as $\vec{\tau}$). Thus the components of a spacetime coordinate x are

$$x^\mu = (t, \vec{x}) = (x^0, x^a);$$

while the components of a Lorentz vector e are

$$e^I = (e^0, e^i).$$

- η_{IJ} is the Minkowski metric, with signature $[-, +, +, +]$. The indices $I, J, \ldots$ are raised and lowered with η_{IJ}. δ_{ij} is the Kronecker delta, or the R^3 metric. The indices $i, j, \ldots$ are raised and lowered with δ_{ij}.

- For reasons explained at the beginning of Chapter 2, I call "gravitational field" the tetrad field $e^I_\mu(x)$, instead of the metric tensor $g_{\mu\nu}(x) = \eta_{IJ}\, e^I_\mu(x) e^J_\nu(x)$.

- ϵ_{IJKL}, or $\epsilon_{\mu\nu\rho\sigma}$, is the completely antisymmetric object with $\epsilon_{0123} = 1$. Similarly for ϵ_{abc}, or ϵ_{ijk}, in 3d. The Hodge star is defined by

$$F^*_{IJ} = \frac{1}{2}\epsilon_{IJKL}\, F^{KL}$$

in flat space, and by the same equation, where $F_{IJ}e^I_\mu e^J_\nu = F_{\mu\nu}$ and $F^*_{IJ}e^I_\mu e^J_\nu = F^*_{\mu\nu}$ in the presence of gravity. Equivalently,

$$F^*_{\mu\nu} = \sqrt{-\det g}\,\frac{1}{2}\epsilon_{\mu\nu\rho\sigma}\, F^{\rho\sigma} = |\det e|\,\frac{1}{2}\epsilon_{\mu\nu\rho\sigma}\, F^{\rho\sigma}.$$

- Symmetrization and antisymmetrization of indices is defined with a half: $A_{(ab)} = \frac{1}{2}(A_{ab} + A_{ba})$ and $A_{[ab]} = \frac{1}{2}(A_{ab} - A_{ba})$.

- I call "curve" on a manifold M, a map

$$\gamma : I \to M$$
$$s \mapsto \gamma^a(s),$$

where I is an interval of the real line R (possibly the entire R.) I call "path" an oriented unparametrized curve, namely an equivalence class of curves under change of parametrization $\gamma^a(s) \mapsto \gamma'^a(s) = \gamma^a(s'(s))$, with $ds'/ds > 0$.

• An orthonormal basis in the Lie algebras $su(2)$ and $so(3)$ is chosen once and for all and these algebras are identified with R^3. For $so(3)$, the basis vectors $(v_i)^j{}_k$ can be taken proportional to $\epsilon_i{}^j{}_k$; for $su(2)$, the basis vectors $(v_i)^A{}_B$ can be taken proportional to the Pauli matrices, see Appendix A1. Thus, an algebra element ω in $su(2) \sim so(3)$ has components ω^i.

• For any antisymmetric quantity v^{ij} with two 3d indices i, j, I use also the one-index notation

$$v^i = \tfrac{1}{2}\,\epsilon^i{}_{jk}\,v^{jk}, \qquad v^{ij} = \epsilon^{ij}{}_k\,v^k;$$

the one-index and the two-indices notation are considered as defining the same object. For instance the $SO(3)$ connections ω^{ij} and A^{ij} are equivalently denoted ω^i and A^i.

Symbols. Here is a list of symbols, with their name and the equation, chapter or section where they are introduced or defined.

A	area	Section 2.1.4
A	Yang–Mills connection	Equation (2.30)
$A, A^i_\mu(x)$	selfdual 4d gravitational connection	Equation (2.19)
$A, A^i_a(\vec{x})$	selfdual or real 3d gravitational connection	Sections 4.1.1, 4.2
$\mathcal{C}$	relativistic configuration space	Section 3.2.1
D_μ	covariant derivative	Equation (2.31)
$Diff^*$	extended diffeomorphism group	Section 6.2.2
$e^I_\mu(x)$	gravitational field	Equation (2.1)
e	determinant of e^I_μ	
e	edge (of spinfoam)	Section 9.1
$E, E^a_i(\vec{x})$	gravitational electric field	Section 4.1.1
f	face (of spinfoam)	Section 9.1
F	curvature two-form	Section 2.1.1
g or U	group element	
G	Newton constant	
$\mathcal{G}$	space of boundary data	Sections 3.2.5– 3.3.3
h_γ	$U(A, \gamma)$	Section 7.1
H	relativistic hamiltonian	Section 3.2
H_0	nonrelativistic (conventional) hamiltonian	Section 3.2
$\mathcal{H}$	quantum state space	Chapter 5
$\mathcal{H}_0$	nonrelativistic quantum state space	Chapter 5

i_n	intertwiner on spin network node n	Section 6.3
i_e	intertwiner on spinfoam edge e	Chapter 9
j	irreducible representation (for $SU(2)$: spin)	
j_l	spin associated to spin network link l	Section 6.2.1
j_f	representation associated to spinfoam face f	Chapter 9
$\mathcal{K}$	kinematical quantum state space	Section 5.2
$\mathcal{K}_0$	$SU(2)$ invariant quantum state space	Section 6.2.3
$\mathcal{K}_{\text{diff}}$	diff-invariant quantum state space	Section 6.2.3
K	boundary quantum space	Sections 5.1.4, 5.3.5
l	link (of spin network)	Section 9.1
l_P	Planck length, $\sqrt{\hbar G c^{-3}}$	
$\mathbf{L}$	length	Section 2.1.4
M	spacetime manifold	
n	node (of spin network)	Section 9.1
p_a	relativistic momenta (including p_t)	Section 3.2
p_t	momentum conjugate to t	Section 3.2
P	the "projector" operator	Section 5.2
P_G	group G projector	Equation (9.117)
P_H	subgroup H projector	Equation (9.119)
$\mathcal{P}$	transition probability	Chapter 5
$\mathcal{P}$	path ordered	Equation (2.81)
q^a	partial observables	Section 3.2
$R^I{}_{J\,\mu\nu}(x)$	curvature	Equation (2.8)
$R^{(j)\alpha}{}_\beta(g)$	matrix of group element g in representation j	
$\mathcal{R}$	3d region	Section 2.1.4
s	s-knot: abstract spin network	Equation (6.4.1)
$|s\rangle$	s-knot state	Equation (6.4.1)
S_{BH}	black-hole entropy	Section 8.2
S	embedded spin network	Section 6.3
$|\mathsf{S}\rangle$	spin network state	Section 6.3.1
$\mathcal{S}$	2d surface	Section 2.1.4
$\mathcal{S}$	space of fast decrease functions	Chapter 5
$\mathcal{S}_0$	space of tempered distributions	Chapter 5
$S[\tilde{\gamma}]$	action functional	Section 3.2
$S(q^a)$	Hamilton–Jacobi function	Section 3.2.2
$S(q^a, q_0^a)$	Hamilton function	Section 3.2.5

t_ρ	thermal time	Sections 3.4, 5.5.1
T	target space of a field theory	Section 3.3.1
U or g	group element	
$U(A, \gamma)$	holonomy	Section 2.1.5
v	vertex (of spinfoam)	Section 9.1
$\mathbf{V}$	volume	Section 2.1.4
$W(q^a, q'^a)$	propagator	Chapter 5
W	transition amplitudes, propagator	Section 5.2
x	4d spacetime coordinates	
$\vec{x}$	3d coordinates	
Z	partition function	Chapter 9
α	loop, closed path	
β	inverse temperature	Section 3.4
γ	path	
γ	motion (in $\mathcal{C}$)	Section 3.2.1
γ	Immirzi parameter	Section 4.2.3
$\tilde{\gamma}$	motion in Ω	Section 3.2
Γ	relativistic phase space	Section 3.2.1
Γ	graph	Section 6.2
Γ	two-complex	Chapter 9
θ	Poincaré–Cartan form on Σ	Section 3.2.2
$\tilde{\theta}$	Poincaré form on Ω	Equation (3.9)
$\eta_{IJ}, \eta_{\mu\nu}$	Minkowski metric $= \text{diag}[-1,1,1,1]$	
λ	cosmological constant	Equation (2.11)
λ	gauge parameter	Section 2.1.3
ρ	statistical state	Sections 3.4, 5.5.1
Σ	constraint surface $H = 0$	Section 3.2.2
σ, Σ	3d boundary surface	Chapter 4
σ	spinfoam	Chapter 9
$\phi(x)$	scalar field	Equation (2.32)
$\psi(x)$	fermion field	Equation (2.35)
ω	presymplectic form on Σ	Section 3.2.2
$\omega^I_{\mu J}(x)$	spin connection	Equation (2.2)
$\tilde{\omega}$	symplectic form on Ω	Section 3.2.2
Ω	space of observables and momenta	Sections 3.2–3.3.2
$\{6j\}$	Wigner $6j$ symbol	Equation (9.33)
$\{10j\}$	Wigner $10j$ symbol	Equation (9.103)
$\{15j\}$	Wigner $15j$ symbol	Equation (9.56)
$|0\rangle$	covariant vacuum in K	Sections 5.1.4, 5.3.5
$|0_t\rangle$	dynamical vacuum in $\mathcal{K}_t$	Sections 5.1.4, 5.3.2
$|0_{\mathrm{M}}\rangle$	Minkowski vacuum in $\mathcal{H}$	Sections 5.1.4, 5.3.1

• *The name of the theory.* Finally, a word about the name of the quantum theory of gravity described in this book. The theory is known as *"loop quantum gravity"* (LQG), or sometimes *"loop gravity"* for short. However, the theory is also designated in the literature using a variety of other names. I list here these other names, and the variations of their use, for the benefit of the disoriented reader.

– *"Quantum spin dynamics"* (QSD) is used as a synonym of LQG. Within LQG, it is sometimes used to designate in particular the dynamical aspects of the hamiltonian theory.

– *"Quantum geometry"* is also sometimes used as a synonym of LQG. Within LQG, it is used to designate in particular the kinematical aspects of the theory. The expression *"quantum geometry"* is generic: it is also widely used in other approaches to quantum spacetime, in particular dynamical triangulations [21] and noncommutative geometry.

– *"Nonperturbative quantum gravity,"* *"canonical quantum gravity"* and *"quantum general relativity"* (QGR) are often used to designate LQG, although their proper meaning is wider.

– The expression *"Ashtekar approach"* is still used sometimes to designate LQG: it comes from the fact that a key ingredient of LQG is the reformulation of classical GR as a theory of connections, developed by Abhay Ashtekar.

– In the past, LQG was also called *"the loop representation of quantum general relativity."* Today, *"loop representation"* and *"connection representation"* are used within LQG to designate the representations of the states of LQG as functionals of loops (or spin networks) and as functionals of the connection, respectively. The two are related in the same manner as the energy ($\psi_n = \langle n|\psi\rangle$) and position ($\psi(x) = \langle x|\psi\rangle$) representations of the harmonic oscillator states.

Part I
Relativistic foundations

*I know that I am mortal, and the creature of a
day...*
*but when I search out the massed wheeling circles
of the stars, my feet no longer touch the earth:
side by side with Zeus himself, I drink my fill of
ambrosia, food of the gods...*

Claudius Ptolemy, *Mathematical Syntaxis*

1

General ideas and heuristic picture

The aim of this chapter is to introduce the general ideas on which this book is based and to present the picture of quantum spacetime that emerges from loop quantum gravity, in a heuristic and intuitive manner. The style of the chapter is therefore conversational, with little regard for precision and completeness. In the course of the book the ideas and notions introduced here will be made precise, and the claims will be justified and formally derived.

1.1 The problem of quantum gravity

1.1.1 Unfinished revolution

Quantum mechanics (QM) and general relativity (GR) have greatly widened our understanding of the physical world. A large part of the physics of the last century has been a triumphant march of exploration of new worlds opened up by these two theories. QM led to atomic physics, nuclear physics, particle physics, condensed matter physics, semiconductors, lasers, computers, quantum optics... GR led to relativistic astrophysics, cosmology, GPS technology... and is today leading us, hopefully, towards gravitational wave astronomy.

But QM and GR have destroyed the coherent picture of the world provided by prerelativistic classical physics: each was formulated in terms of assumptions contradicted by the other theory. QM was formulated using an external time variable (the t of the Schrödinger equation) or a fixed, nondynamical background spacetime (the spacetime on which quantum field theory is defined). But this external time variable and this fixed background spacetime are incompatible with GR. In turn, GR was formulated in terms of riemannian geometry, assuming that the metric is a smooth and deterministic dynamical field. But QM requires that any dynamical field be quantized: at small scales it manifests itself in discrete quanta and is governed by probabilistic laws.

3

We have learned from GR that spacetime is dynamical and we have learned from QM that any dynamical entity is made up of quanta and can be in probabilistic superposition states. Therefore at small scales there should be quanta of space and quanta of time, and quantum superposition of spaces. But what does this mean? We live in a spacetime with quantum properties: a *quantum spacetime*. What is quantum spacetime? How can we describe it?

Classical prerelativistic physics provided a coherent picture of the physical world. This was based on clear notions such as *time, space, matter, particle, wave, force, measurement, deterministic law,* ... This picture has partially evolved (in particular with the advent of field theory and special relativity) but it has remained consistent and quite stable for three centuries. GR and QM have modified these basic notions in depth. GR has modified the notions of space and time; QM the notions of causality, matter, and measurement. The novel, modified notions do not fit together easily. The new coherent picture is not yet available. With all their immense empirical success, GR and QM have left us with an understanding of the physical world which is unclear and badly fragmented. At the foundations of physics there is today confusion and incoherence.

We want to combine what we have learnt about our world from the two theories and to find a new synthesis. This is a major challenge – perhaps the major challenge – in today's fundamental physics. GR and QM have opened a revolution. The revolution is not yet complete.

With notable exceptions (Dirac, Feynman, Weinberg, DeWitt, Wheeler, Penrose, Hawking, 't Hooft, among others) most of the physicists of the second half of the last century have ignored this challenge. The urgency was to apply the two theories to larger and larger domains. The developments were momentous and the dominant attitude was pragmatic. Applying the new theories was more important than understanding them. But an overly pragmatic attitude is not productive in the long run. Towards the end of the twentieth century, the attention of theoretical physics has been increasingly focusing on the challenge of merging the conceptual novelties of QM and GR.

This book is the account of an effort to do so.

1.1.2 How to search for quantum gravity?

How to search for this new synthesis? Conventional field quantization methods are based on the weak-field perturbation expansion. Their application to GR fails because it yields a nonrenormalizable theory. Perhaps this is not surprising: GR has changed the notions of space and time too radically to docilely agree with flat space quantum field theory. Something else is needed.

In science there are no secure recipes for discovery and it is important to explore different directions at the same time. Currently, a quantum theory of gravity is sought along various paths. The two most developed are loop quantum gravity, described in this book, and string theory. Other research directions include dynamical triangulations, noncommutative geometry, Hartle's quantum mechanics of spacetime (this is not really a specific quantum theory of gravity, but rather a general theoretical framework for general-relativistic quantum theory), Hawking's euclidean sum over geometries, quantum Regge calculus, Penrose's twistor theory, Sorkin's causal sets, 't Hooft's deterministic approach and Finkelstein's theory. The reader can find ample references in the general introductions to quantum gravity mentioned in the note at the end of this chapter. Here, I sketch only the general ideas that motivate the approach described in this book, plus a brief comment on string theory, which is currently the most popular alternative to loop gravity.

Our present knowledge of the basic structure of the physical universe is summarized by GR, quantum theory and quantum field theory (QFT), together with the particle-physics standard model. This set of fundamental theories is inconsistent. But it is characterized by an extraordinary empirical success, nearly unique in the history of science. Indeed, currently there is no evidence of any observed phenomenon that clearly escapes, questions or contradicts this set of theories (or a minor modification of the same, to account, say, for a neutrino mass or a cosmological constant). This set of theories becomes meaningless in certain physical regimes. In these regimes, we expect the predictions of quantum gravity to become relevant and to differ from the predictions of GR and the standard model. These regimes are outside our experimental or observational reach, at least so far. Therefore, we have no direct empirical guidance for searching for quantum gravity – as, say, atomic spectra guided the discovery of quantum theory.

Since quantum gravity is a theory expected to describe regimes that are so far inaccessible, one might worry that anything could happen in these regimes, at scales far removed from our experience. Maybe the search is impossible because the range of the possible theories is too large. This worry is unjustified. If this was the problem, we would have plenty of complete, predictive and coherent theories of quantum gravity. Instead, the situation is precisely the opposite: we haven't any. The fact is that we do have plenty of information about quantum gravity, because we have QM and we have GR. Consistency with QM and GR is an extremely strict constraint.

A view is sometime expressed that some totally new, radical and wild hypothesis is needed for quantum gravity. I do not think that this is the case. Wild ideas pulled out of the blue sky have never made science advance. The radical hypotheses that physics has successfully adopted have always been reluctantly adopted because they were forced upon us by new empirical data – Kepler's ellipses, Bohr's quantization, ... – or by stringent theoretical deductions – Maxwell's inductive current, Einstein's relativity ... (see Appendix C). Generally, arbitrary novel hypotheses lead nowhere.

In fact, today we are precisely in one of the typical situations in which theoretical physics has worked at its best in the past. Many of the most striking advances in theoretical physics have derived from the effort of finding a common theoretical framework for two basic and apparently conflicting discoveries. For instance, the aim of combining the keplerian orbits with galilean physics led to newtonian mechanics; combining Maxwell theory with galilean relativity led to special relativity; combining special relativity and nonrelativistic quantum theory led to the theoretical discovery of antiparticles; combining special relativity with newtonian gravity led to general relativity, and so on. In all these cases, major advances have been obtained by "taking seriously"[1] apparently conflicting theories, and exploring the implications of holding the key tenets of both theories for true. Today we are precisely in one of these characteristic situations. We have learned two new very general "facts" about Nature, expressed by QM and GR: we have "just" to figure out what they imply, taken together. Therefore, the question we have to ask is: what have we really learned about the world from QM and from GR? Can we combine these insights into a coherent picture? What we need is a conceptual scheme in which the insights obtained with GR and QM fit together.

This view is *not* the majority view in theoretical physics, at present. There is consensus that QM has been a conceptual revolution, but many do not view GR in the same way. According to many, the discovery of GR has been just the writing of one more field theory. This field theory is, furthermore, likely to be only an approximation to a theory we do not yet know. According to this opinion, GR should not be taken too seriously as a guidance for theoretical developments.

I think that this opinion derives from a confusion: the confusion between the specific form of the Einstein–Hilbert action and the modification of the notions of space and time engendered by GR. The Einstein–Hilbert action might very well be a low-energy approximation of a high-energy theory. But the modification of the notions of space and time does not depend on the specific form of the Einstein–Hilbert action. It depends on its diffeomorphism invariance and its background independence. These properties

---

[1]In [22], Gell-Mann says that the main lesson to be learnt from Einstein is "to 'take very seriously' ideas that work, and see if they can be usefully carried much further than the original proponent suggested."

(which are briefly illustrated in Section 1.1.3 below, and discussed in detail in Chapter 2) are most likely to hold in the high-energy theory as well. One should not confuse the details of the dynamics of GR with the modifications of the notions of space and time that GR has determined. If we make this confusion, we underestimate the radical novelty of the physical content of GR. The challenge of quantum gravity is precisely to fully incorporate this radical novelty into QFT. In other words, the task is to understand what is a general-relativistic QFT, or a background-independent QFT.

Today many physicists prefer disregarding or postponing these foundational issues and, instead, choose to develop and adjust current theories. The most popular strategy towards quantum gravity, in particular, is to pursue the line of research grown in the wake of the success of the standard model of particle physics. The failure of perturbative quantum GR is interpreted as a replay of the failure of Fermi theory.[2] Namely, as an indication that we must modify GR at high energy. With the input of the grand-unified-theories (GUTs), supersymmetry, and the Kaluza–Klein theory, the search for a high-energy correction of GR free from bad ultraviolet divergences has led to higher derivative theories, supergravity, and finally to string theory.

Sometimes the claim is made that the quantum theory of gravity has already been found and it is string theory. Since this is a book about quantum gravity without strings, I should say a few words about this claim. String theory is based on a physical hypothesis: elementary objects are extended, rather than particle-like. This hypothesis leads to a very rich unified theory, which contains much phenomenology, including (with suitable inputs) fermions, Yang–Mills fields and gravitons, and is expected by many to be free of ultraviolet divergences. The price to pay for these theoretical results is a gigantic baggage of additional physics: supersymmetry, extra dimensions, an infinite number of fields with arbitrary masses and spins, and so on.

So far, nothing of this new physics shows up in experiments. Supersymmetry, in particular, has been claimed to be on the verge of being discovered for years, but hasn't shown up. Unfortunately, so far the theory can accommodate any disappointing experimental result because it is hard to derive precise new quantitative physical predictions, with which the theory could be falsified, from the monumental mathematical apparatus of the theory. Furthermore, even recovering the real world is not easy within the theory: the search for a compactification leading to the

---

[2]Fermi theory was an empirically successful but nonrenormalizable theory of the weak interactions, just as GR is an empirically successful but nonrenormalizable theory of the gravitational interaction. The solution has been the Glashow–Weinberg–Salam electroweak theory, which corrects Fermi theory at high energy.

standard model, with its families and masses and no instabilities, has not yet succeeded, as far as I know. It is clear that string theory is a very interesting hypothesis, but certainly not an established theory. It is therefore important to pursue alternative directions as well.

String theory is a direct development of the standard model and is deeply rooted in the techniques and the conceptual framework of flat space QFT. As I shall discuss in detail throughout this book, many of the tools used in this framework – energy, unitary time evolution, vacuum state, Poincaré invariance, S-matrix, objects moving in a spacetime, Fourier transform, ... – no longer make sense in the quantum gravitational regime, in which the gravitational field cannot be approximated by a background spacetime – perhaps not even asymptotically.[3] Therefore string theory does not address directly the main challenge of quantum gravity: understanding what is a background-independent QFT. Facing this challenge directly, before worrying about unification, leads, instead, to the direction of research investigated by loop quantum gravity.[4]

The alternative to the line of research followed by string theory is given by the possibility that the failure of perturbative quantum GR is *not* a replay of Fermi theory. That is, it is not due to a flaw of the GR action, but, instead, it is due to the fact that the conventional weak-field quantum perturbation expansion cannot be applied to the gravitational field.

This possibility is strongly supported a posteriori by the results of loop quantum gravity. As we shall see, loop quantum gravity leads to a picture of the short-scale structure of spacetime extremely different from that of a smooth background geometry. (There are hints in this direction from string theory calculations as well [25].) Spacetime turns out to have a nonperturbative, quantized, discrete structure at the Planck scale, which is explicitly described by the theory. The ultraviolet divergences are cured by this structure. The ultraviolet divergences that appear in the perturbation expansion of conventional QFT are a consequence of the fact that

---

[3]To be sure, the development of string theory has incorporated many aspects of GR, such as curved spacetimes, horizons, black holes and relations between different backgrounds. But this is far from a background-independent framework, such as the one realized by GR in the classical context. GR is not about physics on a curved spacetime, or about relations between different backgrounds: it is about the dynamics of spacetime. A background-independent fundamental definition of string theory is being actively searched for along several directions, but so far the definition of the theory and all calculations rely on background metric spaces.

[4]It has been repeatedly suggested that loop gravity and string theory might merge, because loop gravity has developed precisely the background-independent QFT methods that string theory needs [23]. Also, excitations over a weave (see Section 6.7.1) have a natural string structure in loop gravity [24].

we erroneously replace this discrete Planck-scale structure with a smooth background geometry.

If this is physically correct, ultraviolet divergences do not require the heavy machinery of string theory to be cured. On the other hand, the conventional weak-field perturbative methods cannot be applied, because we cannot work with a fixed smooth background geometry. We must therefore adapt QFT to the full conceptual novelty of GR, and in particular to the change in the notion of space and time induced by GR. What are these changes? I sketch an answer below, leaving a complete discussion to Chapter 2.

1.1.3 The physical meaning of general relativity

GR is the discovery that spacetime and the gravitational field are the same entity. What we call "spacetime" is itself a physical object, in many respects similar to the electromagnetic field. We can say that GR is the discovery that there is no spacetime at all. What Newton called "space," and Minkowski called "spacetime," is unmasked: it is nothing but a dynamical object – the gravitational field – in a regime in which we neglect its dynamics.

In newtonian and special-relativistic physics, if we take away the dynamical entities – particles and fields – what remains is space and time. In general-relativistic physics, if we take away the dynamical entities, nothing remains. The space and time of Newton and Minkowski are re-interpreted as a configuration of one of the fields, the gravitational field. This implies that physical entities – particles and fields – are not immersed in space, and moving in time. They do not live on spacetime. They live, so to say, on one another.

It is as if we had observed in the ocean many animals living on an island: animals on the island. Then we discover that the island itself is in fact a great whale. So the animals are no longer on the island, just animals on animals. Similarly, the Universe is not made up of fields on spacetime; it is made up of fields on fields. This book studies the far-reaching effect that this conceptual shift has on QFT.

One consequence is that the quanta of the field cannot live in spacetime: they must build "spacetime" themselves. This is precisely what the quanta of space do in loop quantum gravity.

We may continue to use the expressions "space" and "time" to indicate aspects of the gravitational field, and I do so in this book. We are used to this in classical GR. But in the quantum theory, where the field has quantized "granular" properties and its dynamics is quantized and therefore only probabilistic, most of the "spatial" and "temporal" features of the gravitational field are lost.

Therefore, to understand the quantum gravitational field we must abandon some of the emphasis on geometry. Geometry represents the classical gravitational field, but not quantum spacetime. This is not a betrayal of Einstein's legacy: on the contrary, it is a step in the direction of "relativity" in the precise sense meant by Einstein. Alain Connes has described beautifully the existence of two points of view on space: the geometric one, centered on space points, and the algebraic, or "spectral" one, centered on the algebra of dual spectral quantities. As emphasized by Alain, quantum theory forces us to a complete shift to this second point of view, because of noncommutativity. In the light of quantum theory, continuous spacetime cannot be anything else than an approximation in which we disregard quantum noncommutativity. In loop gravity, the physical features of space appear as spectral properties of quantum operators that describe our (the observers') interactions with the gravitational field.

The key conceptual difficulty of quantum gravity is therefore to accept the idea that we can do physics in the absence of the familiar stage of space and time. We need to free ourselves from the prejudices associated with the habit of thinking of the world as "inhabiting space" and "evolving in time". Chapter 3 describes a language for describing mechanical systems in this generalized conceptual framework.

This absence of the familiar spacetime "stage" is called the *background independence* of the classical theory. Technically, it is realized by the gauge invariance of the action under (active) diffeomorphisms. A diffeomorphism is a transformation that smoothly drags all dynamical fields and particles from one region of the four-dimensional manifold to another (the precise definition of these transformations is given in Chapter 2). In turn, gauge invariance under diffeomorphism (or *diffeomorphism invariance*) is the consequence of the combination of two properties of the action: its invariance under arbitrary changes of coordinates and the fact that there is no nondynamical "background" field.

1.1.4 Background-independent quantum field theory

Is quantum mechanics[5] compatible with the general-relativistic notions of space and time? It is, provided that we choose a sufficiently general formulation. For instance, the Schrödinger picture is only viable for theories where there is a global observable time variable t; this conflicts with GR, where no such variable exists. Therefore, the Schrödinger picture makes little sense in a background-independent context. But there

---

[5]I use the expression "quantum mechanics" to indicate the theory of all quantum systems, with a finite or infinite number of degrees of freedom. In this sense QFT is part of quantum mechanics.

are formulations of quantum theory that are more general than the Schrödinger picture. In Chapter 5, I describe a formulation of QM sufficiently general to deal with general-relativistic systems. (For another relativistic formulation of QM, see [26].) Formulations of this kind are sometimes denoted "generalized quantum mechanics." I prefer to use "quantum mechanics" to denote any formulation of quantum theory, irrespective of its generality, just as "classical mechanics" is used to designate formalisms with different degrees of generality, such as Newton's, Lagrange's, Hamilton's or symplectic mechanics.

On the other hand, most of the conventional machinery of perturbative QFT is profoundly incompatible with the general-relativistic framework. There are many reasons for this:

- The conventional formalism of QFT relies on Poincaré invariance. In particular, it relies on the notion of energy and on the existence of the nonvanishing hamiltonian operator that generates unitary time evolution. The vacuum, for instance, is the state that minimizes the energy. Generally, there is no global Poincaré invariance, no general notion of energy and no nonvanishing hamiltonian operator in a general-relativistic theory.

- At the root of conventional QFT is the physical notion of particle. The theoretical experience with QFT on curved spacetime [27] and on the relation between acceleration and temperature in QFT [28] indicates that in a generic gravitational situation the notion of particle can be quite delicate. (This point is discussed in Section 5.3.4.)

- Consider a conventional renormalized QFT. The physical content of the theory can be expressed in terms of its n-point functions $W(x_1, \ldots, x_n)$. The n-point functions reflect the invariances of the classical theory. In a general-relativistic theory, invariance under a coordinate transformation $x \to x' = x'(x)$ implies immediately that the n-point functions must satisfy

$$W(x_1, \ldots, x_n) = W(x'(x_1), \ldots, x'(x_n)) \qquad (1.1)$$

and therefore (if the points in the argument are distinct) it must be a constant! That is,

$$W(x_1, \ldots, x_n) = constant. \qquad (1.2)$$

Clearly, we are immediately in a very different framework from conventional QFT.

- Similarly, the behavior for small $|x - y|$ of the two-point function of a conventional QFT

$$W(x, y) = \frac{constant}{|x - y|^d}, \qquad (1.3)$$

expresses the short-distance structure of the QFT. More generally, the short-distance structure of the QFT is reflected in the operator product expansion

$$O(x)O'(y) = \sum_n \frac{O_n(x)}{|x-y|^n}. \tag{1.4}$$

Here $|x - y|$ is the distance measured in the spacetime metric. On flat space for instance $|x-y|^2 = \eta_{\mu\nu}(x^\mu - y^\mu)(x^\nu - y^\nu)$. In a general-relativistic context these expressions make no sense, since there is no background Minkowski (or other) metric $\eta_{\mu\nu}$. In its place, there is the gravitational field, namely the quantum field operator itself. But then, if standard operator product expansion becomes meaningless, the short-distance structure of a quantum gravitational theory must be profoundly different from that of conventional QFT. As we shall see in Chapter 7 this is precisely the case.

There is a tentative escape strategy to circumvent these difficulties: write the gravitational field $e(x)$ as the sum of two terms

$$e(x) = e_{\text{background}}(x) + h(x); \tag{1.5}$$

where $e_{\text{background}}(x)$ is a background field configuration. This may be Minkowski, or any other. Assume that $e_{\text{background}}(x)$ defines spacetime, namely it defines location and causal relations. Then consider $h(x)$ as the gravitational field, governed by a QFT on the spacetime background defined by $e_{\text{background}}$. For instance the field operator $h(x)$ is assumed to commute at spacelike separations, where spacelike is defined in the geometry determined by $e_{\text{background}}(x)$. As a second step one may then consider conditions on $e_{\text{background}}(x)$ or relations between the formulations of the theory defined by different choices of $e_{\text{background}}(x)$. This escape strategy leads to three orders of difficulties. (i) Conventional perturbative QFT of GR based on (1.5) leads to a nonrenormalizable theory. To get rid of the uncontrollable ultraviolet divergences one has to resort to the complications of string theory. (ii) As mentioned, loop quantum gravity shows that the structure of spacetime at the Planck scale is discrete. Therefore physical spacetime has no short-distance structure at all. The unphysical assumption of a smooth background $e_{\text{background}}(x)$ implicit in (1.5) may be precisely the cause of the ultraviolet divergences. (iii) The separation of the gravitational field from spacetime is in strident contradiction with the very physical lesson of GR. If GR is of any guide in searching for a quantum theory of gravity, the relevant spacetime geometry is the one determined by the full gravitational field $e(x)$, and the separation (1.5) is misleading.

A formulation of quantum gravity that does not take the escape strategy (1.5) is a *background-independent*, or general covariant QFT. The main aim of this book is develop the formalism for background-independent QFT.

1.2 Loop quantum gravity

I sketch here the physical picture of quantum spacetime that emerges from loop quantum gravity (LQG). The basic ideas and assumptions on which LQG is based are the following.

(i) Quantum mechanics and general relativity. QM, suitably formulated to be compatible with general covariance, is assumed to be correct. The Einstein equations may be modified at high energy, but the general-relativistic notions of space and time are assumed to be correct. The motivation for these two assumptions is the extraordinary empirical success they have had so far, and the absence of any contrary empirical evidence.

(ii) Background independence. LQG is based on the idea that the quantization strategy based on the separation (1.5) is *not* appropriate for describing the quantum properties of spacetime.

To this we can add:

(iii) No unification. Nowadays, a fashionable idea is that the problem of quantizing gravity has to be solved together with the problem of finding a unified description of all interactions. LQG is a solution of the first problem, not the second.[6]

(iv) Four spacetime dimensions and no supersymmetry. LQG is compatible with these possibilities, but there is nothing in the theory that *requires* higher dimensions or supersymmetry. Higher spacetime dimensions and supersymmetry are interesting theoretical ideas, which, as many other interesting theoretical ideas, can be physically wrong. In spite of 15 years of search, numerous preliminary announcements of discovery then turned out to be false,

---

[6] A motivation for the idea that these two issues are connected is the expectation that we are "near the end of physics." Unfortunately, the expectation of being "near the end of physics" has been present all along the three centuries of the history of modern physics. In the present situation of deep conceptual confusion on the fundamental aspects of the world, I see no sign indicating that we are close to the end of our discoveries about the physical world. When I was a student, it was fashionable to claim that the problem of finding a theory of the strong interactions had to be solved together with the problem of getting rid of renormalization theory. Nice idea. But wrong.

and despite repeated proclamations that supersymmetry was going to be discovered "next year", so far empirical evidence has been solidly and consistently *against* supersymmetry. This might change, but as scientists we must take the indications of the experiments seriously.

On the basis of these assumptions, LQG is a straightforward quantization of GR with its conventional matter couplings. The program of LQG is therefore conservative, and of small ambition. The physical inputs of the theory are just QM and GR, well-tested physical theories. No major additional physical hypothesis or assumption is made (such as: elementary objects are strings, space is made by individual discrete points, quantum mechanics is wrong, GR is wrong, supersymmetry, extra dimensions, ...). No claim of being the final "Theory Of Everything" is made.

On the other hand, LQG has a radical and ambitious side: to merge the conceptual insight of GR into QM. In order to achieve this, we have to give up the familiar notions of space and time. The space continuum "on which" things are located and the time "along which" evolution happens are semiclassical approximate notions in the theory. In LQG, this radical step is assumed in its entirety.

LQG does not make use of most of the familiar tools of conventional QFT because these become inadequate in a background-independent context. It only makes use of the general tools of quantum theory: a Hilbert space of states, operators related to the measurement of physical quantities, and transition amplitudes that determine the probability outcome of measurements of these quantities. Hilbert space of states and operators associated to physical observables are obtained from classical GR following a rather standard quantization strategy. A quantization strategy is a technique for searching for a solution to a well-posed inverse problem: finding a quantum theory with a given classical limit. The inverse problem could have many solutions. As noticed, presently the difficulty is not to discriminate among many complete and consistent quantum theories of gravity. We would be content with one.

1.2.1 Why loops?

Among the technical choices to make in order to implement a quantization procedure is which algebra of field functions to promote to quantum operators. In conventional QFT, this is generally the canonical algebra formed by the positive and negative frequency components of the field modes. The quantization of this algebra leads to the creation and annihilation operators a and $a^\dagger$. The characterization of the positive and negative frequencies requires a background spacetime.

In contrast to this, what characterizes LQG is the choice of a different algebra of basic field functions: a noncanonical algebra based on the holonomies of the gravitational connection. The holonomy (or "Wilson loop") is the matrix of the parallel transport along a closed curve.

The idea that holonomies are the natural variables in a gauge theory has a long history. In a sense, it can be traced back to the very origin of gauge theory, in the physical intuition of Faraday. Faraday understood electromagnetic phenomena in terms of "lines of force." Two key ideas underlie this intuition. First, that the relevant physical variables fill up space; this intuition by Faraday is the origin of field theory. Second, that the relevant variables do not refer to what happens at a point, but rather refer to the relation between different points connected by a line. The mathematical quantity that expresses this idea is the holonomy of the gauge potential along the line. In the Maxwell case, for instance, the holonomy $U(A, \alpha)$ along a loop α is simply the exponential of the line integral along α of the three-dimensional Maxwell potential A:

$$U(A, \alpha) = e^{\oint_\alpha A} = \exp\left\{ \int_0^{2\pi} ds\, A_a(\alpha(s))\, \frac{d\alpha^a(s)}{ds} \right\}. \tag{1.6}$$

In LQG, the holonomy becomes a quantum operator that creates "loop states." In the loop representation formulation of Maxwell theory, for instance, a loop state $|\alpha\rangle$ is a state in which the electric field vanishes everywhere except along a single Faraday line α. More precisely, it is an eigenstate of the electric field with eigenvalue

$$\vec{E}_\alpha(\vec{x}) = \oint ds\, \frac{d\vec{\alpha}(s)}{ds}\, \delta^3(\vec{x}, \vec{\alpha}(s)), \tag{1.7}$$

where $s \mapsto \vec{\alpha}(s)$ is the Faraday line in space. This electric field vanishes everywhere except on the loop α itself, and at every point of α it is tangent to the loop, see Figure 1.1. Notice that the vector distribution field $\vec{E}(\vec{x})$ defined in (1.7) is divergenceless, that is, it satisfies Coulomb law

$$\text{div}\, \vec{E}_\alpha(\vec{x}) = 0 \tag{1.8}$$

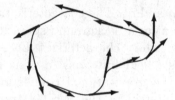

Fig. 1.1   A loop α and the distributional electric field configuration $\vec{E}_\alpha$ (represented by the arrows).

in the sense of distributions. In fact, for any smooth function f we have

$$[\text{div } \vec{E}_\alpha](f) = \int d^3x \, f(\vec{x}) \, \text{div } \vec{E}_\alpha(\vec{x})$$

$$= \int d^3x \, f(\vec{x}) \, \frac{\partial}{\partial x^a} \oint ds \, \frac{d\alpha^a(s)}{ds} \, \delta^3(\vec{x}, \vec{\alpha}(s))$$

$$= - \oint ds \, \frac{d\alpha^a(s)}{ds} \, \frac{\partial}{\partial \alpha^a} f(\alpha(s))$$

$$= - \oint_\alpha df = - \oint ds \, \frac{d}{ds} \, f(\alpha(s)) = 0. \tag{1.9}$$

Indeed, intuitively, Coulomb law requires precisely that an electric field at a point "continues" in the direction of the field itself, namely that it defines Faraday lines. The state $|\alpha\rangle$ is therefore a sort of minimal quantum excitation satisfying (1.8): it is an elementary quantum excitation of a single Faraday line.

The idea that a Yang–Mills theory is truly a theory of these loops has been around for as long as such theories have been studied. Mandelstam, Polyakov, and Wilson, among many others, have long argued that loop excitations should play a major role in quantum Yang–Mills theories, and that we must get to understand quantum Yang–Mills theories in terms of these excitations. In fact, much of the development of string theory has been influenced by this idea.

In *lattice* Yang–Mills theory, namely, in the approximation to Yang–Mills theory where spacetime is replaced by a fixed lattice, loop states have finite norm. In fact, certain finite linear combinations of loop states, called "spin network" states, form a well-defined and well-understood orthonormal basis in the Hilbert space of a lattice gauge theory.

However, in a QFT theory over a *continuous* background, the idea of formulating the theory in terms of loop-like excitations has never proved fruitful. The difficulty is essentially that loop states over a background are "too singular" and "too many." The quantum Maxwell state $|\alpha\rangle$ described above, for instance, has infinite norm; and an infinitesimal displacement of a loop state over the background spacetime produces a distinct, independent, loop state, yielding a continuum of loop states. Over a continuous background, the space spanned by the loop states is far "too big" for providing a basis of the (separable) Hilbert space of a QFT.

However, loop states are not too singular, nor too many, in a *background-independent* theory. This is the key technical point on which LQG relies. The intuitive reason is as follows. Spacetime itself is formed by loop-like states. Therefore the position of a loop state is relevant only *with respect to other loops*, and not with respect to the background. An infinitesimal (coordinate) displacement of a loop state does not produce a distinct quantum state, but only a gauge equivalent representation of the same physical state! Only a finite displacement carrying the loop state across another loop produces a physically different state. Therefore, the size of the space of the loop states is dramatically reduced by diffeomorphism invariance: most of it is just gauge! Equivalently, we can think that a single loop has an intrinsic Planck-size "thickness."

Therefore, in a general-relativistic context the loop basis becomes viable. The state space of the theory, called $\mathcal{K}_{\text{diff}}$, is a separable Hilbert space spanned by loop states. More precisely, as we shall see in Chapter 6, $\mathcal{K}_{\text{diff}}$ admits an orthonormal basis of spin network states, which are formed by finite linear combinations of loop states, and are defined precisely as the spin network states of a lattice Yang–Mills theory. This Hilbert space and the field operators that act on it are described in Chapter 6. They form the basis of the mathematical structure of LQG.

Therefore LQG is the result of the convergence of two lines of thinking, each characteristic of twentieth-century theoretical physics. On the one hand, the intuition of Faraday, Yang and Mills, Wilson, Mandelstam, Polyakov, and others, that forces are described by lines. On the other hand, the Einstein–Wheeler–DeWitt intuition of background independence and background-independent quantum states. Truly remarkably, each of these two lines of thinking is the solution of the blocking difficulty of the other. On the one hand, the traditional nonviability of the loop basis in the continuum disappears because of background independence. On the other hand, the traditional difficulty of controlling diffeomorphism-invariant quantities comes under control thanks to the loop basis.

Even more remarkably, the spin network states generated by this happy marriage turn out to have a surprisingly compelling geometric interpretation, which I sketch below.

1.2.2 Quantum space: spin networks

Physical systems reveal themselves by interacting with other systems. These interactions may happen in "quanta": energy is exchanged with an oscillator of frequency ν in discrete packets, or quanta, of size $E = h\nu$. If the oscillator is in the nth energy eigenstate, we say that there are n quanta in it. If the oscillator is a mode of a free field, we say that there are n "particles" in the field. Therefore we can view the electromagnetic field as made up of its quanta, the photons. What are the quanta of the gravitational field? Or, since the gravitational field is the same entity as spacetime, what are the quanta of space?

The properties of the quanta of a system are determined by the spectral properties of the operators representing the quantities involved in our interaction with the system. The operator associated with the energy of the oscillator, for instance, has a discrete spectrum, and the number of quanta n labels its eigenvalues. The set of its eigenstates form a basis in the state space of the quantum system: this fact allows us to view each state of the system as a quantum superposition of states $|n\rangle$ formed by n quanta. To understand the quantum properties of space, we have therefore to consider the spectral problem of the operators associated with the quantities involved in our interaction with space itself. The most direct interaction we have with the gravitational field is via the geometric structure of the physical space. A measurement of length, area, or volume is, in fact, according to GR, a measurement of a local property of the gravitational field.

For instance, the volume $\mathbf{V}$ of a physical region $\mathcal{R}$ is

$$\mathbf{V} = \int_{\mathcal{R}} \mathrm{d}^3 x \, |\det e(x)|, \tag{1.10}$$

where $e(x)$ is the (triad matrix representing the) gravitational field. In quantum gravity, $e(x)$ is a field operator, and $\mathbf{V}$ is therefore an operator as well.

The volume $\mathbf{V}$ is a nonlinear function of the field e and the definition of the volume operator implies products of local operator-valued distributions. This can be achieved as a limit, using an appropriate regularization procedure. The development of regularization procedures that remain meaningful in the absence of a background metric is a major technical tool on which LQG is based. Using these techniques, a well-defined self-adjoint operator $\mathbf{V}$ can be defined. The computation of its spectral properties is then one of the main results of LQG, and will be derived in Section 6.6.5.

The spectrum of $\mathbf{V}$ turns out to be discrete. Therefore the spacetime volume manifests itself in quanta, of definite volume size, given by the eigenvalues of the volume operator. These quanta of space can be intuitively thought of as quantized "grains" of space or "atoms of space." The first intuitive picture of quantum space is therefore that of "grains of space." These have quantized amounts of volume, determined by the spectrum of the operator $\mathbf{V}$.

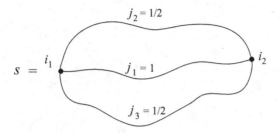

Fig. 1.2 A simple spin network.

The next element of the picture is the information on which grain is adjacent to which. Adjacency (being contiguous, being in touch, being nearby) is the basis of spatial relations. If two spacetime regions are adjacent, that is, if they touch each other, they are separated by a surface $\mathcal{S}$. Let $\mathbf{A}$ be the area of the surface $\mathcal{S}$. Area also is a function of the gravitational field, and is therefore represented by an operator, like volume. The spectral problem for this operator has been solved in LQG, as well. It is discussed in detail in Section 6.6.2. This spectrum turns out also to be discrete. Intuitively, the grains of space are separated by "quanta of area." The principal series of the eigenvalues of the area, for instance, is labeled by multiplets of half-integers $j_i, i = 1, \ldots, n$ and turns out to be given by

$$\mathbf{A} = 8\pi\gamma \, \hbar G \sum_i \sqrt{j_i(j_i + 1)}, \tag{1.11}$$

where γ, the Immirzi parameter, is a free dimensionless constant of the theory.

Consider a quantum state of space $|s\rangle$ formed by N "grains" of space, some of which are adjacent to one another. Represent this state as an abstract graph Γ with N nodes. (By abstract graph I mean here an equivalence class under smooth deformations of graphs embedded in a 3-manifold.) The nodes of the graph represent the grains of space; the links of the graph link adjacent grains and represent the surfaces separating two adjacent grains. The quantum state is then characterized by the graph Γ and by labels on nodes and on links: the label i_n on a node n is the quantum number of the volume and the label j_l on a link l is the quantum number of the area.

A graph with these labels is called an (abstract) "spin network" $s = (\Gamma, i_n, j_l)$, see Figure 1.2. In Section 6.3.1, we will see that the quantum numbers i_n and j_l are determined by the representation theory of the local gauge group $(SU(2))$. More precisely, j_l labels unitary irreducible representations and i_n labels a basis in the space of the intertwiners between the representations adjacent to the node n. The area of a surface cutting

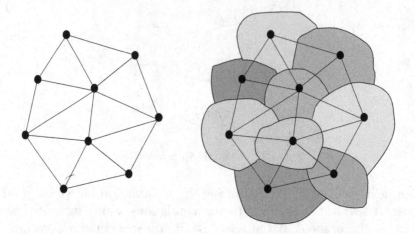

Fig. 1.3 The graph of an abstract spinfoam and the ensemble of "chunks of space," or quanta of volume, it represents. Chunks are adjacent when the corresponding nodes are linked. Each link cuts one elementary surface separating two chunks.

n links of the spin network with labels $j_i, i = 1, \ldots, n$ is then given by (1.11).

As shown in Section 6.3.1, the (kinematical) Hilbert space $\mathcal{K}_{\text{diff}}$ admits a basis labeled precisely by these spin networks. This is a basis of states in which certain area and volume operators are diagonal. Its physical interpretation is the one sketched in Figure 1.3: a spin network state $|s\rangle$ describes a quantized three-geometry.

A loop state $|\alpha\rangle$ is a spin network state in which the graph Γ has no nodes; namely, is a single loop α, and is labeled by the fundamental representation of the group. In such a state, the gravitational field has support just on the loop α itself, as the electric field in (1.7).

In LQG, physical space is a quantum superposition of spin networks, in the same sense that the electromagnetic field is a quantum superposition of n-photon states. The first and basic prediction of the (free) QFT of the electromagnetic field is the existence of the photons, and the specific quantitative prediction of the energy and the momentum of the photons of a given frequency. Similarly, the first prediction of LQG is the existence of the quanta of area and volume, and the quantitative prediction of their spectrum.

The theory predicts that any sufficiently accurate measurement of area or volume would measure one of these spectral values. So far, verifying this prediction appears to be outside our technological capacities.

Where is a spin network? A spin network state does not have a position. It is an abstract graph, not a graph immersed in a spacetime manifold.

Only abstract combinatorial relations defining the graph are significant, not its shape or its position in space.

In fact, a spin network state is not *in* space: it *is* space. It is not localized with respect to something else: something else (matter, particles, other fields) might be localized with respect to it. To ask "where is a spin network" is like asking "where is a solution of the Einstein equations." A solution of the Einstein equations is not "somewhere": it is the "where" with respect to which anything else can be localized. In the same way, the other dynamical objects, such as Yang–Mills and fermion fields, live on the spin network state.

This is a consequence of diffeomorphism invariance. Technically, spin network states are first defined as graphs embedded in a three-dimensional manifold; then the implementation of the diffeomorphism gauge identifies two graphs that can be deformed into each other. They are gauge equivalent. This is like identifying two solutions of the Einstein equations that are related by a change of coordinates. Spin networks embedded in a manifold are denoted S and called "embedded spin networks"; equivalence classes of these under diffeomorphisms are denoted s, and are called "abstract spin networks", or s-knots. A quantum state of space is determined by an s-knot.[7]

The fact that spin networks do not live *in* space, but rather *are* space, has far-reaching consequences. Space itself turns out to have a discrete and combinatorial character. Notice that this is not imposed on the theory, or assumed. It is the result of a completely conventional quantum mechanical calculation of the spectrum of the physical quantities that describe the geometry of space. Since there is no spatial continuity at short scale, there is (literally!) no room in the theory for ultraviolet divergencies. The theory effectively cuts itself off at the Planck scale. Space is effectively granular at the Planck scale, and there is no infinite ultraviolet limit.

Chapter 7 describes how Yang–Mills and fermion fields can be coupled to the theory. This can be obtained by enriching the structure of the spin networks s. In the case of a Yang–Mills theory with gauge group G, for instance, links carry an additional quantum number, labeling irreducible representations G. The spin network itself behaves like the lattice of lattice Yang–Mills theory. In quantum gravity, therefore, the lattice itself becomes a dynamical variable. But notice a crucial difference with respect to conventional lattice Yang–Mills theory: the lattice size is not to be scaled down to zero: it has physical Planck size.

In summary, spin networks provide a mathematically well-defined and physically compelling description of the kinematics of the quantum gravitational field. They also provide a well-defined picture of the small-scale

---

[7]The expression "spin network" is used in the literature to designate both the embedded and the abstract ones, as well as to designate the quantum states they label.

structure of space. It is remarkable that this novel picture of space emerges simply from the combination of old Yang–Mills theory ideas with general-relativistic background independence.

1.2.3 Dynamics in background-independent QFT

The dynamics of the quantum gravitational field can be described giving amplitudes $W(s)$ for spin network states. Let me illustrate here, in a heuristic manner, the physical interpretation of these amplitudes and the way they are defined in the theory. A major feature of this book is that it is based on a general-relativistic way of thinking about observables + evolution. This section sketches this view and may be somewhat harder to follow than the previous ones.

Interpretation of the amplitude $W(s)$. The quantum dynamics of a particle is entirely described by the transition probability amplitudes

$$W(x, t, x', t') = \langle x | e^{-\frac{i}{\hbar} H_0 (t - t')} | x' \rangle = \langle x, t | x', t' \rangle, \qquad (1.12)$$

where $|x, t\rangle$ is the eigenstate of the Heisenberg position operator $x(t)$ with eigenvalue x, H_0 is the hamiltonian operator and $|x\rangle = |x, 0\rangle$. The propagator $W(x, t; x', t')$ depends on two events (x, t) and (x', t') that bound a finite portion of a classical trajectory. The space of the pairs of events (x, t, x', t') is called $\mathcal{G}$ in this book.

A physical experiment consists of a preparation at time t' and a measurement at time t. Say that in a particular experiment we have localized the particle in x' at t' and then found it in x at time t. The set (x, t, x', t') represents the complete set of data of a specific complete observational set up, including preparation and measurement. The space $\mathcal{G}$ is the space of these data sets. In the quantum theory, we associate the complex amplitude $W(x, t, x', t')$, which is a function on $\mathcal{G}$, with any such data set. As emphasized by Feynman, this amplitude codes the full quantum dynamics. Following Feynman, we can compute $W(x, t, x', t')$ with a sum-over-paths that take the values x and x' at t and t', respectively.

If we measure a different observable than position, we obtain states different from the states $|x\rangle$. Let $|\psi_{in}\rangle$ be the state prepared at time t', and let $|\psi_{out}\rangle$ be the state measured at time t. The amplitude associated to these measurements is

$$A = \langle \psi_{out} | e^{-\frac{i}{\hbar} H_0 (t - t')} | \psi_{in} \rangle. \qquad (1.13)$$

The pair of states (ψ_{in}, ψ_{out}) determines a state $\psi = |\psi_{in}\rangle \otimes \langle \psi_{out}|$ in the space $\mathcal{K}_{t,t'}$ which is the tensor product of the Hilbert space of the initial states and (the dual of) the Hilbert space of the final states. The propagator defines a (possibly generalized) state $|0\rangle$ in $\mathcal{K}_{t,t'}$, by $\langle 0|(|x'\rangle \otimes \langle x|) = W(x,t,x',t')$. The amplitude (1.13) can be written simply as

$$A = \langle 0|\psi\rangle. \tag{1.14}$$

Therefore we can express the dynamics from t' to t in terms of a single state $|0\rangle$ in a Hilbert space $\mathcal{K}_{t,t'}$ that represents outcomes of measurements on t' *and* t. The state $|0\rangle$ is called the covariant vacuum, and should not be confused with the state of minimal energy.

Let us extend this idea to field theory. In field theory, the analog of the data set (x,t,x',t'), is the couple $[\Sigma, \varphi]$, where Σ is a 3d surface bounding a finite spacetime region, and φ is a field configuration on Σ. These data define a set of events $(x \in \Sigma, \varphi(x))$ that bound a finite portion of a classical configuration of the field, just as (x,t,x',t') bound a finite portion of the classical trajectory of the particle. The data from a local experiment (measurements, preparation, or just assumptions) must in fact refer to the state of the system on the entire boundary of a finite spacetime region. The field theoretical space $\mathcal{G}$ is therefore the space of surfaces Σ and field configurations φ on Σ. Quantum dynamics can be expressed in terms of an amplitude $W[\Sigma, \varphi]$. Following Feynman's intuition, we can formally define $W[\Sigma, \varphi]$ in terms of a sum over bulk field configurations that take the value φ on the boundary Σ. In fact, in Section 5.3, I argue that the functional $W[\Sigma, \varphi]$ captures the dynamics of a QFT.

Notice that the dependence of $W[\Sigma, \varphi]$ on the geometry of Σ codes the spacetime position of the measuring apparatus. In fact, the relative position of the components of the apparatus is determined by their physical distance and the physical time lapsed between measurements, and these data are contained in the metric of Σ.

Consider now a background-independent theory. Diffeomorphism invariance implies immediately that $W[\Sigma, \varphi]$ is independent of Σ. This is the analog of the independence of $W(x,y)$ from x and y, mentioned in Section 1.1.4. Therefore, in gravity W depends only on the boundary value of the fields. However, the fields include the gravitational field, and the gravitational field determines the spacetime geometry. Therefore, the dependence of W on the fields is still sufficient to code the relative distance and time separation of the components of the measuring apparatus!

What is happening is that in background-dependent QFT we have two kinds of measurements: those that determine the distances of the parts of the apparatus and the time lapsed between measurements, and the actual measurements of the fields' dynamical variables. In quantum gravity, instead, distances and time separations are on an equal footing with the dynamical fields. This is the core of the general-relativistic revolution, and the key for background-independent QFT.

We need one final step. Notice from (1.12) that the argument of W is not the classical quantity, but rather the eigenstate of the corresponding operator. The eigenstates of the gravitational field are spin networks. Therefore in quantum gravity the argument of W must be a spin network, representing the possible outcome of a measurement of the gravitational field (or the geometry) on a closed 3d surface. Thus, in quantum gravity physical amplitudes must be expressed by amplitudes of the form $W(s)$. These give the correlation probability amplitude associated with the outcome s in a measurement of a geometry, just as $W(x, t, x', t')$ does for a particle.

A particularly interesting case is when we can separate the boundary surface in two components, then $s = s_{\text{out}} \cup s_{\text{in}}$. In this case, $W(s_{\text{out}}, s_{\text{in}})$ can be interpreted as the probability amplitude of measuring the quantum three-geometry s_{out} if s_{in} was observed.

Notice that a spin network s_{in} is the analog of (x, t), not just x alone. The time variable is mixed up with the physical variables (Chapter 3). The notion of unitary quantum evolution in time is ill defined in this context, but probability amplitudes remain well defined and physically meaningful (Chapter 5). The quantum dynamical information of the theory is entirely contained in the spin network amplitudes $W(s)$. Given a configuration of space and matter, these amplitudes determine a correlation probability of observing it.

Calculation of the amplitude $W(s)$. In the relativistic formulation of classical hamiltonian theory, dynamics is governed by the relativistic hamiltonian H.[8] This is discussed in detail in Chapter 3. The quantum dynamics is governed by the corresponding quantum operator H. In quantum gravity, H is defined on the space of the spin networks. There is no external time variable t in the theory, and the quantum dynamical equation which replaces the Schrödinger equation is the equation $H\Psi = 0$, called the Wheeler–DeWitt equation. The space of the solutions of the Wheeler–DeWitt equation is denoted $\mathcal{H}$. There is an operator $P : \mathcal{K}_{\text{diff}} \to \mathcal{H}$ that projects $\mathcal{K}_{\text{diff}}$ on the solutions of the Wheeler–DeWitt equation (for a mathematically more precise statement, see Section 5.2).

The transition amplitudes $W(s, s')$ are the matrix elements of the operator P. They define the physical scalar product, namely the scalar product of the space $\mathcal{H}$

$$W(s, s') = \langle s|P|s'\rangle_{\mathcal{K}_{\text{diff}}} = \langle s|s'\rangle_{\mathcal{H}}. \tag{1.15}$$

Thus, the transition amplitude between two states is simply their physical scalar product (Chapter 5). More generally, there is a preferred state $|\emptyset\rangle$

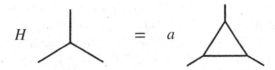

Fig. 1.4   Scheme of the action of H on a node of a spin network.

in $\mathcal{K}_{\mathrm{diff}}$ which is formed by no spin networks. It represents a space with zero volume, or, more precisely, no space at all. The covariant vacuum state which defines the dynamics of the theory is defined by $|0\rangle = P|\emptyset\rangle$. The amplitude of a spin network is defined by

$$W(s) = \langle 0|s\rangle = \langle \emptyset|P|s\rangle \ . \tag{1.16}$$

The construction of the operator H is a major task in LQG. It is delicate and it requires a nontrivial regularization procedure in order to deal with operator products. Chapter 7 is devoted to this construction. Remarkably, the limit in which the regularization is removed exists precisely thanks to diffeomorphism invariance (Section 7.1). This is a second major payoff of background independence. At present, more than one version of the operator H has been constructed, and it is not yet clear which variant (if any!) is correct. The remarks that follow refer to all of them.

The most remarkable aspect of the hamiltonian operator H is that it acts only on the nodes. A state labeled by a spin network without nodes – that is, in which the graph Γ is simply a collection of nonintersecting loops – is a solution of the Wheeler–DeWitt equation. In fact, the unexpected fact that exact solutions of the Wheeler–DeWitt equation could be found at all was the first major surprise that raised interest in LQG in the first place, in the late 1980s.

Acting on a generic state $|s\rangle$, the action of the operator H turns out to be discrete and combinatorial: the topology of the graph is changed and the labels are modified in the vicinity of a node. A typical example of the action of H on a node is illustrated in Figure 1.4: the action on a node splits the node into three nodes and multiplies the state by a number a (that depends on the labels of the spin network around the node). Labels of links and nodes are not indicated in the figure.

Notice the various manners in which the spin network basis is effective in quantum gravity. The states in the spin network basis

(i)   diagonalize area and volume;

(ii) control diff-invariance: diffeomorphism equivalence classes of states are labeled by the s-knots;

(iii) simplify the action of H, reducing it to a combinatorial action on the nodes.

The construction of the hamiltonian operator H completes the definition of the general formalism of LQG in the case of pure gravity. This is
extended to matter couplings in Chapter 7. In Chapter 8, I describe some
of the most interesting applications of the theory. In particular, I illustrate
the application of LQG to cosmology (control of the classical initial singularity, inflation) and to black-hole physics (entropy, emitted spectrum).
I also mention some of its tentative applications in astrophysics.

1.2.4 Quantum spacetime: spinfoam

To be able to compute all the predictions of a theory, it is not sufficient to
have the general definition of the theory. A road towards the calculation
of transition amplitudes in quantum gravity is provided by the spinfoam
formalism.

Following Feynman's ideas, we can give $W(s, s')$ a representation as a
sum-over-paths. This representation can be obtained in various manners.
In particular, it can be intuitively derived from a perturbative expansion,
summing over different histories of sequences of actions of H that send s'
into s.

A path is then the "world-history" of a graph, with interactions happening at the nodes. This world-history is a two-complex, as in Figure
1.5, namely a collection of faces (the world-histories of the links); faces
join at edges (the world-histories of the nodes); in turn, edges join at
vertices. A vertex represents an individual action of H. An example of a
vertex, corresponding to the action of H of Figure 1.4, is illustrated in
Figure 1.6. Notice that on moving from the bottom to the top, a section
of the two-complex goes precisely from the graph on the left-hand side of
Figure 1.4 to the one on the right-hand side. Thus, a two-complex is like
a Feynman graph, but with one additional structure. A Feynman graph is
composed by vertices and edges, a spinfoam by vertices, edges and faces.

Faces are labeled by the area quantum numbers j_l and edges by the
volume quantum numbers i_n. A two-complex with faces and edges labeled in this manner is called a "spinfoam" and denoted σ. Thus, a spinfoam is a Feynman graph of spin networks, or a world-history of spin
networks. A history going from s' to s is a spinfoam σ bounded by s'
and s.

In the perturbative expansion of $W(s, s')$, there is a term associated
with each spinfoam σ bounded by s and s'. This term is the amplitude of
σ. The amplitude of a spinfoam turns out to be given by (a measure term
$\mu(\sigma)$ times) the product over the vertices v of a vertex amplitude $A_v(\sigma)$.
The vertex amplitude is determined by the matrix element of H between
the incoming and the outgoing spin networks and is a function of the labels

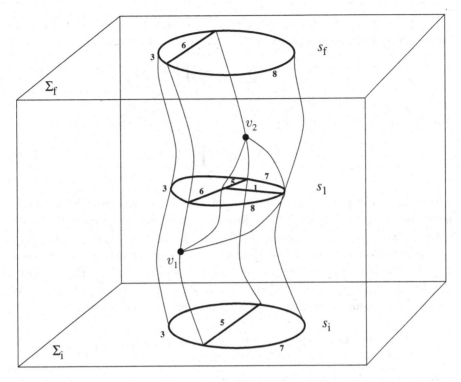

Fig. 1.5 A spinfoam representing the evolution of an initial spin network s_i to a final spin network s_f, via an intermediate spin network s_1. Here v_1 and v_2 are the interaction vertices.

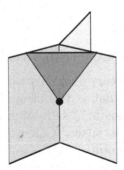

Fig. 1.6 The vertex of a spinfoam.

of the faces and the edges adjacent to the vertex. This is analogous to the amplitude of a conventional Feynman vertex, which is determined by the matrix element of the hamiltonian between the incoming and outgoing states.

The physical transition amplitudes $W(s, s')$ are then obtained by summing over spinfoams bounded by the spin networks s and s'

$$W(s, s') \sim \sum_{\substack{\sigma \\ \partial\sigma = s \cup s'}} \mu(\sigma) \prod_v A_v(\sigma).$$ (1.17)

More generally, for a spin network s representing a closed surface,

$$W(s) \sim \sum_{\substack{\sigma \\ \partial\sigma = s}} \mu(\sigma) \prod_v A_v(\sigma).$$ (1.18)

In general, the Feynman path integral can be derived from Schrödinger theory by exponentiating the hamiltonian operator, but it can also be directly interpreted as a sum over classical trajectories of the particle. Similarly, the spinfoam sum (1.17) can be interpreted as a sum over spacetimes. That is, the sum (1.17) can be seen as a concrete and mathematically well-defined realization of the (ill-defined) Wheeler–Misner–Hawking representation of quantum gravity as a sum over four-geometries

$$W(^3g, {}^3g') \sim \int_{\partial g = {}^3g \cup {}^3g'} [Dg] \, e^{\frac{i}{\hbar} S[g]}.$$ (1.19)

Because of their foamy structure at the Planck scale, spinfoams can be viewed as a mathematically precise realization of Wheeler's intuition of a spacetime "foam." In Chapter 9, I describe various concrete realizations of (1.17), as well as the possibility of directly relating (1.17) with a discretization of (1.19).

1.3 Conceptual issues

The search for a quantum theory of gravity raises questions such as: What is space? What is time? What is the meaning of "being somewhere"? What is the meaning of "moving"? Is motion to be defined with respect to objects or with respect to space? Can we formulate physics without referring to time or to spacetime? And also: What is matter? What is causality? What is the role of the observer in physics?

Questions of this kind have played a central role in periods of major advances in physics. For instance, they played a central role for Einstein, Heisenberg, Bohr and their colleagues; but also for Descartes, Galileo, Newton and their contemporaries, and for Faraday, Maxwell and their colleagues. Today, this manner of posing problems is often regarded as "too philosophical" by many physicists.

Indeed, most physicists of the second half of the twentieth century have viewed questions of this nature as irrelevant. This view was appropriate for the problems they were facing: one does not need to worry about first principles in order to apply the Schrödinger equation to the helium atom, to understand how a neutron star holds together, or to find out the symmetry group governing the strong interactions. During this period, physicists lost interest in general issues. As someone has said during this period, "do not ask what the theory can do for you; ask what you can do for the theory." That is, do not ask foundational questions, just keep developing and adjusting the theory you happen to find in front of you. When the basics are clear and the issue is problem-solving within a given conceptual scheme, there is no reason to worry about foundations: the problems are technical and the pragmatical approach is the most effective one.

Today the kind of difficulties that we face have changed. To understand quantum spacetime, we have to return, once more, to those foundational issues. We have to find new answers to the old foundational questions. The new answers have to take into account what we have learned with QM and GR. This conceptual approach is not the one of Weinberg and Gell-Mann, but it is the one of Newton, Maxwell, Einstein, Bohr, Heisenberg, Faraday, Boltzmann, and many others. It is clear from the writings of the latter that they discovered what they did discover by thinking about general foundational questions. The problem of quantum gravity will not be solved unless we reconsider these questions.

Several of these questions are discussed in the text. Here I only comment on one of these conceptual issues: the role of the notion of time.

1.3.1 Physics without time

The transition amplitudes $W(s, s')$ do not depend explicitly on time. This is to be expected, because the physical predictions of classical GR do not depend explicitly on the time coordinate t either. The theory predicts correlations between physical variables, not the way physical variables evolve with respect to a preferred time variable. But what is the meaning of a physical theory in which the time variable t does not appear?

Let me tell a story. It was Galileo Galilei who first realized that the physical motion of objects on Earth could be described by mathematical laws expressing the evolution of observable quantities $A, B, C, \ldots$ in time. That is, laws for the functions $A(t), B(t), C(t), \ldots$ A crucial contribution by Galileo was to find an effective way to measure the time variable t, and therefore provide an operational meaning to these functions. In fact, Galileo gave a decisive contribution to the discovery of the modern clock by realizing, as a young man, that the small oscillations of a pendulum "take equal time." The story goes that Galileo was staring at the slow oscillations of the big chandelier that can still be seen in the marvelous

Cathedral of Pisa.[9] He checked the period of the oscillations against his pulse and realized that the same number of pulses lapsed during any oscillation of the chandelier. This was the key insight, the basis of the modern clock: today virtually every clock contains an oscillator. Later in life, Galileo used a clock to discover the first quantitative terrestrial physical law in his historic experiments on descent down inclines.

Now, the puzzling part of the story is that while Galileo checked the pendulum against his pulse, not long afterwards doctors were checking their patient's pulse against a pendulum. What is the actual meaning of the pendulum periods taking "equal time"? An equal amount of t lapses in any oscillation: how do we know this, if we can access t only via another pendulum?

It was Newton who cleared up the issue conceptually. Newton *assumes* that an *unobservable* quantity t exists, which flows ("absolute and equal to itself"). We write equations of motion in terms of this t, but we cannot truly access t: we can build clocks that give readings $T_1(t)$, $T_2(t), \ldots$ that, according to our equations, approximate t with the precision we want. What we actually measure is the evolution of other variables against clocks, namely $A(T_1)$, $B(T_1)$. Furthermore, we can check clocks against one another by measuring the functions $T_1(T_2)$, $T_2(T_3), \ldots$ The fact that all these observations agree with what we compute using evolution equations in t gives us confidence in the method. In particular, it gives us confidence that to assume the existence of the *unobservable* physical quantity t is a useful and reasonable thing to do.

Simply: the usefulness of this assumption is lost in quantum gravity. The theory allows us to calculate the relations between observable quantities, such as $A(B)$, $B(C)$, $A(T_1)$, $T_1(A), \ldots$, which is what we see. But it does not give us the evolution of these observable quantities in terms of an unobservable t, as Newton's theory and special relativity do. In a sense, this simply means that there are no good clocks at the Planck scale.

Of course, in a specific problem we can choose one variable, decide to treat it as the independent variable, and call it "the" time. For instance a certain clock time, a certain proper time along a certain particle history, etc. The choice is largely arbitrary and generally it is only locally meaningful. A generally covariant theory does not choose a preferred time variable.

Here are two examples to illustrate this arbitrariness.

- Imagine we throw a precise clock upward and compare its lapsed reading t_f when it lands back, with the lapsed reading t_e of a clock remaining on the Earth. GR predicts that the two clocks read differently, and provides a quantitative relation between t_f

---

[9]Nice story. Too bad the chandelier was hung there a few decades after Galileo's discovery.

and t_e. Is this about the observable t_f evolving in the physical time t_e, or about the observable t_e evolving in the physical time t_f?[10]

- The cosmological context is often indicated as one in which a natural choice of time is available: the cosmological time t_c is the proper time from the Big Bang along the galaxies' worldlines. But an event A happening on Andromeda at the same t_c as ours happens *much later* than an event B on Andromeda simultaneous to us in the sense of Einstein's definition of simultaneity.[11] So, what is happening "right now" on Andromeda? A or B? Furthermore, the real world is not truly homogeneous: when two galaxies, having two different ages relative to the Big Bang, or two different masses, me merge, which of the two has the right time?

So long as we remain within *classical* general relativity, a given gravitational field has the structure of a pseudo-riemannian manifold. Therefore, the dynamics of the theory has no preferred time variable, but we nevertheless have a notion of spacetime for each given solution. But in quantum theory there are no classical field configurations, just as there are no trajectories of a particle. Thus, in quantum gravity the notion of spacetime disappears in the same manner in which the notion of trajectory disappears in the quantum theory of a particle. A single spinfoam can be thought of as representing a spacetime, but the history of the world is not a single spinfoam: it is a sum over spinfoams.

The theory is conceptually well defined without making use of the notion of time. It provides probabilistic predictions for correlations between the physical quantities that we can observe. In principle, we can check these predictions against experiments.[12] Furthermore, the theory provides a clear and intelligible picture of the quantum gravitational field, namely of a "quantum geometry."

Thus, there is no background "spacetime", forming the stage on which things move. There is no "time" along which everything flows. The world in which we happen to live can be understood without using the notion of time.

Bibliographical notes

The fact that perturbative quantum general relativity is nonrenormalizable has been long believed, but was proven only in 1986 by Goroff and Sagnotti [29].

---

[10]If you are tempted to say that the lapsed reading t_e of the clock remaining on Earth gives the "true time" recall that the pseudo-riemannian distance between the two events at which the clocks meet is t_f, not t_e: it is the clock going up and down that follows a geodesic.

[11]Thanks to Marc Lachieze-Rey for this observation.

[12]The special properties of a time variable may emerge only macroscopically. This is discussed in Sections 3.4 and 5.5.1.

For an orientation on current research on quantum gravity see, for instance, the review papers [30–33]. An interesting panoramic of points of view on the problem is in the various contributions to the book [34]. I have given a critical discussion on the present state of spacetime physics in [35–37]. A historical account of development of quantum gravity is given in Appendix B.

As a general introduction to quantum gravity – a subject where nothing yet is certain – the student eager to learn is strongly advised to study also the classic reviews, which are rich in ideas and present different points of view, such as John Wheeler 1967 [38], Steven Weinberg 1979 [39], Stephen Hawking 1979 and 1980 [40,41], Karel Kuchar 1980 [42], and Chris Isham's magistral syntheses [43–45]. On string theory, classic textbooks are Green, Schwarz and Witten, and Polchinksi [46]. For a discussion of the difficulties of string theory and a comparison of the results of strings and loops, see [47], written in the form of a dialog, and [48]. For a fascinating presentation of Alain Connes' vision, see [49]. Lee Smolin's popular-science book [50] provides a readable and enjoyable introduction to LQG.

LQG has inspired novels and short stories. *Blue Mars*, by Kim Stanley Robinson [51], contains a description of the future evolution and merging of loop gravity and strings. I recommend the science fiction novel *Schild Ladder* by Greg Egan [52], which opens with one of the clearest presentations of the picture of space given by loop gravity (Greg is a talented writer and also a scientist who is contributing to the development of LQG), and, for those who can read Italian, *Anna prende il volo*, by Enrico Palandri [53], a charming novel with a gentle meditation on the meaning of the disappearance of time. Literature has the capacity of delicately merging the novel hard views that science develops into the common discourse of our civilization.

2

General Relativity

Lev Landau has called GR "the most beautiful" of the scientific theories. The theory is first of all a description of the gravitational force. Nowadays it is very extensively supported by terrestrial and astronomical observations, and so far it has never been questioned by an empirical observation.

But GR is far more than that. It is a complete modification of our understanding of the basic grammar of nature. This modification does not apply solely to gravitational interaction: it applies to all aspects of physics. In fact, the extent to which Einstein's discovery of this theory has modified our understanding of the physical world and the full reach of its consequences have not yet been completely unraveled.

This chapter is not an introduction to GR, nor an exhaustive description of the theory. For this, I refer the reader to the classic textbooks on the subject. Here, I give a short presentation of the formalism in a compact and modern form, emphasizing the reading of the theory which is most useful for quantum gravity. I also discuss in detail the physical and conceptual basis of the theory, and the way it has modified our understanding of the physical world.

2.1 Formalism

2.1.1 Gravitational field

Let M be the "spacetime" four-dimensional manifold. Coordinates on M are written as $x, x', \ldots$, where $x = (x^\mu) = (x^0, x^1, x^2, x^3)$. Indices $\mu, \nu, \ldots = 0, 1, 2, 3$ are spacetime tangent indices.

- *The gravitational field e is a one-form*

$$e^I(x) = e^I_\mu(x) \, \mathrm{d}x^\mu \tag{2.1}$$

with values in Minkowski space. Indices $I, J, \ldots = 0, 1, 2, 3$ label the components of a Minkowski vector. They are raised and lowered with the Minkowski metric η_{IJ}.

I call "gravitational field" the tetrad field rather than Einstein's metric field $g_{\mu\nu}(x)$. There are three reasons for this: (i) the standard model cannot be written in terms of g because fermions require the tetrad formalism; (ii) the tetrad field e is nowadays more utilized than g in quantum gravity; and (iii) I think that e represents the gravitational field in a more conceptually clean way than g (see Section 2.2.3.) The relation with the metric formalism is given in Section 2.1.5.

- *The spin connection* ω is a one-form with values in the Lie algebra of the Lorentz group $so(3,1)$

$$\omega^I{}_J(x) = \omega^I{}_{\mu J}(x)\, \mathrm{d}x^\mu, \qquad (2.2)$$

where $\omega^{IJ} = -\omega^{JI}$. It defines a covariant partial derivative D_μ on all fields that have Lorentz (I) indices:

$$D_\mu v^I = \partial_\mu v^I + \omega^I{}_{\mu J}\, v^J, \qquad (2.3)$$

and a gauge-covariant exterior derivative D on forms. For instance, for a one-form u^I with a Lorentz index,

$$Du^I = du^I + \omega^I{}_J \wedge u^J. \qquad (2.4)$$

The torsion two-form is defined as

$$T^I = De^I = de^I + \omega^I{}_J \wedge e^J. \qquad (2.5)$$

A tetrad field e determines uniquely a torsion-free spin connection $\omega = \omega[e]$, called compatible with e, by

$$T^I = de^I + \omega[e]^I{}_J \wedge e^J = 0. \qquad (2.6)$$

The explicit solution of this equation is given below in (2.91) or (2.92).

- *The curvature* R of ω is the Lorentz algebra valued two-form[1]

$$R^I{}_J = R^I{}_{J\mu\nu}\, \mathrm{d}x^\mu \wedge \mathrm{d}x^\nu, \qquad (2.7)$$

defined by[2]

$$R^I{}_J = d\omega^I{}_J + \omega^I{}_K \wedge \omega^K{}_J. \qquad (2.8)$$

---

[1] Generally I write spacetime indices $\mu\nu$ before internal Lorentz indices IJ. But for the curvature I prefer to stay closer to Riemann's notation.

[2] Sometimes the curvature of a connection $\omega^I{}_J$ is written as $R^I{}_J = D\omega^I{}_J$. If we naively use the definition (2.4) for D, we get an extra 2 in the quadratic term. The point is that the indices on the connection are not vector indices. That is, (2.4) defines the action of D on sections of a vector bundle, and a connection is not a section of a vector bundle.

We have then immediately from (2.4)

$$D^2 u^I = R^I{}_J \wedge u^J, \tag{2.9}$$

and from this equation and (2.6):

$$R^I{}_J \wedge e^J = 0. \tag{2.10}$$

A region where the curvature is zero is called "flat." Equations (2.5) and (2.8) are called the Cartan structure equations.

- *The Einstein equations* "in vacuum" are

$$\epsilon_{IJKL} \left(e^I \wedge R^{JK} - \frac{2}{3} \lambda \, e^I \wedge e^J \wedge e^K \right) = 0. \tag{2.11}$$

The equation (2.6) relating e and ω and the Einstein equations (2.11) are the field equations of GR in the absence of other fields. They are the Euler–Lagrange equations of the action

$$S[e, \omega] = \frac{1}{16\pi G} \int \epsilon_{IJKL} \left(\frac{1}{4} e^I \wedge e^J \wedge R[\omega]^{KL} - \frac{1}{12} \lambda \, e^I \wedge e^J \wedge e^K \wedge e^L \right) \tag{2.12}$$

where G is the Newton constant[3], and λ is the cosmological constant which I often set to zero below.

- *Inverse tetrad.* Using the matrix $e^\mu_I(x)$, defined to be the inverse of the matrix $e^I_\mu(x)$, we define the Ricci tensor

$$R^I_\mu = R^{IJ}{}_{\mu\nu} \, e^\nu_J, \tag{2.13}$$

and the Ricci scalar

$$R = R^I_\mu \, e^\mu_I, \tag{2.14}$$

and write the vacuum Einstein equations (2.11) as

$$R^I_\mu - \frac{1}{2} R e^I_\mu + \lambda e^I_\mu = 0. \tag{2.15}$$

---

[3]The constant $16\pi G$ has no effect on the classical equations of motion (2.11). However, it governs the strength of the interaction with the matter fields described below, and it also determines the quantum properties of the system. In this it is similar to the mass constant m in front of a free-particle action: the classical equations of motion ($\ddot{x} = 0$) do not depend on m, but the quantum dynamics of the particle does. For instance, the rate at which a wave packet spreads depends on m. Similarly, we will see that the quanta of pure gravity are governed by this constant.

- *Second-order formalism.* Replacing ω with $\omega[e]$ in (2.12) we get the equivalent action

$$S[e] = \frac{1}{16\pi G} \int \epsilon_{IJKL} \left(\frac{1}{4} e^I \wedge e^J \wedge R[\omega[e]]^{KL} - \frac{1}{12} \lambda \, e^I \wedge e^J \wedge e^K \wedge e^L \right). \quad (2.16)$$

The formalisms in (2.12) where e and ω are independent is called the first-order formalism. The two formalism are not equivalent in the presence of fermions; we do not know which one is physically correct, because the effect of gravity on single fermions is hard to measure.

- *Selfdual formalism.* Consider the selfdual "projector" P^i_{IJ} given by

$$P^i_{jk} = \frac{1}{2} \epsilon^i{}_{jk}, \qquad P^i_{0j} = -P^i_{j0} = \frac{i}{2} \delta^i_j, \quad (2.17)$$

where $i = 1, 2, 3$.[4] This verifies the two properties

$$\frac{1}{2} \epsilon_{IJ}{}^{KL} P^i_{KL} = i P^i_{IJ}, \quad P^I_i{}^J P^i_{KL} = P^{IJ}{}_{KL} \equiv \frac{1}{2} \delta^I_{[K} \, \delta^L_{J]} + \frac{i}{4} \epsilon^{IJ}{}_{KL}, \quad (2.18)$$

where $P^{IJ}{}_{KL}$ is the projector on selfdual tensors. Define the complex $SO(3)$ connection

$$A^i_\mu = P^i_{IJ} \, \omega^{IJ}_\mu. \quad (2.19)$$

Equivalently,

$$A^i = \omega^i + i \omega^{0i}. \quad (2.20)$$

(We write $\omega^i = 1/2 \epsilon^i{}_{jk} \omega^{jk}$. See pg. xxii.) We can use the complex selfdual connection A^i (three complex one-forms), instead of the real connection $\omega^I{}_J$ (six real one-forms), as the dynamical variable for GR. (This is equivalent to describing a system with two real degrees of freedom x and y in terms of a single complex variable $z = x + iy$.) In terms of A^i, the vacuum Einstein equations read

$$P_{iIJ} \, e^I \wedge \left(F^i - \frac{2}{3} \lambda \, P^i_{KL} e^K \wedge e^L \right) = 0, \quad (2.21)$$

where $F^i = dA^i + \epsilon^i_{jk} A^j A^k$ is the curvature of A.[5] These are the Euler–Lagrange equations of the action

$$S[e, A] = \frac{1}{16\pi G} \int \left(-i P_{iIJ} \, e^I \wedge e^J \wedge F^i - \frac{1}{12} \lambda \, \epsilon_{IJKL} \, e^I \wedge e^J \wedge e^K \wedge e^L \right), \quad (2.22)$$

which differs from the action (2.12) by an imaginary term that does not change the equations of motion. The selfdual formalism is often used in canonical quantization, because it simplifies the form of the hamiltonian theory. If we replace the imaginary unit i in (2.17) with a real parameter γ, (2.22) is called the Holst action [54], and gives rise to the Ashtekar-Barbero-Immirzi formalism. γ is called the Immirzi parameter.

Plebanski formalism. The Plebanski selfdual two-form is defined as

$$\Sigma^i = P^i_{IJ} \, e^I \wedge e^J. \quad (2.23)$$

That is,

$$\Sigma^1 = e^2 \wedge e^3 + i \, e^0 \wedge e^1 \quad (2.24)$$

---

[4] The complex Lorentz algebra splits into two complex $so(3)$ algebras, called the selfdual and anti-selfdual components: $so(3,1;C) = so(3;C) \oplus so(3;C)$. The projector (2.17) reads out the selfdual component.

[5] Because of the split mentioned in the previous footnote, the curvature of the selfdual component of the connection is the selfdual component of the curvature.

and so on cyclically. A straightforward calculation shows that Σ satisfies

$$D\Sigma^i \equiv d\Sigma^i + A^i{}_j \wedge \Sigma^j = 0. \tag{2.25}$$

where we write $A^i{}_j = \epsilon^i{}_{jk} A^k$. See pg. xxii. The algebraic equations for a triplet of complex two-forms Σ^i

$$3\,\Sigma^i \wedge \Sigma^j = \delta^{ij}\,\Sigma_k \wedge \Sigma^k = -\delta^{ij}\,\overline{\Sigma}_k \wedge \overline{\Sigma}^k, \qquad \Sigma^i \wedge \overline{\Sigma}^j = 0 \tag{2.26}$$

are solved by (2.23), where e^I is an arbitrary real tetrad. The GR action can thus be written as

$$S[\Sigma, A] = \frac{-i}{16\pi G} \int \left(\Sigma_i \wedge F^i + \frac{1}{3}\lambda\,\Sigma_k \wedge \Sigma^k\right), \tag{2.27}$$

where Σ^i satisfies the Plebanski constraints (2.26). The Plebanski formalism is often used as a starting point for spinfoam models.

2.1.2 "Matter"

In the general-relativistic parlance, "matter" is anything which is not the gravitational field. As far as we know, the world is made up of the gravitational field, Yang–Mills fields, fermion fields and, presumably, scalar fields.

- *Maxwell.* The electromagnetic field is described by the one-form field A, the Maxwell potential,

$$A(x) = A_\mu(x)\,dx^\mu. \tag{2.28}$$

Its curvature is the two-form $F = dA$, with components $F_{\mu\nu} = \partial_\mu A_\nu - \partial_\nu A_\mu$. Its dynamics is governed by the action

$$S_M[e, A] = \frac{1}{4} \int F^* \wedge F. \tag{2.29}$$

- *Yang–Mills.* The above generalizes to a nonabelian connection A in a Yang–Mills group G. A defines a gauge covariant exterior derivative D and curvature F. The action is

$$S_{YM}[e, A] = \frac{1}{4} \int \text{tr}[F^* \wedge F], \tag{2.30}$$

where tr is a trace on the algebra.

- *Scalar.* Let $\varphi(x)$ be a scalar field, possibly with values in a representation of G. The Yang–Mills field A defines the covariant partial derivative

$$D_\mu\varphi = \partial_\mu\varphi + A^A_\mu L_A\varphi, \tag{2.31}$$

where L_A are the generators of the gauge algebra in the representations to which φ belongs. The action that governs the dynamics of the field is

$$S_{sc}[e, A, \varphi] = \int d^4x\, e\,\left(\eta^{IJ}\, e^\mu_I\, \overline{D_\mu\varphi}\, e^\nu_J\, D_\nu\varphi + V(\varphi)\right), \tag{2.32}$$

where e is the determinant of e^I_μ and $V(\varphi)$ is a self-interaction potential.

• *Fermion.* A fermion field ψ is a field in a spinor representation of the Lorentz group, possibly with values in a representation of G. The spin connection ω and the Yang–Mills field A define the covariant partial derivative

$$\mathrm{D}_\mu\psi = \partial_\mu\psi + \omega^I_{\mu J}L^J_I\psi + A^A_\mu L_A\psi, \tag{2.33}$$

where L^J_I and L_A are the generators of the Lorentz and gauge algebras in the representations to which ψ belongs. Define

$$\slashed{D}\psi = \gamma^I e^\mu_I \,\mathrm{D}_\mu\psi, \tag{2.34}$$

where γ^I are the standard Dirac matrices. The action that governs the dynamics of the fermion field is

$$S_\mathrm{f}[e,\omega,A,\varphi,\psi] = \int \mathrm{d}^4x \; e \left(\bar\psi \slashed{D}\psi + Y(\varphi,\bar\psi,\psi)\right) + \text{complex conjugate}, \tag{2.35}$$

where the second term is a polynomial interaction potential with a scalar field.

• *The "lagrangian of the world": the standard model.* As far as we know, the world can be described in terms of a set of fields e,ω,A,ψ,φ, where $G = SU(3) \times SU(2) \times U(1)$ and ψ and φ are in suitable multiplets, and is governed by the action

$$\begin{aligned}
S[e,\omega,A,\psi,\varphi] &= S_\mathrm{GR}[e,\omega] + S_\mathrm{YM}[e,A] + S_\mathrm{f}[e,\omega,A,\psi] + S_\mathrm{sc}[e,A,\varphi]\\
&= S_\mathrm{GR}[e,\omega] + S_\mathrm{matter}[e,\omega,A,\varphi,\psi],
\end{aligned} \tag{2.36}$$

with suitable polynomials V and Y. The equations of motion that follow from this action by varying e are the Einstein equations (2.11) with a source term, namely

$$\epsilon_{IJKL}\left(e^I \wedge R^{JK} - \frac{2}{3}\lambda\, e^I \wedge e^J \wedge e^K\right) = 2\pi G\, T_L, \tag{2.37}$$

where the energy-momentum three-form

$$T_I = \frac{\det e}{3!}T^\mu_I\, \epsilon_{\mu\nu\rho\sigma}\mathrm{d}x^\nu \wedge \mathrm{d}x^\rho \wedge \mathrm{d}x^\sigma \tag{2.38}$$

is defined by

$$T_I(x) = \frac{\delta S_\mathrm{matter}}{\delta e^I(x)}. \tag{2.39}$$

Equivalently, the Einstein equations (2.37) can be written as

$$R^I_\mu - \frac{1}{2}Re^I_\mu + \lambda e^I_\mu = 8\pi G\, T^I_\mu. \tag{2.40}$$

$T_\mu^I(x)$ is called the energy-momentum tensor. It is the sum of the individual energy-momentum tensors of the various matter terms.[6]

- *Particles.* The trajectory $x^\mu(s)$ of a point particle is an approximate notion. Macroscopic objects have finite size and elementary particles are quantum entities and therefore have no trajectories. At macroscopic scales, the notion of a point-particle trajectory is nevertheless very useful.

  In the absence of nongravitational forces, the equations of motion for the worldline $\gamma : s \mapsto x^\mu(s)$ of a particle are determined by the action

$$S[e, \gamma] = m \int ds \sqrt{-\eta_{IJ} v^I(s) v^J(s)}, \tag{2.41}$$

where

$$v^I(s) = e_\mu^I(x(s)) v^\mu(s), \tag{2.42}$$

and v^μ is the particle velocity

$$v^\mu(s) = \dot{x}^\mu(s) \equiv \frac{dx^\mu(s)}{ds}. \tag{2.43}$$

This action is independent of the way the trajectory is parametrized, and therefore determines the path, not its parametrization. With the parametrization choice $v_I v^I = -1$, the equations of motion are

$$\ddot{x}^\mu = -\Gamma_{\nu\rho}^\mu \, \dot{x}^\nu \dot{x}^\rho, \tag{2.44}$$

where

$$\Gamma_{\mu\nu}^\sigma = e_j^\rho e^{J\sigma} (e_{\rho I} \partial_{(\mu} e_{\nu)}^I + e_{\nu I} \partial_{[\mu} e_{\rho]}^I + e_{\mu I} \partial_{[\nu} e_{\rho]}^I) \tag{2.45}$$

is called the Levi–Civita connection. In an arbitrary parametrization the equations of motion are

$$\ddot{x}^\mu + \Gamma_{\nu\rho}^\mu \, \dot{x}^\nu \dot{x}^\rho = I(s) \, \dot{x}^\mu, \tag{2.46}$$

where $I(s)$ is an arbitrary function of s.

Minkowski solution. Consider a regime in which we can assume that the Newton constant G is small, that is, a regime in which we can neglect the effect of matter on the gravitational field. Assume also that, within our approximation, the cosmological constant λ is negligible. The Einstein equations (2.11) then admit (among many others) the particularly interesting solution

$$e_\mu^I(x) = \delta_\mu^I, \qquad \omega_{\mu J}^I(x) = 0, \tag{2.47}$$

which is called the Minkowski solution. This solution is everywhere flat.

Assume that the gravitational field is in this configuration. What are the equations of motion of the matter interacting with this particular

---

[6] The energy-momentum tensor defined as the variation of the action with respect to the gravitational field may differ by a total derivative from the one conventional in Minkowski space defined as the Noether current of translations.

gravitational field? These are easily obtained by inserting the Minkowski solution (2.47) into the matter action (2.36)

$$S[A, \varphi, \psi] = S_{\text{matter}}[e = \delta, \omega = 0, A, \varphi, \psi]. \qquad (2.48)$$

The action $S[A, \varphi, \psi]$ is the action of the standard model used in high-energy physics. This action is usually written in terms of the spacetime Minkowski metric $\eta_{\mu\nu}$. This metric is obtained from the Minkowski value (2.47) of the tetrad field. For instance, in the action of a scalar field (2.32), the combination $\eta^{IJ}e_I^\mu(x)e_J^\nu(x)$ becomes

$$\eta^{IJ}e_I^\mu(x)e_J^\nu(x) = \eta^{IJ}\delta_I^\mu \, \delta_J^\nu = \eta^{\mu\nu} \qquad (2.49)$$

on this solution.

The Minkowski metric $\eta_{\mu\nu}$ of special-relativistic physics is nothing but a particular value of the gravitational field. It is one of the solutions of the Einstein equation, within a certain approximation.

2.1.3 Gauge invariance

The general definition of a system with a gauge invariance, and the one which is most useful for understanding the physics of gauge systems, is the following, which is due to Dirac. Consider a system of evolution equations in an evolution parameter t. The system is said to be "gauge" invariant if evolution is under-determined, that is, if there are two distinct solutions that are equal for t less than a certain $\hat{t}$; see Figure 2.1. These two solutions are said to be "gauge equivalent." Any two solutions are said to be gauge equivalent if they are gauge equivalent (as above) to a third solution. The gauge group $\mathcal{G}$ is a group that acts on the physical fields and maps gauge-equivalent solutions into one another. Since classical physics is deterministic, under-determined evolution equations are physically consistent only under the stipulation that only quantities invariant under gauge transformations are physical predictions of the theory. These quantities are called the gauge-invariant observables.

The equations of motion derived by the action (2.36) are invariant under three groups of gauge transformations: (i) local Yang–Mills gauge transformations, (ii) local Lorentz transformations and (iii) diffeomorphism transformations. They are described below. Gauge-invariant observables must be invariant under these three groups of transformations.

(i) **Local G transformations.** G is the Yang–Mills group. A local G transformation is labeled by a map $\lambda : M \to G$. It acts on φ, ψ and the connection A in the

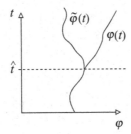

Fig. 2.1   Dirac definition of gauge: two different solutions of the equations of motion must be considered gauge equivalent if they are equal for $t < \hat{t}$.

well-known form, while e and ω are invariant,

$$\lambda: \quad \varphi(x) \mapsto R_\varphi(\lambda(x))\, \varphi(x), \tag{2.50}$$

$$\psi(x) \mapsto R_\psi(\lambda(x))\, \psi(x), \tag{2.51}$$

$$A_\mu(x) \mapsto R(\lambda(x))\, A_\mu(x) + \lambda(x)\partial_\mu \lambda^{-1}(x), \tag{2.52}$$

$$e_\mu^I(x) \mapsto e_\mu^I(x), \tag{2.53}$$

$$\omega_{\mu J}^I(x) \mapsto \omega_{\mu J}^I(x). \tag{2.54}$$

Here R_φ and R_ψ are the representations of G to which φ and ψ belong and R is the adjoint representation.

(ii) Local Lorentz transformations. A local Lorentz transformation is labeled by a map $\lambda: M \to SO(3,1)$. It acts on φ, ψ and the connection ω precisely as a Yang–Mills local transformation with Yang–Mills group $G = SO(3,1)$. Scalars φ belong to the trivial representation; fermions ψ belong to the spinor representations S. The gravitational field e transforms in the fundamental representation. Explicitly, writing an element of $SO(3,1)$ as $\lambda^I{}_J$, we have

$$\lambda: \quad \varphi(x) \mapsto \varphi(x), \tag{2.55}$$

$$\psi(x) \mapsto S(\lambda(x))\, \psi(x), \tag{2.56}$$

$$A_\mu(x) \mapsto A_\mu(x), \tag{2.57}$$

$$e_\mu^I(x) \mapsto \lambda^I{}_J(x)\, e_\mu^J(x), \tag{2.58}$$

$$\omega_{\mu J}^I(x) \mapsto \lambda^I{}_K(x)\omega_{\mu L}^K(x)\lambda^L{}_J(x) + \lambda_K^I(x)\partial_\mu\lambda^K{}_J(x). \tag{2.59}$$

(iii) Diffeomorphisms. Third, and most important, is the invariance under diffeomorphisms. A diffeomorphism gauge transformation is labeled by a smooth invertible map $\phi: M \to M$ (that is, by a "diffeomorphism" of M).[7] It acts *nonlocally* on all the fields, by pulling them back, according to their form char-

---

[7]There is an unfortunate terminological imprecision. A map $\phi: M \to M$ is called a diffeomorphism. The associated transformations (2.60)–(2.64) on the fields are also often loosely called a diffeomorphism (also in this book), instead of diffeomorphism gauge transformations. This tends to generate confusion.

acter: φ and ψ are zero forms, e, ω and A are one-forms:[8]

$$\phi: \qquad \varphi(x) \mapsto \varphi(\phi(x)), \tag{2.60}$$

$$\psi(x) \mapsto \psi(\phi(x)), \tag{2.61}$$

$$A_\mu(x) \mapsto \frac{\partial \phi^\nu(x)}{\partial x^\mu} A_\nu(\phi(x)), \tag{2.62}$$

$$e_\mu^I(x) \mapsto \frac{\partial \phi^\nu(x)}{\partial x^\mu} e_\nu^I(\phi(x)), \tag{2.63}$$

$$\omega_{\mu J}^I(x) \mapsto \frac{\partial \phi^\nu(x)}{\partial x^\mu} \omega_{\nu J}^I(\phi(x)). \tag{2.64}$$

These three groups of transformations send solutions of the equations of motion into other solutions of the equations of motion. They are gauge transformations because we can take these transformations to be the identity before a given coordinate time $\hat{t}$ and different from the identity afterwards. Therefore they are responsible for the under-determination of the evolution equations. Following Dirac's argument given above, physical predictions of the theory must be given by quantities invariant under all three of these transformations.

In particular, let a local quantity in spacetime be a quantity dependent on a fixed given point x. Notice that such a quantity cannot be invariant under a diffeomorphism. Therefore no local quantity in spacetime (in this sense) is a gauge-invariant observable in GR. The meaning of this fact and the far-reaching consequences of diffeomorphism invariance are discussed below in Section 2.3.2.

2.1.4 Physical geometry

At each point x of the spacetime manifold M, the gravitational field $e_\mu^I(x)$ defines a map from the tangent space $T_x M$ to Minkowski space. The map sends a vector v^μ in $T_x M$ into the Minkowski vector $u^I = e_\mu^I(x)v^\mu$. The Minkowski length $|u| = \sqrt{-u \cdot u} = \sqrt{-\eta_{IJ}u^I u^J}$ defines a norm $|v|$ of the tangent vector v^μ

$$|v| \equiv |u| = \sqrt{-\eta_{IJ}(e_\mu^I(x)v^\mu)\,(e_\nu^J(x)v^\nu)}; \tag{2.65}$$

---

[8]Under this definition, internal Lorentz, spinor, and gauge indices do not transform under a diffeomorphism. Alternatively, one should consider fiber-preserving diffeomorphisms of the Lorentz and gauge bundle. This alternative can be viewed as mathematically more clean and physically more attractive, because it makes more explicit the fact that local inertial frames or local gauge choices at different spacetime points cannot be identified (see later). However, the mathematical description of a diffeomorphism becomes more complicated, while the two choices are ultimately physically equivalent, due to the gauge invariance under local Lorentz and gauge transformations. The proper mathematical transformation of a spinor under diffeomorphisms is discussed in [55] and [56].

$|v|$ is called the "physical length" of the tangent vector v. The tangent vector v is called timelike (spacelike or lightlike) if u is timelike (spacelike or lightlike).

This fact allows us to assign a size to any d-dimensional surface in M. At any point x on the surface, the gravitational field maps the tangent space of the surface into a surface in Minkowski space. This surface carries a volume form, which can be pulled back to the tangent space of x and then to the surface itself, and integrated. In particular:

The length $\mathbf{L}$ of a curve $\gamma : s \mapsto x^\mu(s)$ is the line integral of the norm of its tangent

$$\mathbf{L}[e,\gamma] = \int |d\gamma| = \int ds \, |u(s)| = \int ds \, \sqrt{-\eta_{IJ} \, u^I(s) \, u^J(s)}, \quad (2.66)$$

where

$$u^I(s) = e^I_\mu(\gamma(s)) \, \frac{dx^\mu(s)}{ds}. \quad (2.67)$$

This can be written as the line integral of the norm of the one-form $e^I(x) = e^I_\mu(x)dx^\mu$ along γ:

$$\mathbf{L}[e,\gamma] = \int_\gamma |e| . \quad (2.68)$$

The length is independent of the parametrization and the orientation of γ. A curve is called timelike if its tangent is everywhere timelike. Notice that the action of a particle (2.41) is nothing but the length of its path in spacetime

$$S[e,\gamma] = m \, \mathbf{L}[e,\gamma]. \quad (2.69)$$

The area $\mathbf{A}$ of a two-dimensional surface $\mathcal{S} : \sigma = (\sigma^i) \mapsto x^\mu(\sigma^i)$, $i = 1, 2$, immersed in M, is

$$\mathbf{A}[e,\mathcal{S}] = \int |d^2\mathcal{S}| = \int_\mathcal{S} d^2\sigma \, \sqrt{\det(u_i \cdot u_j)}, \quad (2.70)$$

where

$$u^I_i(\sigma) = e^I_\mu(\gamma(\sigma)) \, \frac{\partial x^\mu(\sigma)}{\partial \sigma^i}, \quad (2.71)$$

and the determinant is over the i, j indices. That is,

$$\mathbf{A}[e,\mathcal{S}] = \int d^2\sigma \, \sqrt{(u_1 \cdot u_1)(u_2 \cdot u_2) - (u_1 \cdot u_2)^2}. \quad (2.72)$$

A surface is called spacelike if its tangents are all spacelike.

The volume $\mathbf{V}$ of a three-dimensional region $\mathcal{R} : \sigma = (\sigma^i) \mapsto x^\mu(\sigma^i)$, $i = 1, 2, 3$, immersed in M, is

$$\mathbf{V}[e, \mathcal{R}] = \int |\mathrm{d}^3\mathcal{R}| = \int_{\mathcal{R}} \mathrm{d}^3\sigma \sqrt{n \cdot n}, \tag{2.73}$$

where

$$n_I = \epsilon_{IJKL} \, u_1^J u_2^K u_3^L \tag{2.74}$$

is normal to the surface. A region is called spacelike if n is everywhere timelike.

The quantities $\mathbf{L}$, $\mathbf{A}$ and $\mathbf{V}$ are particular functions of the gravitational field e. The reason they have these geometric names is discussed below in Section 2.2.3.

2.1.5 Holonomy and metric

In GR, quantities close to observations, such as lengths and areas, are nonlocal in the sense that they depend on finite but extended regions in spacetime, such as lines and surfaces. Another natural nonlocal quantity which plays a central role in the quantum theory is the holonomy U of the gravitational connection (ω, or its selfdual part A) along a curve γ.

Definition of the holonomy. Given a connection A in a group G over a manifold M, the holonomy is defined as follows. Let a curve γ be a continuous, piecewise smooth map from the interval $[0, 1]$ into M,

$$\gamma : [0, 1] \longrightarrow M, \tag{2.75}$$
$$s \longmapsto x^\mu(s). \tag{2.76}$$

The holonomy, or parallel propagator, $U[A, \gamma]$ of the connection A along the curve γ is the element of G defined by

$$U[A, \gamma](0) = \mathbb{1} , \tag{2.77}$$

$$\frac{\mathrm{d}}{\mathrm{d}s} U[A, \gamma](s) - \dot{\gamma}^\mu(s) \, A_\mu\big(\gamma(s)\big) \, U[A, \gamma](s) = 0, \tag{2.78}$$

$$U[A, \gamma] = U[A, \gamma](1), \tag{2.79}$$

where $\dot{\gamma}^\mu(s) \equiv \mathrm{d}x^\mu(s)/\mathrm{d}s$ is the tangent to the curve. (In the mathematical literature, the term "holonomy" is generally used for closed curves only. In the quantum gravity literature, it is commonly employed for open curves as well.) The formal solution of this equation is

$$U[A, \gamma] = \mathcal{P} \exp \int_0^1 \mathrm{d}s \, \dot{\gamma}^\mu(s) \, A_\mu^i\big(\gamma(s)\big) \, \tau_i \equiv \mathcal{P} \exp \int_\gamma A , \tag{2.80}$$

where τ_i is a basis in the Lie algebra of the group G and the path ordered $\mathcal{P}$ is defined by the power series expansion

$$\mathcal{P}\exp\int_0^1 ds\, A\big(\gamma(s)\big)$$

$$=\sum_{n=0}^{\infty}\int_0^1 ds_1\int_0^{s_1} ds_2\cdots\int_0^{s_{n-1}} ds_n\, A\big(\gamma(s_n)\big)\cdots A\big(\gamma(s_1)\big). \quad (2.81)$$

The connection A is a rule that defines the meaning of parallel-transporting a vector in a representation R of G from a point of M to a nearby point: the vector v at x is defined to be parallel to the vector $v + R(A_a dx^\mu)v$ at $x + dx$. A vector is parallel-transported along γ to the vector $R(U(A,\gamma))v$.

An important property of the holonomy is that it transforms homogeneously under the gauge transformation (2.52) of A. That is: $U[A_\lambda,\gamma] = \lambda(x_f^\gamma)U[A,\gamma]\lambda^{-1}(x_i^\gamma)$, where $x_{i,f}^\gamma$ are the initial and final points of γ.

A technical remark that we shall need later on: the holonomy of any curve γ is well defined even if there are (a finite number of) points where γ is nondifferentiable and A is ill defined. The reason is that we can break γ into components where everything is differentiable and define the holonomy of γ as the product of the holonomies of the components, which are well defined by continuity.

Physical interpretation of the holonomy. Consider two left-handed neutrinos that meet at the spacetime point A, separate and then meet again at the spacetime point B. Assume their spins are parallel at A and evolve under the sole influence of the gravitational field. What is their relative spin at B? A left-handed neutrino lives in the selfdual representation of the Lorentz group and therefore its spin is parallel-transported by the selfdual connection A. Let γ_1 and γ_2 be the worldlines of the two neutrinos from A to B and let $\gamma = \gamma_2^{-1} \circ \gamma_1$ be the loop formed by the two worldlines. If the first neutrino has spin ψ at B, the second has spin $\psi' = U(A,\gamma)\psi$. By having the two neutrinos interact, we can in principle measure a quantity such as $\alpha = 2\mathrm{Re}\langle\psi|\psi'\rangle$, which (assuming $|\psi| = 1$) gives the trace of the holonomy $\alpha = \mathrm{tr}\, U[A,\gamma]$.

Metric notation. Einstein wrote GR in terms of the metric field. Here I give the translation to metric variables. Notice, however, that this is necessarily incomplete, since the fermion equations of motion cannot be written in terms of the metric field.

The metric field g is a symmetric tensor field defined by

$$g_{\mu\nu}(x) = e_\mu^I(x)\, e_\nu^J(x)\, \eta_{IJ}. \tag{2.82}$$

At each point x of M, g defines a scalar product in the tangent space $T_x M$

$$(u,v) = g_{\mu\nu}(x)u^\mu v^\nu, \qquad u,v \in T_x M, \tag{2.83}$$

and therefore maps $T_x M$ into $T_x^* M$. In other words, $g_{\mu\nu}$ and its inverse $g^{\mu\nu}$ can be used to raise and lower tangent indices. The fact that $e_I^\mu(x) \equiv \eta_{IJ}g^{\mu\nu}e_\nu^J(x)$ is the inverse matrix of $e_\mu^I(x)$ is then a result, not a definition.

The metric-preserving linear connection Γ is the field $\Gamma_{\mu\nu}^\rho(x)$ defined by

$$\Gamma_{\mu\nu}^\rho = e^\rho{}_I(\partial_\mu e_\nu^I + \omega_{\mu J}^I\, e_\nu^J). \tag{2.84}$$

It defines a covariant partial derivative D_μ on all fields that have tangent (μ) indices

$$D_\mu v^\nu = \partial_\mu v^\nu + \Gamma_{\mu\rho}^\nu v^\rho. \tag{2.85}$$

Together with ω, it defines a covariant partial derivative D_μ on all objects that have Lorentz as well as tangent indices. In particular, notice that (2.84) yields immediately

$$D_\mu e_\nu^I = \partial_\mu e_\nu^I + \omega_{\mu J}^I\, e_\nu^J - \Gamma_{\mu\nu}^\rho\, e_\rho^I = 0. \tag{2.86}$$

The antisymmetric part $T_{\mu\nu}^\rho = \Gamma_{\mu\nu}^\rho - \Gamma_{\nu\mu}^\rho$ of the linear connection gives the torsion $T^I = e_\rho^I T_{\mu\nu}^\rho dx^\mu dx^\nu$, defined in (2.5).

The Levi–Civita connection is the (metric-preserving) linear connection determined by e and $\omega[e]$. That is, it is defined by

$$\partial_\mu e_\nu^I + \omega[e]_{\mu J}^I\, e_\nu^J - \Gamma_{\mu\nu}^\rho\, e_\rho^I = 0, \tag{2.87}$$

whose solution is (2.45). It is torsion-free. Notice that the antisymmetric part of this equation is the first Cartan structure equation with vanishing torsion, namely (2.6), which is sufficient to determine $\omega[e]$ as a function of e.

The Levi–Civita connection is uniquely determined by g: it is the unique torsion-free linear connection that is metric preserving, namely that satisfies

$$D_\mu g_{\nu\rho} = 0, \tag{2.88}$$

or equivalently,

$$\partial_\mu g_{\nu\rho} - \Gamma_{\mu\nu}^\sigma g_{\sigma\rho} - \Gamma_{\mu\rho}^\sigma g_{\nu\sigma} = 0. \tag{2.89}$$

This equation is solved by (2.45), or

$$\Gamma_{\mu\nu}^\rho = \frac{1}{2}g^{\rho\sigma}(\partial_\mu g_{\sigma\nu} + \partial_\nu g_{\mu\sigma} - \partial_\sigma g_{\mu\nu}). \tag{2.90}$$

Notice that equations (2.87) and (2.90) allow us to write the explicit solution of the GR equation of motion (2.6)

$$\omega[e]_{\mu J}^I = e_J^\nu(\partial_\mu e_\nu^I - \Gamma_{\mu\nu}^\rho e_\rho^I), \tag{2.91}$$

where Γ is given by (2.90) and g by (2.82). Explicitly, this gives, with a bit of algebra,

$$\omega[e]_\mu^{IJ} = 2\,e^{\nu[I}\partial_{[\mu}e_{\nu]}{}^{J]} + e_{\mu K}e^{\nu I}e^{\sigma J}\partial_{[\sigma}e_{\nu]}{}^K. \tag{2.92}$$

The Riemann tensor can be defined via

$$R^\mu{}_{\nu\rho\sigma}\,e_\mu^I = R^I{}_{J\,\rho\sigma}\,e_\nu^J. \tag{2.93}$$

The Ricci tensor is

$$R_{\mu\nu} = R_\mu^I\,e_{I\nu}, \tag{2.94}$$

where R_μ^I is defined in (2.13). The energy-momentum tensor (see footnote 6 after (2.40))

$$T_{\mu\nu} = T_\mu^I\,e_{I\nu}. \tag{2.95}$$

In terms of these quantities, the Einstein equations (2.40) read

$$R_{\mu\nu} - \frac{1}{2}Rg_{\mu\nu} + \lambda g_{\mu\nu} = 8\pi G\,T_{\mu\nu}. \tag{2.96}$$

The Minkowski solution is

$$g_{\mu\nu}(x) = \eta_{\mu\nu}, \tag{2.97}$$

where we see clearly that the spacetime Minkowski metric is nothing but a particular value of the gravitational field. With a straightforward calculation, the action (2.12) reads

$$S[g] = \frac{1}{16\pi G}\int (R+\lambda)\,\sqrt{-\det g}\,\,\mathrm{d}^4x. \tag{2.98}$$

The matter action cannot be written in metric variables.

Riemann geometry. The tensor g equips the spacetime manifold M with a metric structure: it defines a distance between any two points, and this distance is a smooth function on M. (More precisely, it defines a pseudo-metric structure, as distance can be imaginary.) Riemann studied the structure defined by (M,g), called today a riemannian manifold, and defined the Riemann curvature tensor as a generalization of Gauss theory of curved surfaces to an arbitrary number of dimensions. Riemann presented this mathematical theory as a general theory of "geometry" that generalizes Euclidean geometry. Einstein utilized this mathematical theory for describing the physical dynamics of the gravitational field. In retrospect, the reason this was possible is because, as understood by Einstein, the euclidean structure of the physical space in which we live is determined by the local gravitational field. Therefore elementary physical geometry is simply a description of the local properties of the gravitational field, as revealed by matter (rigid bodies) interacting with it. This point is discussed in more detail below in Section 2.2.3.

The basic equations of GR presented in this section do not look too different from the equations of a prerelativistic[9] field theory, such as QED or the standard model. But the similarity can be very misleading. The physical interpretation of a general-relativistic theory is very different from the interpretation of a prerelativistic one. In particular, the meaning of

---

[9] Recall that in this book "relativistic" means *general* relativistic.

the coordinates x^μ is different than in prerelativistic physics, and the gauge-invariant observables are not related to the fields as they are in prerelativistic physics.

The process of understanding the physical meaning of the GR formalism has taken many decades and perhaps it is not entirely concluded yet. For several decades after Einstein's discovery of the theory, for instance, it was not clear whether or not the theory predicted gravitational waves. The prevailing opinion was that wave solutions were only a coordinate artifact and did not represent physical waves capable of carrying energy and momentum, or, as Bondi put it, capable of "boiling a glass of water." This opinion was wrong of course. Einstein himself badly misinterpreted the meaning of the Schwarzschild singularity. Wrong high-precision measurements of the Earth–Moon distance have been in the literature for a while, because of a mistake due to a conceptual confusion between physical and coordinate distance.

I do not want to give the impression that GR is "foggy." Quite the reverse, the fact that in all these and similar instances consensus has eventually emerged indicates that the conceptual structure of GR is secure. But to understand this conceptual structure, to understand how to use the equations of GR correctly and how to relate the quantities appearing in these equations to the numbers measured in the laboratory or observed by the astronomers, is definitely a nontrivial problem. More generally, the problem is to understand what precisely GR says about the world. Clarity in this respect is essential if we want to understand the quantum physics of the theory.

In order to shed light on this problem, it is illuminating to retrace the conceptual path and the problems that led to the discovery of the theory. This is done in the following Section 2.2. The impatient reader may skip Section 2.2 and jump to Section 2.3, where the interpretation of GR is compactly presented (but impatience slows understanding).

2.2 The conceptual path to the theory

The roots of GR are in two distinct problems. Einstein's genius was to understand that the two problems solve each other.

2.2.1 *Einstein's first problem: a field theory for the newtonian interaction*

It was Newton who discovered dynamics. But to a large extent it was Descartes who, a generation earlier, fixed the general rules of the modern science of nature, or the *Scientia Nova* as it was called at the time. One of

Descartes' prescriptions was the elimination of all the "influences from far away" that plagued mediaeval science. According to Descartes, physical interactions happen only between contiguous entities – as in collisions, pushes and pulls. Newton violated this prescription, describing gravity as the instantaneous "action-at-a-distance" of the force

$$F = G \, \frac{m_1 m_2}{d^2}. \tag{2.99}$$

Newton did not introduce action-at-a-distance with a light heart; he calls it "repugnant." His violation of the cartesian prescriptions was one of the reasons for the strong initial opposition to newtonianism. For many, his law of gravitation sounded too much like the discredited "influences from the stars" of the ineffective science of the Middle Ages. But the empirical success of Newton's dynamics and gravitational theory was so immense that most worries about action-at-a-distance dissipated.

Two centuries later, it is another Briton who finds the way to address the problem afresh, in an effort to understand electric and magnetic forces. Faraday introduces a new notion,[10] which is going to revolutionize modern physics: the notion of *field*. For Faraday, the field is a set of lines filling space. The Faraday lines begin and end on charges; in the absence of charges, each line closes, forming a *loop*. In his wonderful book, which is one of the pillars of modern physics and has virtually no equations, Faraday discusses whether the field is a real physical entity.[11] Maxwell formalizes Faraday's powerful physical intuition into a beautiful

---

[10] Many ideas of modern science have been resuscitated from hellenistic science [57]. Is the Faraday–Maxwell notion of field a direct descendant of the notion of $\pi\nu\epsilon\hat{\nu}\mu\alpha$ (pneuma), that appears for instance in Hipparchus as the carrier of the attraction of the Moon on the oceans, causing the tides, and which also appears in contexts related to magnetism [58]? Did Faraday know this notion?

[11] "With regards to the great point under consideration, it is simply: whether the lines of force have a physical existence or not. ... I think that the physical nature of the lines must be granted." [59] Strictly speaking, we can translate the problem in modern terms as to whether the field has degrees of freedom independent from the charges or not. But this doesn't diminish the ontological significance of Faraday's question, which seems to me transparent in these lines. Faraday's continuation is lovely: "And though I should not have raised the argument unless I had thought it both important and likely to be answered ultimately in the affirmative, I still hold the opinion with some hesitation, with as much, indeed, as accompanies any conclusion I endeavor to draw respecting points in the very depths of science." I think that Faraday's greatness shines in this "hesitation," which betrays his full awareness of the importance of the step he is taking (virtually all of modern fundamental physics comes out of these lines) as well as the full awareness of the risk of taking any major novel step.

mathematical theory – a field theory. At each spacetime point, Maxwell electric and magnetic fields represent the tangent to the Faraday line. There is no action-at-a-distance in the theory: the Coulomb description of the electric force between two charges, namely the instantaneous action-at-a-distance law

$$F = k \frac{q_1 \, q_2}{d^2}, \tag{2.100}$$

is understood to be correct only in the static limit. A charge q_1 at distance d from another charge q_2 does not produce an instantaneous force on q_2, because if we move q_1 rapidly away, it takes a time $t = d/c$ before q_2 begins to feel any change. This is the time the interaction takes to move across space at a finite speed, in a manner remarkably consistent with Descartes' prescription.

When Einstein studies physics, Maxwell theory is only three decades old. In his writings, Einstein rhapsodizes on the beauty of Maxwell theory and the profound impression it made upon him. Given the formal similarity of the Newton and Coulomb forces (2.99) and (2.100), it is completely natural to suspect that (2.99) also is only correct in the static limit. Namely, that the gravitational force is not instantaneous either: if a neutron star, rushing at great speed from the deep sky, smashed away the Sun, it would take a finite time before any effect be felt on Earth. That is, it is natural to suspect that there is a field theory behind Newton theory as well. Einstein set out to find this field theory. GR is what he found.

Special relativity. In fact, the need for a field theory behind Newton law (2.99) is not just *suggested* by the Coulomb–Maxwell analogy: it is indirectly *required* by Maxwell theory. The reason is that Maxwell theory not only eliminated the apparent action-at-a-distance of Coulomb law (2.100), but it also led to a reorganization of the notions of space and time which, in turn, renders *any* action-at-a-distance inconsistent. This reorganization of the notions of space and time is special relativity, a key step towards GR.

In spite of its huge empirical success, Maxwell theory had an apparent flaw if taken as a fundamental theory:[12] it is not galilean invariant. Galilean invariance is a consequence of the equivalence of inertial frames – at least it had always been understood as such. Inertial frame equivalence, or the fact that velocity is a relative notion, is one of the pillars of dynamics. The story goes that in the silent halls of Warsaw's University, an

---

[12]Rather than as a phenomenological theory of the disturbances of a mechanical ether whose dynamics is still to be found.

old and grave professor stormed out of his office like a madman, shouting "Eureka! Eureka! The new Archimedes is born!", when he saw Einstein's 1905 paper, offering the solution of this apparent contradiction. The way Einstein solves the problem is an example of theoretical thinking at its best. I think it should be kept in mind as an exemplar, when we consider the apparent contradictions between GR and QM.

Einstein *maintains* his confidence in the galilean discovery that physics is the same in all moving inertial frames and *also* maintains his confidence that Maxwell equations are correct, in spite of the apparent contradiction. He realizes that there is contradiction only because we implicitly hold a *third* assumption. By dropping this third assumption, the contradiction disappears. The third assumption regards the notion of time. It is the idea that it is always meaningful to say which of two distant events, A and B, happens first. Namely, that simultaneity is well defined in a manner independent of the observer. Einstein observes that this is a prejudice we have on the structure of reality. We can drop this prejudice and accept the fact that the temporal ordering of distant events may have no meaning. If we do so, the picture returns to consistency.

The success of special relativity was rapid, and the theory is today widely empirically supported and universally accepted. Still, I do not think that special relativity has really been fully absorbed even now: the large majority of cultivated people, as well as a surprisingly high number of theoretical physicists still believe, deep in their heart, that there is something happening "right now" on Andromeda; that there is a single universal time ticking away the life of the Universe. Do you, my reader?

An immediate consequence of special relativity is that action-at-a-distance is not just "repugnant" as Newton felt: it is a nonsense. There is no (reasonable) sense in which we can say that the force due to the mass m_1 acts on the mass m_2 "instantaneously." If special relativity is correct, (2.99) is not just *likely* to be the static limit of a field theory: it *has* to be the static limit of a field theory. When the neutron star hits the Sun, there is no "now" at which the Earth could feel the effect. The information that the Sun is no longer there must travel from Sun to Earth across space, carried by an entity. This entity is the gravitational field.

Maxwell → Einstein. Therefore, shortly after having worked out the key consequences of special relativity, Einstein attacks what is obviously the next problem: searching the field theory that gives (2.99) in the static limit. His aim is to do for (2.99) what Faraday and Maxwell had done for (2.100). The result in brief is the following, expressed in modern language.

Maxwell's solution to the problem is to introduce the one-form field $A_\mu(x)$.

The force on the particles is

$$\ddot{x}^\mu = eF^\mu{}_\nu \, \dot{x}^\nu, \qquad (2.101)$$

where F is constructed with the first derivatives of A.

A satisfies the (Maxwell) field equations

$$\partial_\mu F^{\nu\mu} = J^\nu, \qquad (2.102)$$

a system of second-order partial differential equations for A, with the charge current J^ν as source.

More generally, the field equations can be obtained as Euler–Lagrange equations of the action

$$S[A, \text{matt}] = \frac{1}{4} \int F^* \wedge F$$
$$+ S_{\text{matt}}[A, \text{matt}], \quad (2.103)$$

where F is the curvature of A.

S_{matt} is obtained from the matter action by replacing derivatives with covariant derivatives.

It follows that the source of the field equations is

$$J^\mu = \frac{\delta}{\delta A_\mu} S_{\text{matt}}[A, \text{matt}]. \qquad (2.104)$$

Einstein's solution is to introduce the field $e_\mu^I(x)$, a one-form with value in Minkowski space.

The force on the particles is (eq. (2.44))

$$\ddot{x}^\mu = -\Gamma^\mu_{\nu\rho} \, \dot{x}^\nu \dot{x}^\rho, \qquad (2.105)$$

where Γ is constructed with the first derivatives of e (equation (2.45)).

e satisfies the (Einstein) field equations (eq. (2.37), here with $\lambda = 0$)

$$R_\mu^I - \frac{1}{2} e_\mu^I R = 8\pi G \, T_\mu^I, \qquad (2.106)$$

a system of second order partial differential equations for e, with the energy momentum tensor T_μ^I as source.

More generally, the field equations can be obtained as Euler–Lagrange equations of the action ((2.36) in second order form)

$$S[e, \text{matt}] = \frac{1}{16\pi G} \int e^I \wedge e^J \wedge R^{KL} \epsilon_{IJKL}$$
$$+ S_{\text{matt}}[e, \text{matt}], \qquad (2.107)$$

where R is the curvature of the connection ω compatible with e.

S_{matt} is obtained from the matter action by replacing derivatives with covariant derivatives and the Minkowski metric with the gravitational metric.

It follows that the source of the field equations is (2.37)

$$T_\mu^I = \frac{\delta}{\delta e_\mu^I} S_{\text{matt}}[e, \text{matt}]. \qquad (2.108)$$

The structural similarity between the theories of Maxwell and Einstein theories is evident. However, this is only half of the story.

2.2.2 Einstein's second problem: relativity of motion

To understand Einstein's second problem, we have to return again to the origin of modern physics. In the western culture there are two traditional ways of understanding what is "space": as an *entity* or as a *relation*.

"Space is an entity" means that space still exists when there is nothing else besides space. It exists by itself, and objects move in it. This is the way Newton describes space, and is called absolute space. It is also the way spacetime (rather than space) is understood in

special relativity. Although considered since ancient times (in the democritean tradition), this way of understanding space was not the traditional dominant view in western culture. The dominant view, from Aristotle to Descartes, was to understand space as a relation.

"Space is a relation" means that the world is made up of physical objects, or physical entities. These objects have the property that they can be in touch with one another, or not. Space is this "touch", or "contiguity", or "adjacency" relation between objects. Aristotle, for instance, defines the spatial location of an object as the (internal) boundary of the set of the objects that surround it. This is relational space.

Strictly connected to these two ways of understanding space, there are two ways of understanding motion.

"Absolute motion." If space is an entity, motion can be defined as going from one part of space to another part of space. This is how Newton defines motion.

"Relative motion." If space is a relation, motion can only be defined as going from the contiguity of one object to the contiguity of another object. This is how Descartes[13]. and Aristotle[14] define motion.

For a physicist, the issue is which of these two ways of thinking about space and motion allows a more effective description of the world.

For Newton, space is absolute and motion is absolute.[15] This is a second violation of cartesianism. Once more, Newton does not take this step

---

[13] "We can say that movement is the transference of one part of matter or of one body, from the vicinity of those bodies immediately contiguous to it, and considered at rest, into the vicinity of some others". (Descartes, *Principia Philosophiae*, Section II-25, p. 51) [60]

[14] Aristotle insists that motion is relative. He illustrates the point with the example of a man walking on a boat. The man moves with respect to the boat, which moves with respect to the water of the river, which moves with respect to the ground . . . Aristotle's relationalism is tempered by the fact that there are preferred objects that can be used as a preferred reference: the Earth at the center of the Universe, and the celestial spheres, in particular one of the fixed stars. Thus, we can say that something is moving "in absolute terms" if it moves with respect to the Earth. However, there are *two* preferred frames in ancient cosmology: the Earth *and* the fixed stars, and the two rotate with respect to each other. The thinkers of the Middle Ages did not miss this point, and discussed at length whether the stars rotate around the Earth or the Earth rotates under the stars. Remarkably, in the fourteenth century Buridan concluded that neither view is more true than the other on grounds of reason and Oresme studied the rotation of the Earth, more than a century before Copernicus.

[15] "So, it is necessary that the definition of places, and hence local motion, be referred to some motionless thing such as extension alone or *space*, in so far as space is seen truly distinct from moving bodies." [61]. This is in open contrast with Descartes, definition, given in footnote 13.

with a light heart: he devotes a long initial section of the *Principia* to explain the reasons of his choice. The strongest argument in Newton's favor is entirely a posteriori: his theoretical construction works extraordinarily well. Cartesian physics was never as effective. But this is not Newton's argument. Newton resorts to empirical evidence, discussing a famous experiment with a bucket.

Newton's bucket. Consider a "bucket full of water, hung by a long cord, so often turned about that the cord is strongly twisted." Whirl the bucket, so that it starts rotating and the cord untwisting. At first

(i) *the bucket rotates (with respect to us) and the water remains still. The surface of the water is flat.*

Then the motion of the bucket is transmitted to the water by friction and thus the water starts rotating together with the bucket. At some time

(ii) *the water and the bucket rotate together. The surface of the water is no longer flat: it is concave.*

We know from experience that the concavity of the water is caused by rotation. Rotation with respect to what? Newton's bucket experiment shows something subtle about this question. If motion is change of place with respect to the surrounding objects, as Descartes demands, then we must say that in (i) water rotates (with respect to the bucket, which surrounds it), while in (ii) water is still (with respect to the bucket). But, observes Newton, the concavity of the surface appears in (ii), not in (i). It appears when the water is still with respect to the bucket, not when the water moves with respect to the bucket. Therefore the rotation that produces the physical effect is not the rotation with respect to the bucket. It is the rotation with respect to...what?

It is rotation with respect to space itself, answers Newton. The concavity of the water surface is an effect of the absolute motion of the water: the motion with respect to absolute space, not to the surrounding bodies. This, claims Newton, proves the existence of absolute space.

Newton's argument is subtle, and for three centuries, nobody had been able to defeat it. To understand it correctly, we should lay to rest a common misunderstanding. Relationalism, namely the idea that motion can be defined only in relation to other objects, should not be confused with galilean relativity. Galilean relativity is the statement that "rectilinear uniform motion" is a priori indistinguishable from stasis. Namely that velocity (just velocity!) is relative to other bodies. Relationalism, on the other hand, holds that *any* motion (however zig-zagging) is a priori indistinguishable from stasis. The very formulation of galilean relativity assumes a nonrelational definition of motion: "rectilinear and uniform" with respect to what?

Now, when Newton claimed that motion with respect to absolute space is real and physical, he, in a sense, overdid it, insisting that even rectilinear uniform motion is absolute. This caused a painful debate, because there are no physical effects of inertial motion, and therefore the bucket argument fails for this particular class of motions.[16] Therefore, inertial motion and velocity are to be considered relative in newtonian mechanics.

What Newton needed for the foundation of dynamics – and what we are discussing here – is not the relativity of inertial motion: it is whether *accelerated* motion, exemplified by the rotation of the water in the bucket, is relative or absolute. The question here is not whether or not there is an absolute space with respect to which velocity can be defined. The question is whether or not there is an absolute space with respect to which *acceleration* can be defined. Newton's answer, supported by the bucket argument, was positive. Without this answer, Newton's main law

$$\vec{F} = m\vec{a} \tag{2.109}$$

wouldn't even make sense.

Opposition to Newton's absolute space was even stronger than opposition to his action-at-a-distance. Leibniz and his school argued fierily against Newton absolute motion and Newton's use of absolute acceleration.[17] Doubts never really disappeared down through subsequent centuries and a lingering feeling remained that something was wrong in Newton's argument. At the end of the nineteenth century, Ernst Mach returned to the issue suggesting that Newton's bucket argument could be wrong because the water does not rotate with respect to absolute space, it rotates with respect to the full matter content of the Universe. I will comment on this idea and its influence on Einstein in Section 2.4.1. But, as for action-at-a-distance, the immense empirical triumph of newtonianism could not be overcome.

Or could it? After all, in the early twentieth century, 43 seconds of arc in Mercury's orbit were observed, which Newton's theory didn't seem to be able to account for...

Generalize relativity. Einstein was impressed by galilean relativity. The velocity of a single object has no meaning; only the velocity of objects with respect to one another is meaningful. Notice that, in a sense, this is a failure of Newton's program of revealing the "true motions." It is a minor, but significant failure. For Einstein, this was a hint that there is something wrong in the newtonian (and special-relativistic) conceptual scheme.

---

[16]Newton is well aware of this point, which is clearly stated in the Corollary V of the *Principia*, but he chooses to ignore it in the introduction to *Principia*. I think he did this just to simplify his argument, which was already hard enough for his contemporaries.

[17]Leibniz had other reasons of complaint with Newton. The two were fighting over the priority for the invention of calculus – scientists' frailties remain the same in all centuries.

In spite of its immense empirical success, Newton's idea of an absolute space has something deeply disturbing in it. As Leibniz, Mach, and
many others emphasized, space is a sort of extrasensorial entity that acts
on objects but cannot be acted upon. Einstein was convinced that the
idea of such an absolute space was wrong. There can be no absolute
space, no "true motion." Only relative motion, and therefore relative acceleration, must be physically meaningful. Absolute acceleration should
not enter physical equations. With special relativity, Einstein had succeeded in vindicating galilean relativity of velocities from the challenge of
Maxwell theory. He was then convinced that he could vindicate the entire
aristotelian–cartesian relativity of motion. In Einstein's terms, "the laws
of motion should be the same in all reference frames, not just in the inertial frames." Things move with respect to one another, not with respect to
an absolute space; there cannot be any physical effect of absolute motion.

According to many contemporary physicists, this is excessive weight
given to "philosophical" thinking, which should not play a role in physics.
But Einstein's achievements in physics are far more effective than the ones
obtained by these physicists.

2.2.3 The key idea

The question addressed in Newton's bucket experiment is the following.
The rotation of the water has a physical effect – the concavity of the
water surface: with respect to *what* does the water "rotate?" Newton
argues that the relevant rotation is not the rotation with respect to the
surrounding objects (the bucket), therefore it is rotation with respect to
absolute space. Einstein's new answer is simple and fulgurating:

The water rotates with respect to a local physical entity: the gravitational field.

It is the gravitational field, not Newton's inert absolute space, that tells
objects if they are accelerating or not, if they are rotating or not. There
is no inert background entity such as newtonian space: there are only
dynamical physical entities. Among these are the fields. Among the fields
is the gravitational field.

The flatness or concavity of the water surface in Newton's bucket is not
determined by the motion of the water with respect to absolute space.
It is determined by the physical interaction between the water and the
gravitational field.

The two lines of Einstein's thinking about gravity (finding a field theory for the newtonian interaction, and getting rid of absolute acceleration)
meet here. Einstein's key idea is that Newton has mistaken the gravitational field for an absolute space.

What leads Einstein to this idea? Why should newtonian acceleration be defined with respect to the gravitational field? The answer is given by the special properties of the gravitational interaction.[18] These can be revealed by a thought experiment called Einstein's elevator. I present below a modern and more realistic version of Einstein's elevator argument.

An "elevator" argument: newtonian cosmology. Here is a simple physical situation that illustrates that inertia and gravity are the same thing. The model is simple, but completely realistic. It leads directly to the physical intuition underlying GR.

In the context of newtonian physics, consider a universe formed by a very large spherical cloud of galaxies. Assume that the galaxies are – and remain – uniformly distributed in space, with a time-dependent density $\rho(t)$, and that they attract each other gravitationally. Let C be the center of the cloud. Consider a galaxy A (say, ours) at a distance $r(t)$ from the center C. As is well known, the gravitational force on A due to the galaxies outside a sphere of radius r around C cancels out, and the gravitational force due to the galaxies inside this sphere is the same as the force due to the same mass concentrated in C. Therefore the gravitational force on A is

$$F = -G\frac{m_A \frac{4}{3}\pi r^3(t)\, \rho(t)}{r^2(t)}, \tag{2.110}$$

or

$$\frac{d^2 r}{dt^2} = -G\frac{4}{3}\pi\, r(t)\rho(t). \tag{2.111}$$

If the density remains spatially constant, it scales uniformly as r^{-3}. That is, $\rho(t) = \rho_0 r^{-3}(t)$, where ρ_0 is a constant equal to the density at $r(t) = 1$. Therefore,

$$\frac{d^2 r}{dt^2} = -\frac{4}{3}\pi G\rho_0 \frac{1}{r^2(t)} = -\frac{c}{r^2(t)}, \tag{2.112}$$

where

$$c = \frac{4\pi G\rho_O}{3} \tag{2.113}$$

is a constant. Equation (2.112) is the Friedmann cosmological equation which governs the expansion of the universe. (It is the same equation that one obtains from full GR in the spatially flat case.)

In the newtonian model we are considering, the galaxy C is in the center of the universe and defines an inertial frame, while the galaxy A is not in the center, and is not inertial. Assume that the cloud is so large that its boundary cannot be observed from C or A. If you are in one of these two galaxies, how can you tell in which you are? That is, how can you tell whether you are in the inertial reference frame C or in the accelerated frame A?

The answer is, very remarkably, that you cannot. Since the entire cloud expands or contracts uniformly, the picture of the local sky looks uniformly expanding or contracting precisely in the same manner from all galaxies. But you cannot detect if you are in the inertial galaxy C or in the accelerated galaxy A by local experiments either! Indeed, to detect if you are in an accelerated frame you have to observe inertial forces,

---

[18] Gravity is "special" in the sense that newtonian absolute space is a configuration of the gravitational field. Once we get rid of the notion of absolute space, the gravitational interaction is no longer particularly special. It is one of the fields forming the world. But it is a very different world from that of Newton and Maxwell.

such as the ones that make the water surface of Newton's bucket concave. The A frame acceleration is

$$\vec{a} = \frac{c}{r^2(t)}\vec{u}, \tag{2.114}$$

where $\vec{u}$ is a unit vector pointing towards C. Therefore, there is an inertial force

$$\vec{F}_{\text{inertial}} = -\frac{c}{r^2(t)}\vec{u} \tag{2.115}$$

on all moving masses. This is the force that should allow us to detect that the frame is not inertial. However, all masses feel, besides the local forces $\vec{F}_{\text{local}}$, also the cosmological gravitational pull towards C,

$$\vec{F}_{\text{cosmological}} = \frac{c}{r^2(t)}\vec{u}, \tag{2.116}$$

so that their motion in the accelerated A frame is governed by

$$m\vec{a} = \vec{F}_{\text{local}} + \vec{F}_{\text{inertial}} + \vec{F}_{\text{cosmological}} \tag{2.117}$$

$$= \vec{F}_{\text{local}} \tag{2.118}$$

because (2.115) and (2.116) cancel out exactly. Therefore, the local dynamics in A looks precisely as if it were inertial. The parabola of a falling stone in A, seen from the accelerated A frame, looks as a straight line. There is no way of telling if you are the center, and no way of telling if you are inertial or not.

How do we interpret this impossibility of detecting the inertial frame? According to newtonian physics, the dynamics in C or A should be completely different. But this difference is not physically observable. In the newtonian conceptual scheme, A is noninertial, there are gravitational forces and inertial forces, but there is a sort of conspiracy that hides both of them. In fact, the situation is completely general: in a sufficiently small region, inertial and gravitational forces cancel to any accuracy in a free-falling reference system.[19] It is clear that there should be a better way of understanding this physical situation, without resorting to all these unobservable forces.

The better way is to drop the newtonian preferred *global* frame, and to realize each galaxy has its own *local* inertial reference frame. We can define local inertial frame by the absence of observable inertial effects, as in newtonian physics. Each galaxy then has its local inertial frame. These

---

[19] This is the equivalence principle. By the way, Newton, the genius, knew it: "If bodies, moved among themselves, are urged in the direction of parallel lines by equal accelerative forces, they will all continue to move among themselves, after the same manner as if they had not been urged by those forces." (Newton, *Principia*, Corollary VI to the "Laws of Motion") [62]. Newton uses this corollary for computing the complicated motion of the Moon in the Solar System. In the frame of the Earth, inertial forces and the solar gravity cancel out with good approximation, and the Moon follows a keplerian orbit.

frames are determined by the gravitational force. That is, it is gravity that determines, at each point, what is inertial. Inertial motion is such with respect to the local gravitational field, not with respect to absolute space.

Gravity determines then the way the frames of different galaxies fall with respect to one another. The gravitational field expresses the relation between the various inertial frames. It is the gravitational field that determines inertial motion. Newton's true motion is not motion with respect to absolute space: it is motion with respect to a frame determined by the gravitational field. It is motion relative to the gravitational field. Equation (2.109) governs the motion of objects with respect to the gravitational field.

The form of the gravitational field. Recall that Einstein's problem was to describe the gravitational field. The discussion above indicates that the gravitational field can be viewed as the field that determines, at each point of spacetime, the preferred frames in which motion is inertial. Let us write the mathematics that expresses this intuition.

Return to the cloud of galaxies. Since we have dropped the idea of a global inertial reference system, let us coordinatize events in the cloud with *arbitrary* coordinates $x = (x^\mu)$. The precise physical meaning of these coordinates is discussed in detail in the next section. Let x_A^μ be coordinates of a particular event A, say in our galaxy. Since these coordinates are arbitrarily chosen, motion described in the coordinates x^μ is, in general, not inertial in our galaxy. For instance, particles free from local forces do not follow straight lines. But we can find a locally inertial reference frame around A. Let us denote the coordinates it defines as X^I, and take the event A as the origin so that $X^I(A) = 0$. The coordinates X^I can be expressed as functions

$$X^I = X^I(x) \tag{2.119}$$

of the arbitrary coordinates x. In the x coordinates, the noninertiality of the motion in A is gravity. Gravity in A is the information of the change of coordinates that takes us to inertial coordinates. This information is contained in the functions (2.119). But only the value of these functions in a small neighborhood around A is relevant; because, if we move away, the local inertial frame will change. Therefore we can Taylor-expand (2.119) and keep only the first nonvanishing term. As $X^I(A) = 0$, to first nonvanishing order we have

$$X^I(x) = e_\mu^I(x_A)\, x^\mu, \tag{2.120}$$

where we have defined

$$e^I_\mu(x_A) = \left.\frac{\partial X^I(x)}{\partial x^\mu}\right|_{x=x(A)}. \tag{2.121}$$

The quantity $e^I_\mu(x_A)$ contains all the information we need to know the local inertial frame in A. The construction can be repeated at each point x. The quantity

$$e^I_\mu(x) = \left.\frac{\partial X^I(x)}{\partial x^\mu}\right|_{x}, \tag{2.122}$$

where X^I are now inertial coordinates at x, is the gravitational field at x. This is the form of the field introduced in Section 2.1.1.

The gravitational field $e^I_\mu(x)$ is therefore the jacobian matrix of the change of coordinates from the x coordinates to the coordinates X^I that are locally inertial at x. The field $e^I_\mu(x)$ is also called the "tetrad" field, from the Greek word for "four", or the "soldering form", because it "solders" a Minkowski vector bundle to the tangent bundle, or, following Cartan, the "moving frame", although there is nothing moving about it.

Transformation properties. If the coordinate system X^I defines a local inertial system at a given point, so does any other local coordinate system $Y^J = \Lambda^J_{\ I} X^I$, where Λ is a Lorentz transformation. Therefore the index I of $e^I_\mu(x)$ transforms as a Lorentz index under a local Lorentz transformation, and the two fields $e^I_\mu(x)$ and

$$e'^J_\mu(x) = \Lambda^J_{\ I}(x) e^I_\mu(x) \tag{2.123}$$

represent the same physical gravitational field. Thus, this description of gravity has a local Lorentz gauge invariance.

What happens if instead of using the physical coordinates x we had chosen coordinates $y = y(x)$? The chain rule determines the field $e'^I_\nu(y)$ that we would have found had we used coordinates y

$$e'^I_\nu(y) = \frac{\partial x^\mu(y)}{\partial y^\nu} e^I_\mu(x(y)). \tag{2.124}$$

The transformation properties (2.123) and (2.124) are precisely the transformation properties (2.58) and (2.63) under which the GR action is invariant.

These transformation laws are also the ones of a one-form field valued in a vector bundle P over the spacetime manifold M, whose fiber is Minkowski space $\mathcal{M}$, associated with a principal $SO(3,1)$ Lorentz bundle. This is a natural geometric setting for the gravitational field. The connection ω defined in Section 2.1.1 is a connection of this bundle. This setting realizes the physical picture of a patchwork of Minkowski spaces, suggested by the cloud of galaxies, carrying Lorentz frames at each galaxy. More precisely, the gravitational field can be viewed as map $e : TM \to P$ that sends tangent vectors to Lorentz vectors.

Matter. Finally, consider a particle moving in spacetime along a worldline $x^\mu(\tau)$. If a particle has velocity $v^\mu = dx^\mu/d\tau$ at a point x, its velocity in local Minkowski coordinates X^I at x is

$$u^I = \left.\frac{\partial X^I(x)}{\partial x^\mu}\right|_x v^\mu = e^I_\mu(x)v^\mu. \tag{2.125}$$

In this local Minkowski frame, the infinitesimal action along the trajectory is

$$dS = m\sqrt{-\eta_{IJ}u^I u^J}\, d\tau. \tag{2.126}$$

Therefore, the action along the trajectory is the one given in (2.41). The same argument applies to all matter fields: the action is a sum over spacetime of local terms which can be inferred from their Minkowski space equivalent.

Metric geometry. In Section 2.1.4, we saw that the gravitational field e defines a metric structure over spacetime. One is often tempted to give excessive significance to this structure, as if distance was an essential property of reality. But there is no a priori kantian notion of distance needed to understand the world. We could have developed physics without ever thinking about distances, and still have retained the complete predictive and descriptive power of our theories.

What is the physical meaning of the spacetime metric structure? What do we mean when we say that two points are 3 centimeters apart, or two events are 3 seconds apart?

The answer is in the dynamics of matter interacting with the gravitational field. Let us first consider Minkowski space. Consider two objects A and B that are 3 centimeters apart. This means that if we put a ruler between the two points, the part of the ruler that fits between the two is marked 3 cm. The shape of the ruler is determined by the Maxwell and Schrödinger equations at the atomic level. These equations contain the Minkowski tensor η_{IJ}. They have stable solutions in which the molecules maintain positions (better: vibrate around equilibrium positions) at a fixed "distance" L from one another. L is determined by the constants in these equations. This means that the molecules maintain positions at points with coordinate distances Δx^I such that

$$\eta_{IJ}\Delta x^I \Delta x^J = L^2. \tag{2.127}$$

We exploit this peculiar behavior of condensed matter for coordinatizing spacetime locations. That is, "distance" is nothing but a convenient manner for labeling locations determined by material objects (the ruler), whose dynamics is governed by certain equations. We could avoid mentioning distance by saying a number $N = 3\,[cm]/L$ of molecules, obeying the Maxwell and Schrödinger equations with given initial values, fit between A and B.

Consider now the same situation in a gravitational field e. Again, the fact that two points A and B are 3 centimeters apart means that we can fit the N molecules of the ruler between A and B. But now the dynamics of the molecules is determined by their interaction with the gravitational field. The Maxwell and Schrödinger equations have stable solutions in which the molecules keep themselves at coordinate distances Δx^μ such that

$$\eta_{IJ}e_\mu^I(x)e_\nu^J(x)\Delta x^\mu \Delta x^\nu = L^2. \tag{2.128}$$

Thus, a measure of distance is a measurement of the local gravitational field, performed exploiting the peculiar way matter interacts with gravity.

The same is true for temporal intervals. Consider two events A and B that happen in time. The meaning that 3 seconds have elapsed between A and B is that a second-ticking clock has ticked three times in this time interval. The physical system that we use as a clock interacts with the gravitational field. The pace of the clock is determined by the local value of e. Thus, a clock is nothing but a device measuring an extensive function of the gravitational field along a worldline going from A to B.

Imagine that a particle falls along a timelike geodesic from A to B. We know from special relativity that the increase of the action of the particle in the particle frame is

$$dS = m dt, \tag{2.129}$$

where m is the particle mass. Therefore, a clock comoving with the particle will measure the quantity

$$T = \frac{1}{m}S = \int_A^B d\tau \sqrt{-\eta_{IJ}e_\mu^I e_\nu^J \dot{x}^\mu \dot{x}^\nu}. \tag{2.130}$$

Thus, a clock is a device for measuring a function T of the gravitational field. In general, any metric measurement is nothing but a measurement of a nonlocal function of the gravitational field.

This is true in an arbitrary gravitational field e as well as in flat space. In flat space, we can use these measurements for determining positions with respect to the gravitational field. Since the flat-space gravitational field is Newton absolute space, these measurements locate points in spacetime.

2.2.4 Active and passive diffeomorphisms

Before getting to the last and main step in Einstein's discovery of GR, we need the notion of active diffeomorphism. I introduce this notion with an example.

Consider the surface of the Earth, and call it M. At each point $P \in M$ on Earth, say the city of Paris, there is a certain temperature $T(P)$. The temperature is a scalar function $T : M \rightarrow R$ on the Earth's surface. Imagine a simplified model of weather evolution in which the only factor determining temperature change was the displacement of air due to wind. By this I mean the following. Fix a time interval: say we call T the temperature on May 1st, and $\tilde{T}$ the temperature on May 2nd. During this time interval, the winds move the air which is over a point $Q = \phi(P)$ to the point P. If, say, Q is the French village of Quintin, this means that the winds have blown the air of Quintin to Paris. Assume the temperature $\tilde{T}(P)$ of Paris on May 2nd is equal to the temperature $T(Q)$ of Quintin the day before. The "wind" map ϕ is a map from the

Earth's surface to itself, which associates with each point P the point Q from which the air has been blown by the wind. From May 1st to May 2nd, the temperature field changes then as follows:

$$T(P) \to \tilde{T}(P) = T(\phi(P)). \tag{2.131}$$

Assuming it is smooth and invertible, the map $\phi : M \to M$ is an *active diffeomorphism*. The scalar field T on M is transformed by this active diffeomorphism as in (2.131): it is "dragged" along the surface of the Earth by the diffeomorphism ϕ. Notice that coordinates play no role in all this.

Now imagine that we choose certain geographical coordinates x to coordinatize the surface of the Earth. For instance, latitude and longitude, namely the polar coordinates $x = (\theta, \varphi)$, with $\varphi = 0$ being Greenwich. Using these coordinates, the temperature is represented by a function of the coordinates $T(x)$. The May 1st temperature $T(x)$ and the May 2nd temperature $\tilde{T}(x)$ are related by

$$\tilde{T}(x) = T(\phi(x)). \tag{2.132}$$

For instance, if the wind has blown uniformly westward by $2°20'$ degrees (Quintin is $2°20'$ west of Paris), then

$$\tilde{T}(\theta, \varphi) = T(\theta, \varphi + 2°20'). \tag{2.133}$$

Of course, there is nothing sacred about this choice of coordinates. For instance, the French might resent that the origin of the coordinates is Greenwich, and have it pass through Paris, instead. Thus, the French would describe the same temperature field that the British describe as $T(\theta, \varphi)$ by means of different polar coordinates, defined by $\varphi = 0$ being Paris. Since Paris is $2°20'$ degrees East of Greenwich, for the French the temperature field on May 1st is

$$T'(\theta, \varphi) = T(\theta, \varphi + 2°20'). \tag{2.134}$$

This is a change of coordinates, or a *passive diffeomorphism*.

Now the two equations (2.133) and (2.134) look precisely the same. But it would be silly to confuse them. In (2.133), $\tilde{T}(\theta, \varphi)$ is the temperature on May 2nd; while in (2.134), $T'(\theta, \varphi)$ is the temperature on May 1st, but written in French coordinates. In summary, the first equation represents a change in the temperature field due to the wind, the second equation represents a change in convention. The first equation describes an "active diffeomorphism," the second a change of coordinates, also called a "passive diffeomorphism."

Given a manifold M, an active diffeomorphism ϕ is a smooth invertible map from M to M. A scalar field T on M is a map $T : M \to R$. Given an active diffeomorphism ϕ, we define the new scalar field $\tilde{T}$ transformed by ϕ as

$$\tilde{T}(P) = T(\phi(P)). \tag{2.135}$$

Coordinates play no role in this.

A coordinate system x on a d-dimensional manifold M is an invertible differentiable map from (an open set of) M to R^d. Given a field T on M, this map determines the function $t : R^d \to R$ defined by $t(x) = T(P(x))$, called "the field T in coordinates x."[20] A passive diffeomorphism is an

---

[20]In the physics literature, the two maps $T : M \to R$ and $t = T \circ x^{-1} : R^d \to R$,

invertible differentiable map $\phi : R^d \rightarrow R^d$ that defines a new coordinate system x' on M by $x(P) = \phi(x'(P))$. The value of the field T in coordinates x' is given by

$$t'(x') = t(\phi(x')). \tag{2.136}$$

Beware the formal similarity between (2.135) and (2.136).

The above extends immediately to all structures on M. For instance, an active diffeomorphism ϕ carries a one-form field e on M to the new one-form field $\tilde{e} = \phi^* e$, the pull-back of e under ϕ and so on.

In particular, a metric $d : M \times M \rightarrow R^+$ is an assignment of a distance $d(A, B)$ between any two points A and B of M. An active diffeomorphism defines the new metric $\tilde{d}$ given by $\tilde{d}(A, B) \equiv d(\phi^{-1}(A), \phi^{-1}(B))$. The two metrics d and $\tilde{d}$ are isometric but distinct.[21] An equivalence class of metrics under active diffeomorphisms is sometimes called a "geometry." Given a coordinate system, we can represent a (Riemannian) metric d by means of a tensor field on R^d: Riemann's metric tensor $g_{\mu\nu}(x)$ or, equivalently, the tetrad field $e^I_\mu(x)$. Under a change of coordinate system, the same metric is represented by a different $g_{\mu\nu}(x)$ or $e^I_\mu(x)$.

The example of the Earth's temperature given above illustrates a peculiar relation between active and passive diffeomorphisms: given two temperature fields T and $\tilde{T}$ related by an active diffeomorphism, we can always find a coordinate transformation such that in the new coordinates $\tilde{T}$ is represented by the same function as T in the old coordinates. This simple mathematical observation is at the root of Einstein's arguments that I will describe below. (The argument will be essentially that a theory that does not distinguish coordinate systems cannot distinguish fields related by active diffeomorphisms either.)

More precisely, the relation between active and passive diffeomorphisms is as follows. The group of the active diffeomorphisms acts on the space

---

are always indicated with the same symbol, generating confusion between active and passive diffeomorphisms. In this paragraph I use distinct notations. In the rest of the text, however, I shall adhere to the standard notation and indicate the field and its coordinate representation with the same symbol.

[21] Here is an example of isometric but distinct metrics. The 2001 *Shell* road-map says that the distances between New York (NY), Chicago (C) and Kansas City (KC) are: $d(\text{NY, C}) = 100$ miles, $d(\text{C, KC}) = 50$ miles, $d(\text{KC, NY}) = 100$ miles, while the 2002 *Lonely Planet* tourist guide claims that these distances are $\tilde{d}(\text{NY, C}) = 100$ miles, $\tilde{d}(\text{C, KC}) = 100$ miles, $\tilde{d}(\text{KC, NY}) = 50$ miles. Obviously these are not the same distances. But they are isometric: the two are transformed into each other by the active diffeomorphism $\phi(\text{NY}) = \text{C}$, $\phi(\text{C}) = \text{KC}$, $\phi(\text{KC}) = \text{NY}$.

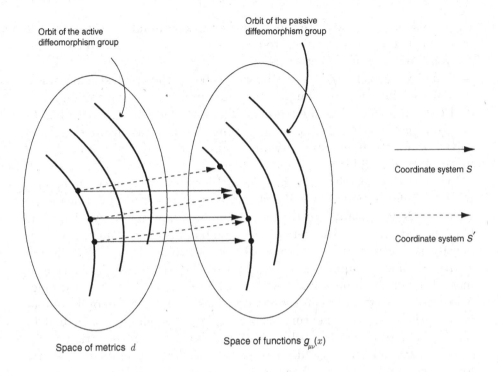

Fig. 2.2   Active and passive diffeomorphisms.

of metrics d. The group of passive diffeomorphisms acts on the space of functions $g_{\mu\nu}(x)$. The orbits of the first group are in natural one-to-one correspondence with the orbits of the second. However, the relation between the individual metrics d and the individual functions $g_{\mu\nu}(x)$ depends on the coordinate system chosen. The situation is illustrated in Figure 2.2.

2.2.5 General covariance

Around 1912, using the idea that any motion is relative, Einstein had found the form of the gravitational field as well as the equations of motions of matter in a given gravitational field. This was already a remarkable achievement, but the field equations for the gravitational field were still missing. In fact, the best part of the story had yet to come.

Two problems remained open: the field equations, and understanding the physical meaning of the coordinates x^{μ} introduced above. Einstein struggled with these two problems during the years 1912–1915, trying several solutions and changing his mind repeatedly. Einstein has called this search his "struggle with the meaning of the coordinates." The struggle was epic. The result turned out to be amazing. In Einstein's words, it was "beyond my wildest expectations."

To increase Einstein's stress, Hilbert, probably the greatest mathematician at the time, was working on the same problem, trying to be first to find the gravitational field equations. The fact that Hilbert, with his far superior mathematical skills, could not find these equations first, testifies to the profound differences between fundamental physical problems and mathematical problems.

In his search for the field equations, Einstein was guided by several pieces of information. First, the static limit of the field equations must yield the Newton law, as the static limit of Maxwell theory yields the Coulomb law. Second, the source of Coulomb law is charge; and the charge density is the temporal component of four-current $J^\mu(x)$, which is the source of Maxwell equations. The source of the Newtonian interaction is mass. Einstein had understood with special relativity that mass is in fact a form of energy and that the energy density is the temporal component of the energy-momentum tensor $T_{\mu\nu}(x)$. Therefore, $T_{\mu\nu}(x)$ had to be the likely source of the field equations. Third, the introduction of the gravitational field was based on the use of arbitrary coordinates, therefore there should be some form of covariance under arbitrary changes of coordinates in the field equations. Einstein searched for covariant second-order equations as relations between tensorial quantities, since these are unaffected by coordinate change. He learned from Riemannian geometry that the only combination of second derivatives of the gravitational field that transforms tensorially is the Riemann tensor $R^\mu_{\nu\rho\sigma}(x)$. This was in fact Riemann's major result. Einstein knew all this in 1912. To derive Einstein's field equations (2.97) from these ideas is a simple calculation, presented in all GR textbooks, and which a good graduate student can today repeat easily. Still, Hilbert couldn't do it, and Einstein got stuck for several years. What was the problem?

The problem was "the meaning of the coordinates." Here is the story.

1 Einstein for general covariance. At first, Einstein demands the field equations for the gravitational field $e^I_\mu(x)$ to be generally covariant on M. This means that if $e^I_\mu(x)$ is a solution, then $e'^I_\nu(y)$ defined in (2.124) should also be a solution. For Einstein, this requirement (unheard of at the time) was the formalization of the idea that the laws of nature must be the same in all reference frames, and therefore in all coordinate systems.

2 Einstein against general covariance. In 1914, however, Einstein convinces himself that the field equations should *not* be generally covariant

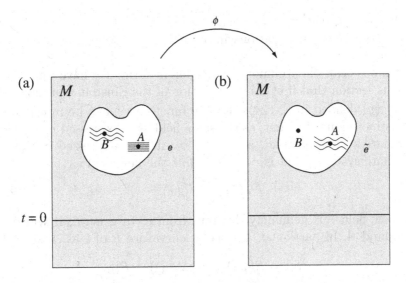

Fig. 2.3   The active diffeomorphism ϕ drags the nonflat (wavy) gravitational field from the point B to the point A.

[63]. Why? Because Einstein rapidly understands the physical consequences of general covariance, and he initially panics in front of them. The story is very instructive, because it reveals the true magic hidden inside GR. Einstein's argument *against* general covariance is the following.[22]

Consider a region of spacetime containing two spacetime points A and B. Let e be a gravitational field in this region. Say that around the point A the field is flat while at the point B it is not (see Figure 2.3(a)). Next, consider a map ϕ from M to M that maps the point A to the point B. Consider the new field $\tilde{e} = \phi^*e$ which is pulled back by this map. The value of the field $\tilde{e}$ at A is determined by the value of e at B, and therefore the field $\tilde{e}$ will not be flat around A (see Figure 2.3(b)).

Now, if e is a solution of the equations of motion, and if the equations of motion are generally covariant, then $\tilde{e}$ is also a solution of the equations of motion. This is because of the relation between active diffeomorphisms and changes of coordinates: we can always find two different coordinate systems on M, say x and y, such that the function $e^I_\mu(x)$ that represents e in the coordinate system x is the same function as the function $\tilde{e}^I_\mu(y)$ that represents $\tilde{e}$ in the coordinate systems y. Since the equations of motion

---

[22] At first, Einstein got discouraged about generally covariant field equations because of a mistake he was making while deriving the static limit: the calculation yielded the wrong limit. But this is of little importance here, given the powerful use that Einstein has been routinely capable of making of general conceptual arguments.

are the same in the two coordinate systems, the fact that this function satisfies the Einstein equations implies that e as well as $\tilde{e}$ are physical solutions.

Let me repeat the argument in a different form. We have found in the previous section that if $e_\mu^I(x)$ is a solution of the Einstein equations, then so is $e'^I_\nu(y)$ defined in (2.124). But the function e'^I_ν can be interpreted in two distinct manners. First, as the same field as e, expressed in a different coordinate system. Second, as a *different* field $\tilde{e}$, expressed in the same coordinate system. That is, we can *define* the new field as

$$\tilde{e}_\mu^I(x) = e'^I_\mu(x). \tag{2.137}$$

This new field $\tilde{e}$ is genuinely different from e. In general, it will not be flat around A. In particular, the scalar curvature $\tilde{R}$ of $\tilde{e}$ at A is

$$\tilde{R}|_A = \tilde{R}(x_A) = R(\phi(x_A)) = R|_B. \tag{2.138}$$

In other words, if the equations of motion are generally covariant they are *also* invariant under active diffeomorphisms.

Given this, Einstein makes the following famous observation.

The "hole" argument: Assume the gravitational field-equations are generally covariant. Consider a solution of these equations in which the gravitational field is e and there is a region H of the universe without matter (the "hole," represented as the white region in Figure 2.3). Assume that inside H there is a point A where e is flat and a point B where it is not flat. Consider a smooth map $\phi : M \to M$ which reduces to the identity outside H, and such that $\phi(A) = B$, and let $\tilde{e} = \phi^* e$ be the pull-back of e under ϕ. The two fields e and $\tilde{e}$ have the same past, are both solutions of the field equations, but have different properties at the point A. Therefore, the field equations do not determine the physics at the spacetime point A. Therefore they are not deterministic. But we know that (classical) gravitational physics is deterministic. Therefore either

   (i) the field equations must not be generally covariant, or

   (ii) there is no meaning in talking about the physical spacetime point A.

On the basis of this argument, Einstein searched for nongenerally covariant field equations for three years, in a frantic race against Hilbert.

3 Einstein's return to general covariance. Then, rather suddenly, in 1915, Einstein published generally covariant field equations. What had happened? Why had Einstein changed his mind? Is there a mistake in the

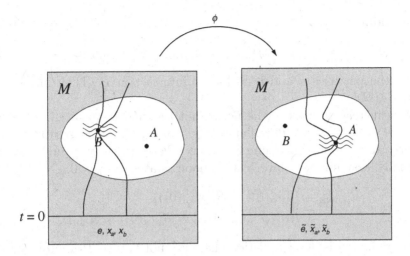

Fig. 2.4   The diffeomorphism moves the nonflat region *as well as the intersection point of the two particles a and b* from the point B to the point A.

hole argument? No, the hole argument is correct. The correct physical conclusion, however, is (ii), not (i). This point hit Einstein like a flash of lightning, the precise conceptual discovery to which all his previous thoughts had led.

Einstein's way out from the difficulty raised by the hole argument is to realize that there is no meaning in referring to "the point A" or "the event A," without further specifications.

Let us follow Einstein's explanation in detail.

Spacetime coincidences. Consider again the solution e of the field equations, but assume that in the universe there are also the two particles a and b. Say that the worldlines $(x_a(\tau_a), x_b(\tau_b))$ of the two particles intersect at the spacetime point B, see Figure 2.4.

Now, for given initial conditions, the worldlines of the particles are determined by the gravitational field. They are geodesics of e or, if other forces are involved, they satisfy the geodesic equation with an additional force term. Consider the field $\tilde{e} = \phi^* e$. The particles' worldlines $(x_a(\tau_a), x_b(\tau_b))$ are *no longer* solutions of the particles' equations of motion in this gravitational field. If the gravitational field is $\tilde{e}$ instead of e, the particles' motions over M will be different. But it is easy to find the motion of the particles determined by $\tilde{e}$, precisely because the complete set of equations of motion is generally covariant. Therefore, an active diffeomorphism on the gravitational field *and* the particles sends solutions into solutions. Thus, the motion of the particles in the field $\tilde{e}$ is given by the worldlines

$$\tilde{x}_a(\tau_a) = \phi^{-1}(x_a(\tau_a)), \qquad \tilde{x}_b(\tau_b) = \phi^{-1}(x_b(\tau_b)). \qquad (2.139)$$

Then, the particles a and b no longer intersect in B. They intersect in $A = \phi^{-1}(B)$!

Now, instead of asking whether or not the field is flat at A, let us ask whether or not the field is flat at the point where the particles meet. Clearly, the result is the same for the two cases (e, x_a, x_b) and $(\tilde{e}, \tilde{x}_a, \tilde{x}_b)$. Formally, assuming the intersection point is at $\tau_a = \tau_b = 0$,

$$\tilde{R}|_{\text{inters}} = \tilde{R}(\tilde{x}_a(0)) = R(\phi(\tilde{x}_a(0)))$$
$$= R(\phi(\phi^{-1}(x_a(0)))) = R(x_a(0)) = R|_{\text{inters}}. \qquad (2.140)$$

This prediction is deterministic. There are not two contradictory predictions, therefore there is determinism, so long as we restrict ourselves to this kind of prediction. Einstein calls "spacetime coincidences" this way of determining points.

Einstein observes that this conclusion is general: the theory does not predict what happens at spacetime points (like newtonian and special-relativistic theories do). Rather, it predicts what happens at locations *determined by the dynamical elements of the theory themselves*. In Einstein's words:

> All our space-time verifications invariably amount to a determination of space-time coincidences. If, for example, events consisted merely in the motion of material points, then ultimately nothing would be observable but the meeting of two or more of these points. Moreover, the results of our measuring are nothing but verifications of such meetings of the material points of our measuring instruments with other material points, coincidences between the hands of a clock and points on the clock dial, and observed point-events happening at the same place at the same time. The introduction of a system of reference serves no other purpose than to facilitate the description of the totality of such coincidences. [64]

The two solutions (e, x_a, x_b) and $(\tilde{e}, \tilde{x}_a, \tilde{x}_b)$ are only distinguished by their localization on the manifold. They are different in the sense that they ascribe different properties to manifold points. However, if we demand that localization is defined only with respect to the fields and particles themselves, then there is nothing that distinguishes the two solutions physically. In fact, concludes Einstein, the two solutions represent the same physical situation. The theory is gauge invariant in the sense of Dirac, under active diffeomorphisms: there is a redundancy in the mathematical formalism; the same physical world can be described by different solutions of the equations of motion.

It follows that localization on the manifold has no physical meaning. The physical picture is completely different from the example of the temperature field on the Earth's surface illustrated in the previous section. In that example, the cities of Paris and Quintin were real distinguishable entities, independent from the temperature field. In GR, general covariance is compatible with determinism only assuming that individual spacetime points have no physical meaning by themselves. It is like having only the temperature field, without the underlying Earth.

What disappears in this step is precisely the background spacetime that Newton believed to have been able to detect with great effort beyond the apparent relative motions.

Einstein's step toward a profoundly novel understanding of nature is achieved. Background space and spacetime are effaced from this new understanding of the world. Motion is entirely relative. Active diffeomorphism invariance is the key to implement this complete relativization. Reality is not made up of particles and fields on a spacetime: it is made up of particles and fields (including the gravitational field), that can only be localized with respect to one another. No more fields on spacetime: just fields on fields. Relativity has become general.

2.3 Interpretation

General covariance makes the relation between formalism and experiment far more indirect than in conventional field theories.

Take Maxwell theory as an example. We assume that there is a background spacetime. We have special objects at our disposal (the walls of the lab, the Earth) that define an inertial frame to a desired approximation. These objects allow us to designate locations relative to background spacetime. We have two kinds of measuring devices: (a) meters and clocks that measure distance and time intervals from these reference objects, and (b) devices that measure the electric and magnetic fields. The reading of the devices (a) gives us x^μ. The reading of the devices (b) gives us $F_{\mu\nu}$. We measure the two and say that the field has the value $F_{\mu\nu}$ at the point x^μ. The theory can predict the value $F_{\mu\nu}$ at the point x^μ.

We cannot do the same in GR. The theory does not predict the value of the field at the point x^μ. So, how do we compare theory and observations?

2.3.1 Observables, predictions and coordinates

As discussed at the end of the previous section, a physical state does not correspond to a solution $e(x)$ of Einstein's equations, but to an equivalence class of solutions under active diffeomorphisms. Therefore, the quantities that the theory predicts are all and only the quantities that are well defined on these equivalence classes. That is, only the quantities that are invariant under diffeomorphisms. These quantities are independent from the coordinates x^μ.

In concrete applications of the theory, these quantities are generally obtained by solving away the coordinates x from solutions to the equations of motion. Here are a few examples.

Solar System. Consider the dynamics of the Solar System. The variables are the gravitational field $e(x)$ and the worldlines of the planets, $x_n(\tau_n)$. Fix a solution $(e(x), x_n(\tau_n))$ to the equations of motion. We want to derive physical predictions from this solution and compare them with observations. Choose for simplicity $\tau_n = x^0$; so that the solution is expressed by $(e(x), \vec{x}_n(x^0))$. Consider the worldline of the Earth. Compute the distance $d_n(x^0)$ between the Earth and the planet n, defined as the proper time elapsed along the Earth's worldline while a null geodesic (a light pulse) leaving the Earth at x_0 travels from Earth to the planet and back.

The functions $(d_n(x^0))$ can be computed from the given solutions to the equations of motion. Consider a space $\mathcal{C}$ with coordinates (d_n). The functions $(d_n(x^0))$ define a curve γ on this space.

We can associate a measuring device with each d_n: a laser apparatus that measures the distance to planet n. These quantities can be measured together. We obtain the event (d_n), which can be represented by a point in $\mathcal{C}$. The theory predicts that this point will fall on the curve γ. A sequence of these events can be compared with the curve γ, and in this way we can test the given solutions to the equations of motion against experience. (In the terminology of Chapter 3, the quantities d_n are partial observables.) Notice that this can be done with arbitrary precision and that distant stars, inertial systems, preferred coordinates, or choice of time variable, play no role.

Clocks. Consider the gravitational field around the Earth. Consider two worldlines. Let the first be the worldline of an object fixed on the Earth's surface. Let the second be the worldline of an object in free fall on a keplerian orbit around the Earth, that is, a satellite. Fix an arbitrary initial point P on the worldline of the orbiting object and let T_1 be the proper time from P along this worldline. Send a light signal from P to the object on Earth; let Q be the point on the Earth's worldline when the signal is received and let T_2 be the proper time from Q along this worldline. Then let $T_2(T_1)$ be the reception proper time on Earth of a signal sent at T_1 proper time in orbit. GR allows us to compute the function $T_2(T_1)$ for any T_1.

It is easy to associate measuring devices to T_1 and T_2: these are a clock on Earth and a clock in orbit. If the orbiting object sends a signal at fixed proper times T_1, the reception times T_2 can be compared with the predictions of the theory. Here T_1 and T_2 are the partial observables. I let you decide which one of the two is the "true time variable."

Solar System with a clock. We can add a clock to the Solar System measurements described above. Fixing arbitrarily an initial event on Earth (a particular eclipse, the birth of Jesus, or the death of John Lennon), we can compute the proper time $T(x^0)$ lapsed from this event along the Earth's worldline. The partial observable T can be added to the partial observables d_n, giving the set (d_n, T) of partial observables. If we do so, it may be convenient to express the correlations (d_n, T) as functions $d_n(T)$. A complete gauge-invariant observable, fully predicted by the theory, is the value $d_n(T)$ of a planet distance at a certain given Earth proper time T from the initial event. Notice that T is not a coordinate. It is a complicated nonlocal function of the gravitational field, to which a measuring device (measuring a partial observable) has been attached. The use of a clock on Earth to determine a local temporal localization is just a matter of convenience.

Binary pulsar. Consider a binary-star system in which one of the two stars is a pulsar. Because of a Doppler effect, the frequency of the pulsing signal oscillates with the orbital period of the system. This fact allows us to count the number of pulses in each orbit. Let N_n be the number of pulses we receive in the nth orbit. A theoretical model of the pulsar allows us to compute the expected decrease in orbital period due to gravitational wave emission and therefore the expected sequence N_n which can be compared with the observed one. Doing this with sufficient care won J.H. Taylor and R.A. Hulse the 1993 Nobel Prize.

Notice that in all these examples the coordinates x^μ have disappeared from the observable quantities. This is true in general. A theoretical model of a physical system is made using coordinates x^μ, but then observable quantities are independent of the coordinates x^μ.[23]

2.3.2 The disappearance of spacetime

In the mathematical formalism of GR we utilize the "spacetime" manifold M, coordinatized by x. However, a state of the universe does not

---

[23] Unless we gauge-fix them to given partial observables, see Section 2.4.6.

correspond to a configuration of fields on M. It corresponds to an equivalence class of field configurations under active diffeomorphisms. An active diffeomorphism changes the localization of the field on M by dragging it around. Therefore, localization on M is just gauge: it is physically irrelevant.

In fact, M itself has no physical interpretation, it is just a mathematical device, a gauge artifact. Pre-general-relativistic coordinates x^μ designate points of the physical spacetime manifold "where" things happen (see a detailed discussion below in Section 2.4.5); in GR there is nothing of the sort. The manifold M cannot be interpreted as a set of physical "events," or physical spacetime points "where" the fields take value. It is meaningless to ask whether or not the gravitational field is flat around the point A of M, because there is no physical entity "spacetime point A." Contrary to Newton and to Minkowski, there are no spacetime points where particles and fields live. There are no spacetime points at all. The Newtonian notions of space and time have disappeared.

In Einstein's words:

> ... the requirement of general covariance takes away from space and
> time the last remnant of physical objectivity ... [64]

Einstein justifies this conclusion in the immediate continuation of this text, which is the paragraph I quoted at the end of the previous section, with the observation that all observations are spacetime coincidences.

In newtonian physics, if we take away the dynamical entities, what remains is space and time. In general-relativistic physics, if we take away the dynamical entities, nothing remains. The space and time of Newton and Minkowski are reinterpreted as a configuration of one of the fields, the gravitational field.

Concretely, this radically novel understanding of spatial and temporal relations is implemented in the theory by the invariance of the field equations under diffeomorphisms. Because of background independence – that is, since there are no nondynamical objects that break this invariance in the theory – diffeomorphism invariance is formally equivalent to general covariance, namely the invariance of the field equations under arbitrary changes of the spacetime coordinates $\vec{x}$ and t.

Diffeomorphism invariance implies that the spacetime coordinates $\vec{x}$ and t used in GR have a different physical meaning to the coordinates $\vec{x}$ and t used in prerelativistic physics. In prerelativistic physics, $\vec{x}$ and t denote localization with respect to appropriately chosen reference objects. These reference objects are chosen in such a way that they make the physical influence of background spacetime manifest. In particular, their motion can be chosen to be inertial. In GR, on the other hand, the spacetime coordinates $\vec{x}$ and t have no physical meaning: physical predictions of GR are independent of the coordinates $\vec{x}$ and t.

A physical theory should not describe the location in space and the evolution in time of dynamical objects. It describes *relative* location and *relative* evolution of dynamical objects. Newton introduced the notion of background spacetime because he needed the acceleration of a particle to be well defined (so that $\vec{F} = m\vec{a}$ could make sense). In the newtonian theory and in special relativity, a particle accelerates when it does so with respect to a fixed spacetime in which the particle moves. In general relativity, a particle (a dynamical object) accelerates when it does so with respect to the local values of the gravitational field (another dynamical object). There is no meaning for the location of the gravitational field, or the location of the particle: only the relative location of the particle with respect to the gravitational field has physical meaning.

What remains of the prerelativistic notion of spacetime is a relation between dynamical entities: we can say that two particles' worldlines "intersect"; that a field has a certain value "where" another field has a certain value; or that we measure two partial observables "together." This is precisely the modern realization of Descartes' notion of *contiguity*, and it is the basis of spatial and temporal notions in GR.

As Whitehead put it, we cannot have spacetime without dynamical entities, anymore than saying that we can have the cat's grin without the cat. The world is made up of fields. Physically, these do not live on spacetime. They live, so to say, on one another. No more fields on spacetime, just fields on fields. It is as outlined in the metaphor in Section 1.1.3, where we no longer had animals on the island, just animals on the whale, animals on animal. Our feet are no longer in space: we have to ride the whale.

2.4 *Complements

I close this chapter by discussing a certain number of issues related to the interpretation of GR.

2.4.1 Mach principles

The ideas of Ernst Mach had a strong influence on Einstein's discovery of GR. Mach presented a number of acute criticisms to Newton's motivations for introducing absolute space and absolute time. In particular, he pointed out that in Newton's bucket argument there is a missing element: he observed that the inertial reference frame (the reference frame with respect to which rotation has detectable physical effects) is also the reference frame in which the fixed stars do not rotate. Mach then suggested that the inertial reference frame is not determined by absolute space, but rather it is determined by the entire matter content of the Universe, including distant stars. He suggested that if we could repeat the experiment with a very massive bucket, the mass of the bucket would affect the inertial frame, and the inertial frame would rotate with the bucket.

In the light of GR, the observation is certainly pertinent and it is clear that the argument may have played a role in Einstein's dismissal of Newton's argument. However, for some reason, the precise relation between Mach's suggestion and GR has generated a vast debate. Mach's suggestion that inertia is determined by surrounding matter has been called "the Mach principle" and much ink has been employed to discuss whether or not GR implements this principle; whether or not "GR is machian." Remarkably, in the literature one finds arguments and proofs in favor as well as against the conclusion that GR is machian. Why this confusion?

Because there is no well-defined "Mach principle." Mach provided a very important but vague suggestion, that Einstein developed into a theory, not a precise statement that can be true or false. Every author that has discussed "the Mach principle" has actually considered a *different* principle. Some of these "Mach principles" are implemented in GR, others are not.

In spite of the confusion, or perhaps thanks to it, the discussion on how machian GR is sheds some light on the physical content of GR. Here I list several versions of the Mach principle that have been considered in the literature, and, for each of these, I comment on whether this particular Mach principle is True or False in GR. In the following, "matter" means any dynamical entity except the gravitational field.

- **Mach principle 1:** *Distant stars can affect the local inertial frame.*
  True. Because matter affects the gravitational field.

- **Mach principle 2:** *The local inertial frame is completely determined by the matter content of the Universe.*
  False. The gravitational field has independent degrees of freedom.

- **Mach principle 3:** *The rotation of the inertial reference frame inside the bucket is in fact dragged by the bucket, and this effect increases with the mass of the bucket.*
  True. In fact, this is the Lense–Thirring effect: a rotating mass drags the inertial frames in its vicinity.

- **Mach principle 4:** *In the limit in which the mass of the bucket is large, the internal inertial reference frame rotates with the bucket.*
  Depends. It depends on the details of the way the limit is taken.

- **Mach principle 5:** *There can be no global rotation of the Universe.*
  False. Einstein believed this to be true in GR, but Gödel's solution is a counter-example.

- **Mach principle 6:** *In the absence of matter, there would be no inertia.*
  False. There are vacuum solutions of the Einstein field equations.

- **Mach principle 7:** *There is no absolute motion, only motion relative to something else; therefore the water in the bucket does not rotate in absolute terms, it rotates with respect to some dynamical physical entity.*
  True. This is the basic physical idea of GR.

- **Mach principle 8:** *The local inertial frame is completely determined by the dynamical fields in the Universe.*
  True. In fact, this is precisely Einstein's key idea.

2.4.2 Relationalism versus substantivalism

In contemporary philosophy of science there is an interesting debate on the interpretation of GR. The two traditional theses about space – absolute and relational – suitably edited to take into account scientific progress, continue under the names of *substantivalism* and *relationalism*. Here I present a few considerations on the issue.

GR changes the notion of spacetime in physics in the sense of relationalism. In pre-relativistic physics, spacetime is a fixed nondynamical entity, in which physics happens. It is a sort of structured container which is the home of the world. In relativistic physics, there is nothing of the sort. There are only interacting fields and particles. The only notion of localization which is present in the theory is relative: dynamical objects can be localized only with respect to one another. This is the notion of space defended by Aristotle and Descartes, against which Newton wrote the initial part of *Principia*. Newton had two points: the physical reality of inertial effects such as the concavity of the water in the bucket, and the immense empirical success of his theory based on absolute space. Einstein provided an alternative interpretation for the cause of the concavity – interaction with the local gravitational field – and a theory based on relational space that has better empirical success than Newton theory. After three centuries, the European culture has returned to a fully relational understanding of space and time.

At the basis of cartesian relationalism is the notion of "contiguity." Two objects are contiguous if they are close to one another. Space is the order of things with respect to the contiguity relation. At the basis of the spacetime structure of GR is essentially the same notion. Einstein's "spacetime coincidences" are analogous to Descartes "contiguity."

A substantivalist position can nevertheless still be defended to some extent. Einstein's discovery is that newtonian spacetime and the gravitational field are the same entity. This can be expressed in two equivalent ways. One states that there is no spacetime; there is only the gravitational field. This is the choice I have made in this book. The second states that there is no gravitational field; it is spacetime that has dynamical properties. This choice is common in the literature. I prefer the first because I find that the differences between the gravitational field and other fields are more accidental than essential. But the choice between the two points of view is only a matter of choice of words, and thus, ultimately, personal taste. If one prefers to keep the name "spacetime" for the gravitational field, then one can still hold a substantivalist position and claim that, according to GR, spacetime is an entity, not a relation. Furthermore, localization can be defined with respect to the gravitational field, and therefore the substantivalist can say that spacetime is an entity that defines localization. For an articulation of this thesis, see, for instance, [65].

However, this is a very weakened substantivalist position. One is free to call "spacetime" anything with respect to which we define position. But to what extent is spacetime different from any arbitrary continuum of objects used to define position? Newton's acute formulation of his substantivalism, already mentioned in footnote 15 above, contains a precise characterization of "space":

> ... so it is necessary that the definition of places, and hence of local motion, be referred to some motionless thing such as extension alone or "space," *in so far as space is seen to be truly distinct from moving bodies.*[24]

The characterizing feature of space is that of being truly distinct from *moving* bodies, that is, in modern terms and after the Faraday–Maxwell conceptual revolution, that of

---

[24] I. Newton, *De Gravitatione et aequipondio fluidorum*, [61].

being truly distinct from dynamical entities such as particles or fields. This is clearly not the case for the spacetime of GR. If the modern substantivalist is happy to give up Newton's strong substantivalism and identify the thesis that "spacetime is an entity" with the thesis that "spacetime is the gravitational field, which is a dynamical entity," then the distinction between substantivalism and relationalism is completely reduced to one of semantics.

When two opposite positions in a long-standing debate have come so close that their distinction is reduced to semantics, one can probably say that the issue is solved. I think one can say that in this sense GR has solved the long-standing issue of the relational versus substantivalist interpretations of space.

2.4.3 Has general covariance any physical content? Kretschmann's objection

Virtually any field theory can be reformulated in a generally covariant form. An example of a generally covariant reformulation of a scalar field theory on Minkowski spacetime is presented below. This fact has led some people to wonder whether general covariance has any physical significance at all. The argument is as follows: if any theory can be formulated in a general covariant language, then general covariance is not a principle that selects a particular class of theories, therefore it has no physical content. This argument was presented by Kretschmann shortly after Einstein's publication of GR. It is heard among some philosophers of science, and sometimes used also by some physicists that dismiss the conceptual novelty of GR.

I think that the argument is wrong. The *non sequitur* is the idea that a formal property that does not restrict the class of admissible theories, has no physical significance. Why should that be? Formalism is flexible, and we can artificially give a theory a certain formal property, especially if we accept byzantine formulations. But it does not follow from this that the use of one formalism or another is irrelevant. Physics is the search for the more effective formalism to read Nature. The relevant question is not whether general covariance restricts the class of admissible theories, but whether GR could have been conceived or understood at all, without general covariance. Let me illustrate this point with the example of rotational invariance.

Kretschmann's objection applied to rotational symmetry. Ancient physics assumed that space has a preferred direction. The "up" and the "down" were considered absolutely defined. This changes with newtonian physics, where space has rotational symmetry: all spatial directions are a priori equivalent, and only contingent circumstances – such as the presence of a nearby mass like the Earth – can make one direction particular. Physicists often say that rotational invariance limits the admissible forces. But strictly speaking, this is not true. Kretschmann's objection applies equally well to rotational invariance: given a theory which is not rotationally invariant, we can reformulate it as a rotationally invariant theory, just by adding some variable. For instance, consider a physical theory T in which all bodies are subject to a force in the z-direction, $F = -g$, where g is a constant (such as gravity). This is a nonrotationally-invariant theory. Now consider another theory T' in which there is a dynamical vector quantity $\vec{v}$, of length unity, and a force $\vec{F} = g\vec{v}$. The theory T' is rotationally invariant, but in each solution the vector $\vec{v}$ will take a particular value, in a particular direction. Calling z this direction we have precisely the same phenomenology as theory T.

The example shows that we can express a nonrotationally invariant theory T in a rotationally invariant formalism T'. Therefore rotational invariance does not *truly* restrict the class of admissible theories. Shall we conclude with Kretschmann that rotational invariance has no physical significance?

Obviously not. Modern physics *has* made real progress with respect to ancient physics in understanding that space is rotationally invariant. Where is the progress? It is in the fact that the discovery of the rotational invariance of space puts us in a far more effective position for understanding Nature. We can say that we have discovered that in general there is no preferred "up" and "down" in the Universe. Equivalently, we can say that a rotationally invariant physical formalism is far more effective for understanding Nature than a nonrotationally invariant one.

There are two key issues here. First, it would have been difficult to find newtonian theory within a conceptual framework in which the "up" and the "down" are considered absolute. Second, reformulating the theory T in the rotationally invariant form T' modifies our understanding of it: we have to introduce the dynamical vector $\vec{v}$. From the point of view of the two theories T and T', the vector $\vec{v}$ is a byzantine construction without much sense. But notice that from the point of view of understanding Nature, the introduction of $\vec{v}$ points to the physically correct direction: we are led to investigate the nature and the dynamics of this vector; $\vec{v}$ is indeed the local gravitational field, and this is precisely the right track towards a more effective understanding of Nature. This is the strength of having understood rotational invariance.

In fact, if there is rotational invariance in the Universe, there should be a rotationally invariant manner of understanding ancient physics, which, in its limited extent, was effective. Theory T' above represents precisely this better understanding of ancient physics. More than that, the reinterpretation itself indicates a new effective way of understanding the world. In conclusion, the fact that the effective but nonrotationally invariant theory T admits the byzantine rotationally invariant formulation T' is not an argument for the physical irrelevance of rotational invariance. Far from that, it is something that is required for us to have confidence in rotational invariance.

On the one hand, rotational invariance is interesting because it *enlarges*, not because it *restricts*, the kind of physics we can naturally describe. On the other hand, rotational invariance *does* drastically reduce the kind of theories that we are *willing* to consider. Not because it forbids us to write certain theories – such as theory T' –, but because if we want to describe a theory such as T, we have to pay a price. Here the introduction of the vector $\vec{v}$. It is up to the theoretician to judge whether this price is worth paying, that is, whether $\vec{v}$ is in fact a physical entity worthwhile considering.

The value of a novel idea or a novel language in theoretical physics is not in the fact that old physics cannot be expressed in the new language. It is simply in the fact that it is more effective for describing reality. A physical theoretical framework is a map of reality. If the symbols of the map are better chosen, the map is more effective. A new language, by itself, rarely truly restricts the kind of theories that can be expressed. But it renders certain theories far simpler and others awkward. It orients our investigation on Nature. This, and nothing else, is scientific knowledge.

Let me come back to general covariance. Like rotational invariance, general covariance is a novel language which expresses a general physical idea about the world. It is possible to express Newtonian physics in a generally covariant language. It is also possible to express GR physics in a nongenerally covariant language (by gauge-fixing the coordinates). But newtonian physics expressed in a covariant language or GR expressed in a noncovariant language are both monsters formulated in a form far more intricate than what is possible. Nobody would have found them.

What Einstein discovered is that two classes of entities previously considered distinct are in fact entities of the same kind. Newton taught us that (an effective way to understand the world is to think that) the world is made up of two clearly *distinct* classes of entities, of very different nature. The first class is formed by space and time. The second class includes all dynamical entities moving in space and in time. In newtonian physics these two classes of entities are different in many respects, and enter the formalism of physical models in very different manners. Einstein has understood that (a more effective way to understand the world is to think that) the world is *not* made up of two distinct kinds of entities. There is only one type of entity: dynamical fields. General covariance is the language for describing a world without distinction between the spacetime entities and the dynamical entities. It is the language that does not assume this distinction.

We can reinterpret prerelativistic physics in a generally covariant language. It suffices to rewrite the newtonian absolute space and absolute time as a dynamical field, and then write generally covariant equations that fix them to their flat-space values. But if we do so, we are not denying the physical content of Einstein's idea. On the contrary, we are simply reinterpreting the world in Einstein's terms. In other words, we are showing the strength, not the weakness, of general covariance. Furthermore, in so doing we introduce a new physical field and we find ourselves in the funny situation of having to write equations of motion for this field that constrain it to a single value. Thus we have a theory where one of the dynamical fields is strangely constrained to a single value. This immediately suggests that perhaps we can relax these equations and allow a full dynamics for this field. If we do so, we are directly on the track of GR. Again, far from showing the physical irrelevance of general covariance, this indicates its enormous cognitive strength.

I think that the mistake behind Kretschmann's argument is an excessively legalistic reading of the scientific enterprise. It is the mistake of taking certain common physicists' statements too literally. Physicists often write that a certain symmetry or a certain principle "uniquely determines" a certain theory. At a close reading, these statements are almost always much exaggerated. The uniqueness only holds under a vast number of other assumptions that are left implicit and which are facts or ideas the physicist considers natural, and does not bother detailing. The typical physicist carelessly dismisses counter-examples by saying that they would be unphysical, implausible, or completely artificial. The connection between general physical ideas, general principles, intuitions, symmetries, is a burning melt of powerful ideas, not the icy demonstration of a mathematical theorem. What is at stake is finding the most effective language for thinking the world, not writing axioms. It is language in formation, not bureaucracy.[25]

---

[25] Historically, the entire issue might be the result of a misunderstanding. Kretschmann attacked Einstein in a virulent form. In particular, he attacked Einstein's coincidences solution of the hole argument. Now Einstein probably learned the idea that coincidences are the only observables precisely from Kretschmann, but didn't give much credit to Kretschmann for this. I suppose this should have made Kretschmann quite bitter. I think that Kretschmann's subtext in saying that general covariance is empty was not that general covariance was no progress with respect to old physics: it was that general covariance was no progress with respect to what he himself had already realized before Einstein.

Generally covariant flat-space field theory. Consider the field theory of a free massless scalar field $\phi(x)$ on Minkowski space. The theory is defined by the action

$$S[\phi] = \int d^4x \; \eta^{\alpha\beta} \partial_\alpha \phi \; \partial_\beta \phi. \tag{2.141}$$

The equation of motion is the flat-space Klein–Gordon equation

$$\eta^{\alpha\beta} \partial_\alpha \partial_\beta \phi = 0, \tag{2.142}$$

and the theory is obviously not generally covariant.

A trivial way to reformulate this theory in generally covariant language is to introduce the tetrad field $e_\mu^\alpha(x)$ and write the equations

$$\partial_\mu(e \, \eta^{\alpha\beta} e_\alpha^\mu e_\beta^\nu \partial_\nu \phi) = 0, \tag{2.143}$$

$$R^\alpha{}_{\beta\mu\nu} = 0. \tag{2.144}$$

The solution of (2.144) is that e is flat. Since the system is covariant we can choose a gauge in which $e_\mu^\alpha(x) = \delta_\mu^\alpha$. In this gauge, (2.143) becomes (2.141).

A more interesting way is as follows. Consider a field theory for five scalar fields $\Phi^A(x)$, where $A = 1, \ldots, 5$. Use the notation

$$V_A = \epsilon_{ABCDE} \; \partial_\mu \Phi^B \partial_\nu \Phi^C \partial_\rho \Phi^D \partial_\sigma \Phi^E \, \epsilon^{\mu\nu\rho\sigma}, \tag{2.145}$$

where $\epsilon^{\mu\nu\rho\sigma}$ and ϵ_{ABCDE} are the 4-dimensional and 5-dimensional completely antisymmetric pseudo-tensors. Consider the theory defined by the action

$$S[\Phi^A] = \int d^4x \; V_5^{-1}(V_4 V_4 - V_3 V_3 - V_2 V_2 - V_1 V_1), \tag{2.146}$$

where V_5 is assumed never to vanish. The theory is invariant under diffeomorphisms. Indeed, V_A transforms as a scalar density (because $\epsilon^{\mu\nu\rho\sigma}$ is a scalar density), hence the integrand is a scalar density and the integral is invariant. For $\alpha = 1, 2, 3, 4$, define the matrix

$$E_\mu^\alpha(x) = \partial_\mu \Phi^\alpha(x), \tag{2.147}$$

its inverse E_α^μ and its determinant E. Varying Φ^5, we obtain the equation of motion

$$\partial_\mu(E \, \eta^{\alpha\beta} E_\alpha^\mu E_\beta^\nu \partial_\nu \Phi^5) = 0. \tag{2.148}$$

This is the massless Klein–Gordon equation (2.143) interacting with a gravitational field E_μ^α. Varying Φ^α we do not obtain independent equations. We obtain the energy-momentum conservation law implied by (2.148). The fact that there is only one independent equation is a consequence of the fact that there is a four-fold gauge invariance. We can choose a gauge in which

$$\Phi^a(x) = x^a. \tag{2.149}$$

We then have immediately $E_\mu^a = \delta_\mu^a$, and (2.148) becomes (2.142). The other four equations are

$$\partial_a(\partial^a \Phi^5 \partial_b \Phi^5 - \frac{1}{2} \delta_b^a \partial_c \Phi^5 \partial_c \Phi^5) = 0. \tag{2.150}$$

Even better, we may not fix the gauge, and consider the gauge-invariant function of four variables $\phi(X^a)$ defined by

$$\phi(\Phi^a(x)) = \Phi^5(x). \tag{2.151}$$

This function satisfies the Minkowski-space Klein–Gordon equation (2.142).

How to interpret such a theory? The theory (2.141) is not generally covariant, therefore its coordinates x are (partial) observables. The theory is defined by five partial observables: four x^μ and ϕ. To interpret the theory we must have measuring procedures associated with these five quantities. The relation between these observables is governed by (2.141). On the other hand, the theory (2.146) is generally covariant; therefore the coordinates x are not observable. The theory is defined by five partial observables: the five ϕ^A. We must have measuring procedures associated with these five quantities. The relation between these observables is governed again by (2.141). Therefore in the two cases we have the same partial observables, identified by $\Phi^A \leftrightarrow (x^a, \phi)$, related by the same equation.

There is only one subtle but important difference between theory (2.146) and theory (2.141). Theory (2.141) separates the five partial observables (x, ϕ) into two sets: the independent ones (x) and the dependent one (ϕ). Theory (2.146) treats the five partial observables Φ^A on an equal footing. Thus, in a strict sense, theory (2.141) contains one extra item of information: a distinction between dependent and independent partial observables. Because of this difference, the two theories reflect two quite different interpretations of the world. The first describes a world's ontology split into spacetime and matter. The second describes a world where the spacetime structure is interpreted as relational.

2.4.4 Meanings of time

The concept of time used in natural language carries many properties. Within a given theoretical framework (say newtonian mechanics) time maintains some of these properties, and loses others. In different theoretical frameworks time has different properties. The best-known example is probably the directionality of time: absent in mechanics, present in thermodynamics. But many other features of time lack in one theory and are present in others. For instance, a property of time in newtonian mechanics is uniqueness: there is a unique time interval between any two events. Conversely, in special relativity there are as many time variables as there are Lorentz observers $(x^0, x'^0, \ldots)$. Another attribute of time in newtonian mechanics is globality: every solution of the equations of motion "passes" through every value of newtonian time t once and only once. In some cosmological models, on the other hand, there is no choice of time variable with such a property: there is "no time," if we demand that being global is an essential property of time. In other words, we use the word "time" to denote quite different concepts, that may or may not include this or that property.

Here I describe a simple classification of possible attributes of time. Below, I identify and list nine properties of time. Then I describe and tabulate ten separate levels of increasing complexity of the notion of time, corresponding to an increasing number of properties. Theories typically fall in one of these levels, according to the set of attributes that the theory ascribes to the notion of time it uses. The ten-fold arrangement is conventional: the main point I intend to emphasize is that a single, clear and pure notion of "time" does not exist.

Properties of time. Consider an infinite set S without any structure. Add to S a topology and a differential structure dx. Thus, S becomes a manifold; assume that this manifold is one-dimensional, and denote the set S together with its differentiable structure as the line $L = (S, dx)$. Next, assume we add a metric structure d to L; denote the resulting metric line as $M = (S, dx, d)$. Next, fix an ordering $<$ (a direction) in M. Denote the resulting oriented line as the affine line $A = (S, dx, d, <)$. Next, fix a preferred point of A as the origin 0; the resulting space is isomorphic to the real line $R = (S, \mathrm{d}x, d, <, 0)$.

The real line R is the traditional metaphor for the idea of time. Time is frequently represented by a variable t in R. The structure of R corresponds to an ensemble of properties that we naturally associate to the notion of time, as follows. (a) The existence of a topology on the set of the time instants, namely the existence of a notion of two time instants being close to each other, and the fact that time is "one-dimensional." (b) The existence of a metric. Namely the possibility of stating that two distinct time intervals are equal in magnitude; time is "metric." (c) The existence of an ordering relation between time instants. Namely, the possibility of distinguishing the past direction from the future direction. (d) The existence of a preferred time instant, the present, the "now." To capture these properties in mathematical language, we describe time as a real line R. An affine line A describes time up to the notion of present; a metric line M describes time up to the notions of present and past/future distinction; a line L describes time up to the notion of metricity.

In newtonian mechanics, we begin by representing time as a variable in R, but then the equations are invariant both under $t \mapsto -t$ and under $t \mapsto t + a$. Thus the theory is actually defined in terms of a variable t in a metric line M. Newtonian mechanics, in fact, incorporates both the notions of topology of the set of time instants and (in a very essential way) the fact that time is metric, but it does not make any use of the notion of present nor the direction of time. This is well known. Note that Newton theory is not inconsistent with the introduction of the notions of a present and of time-directionality: it simply does not make any use of these notions. These notions are not present in Newton theory.

The properties listed above do not exhaust the different ways in which the notion of time enters physical theories; the development of theoretical physics has modified substantially the natural notion of time. A first modification was introduced by special relativity. Einstein's definition of the time coordinate of distant events yields a notion of time which is observer dependent. An invariant structure can be maintained at the price of relaxing the 1d character of time and the 3d character of space in favor of a notion of 4d spacetime. Alternatively, we may say that the notion of a single time is replaced by a three-parameter family of times $t_{\vec{v}}$, one for each Lorentz observer. Therefore, the time we use in special relativity is not unique as is the time in newtonian mechanics. Rather than a single line, we have a three-parameter family of lines (the straight lines through the origin that fill the light cone of Minkowski space). Denote this three-parameter set of lines as M^3.

Times in GR. There are several distinct possibilities of identifying "time" in GR. Each singles out a different notion of time. Each of these notions reduces to the standard nonrelativistic or special-relativistic time in appropriate limits, but each lacks at least some of the properties of nonrelativistic time. The most common ways of identifying time within GR are the following.

Coordinate time x^0. Coordinate time can be arbitrarily rescaled, and does not provide a way of identifying two time intervals as equal in duration. Therefore it is not metric, in the sense defined above. In addition, the possibility of changing the time coordinate freely from point to point implies that there is an infinite-dimensional choice of equally good coordinate times. Finally, unlike prerelativistic time, x^0 is not an observable quantity. Denote the set of all the possible coordinate times as L^∞.

Proper time τ. This notion of time is metric. But it is very different from the notion of

time in special relativity for several reasons. First, it is determined by the gravitational field. Second, we have a different time for each worldline, or, infinitesimally, for every speed at every point. For an infinitesimal timelike displacement
dx^μ at a point x, the infinitesimal time interval is $d\tau = \sqrt{-g_{\mu\nu}(x)\,dx^\mu dx^\nu}$. This
notion of time is a radical departure from the notion of time used in special rela-
tivity because it is determined by the dynamical fields in the theory. A solution
of Einstein's equations defines a point in the phase space Γ of GR. It assigns
a metric structure to every worldline. Therefore this notion of time is given by
a function from the phase space Γ multiplied by the set of the worldlines wl
into the metric structures $d : wl \times wl \mapsto R^+$. Denote this function as m^∞. Call
"internal" a notion of time affected by the dynamics.

Before GR, dynamics could be expressed as evolution in a single time variable which has metric properties and could be measured. In general-relativistic
physics, this concept of time splits into two distinct concepts: we can still view
the dynamics as evolution in a time variable, x^0, but this time has no metric
properties and is not observable; alternatively, there is a notion of time that
has metric properties, τ, but the dynamics of the theory cannot be expressed as
evolution in τ. Is there a way to go around this split and view GR as a dynamical theory in the sense of a theory expressing evolution in an observable metric
time?

Clock time. The dynamics of GR determines how observable quantities evolve with
respect to one another. We can always choose one observable quantity t_c, de-
clare it the independent one, and describe how the other observables evolve as
functions of it. A typical example of this clock time is the radius of a spatially
compact universe in relativistic cosmology R. Formally, clock time is a function
on the extended configuration space $\mathcal{C}$ of the theory (see Chapter 3.) Denote this
notion of time as the clock time $\tau_c : \mathcal{C} \mapsto R$.

Under this definition of time, GR becomes similar to a standard hamiltonian
dynamical theory. A clock time, however, generally behaves as a clock only in
certain states or for a limited amount of time. The radius of the universe, for
instance, fails as a good time variable when the universe recollapses. In general, a clock time lacks temporal globality. In fact, several results are known
concerning obstructions to defining a function t_c that behaves as "a good time"
globally [66].

Notice that some of these relativistic notions of time are, in a sense, opposite to the
prerelativistic case: while in newtonian theory time evolution is captured by a function
from the metric line M (time) to the configuration or phase space, now the notion of
time is captured by a function from the configuration or phase space to the metric line.
This inversion is the mathematical expression of the physical idea that the flow of time
is affected or determined by the dynamics of the system itself.

Finally, none of the ways of thinking of time in classical GR can be uncritically
extended to the quantum regime. Quantum fluctuations of physical clocks, and quantum superposition of different metric structures make the very notion of time fuzzy at
the Planck scale. As will be discussed in the second part of this book, a fundamental
concept of time may be absent in quantum gravity.

Notions of time. Notice that properties of time progressively disappear in going toward
more fundamental physical theories. At the opposite end of the spectrum, there are

properties associated with the notion of time used in the natural languages which are not present in physical theories. They play a role in other areas of natural investigations. I mention these properties for the sake of completeness. These are, for instance, memory, expectations, and the psychological perception of free will.

To summarize, I have identified the following properties of the notion of time.

1 Existence of memory and expectations.
2 Existence of a preferred instant of time: the present, the now.
3 Directionality: the possibility of distinguishing the past from the future direction.
4 Uniqueness: the feature that is lost in special and general relativity, where we cannot identify a preferred time variable.
5 The property of being external: the independence of the notion of time from the dynamical variables of the theory.
6 Spatial globality: the possibility of defining the same time variable in all space points.
7 Temporal globality: the fact that every motion goes through every value of the time variable once and only once.
8 Metricity: the possibility of saying that two time intervals have equal duration.
9 One-dimensionality, namely the possibility of arranging the time instants in a one-dimensional manifold.

This discussion suggests a sequence of notions of time, which I list here in order of decreasing complexity.

Time of natural language. This is the notion of time of everyday language, which includes all the features just listed. This notion of time is not necessarily nonscientific: for instance, any scientific approach to, say, the human brain, should make use of this notion of time.

Time-with-a-present. This is the notion of time that has all the features just listed, including the existence of a preferred instant, the present, but not the notions of memory and expectations, which are notions usually more related to complex systems (brain) than to time itself. The notion of present is generally considered a feature of time itself. This notion of time is the one to which often people refer when they refer to the "flow of time" or Eddington's "vivid perception of the flow of time" [67]. This notion of time can be described by the structure of a parametrized line R.

Thermodynamical time. If we maintain the distinction between a future direction and a past direction, but we give up the notion of present, we obtain the notion of time typical of thermodynamics. Since thermodynamics is the first physical science that appears in this list, this is maybe a good place to emphasize that the notion of present, of the "now," is completely absent from the description of the world in physical terms. This notion of time can be described by the structure of an affine line A.

Newtonian time. In newtonian mechanics there is no preferred direction of time. Notice that in the absence of a preferred direction of time the notions of cause and effect are interchangeable. This notion of time can be described by the structure of a metric line M.

Special-relativistic time. If we give up uniqueness, we have the time used in special relativity: different Lorentz observers have a different notion of time. Special-relativistic time is still external, spatially and temporally global, metrical and one-dimensional, but it is not unique: There is a three-parameter set of quantities that share the status of time. This notion of time can be described by the three-parameter set of metric lines M^3.

Table 2.1 Notions of time.

Time notion	Property	Example	Form
natural language time	memory	brain	?
time-with-a-present	present	biology	R
thermodynamical time	direction	thermodynamics	A
newtonian time	unique	newtonian mechanics	M
special-relativistic time	external	special relativity	M^3
cosmological time	spatially global	cosmological time	m
proper time	temporally global	worldline proper time	m^∞
clock time	metric	clocks in GR	c
parameter time	one-dimensional	coordinate time	L^∞
no-time	none	quantum gravity	none

Cosmological time. By this I indicate a time which is spatially and temporally global, metrical and one-dimensional, but it is not external, namely it is dynamically determined by the theory. Proper time in cosmology is the typical example. It is the most structured notion of time that occurs in GR. Denote it by m.

Proper time. By this I indicate a time which is temporally global, metrical and one-dimensional, but it is not spatially global, as the notion of proper time along worldlines in GR. It can be represented by a function m^∞ defined on the cartesian product of the phase space and the ensemble of the worldlines.

Clock time. By this I indicate a time which is metrical and one-dimensional, but it is not temporally global. A realistic matter clock in GR defines a time in this sense. This notion of time can be described by a function c on the phase space.

Parameter time. By this we mean a notion of time which is not metric and not observable. The typical example is the coordinate-time in GR. Another example of parameter time is the evolution parameter in the parametrized formulation of the dynamics of a relativistic particle. Parameter time is described by an unparametrized line L, or by an infinite set L^∞ of unparametrized lines.

No-time. Finally, this is the bottom level in the analysis; it is not a time concept, but rather I indicate by no-time the idea that a predictive physical theory can be well defined also in the absence of any notion of time.

The list must not be taken rigidly. It is summarized in the Table 2.1.

There is a interesting feature that emerges from the above analysis: the hierarchical arrangement. While some details of this arrangement may be artificial, nevertheless the analysis points to a general fact: moving from theories of "special" objects, like the brain or living beings, toward more general theories that include larger portions of Nature, we make use of a physical notion of time that is less specific and has less determinations. If we observe Nature at progressively more fundamental levels, and we seek for laws that hold in more general contexts, then we discover that these laws require or admit an increasingly weaker notion of time.

This observation suggests that "high level" features of time are not present at the fundamental level, but "emerge" as features of specific physical regimes, like the notion of "water surface" emerges in certain regimes of the dynamics of a combination of water and air molecules (see, for instance, [68]).

Notions of time with more attributes are high-level notions that have no meaning in more general situations. The uniqueness of newtonian time, for instance, makes sense only in the special regime in which we consider an ensemble of bodies moving slowly with respect to each other. Thus, the notion of a unique time is a high-level notion that makes sense only for some regimes in Nature. For general systems, most features of time are genuinely meaningless.

2.4.5 Nonrelativistic coordinates

The precise meaning of the coordinates $x = (\vec{x}, t)$ in newtonian and special-relativistic physics is far from obvious. Let me recall it here, in order to clarify the precise difference between these and the relativistic coordinates.

Newton is well aware that the motions we observe are relative motions, and stresses this point in *Principia*. His point is not that we can directly *observe* absolute motion. His point is that we can *infer* the absolute motion or "true motions," or motion with respect to absolute space, from its physical effects (such as the concavity of the water in the bucket), starting from our observation of relative motions.

For instance, we observe and describe motions with respect to Earth; but from subtle effects, such as Foucault's pendulum, we infer that these are not true motions. The experiment of the bucket is an example of the possibility of revealing true motion (rotation of the water with respect to space), disentangling it from relative motion (rotation with respect to the bucket), by means of an observable effect (the concavity of the water surface).[26]

For Newton, the coordinates $\vec{x}$ that enter his main equation

$$\vec{F} = m \, \frac{\mathrm{d}^2 \vec{x}(t)}{\mathrm{d}t^2} \tag{2.152}$$

are the coordinates of absolute space. However, since we cannot directly observe space, the only way we can coordinatize space points is by using physical objects. The coordinates $\vec{x}$ of the object A moving along the trajectory $\vec{x}(t)$ are therefore defined as distances from a chosen system O of objects, which we call a "reference frame." But then $\vec{x}$ are not the coordinates of absolute space. So, how can equation (2.152) work?

The solution of the difficulty, is to use the capacity of unveiling "true motion" that Newton has pointed out, in order to select the objects forming the reference frame O wisely. There are "good" and "bad" reference frames. The good ones are the ones in which no effect such as the concavity of the water surface of Newton's bucket can be

---

[26] Newton accords deep significance to the fact that we can unveil true motion. He describes relative motion as the way reality is observed by us, and true motion as the way reality might be directly "perceived", or "sensed", by God. This is why Newton calls space – the entity with respect to which true motion happens – the "Sensorium of God": true motion is motion "with respect to God," or "as perceived by God." There is a platonic tone in this idea that reason finds the way to the veiled divine truth beyond appearances. I wouldn't read this as so removed from modernity as it is often portrayed. There isn't all that much difference between Newton's inquiry into God's way of "sensing the world," and the modern search for the most effective way of conceptualizing reality. Newton's God plays a mere linguistical role here: the role of denoting a major enterprise, upgrading our own conceptual structure for understanding reality.

observed, within a desired accuracy. Equation (2.152) is correct, to the desired accuracy, if we use coordinates defined with respect to these good frames. In other words, the physical content of (2.152) is actually quite subtle:

> There exist reference objects O with respect to which the motion of any other object A is correctly described by (2.152).

This is a statement that begins to be meaningful only when a sufficiently large number of moving objects is involved.

Notice also that for this construction to work it is important that the objects O forming the reference frame are not affected by the motion of the object A. There shouldn't be any dynamical interaction between A and O.

Special relativity does not change much of this picture. Since absolute simultaneity makes no sense, if the event A is distant from the clock in the origin, its time t is ill defined. Einstein's idea is to *define* a procedure for assigning a t to distant events, using clocks moving inertially.

> At clock time t_e, send a light signal that reaches the event. Receive the reflected signal back at t_r. The time coordinate of the event is *defined* to be $t_A = \frac{1}{2}(t_e + t_r)$.

It is important to emphasize that this is a useful definition, not a metaphysical statement that the event A happens *"right at the time when"* the observer clock displays t_A.

Special relativity replaces Newton's absolute space and absolute time with a single entity: Minkowski's absolute spacetime, while the notion of inertial system and the meaning of the coordinates are the same as in newtonian mechanics.

Summarizing, these coordinates have the following properties.

(i) Coordinates describe position with respect to physical reference objects (reference frames).

(ii) Space coordinates are defined by the *distance* from the reference bodies. Time coordinates are defined with respect to isochronous clocks.

(iii) Reference objects are appropriately chosen: they are such that the reference system they define is inertial.

(iv) Inertial frames reveal the structure of absolute spacetime itself.

(v) The object A whose dynamics is described by the coordinates does not interact with the reference objects O. There is no dynamical coupling between A and O.

Relativistic coordinates do not have *any* of these properties. The fact that the two are indicated with the same notation x^μ is only an unfortunate historical accident.

2.4.6 Physical coordinates and GPS observables

Instead of working with arbitrary unphysical coordinates x^μ, we can choose to coordinatize spacetime events with coordinates X^μ having an assigned physical interpretation. For instance, we can describe the Universe by giving a name $\vec{X}$ to each galaxy, and choosing X^0 as the proper time from the Big Bang, along the galaxy worldline. If we do so, the defining properties of the coordinates X must be added to the formalism. We must add a certain number of equations for the gravitational field: the equations of motions of the objects used to fix the coordinates (the galaxies, in the example). These additional equations gauge-fix general covariance.

The gauge-fixing can also be partial. For instance, a common choice is

$$\tilde{e}_0^0(X) = 1, \qquad \tilde{e}_0^i(X) = 0, \qquad \tilde{e}_a^0(X) = 0, \qquad (2.153)$$

where $i = 1, 2, 3$ and $a = 1, 2, 3$. This corresponds to partially fixing the coordinates by requiring that X^0 measures proper time, that equal X^0 surfaces are locally instantaneity surfaces in the sense of Einstein for the constant $\vec{X}$ lines and that the local Lorentz frames are chosen so that these lines are still.

If the coordinates are fully specified, the set formed by these physical gauge-fixing equations and the equations of motion has no residual gauge invariance; that is, initial data determine evolution uniquely. This procedure can be implemented in many possible ways, since there are arbitrarily many ways of fixing physical coordinates, and none is a priori better than any other. In spite of this arbitrariness, this procedure is often convenient, when the physical situation suggests a natural coordinate choice, as in the cosmological context mentioned.

Physical coordinates X^μ defined by matter filling space can only be effectively used in the cosmological context because it is only at the cosmological scale that matter fills space. In a system in which there are empty regions, such as the Solar System, these physical coordinates are not available. An interesting alternative choice is provided by the GPS coordinates described below.

The physical coordinates X^μ are partial observables and we can associate measuring devices with them.

Undetermined physical coordinates. Finally, there is a third interpretation of the coordinates of GR, which is intermediate between arbitrary coordinates x^μ and physical coordinates X^μ. Imagine that a region of the universe is filled with certain light objects, which may not be in free fall. We can use these objects to define physical coordinates X^μ, but also choose to ignore the equations of motion of these objects. We obtain a system of equations for the gravitational field and other matter, expressed in terms of coordinates X^μ that are interpreted as the spacetime location of reference objects whose dynamics we *have chosen* to ignore.

This set of equations is under-determined: the same initial conditions can evolve into different solutions. However, the interpretation of such under-determination is simply that we have chosen to neglect part of the equations of motion. Different solutions with the same initial conditions represent the same physical configuration of the fields, but expressed, say, in one case with respect to free-falling reference objects, in the other case with respect to reference objects on which a force has acted at a certain moment, and so on. This procedure has the disadvantage of being useless in quantum theory, where we cannot assume that something is observable and at the same time neglect its dynamics.

In conclusion, one should always be careful in talking about general-relativistic coordinates whether one is referring to

(i)   arbitrary mathematical coordinates x;

(ii)   physical coordinates X with an interpretation as positions with respect to objects whose equations of motion are taken into account;

(iii) physical coordinates with an interpretation as positions with respect to objects whose equations of motion are ignored.

The system of equations of motion is nondeterministic in (i) and (iii), deterministic in (ii). The coordinates are partial observables in (ii) and (iii), but not in (i). Confusion about observability in GR follows from confusing these three different interpretations of the coordinates. The following is an example of physical coordinates.

GPS observables. In the literature there are many attempts to define useful physical coordinates. It is easier to define physical coordinates in the presence of matter than in the context of pure GR. Ideally, we can consider GR interacting with four scalar matter fields. Assume that the configuration of these fields is sufficiently nondegenerate. Then the components of the gravitational field at points defined by given values of the matter fields are gauge-invariant observables. This idea has been developed in a number of variants, such as dust-carrying clocks and others (see [69–71] and references therein). The extent to which the result is realistic or useful is questionable. It is rather unsatisfactory to understand the theory in terms of fields that do not exist, or phenomenological objects such as dust, and it is questionable whether these procedures could make sense in the quantum theory where the aim is to describe Planck scale dynamics. Earlier attempts to write a complete set of gauge-invariant observables are in the context of pure GR [72]. The idea is to construct four scalar functions of the gravitational field (say, scalar polynomials of the curvature), and use these to localize points. The value of a fifth scalar function in a point where the four scalar functions have a given value is a gauge-invariant observable. This works, but the result is mathematically very intricate and physically very unrealistic. It is certainly possible, in principle, to construct detectors of such observables but I doubt any experimenter would get funded for a proposal to build such an apparatus.

There is a simple way out based on GR coupled with a minimal and *very* realistic amount of additional matter. Indeed, this way out is so realistic that it is in fact real: it is essentially already implemented by existing technology, the Global Positioning System (GPS), which is the first technological application of GR or the first large-scale technology that needs to take GR effects into account [73].

Consider a generally covariant system formed by GR coupled with four small bodies. These are taken to have negligible mass; they will be considered as point particles for simplicity, and called "satellites." Assume that the four satellites follow timelike geodesics, that these geodesics meet in a common (starting) point O, and at O they have a given (fixed) speed – the same for all four – and directions as the four vertices of a tetrahedron. The theory might include any other matter. Then (there is a region $\mathcal{R}$ of spacetime for which) we can uniquely associate four numbers $s^\alpha, \alpha = 1, 2, 3, 4$ to each spacetime point p as follows. Consider the past lightcone of p. This will (generically) intersect the four geodesics in four points p_α. The numbers s^α are defined as the distance between p_α and O. (That is: the proper time along the satellites' geodesic.) We can use the s^α as physically defined coordinates for p. The components $g_{\alpha\beta}(s)$ of the metric tensor in these coordinates are gauge-invariant quantities. They are invariant under four-dimensional diffeomorphisms (because these deform the metric as well as the satellites' worldlines). They define a complete set of gauge-invariant observables for the region $\mathcal{R}$.

The physical picture is simple, and its realism is transparent. Imagine that the four "satellites" are in fact satellites, each carrying a clock that measures the proper time along its trajectory, starting at the meeting point O. Imagine also that each satellite broadcasts its local time with a radio signal. Suppose I am at the point p and have an electronic device that simply receives the four signals and displays the four readings, see Figure 2.5. These four numbers are precisely the four physical coordinates s^α defined above. Current technology permits us to perform these measurements with an accuracy well within the relativistic regime [73, 74]. If we then use a rod and a clock, and measure the physical 4-distances between s^α coordinates, we are directly measuring the components of the metric tensor in the physical coordinate system. In the terminology of Chapter 3, the s^α are *partial* observables, while the $g_{\alpha\beta}(s)$ are *complete* observables.

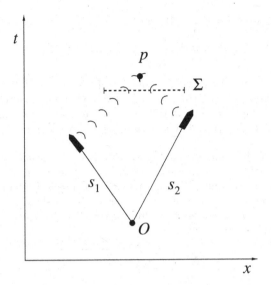

Fig. 2.5 s_1 and s_2 are the GPS coordinates of the point p. Σ is a Cauchy surface with p in its future domain of dependence.

As shown below, the physical coordinates s^α have nice geometrical properties; they are characterized by

$$g^{\alpha\alpha}(s) = 0, \qquad \alpha = 1, \ldots, 4. \tag{2.154}$$

Surprisingly, in spite of the fact that they are defined by what looks like a rather non-local procedure, the evolution equations for $g_{\alpha\beta}(s)$ are local. These evolution equations can be written explicitly using the Arnowitt–Deser–Misner (ADM) variables (see [131] of Chapter 3 for details). Lapse and Shift turn out to be fixed local functions of the three metric.

In what follows, I first introduce the GPS coordinates s^α in Minkowski space. Then I consider a general spacetime. I assume the Einstein summation convention only for couples of repeated indices that are one up and one down. Thus, in (2.154) α is not summed. While dealing with Minkowski spacetime, the spacetime indices μ, ν are raised and lowered with the Minkowski metric. Here I write an arrow over three- as well as four-dimensional vectors. Also, here I use the signature $[+, -, -, -]$, in order to have the same expressions as in the original article on the subject.

Consider a tetrahedron in three-dimensional euclidean space. Let its center be at the origin and its four vertices $\vec{v}^\alpha$, where the vectors $\vec{v}^\alpha$ have unit length $|\vec{v}^\alpha|^2 = 1$ and $\vec{v}^\alpha \cdot \vec{v}^\beta = -1/3$ for $\alpha \neq \beta$. Here $\alpha = 1, 2, 3, 4$ is an index that distinguishes the four vertices, and should not be confused with vector indices. With a convenient orientation, these vertices have cartesian coordinates ($a = 1, 2, 3$)

$$v^{1a} = (0,\ 0,\ 1), \qquad\qquad v^{2a} = (2\sqrt{2}/3,\ 0,\ -1/3), \tag{2.155}$$

$$v^{3a} = (-\sqrt{2}/3,\ \sqrt{2/3},\ -1/3), \quad v^{4a} = (-\sqrt{2}/3,\ -\sqrt{2/3},\ -1/3). \tag{2.156}$$

Let us now go to a four-dimensional Minkowski space. Consider four timelike 4-vectors $\vec{W}^\alpha$, of length unity, $|\vec{W}^\alpha|^2 = 1$, representing the normalized 4-velocities of four particles moving away from the origin in the directions $\vec{v}^\alpha$ at a common speed v. Their

Minkowski coordinates ($\mu = 0, 1, 2, 3$) are

$$W^{\alpha\mu} = \frac{1}{\sqrt{1 - v^2}}\, (1,\ v v^{\alpha a}). \tag{2.157}$$

Fix the velocity v by requiring the determinant of the matrix $W^{\alpha\mu}$ to be unity. (This choice fixes v at about one-half the speed of light; a different choice changes only a few normalization factors in what follows.) The four by four matrix $W^{\alpha\mu}$ plays an important role in what follows. Notice that it is a fixed matrix whose entries are certain given numbers.

Consider one of the four 4-vectors, say $\vec{W} = \vec{W}^1$. Consider a free particle in Minkowski space that starts from the origin with 4-velocity $\vec{W}$. Call it a "satellite." Its worldline l is $\vec{x}(s) = s\vec{W}$. Since $\vec{W}$ is normalized, s is precisely the proper time along the worldline. Consider now an arbitrary point p in Minkowski spacetime, with coordinates $\vec{X}$. Compute the value of s at the intersection between l and the past lightcone of p. This is a simple exercise, giving

$$s = \vec{X} \cdot \vec{W} - \sqrt{(\vec{X} \cdot \vec{W})^2 - |\vec{X}|^2}. \tag{2.158}$$

Now consider four satellites, moving out of the origin at 4-velocity $\vec{W}^\alpha$. If they radio broadcast their position, an observer at the point p with Minkowski coordinates $\vec{X}$ receives the four signals s^α

$$s^\alpha = \vec{X} \cdot \vec{W}^\alpha - \sqrt{(\vec{X} \cdot \vec{W}^\alpha)^2 - |\vec{X}|^2}. \tag{2.159}$$

Introduce (nonlorentzian) general coordinates s^α on Minkowski space, defined by the change of variables (2.159). These are the coordinates read out by a GPS device in Minkowski space. The jacobian matrix of the change of coordinates is given by

$$\frac{\partial s^\alpha}{\partial x^\mu} = W^\alpha_\mu - \frac{W^\alpha_\mu (\vec{X} \cdot \vec{W}^\alpha) - X_\mu}{\sqrt{(\vec{X} \cdot \vec{W}^\alpha)^2 - |\vec{X}|^2}}, \tag{2.160}$$

where W^α_μ and X_μ are $W^{\alpha\mu}$ and X^μ with the spacetime index lowered with the Minkowski metric. This defines the tetrad field $e^\alpha_\mu(s)$

$$e^\alpha_\mu(s(X)) = \frac{\partial s^\alpha}{\partial x^\mu}(X). \tag{2.161}$$

The contravariant metric tensor is given by $g^{\alpha\beta}(s) = e^\alpha_\mu(s) e^{\mu\beta}(s)$. Using the relation $|\vec{W}^\alpha|^2 = 1$, a straightforward calculation shows that

$$g^{\alpha\alpha}(s) = 0, \qquad \alpha = 1, \ldots, 4. \tag{2.162}$$

This equation has the following nice geometrical interpretation. Fix α and consider the one-form field $\omega^\alpha = ds^\alpha$. In s^α coordinates, this one-form has components $\omega^\alpha_\beta = \delta^\alpha_\beta$, and therefore "length" $|\omega^\alpha|^2 = g^{\beta\gamma}\omega^\alpha_\beta\omega^\alpha_\gamma = g^{\alpha\alpha}$. But the "length" of a one-form is proportional to the volume of the (infinitesimal, now) 3-surface defined by the form. The 3-surface defined by ds^α is the surface $s^\alpha = $ *constant*. But $s^\alpha = $ *constant* is the set of points that read the GPS coordinate s^α, namely that receive a radio broadcasting from a same event p_α of the satellite α, namely that are on the future lightcone of p_α. Therefore $s^\alpha = $ *constant* is a portion of this lightcone; it is a null surface and therefore its volume is zero. And so $|\omega^\alpha|^2 = 0$, and $g^{\alpha\alpha} = 0$.

Since the s^α coordinates define $s^\alpha = constant$ surfaces that are null, we denote them as "null GPS coordinates." It is useful to introduce another set of GPS coordinates as well, which have the traditional timelike and spacelike character. We denote these as s^μ, call them "timelike GPS coordinates," and define them by

$$s^\alpha = W^\alpha_\mu s^\mu. \tag{2.163}$$

This is a simple algebraic relabeling of the names of the four GPS coordinates, such that $s^{\mu=0}$ is timelike and $s^{\mu=a}$ is spacelike. In these coordinates, the gauge condition (2.162) reads

$$W^\alpha_\mu W^\alpha_\nu g^{\mu\nu}(s) = 0. \tag{2.164}$$

This can be interpreted geometrically as follows. The (timelike) GPS coordinates are coordinates s^μ such that the four one-form fields

$$\omega^\alpha = W^\alpha_\mu ds^\mu \tag{2.165}$$

are null.

Let us now jump from Minkowski space to full GR. Consider GR coupled with four satellites of negligible mass that move geodesically and whose worldlines emerge from a point O with directions and velocity as above. Locally around O the metric can be taken to be minkowskian; therefore the details of the initial conditions of the satellites' worldlines can be taken as above. The phase space of this system is the one of pure GR plus ten parameters, giving the location of O and the Lorentz orientation of the initial tetrahedron of velocities. The integration of the satellites' geodesics and of the lightcones can be arbitrarily complicated in an arbitrary metric. However, if the metric is sufficiently regular, there will still be a region $\mathcal{R}$ in which the radio signals broadcast by the satellites are received. (In the case of multiple reception, the strongest one can be selected. That is, if the past lightcone of p intersects l more than once, generically there will be one intersection which is at shorter luminosity distance.) Thus, we still have well-defined physical coordinates s^α on $\mathcal{R}$. Equation (2.162) holds in these coordinates, because it depends only on the properties of the light propagation around p. We define also timelike GPS coordinates s^μ by (2.163), and we get condition (2.164) on the metric tensor.

To study the evolution of the metric tensor in GPS coordinates it is easier to shift to ADM variables N, N^a, γ_{ab}. These are functions of the covariant components of the metric tensor, defined in general by

$$ds^2 = g_{\mu\nu}dx^\mu dx^\nu = N^2 dt^2 - \gamma_{ab}(dx^a - N^a dt)(dx^b - N^b dt). \tag{2.166}$$

Equivalently, they are related to the contravariant components of the metric tensor by

$$g^{\mu\nu} v_\mu v_\nu = -\gamma^{ab} v_a v_b + (n^\mu v_\mu)^2, \tag{2.167}$$

where γ^{ab} is the inverse of γ_{ab} and $n^\mu = (1/N, N^a/N)$. Using these variables, the gauge condition (2.164) reads

$$W^\alpha_a W^\alpha_b \gamma^{ab} = (W^\alpha_\mu n^\mu)^2. \tag{2.168}$$

Notice now that this can be solved for the Lapse and Shift as a function of the 3-metric (recall that W^α_μ are fixed numbers), obtaining

$$n^\mu = W^\mu_\alpha q^\alpha, \tag{2.169}$$

where W_α^μ is the inverse of the matrix W_μ^α, and

$$q^\alpha = \sqrt{W_a^\alpha W_b^\alpha \gamma^{ab}}. \tag{2.170}$$

Or, explicitly,

$$N = \frac{1}{W_\alpha^0 q^\alpha}, \qquad N^a = \frac{W_\alpha^a q^\alpha}{W_\alpha^0 q^\alpha}. \tag{2.171}$$

The geometrical interpretation is as follows. We want the one-form ω^α defined in (2.165) to be null, namely its norm to vanish. But in the ADM formalism this norm is the sum of two parts: the norm of the pull-back of ω^α on the constant time ADM surface, which is q^α, given in (2.170), and depends on the three metric; plus the square of the projection of ω^α on n^μ. We can thus obtain the vanishing of the norm by adjusting the Lapse and Shift. We have four conditions (one per α) and we can thus determine Lapse and Shift from the 3-metric. In other words, whatever the 3-metric, we can always adjust Lapse and Shift so that the gauge condition (2.164) is satisfied. But in the ADM formalism, the arbitrariness of the evolution in the Einstein equations is entirely captured by the freedom in choosing Lapse and Shift. Since here Lapse and Shift are uniquely determined by the 3-metric, evolution is determined uniquely if the initial data on a Cauchy surface are known. Therefore, the evolution in the GPS coordinate s^0 of the GPS components of the metric tensor, $g_{\mu\nu}(s)$, is governed by deterministic equations: the ADM evolution equation with Lapse and Shift determined by (2.170)–(2.171). Notice also that evolution is local, since the ADM evolution equations, as well as the (2.170)–(2.171), are local.[27]

How can the evolution of the quantities $g_{\mu\nu}(s)$ be local? The conditions on the null surfaces described in the previous paragraph are nonlocal. Coordinate distances typically yield nonlocality: imagine we define physical coordinates in the Solar System using the cosmological time t_c and the spatial distances x_S, x_E, x_J (at fixed t_c) from, say, the Sun the Earth and Jupiter. The metric tensor $g_{\mu\nu}(t_c, x_S, x_E, x_J)$ in these coordinates is a gauge-invariant observable, but its evolution is highly nonlocal. To see this, imagine that in this moment (in cosmological time), Jupiter is swept away by a huge comet. Then the value of $g_{\mu\nu}(t_c, x_S, x_E, x_J)$ here changes instantaneously, without any local cause: the value of the coordinate x_J has changed because of an event happening far away. What's special about the GPS coordinates that avoids this nonlocality? The answer is that the value of a GPS coordinate at a point p does in fact depend on what happens "far away" as well. Indeed, it depends on what happens to the satellite. However, it only depends on what happened to the satellite when it was broadcasting the signal received in p, and this is in the past of p! If p is in the past domain of dependence of a partial Cauchy surface Σ, then the value of $g_{\mu\nu}(s)$ in p is completely determined by the metric and its derivative on Σ, namely evolution is causal, because the entire information needed to set up the GPS coordinates is in the data in Σ, see Figure 2.5. Explicitly, the $s^\alpha = constant$ surfaces around Σ can be uniquely integrated ahead all the way to p. They certainly can, as they represent just the evolution of a light front! This is how local evolution is achieved by these coordinates.

Summarizing, I have introduced a set of physical coordinates, determined by certain material bodies. Geometrical quantities such as the components of the metric tensor expressed in physical coordinates are of gauge-invariant observables. There is no need to introduce a large unrealistic amount of matter or to construct complicated and unrealistic physical quantities out of the metric tensor. Four particles are sufficient to

---

[27]This does not imply that the full set of equations satisfied by $g_{\mu\nu}(s)$ must be local, since initial conditions on $s^0 = 0$ satisfy four other constraints besides the ADM ones.

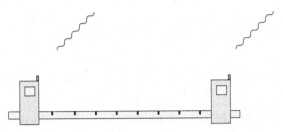

Fig. 2.6 A simple apparatus to measure the gravitational field. Two GPS devices, reading s_L^μ and s_R^μ respectively, connected by a 1 meter rod. If for instance $s_R^\mu = s_L^\mu$ for $\mu = 0, 2, 3$, then the local value of $g_{11}(s)$ is: $g_{11}(s) = (s_R^1 - s_L^1)^{-2} m^2$.

coordinatize a (region of a) four-geometry. Furthermore, the coordinatization procedure is not artificial: it is the real one utilized by existing technology.

The components of the metric tensor in (timelike) GPS coordinates can be measured as follows (see Figure 2.6). Take a rod of physical length L (small with respect to the distance along which the gravitational field changes significantly) with two GPS devices at its ends (reading timelike GPS coordinates). Orient the rod (or search among recorded readings) so that the two GPS devices have the same reading s of all coordinates except for s^1. Let δs^1 be the difference in the two s^1 readings. Then we have along the rod

$$ds^2 = g_{11}(s)\delta s^1 \delta s^1 = L^2. \tag{2.172}$$

Therefore,

$$g_{11}(s) = \left(\frac{L}{\delta s^1}\right)^2. \tag{2.173}$$

Nondiagonal components of $g_{ab}(s)$ can be measured by simple generalizations of this procedure. The $g_{0b}(s)$ are then algebraically determined by the gauge conditions. In a thought experiment, data from a spaceship traveling in a spacetime region could be used to produce a map of values of the GPS components of the metric tensor. Instead of using a rod, which is a rather crude device for measuring distances, one could send a light pulse forward and back between the two GPS devices kept at fixed spatial s^μ coordinates. If T is the (physical) time for flying back and forward measured by a precise clock on one device, then $g_{11}(s) = (cT/2\delta s^1)^2$. This is valid so long as T and L are small compared to the distances over which the $g_{ab}(s)$ change by amounts of the order of the experimental errors.

The individual components of the metric tensor expressed in physical coordinates are measurable. The statement that "the curvature is measurable but the metric is not measurable," which is often heard, is incorrect. Both metric and curvature in physical coordinates are measurable and predictable. Neither metric nor curvature in arbitrary nonphysical coordinates are measurable.

The GPS coordinates are partial observables (see Chapter 3). The complete observables are the quantities $g_{\mu\nu}(s)$, for any given value of the coordinates s^μ. These quantities are diffeomorphism invariant, are uniquely determined by the initial data and in a canonical formulation are represented by functions on the phase space that commute with all constraints.

The GPS observables are a straightforward generalization of Einstein's "spacetime coincidences." In a sense, they are *precisely* Einstein's point coincidences. Einstein's "material points" are just replaced by photons (light pulses): the spacetime point s^α is characterized as the meeting point of four photons designated by the fact of carrying the radio signals s^α.

———

Bibliographical notes

There are many beautiful classic textbooks on GR. Two among the best, offering remarkably different points of view on the theory, are Weinberg [75] and Wald [76]. The first stresses the similarity between GR and flat-space field theory; the second, on the contrary, emphasizes the geometric reading of GR. Here I have followed a third path: I place emphasis on the change of the notions of space and time needed for general-relativistic physics (which affects quantization dramatically), but I put little emphasis on the geometric interpretation of the gravitational field (which is going to be largely lost in the quantum theory).

Relevant mathematics is nicely presented, for instance, in the text by Choquet-Bruhat, DeWitt-Morette and Dillard-Bleick [77] and in [16]. On the large empirical evidence in favor of GR, piled up in the recent years, see Ciufolini and Wheeler [78].

The tetrad formalism and its introduction into quantum gravity are mainly due to Cartan, to Weyl [80], and to Schwinger [80] the first-order formalism to Palatini. The Plebanski two-form was introduced in [81]. The selfdual connection, which is at the root of Ashtekar's canonical theory (see Chapter 4), was introduced by Amitaba Sen [82]. The lagrangian formulation for the selfdual connections was given in [83]. A formulation of GR based on the sole connection is discussed in [84].

Interesting reconstructions of Einstein's path towards GR are in [85,86]. Kretschmann's objection to the significance of general covariance appeared in [87]. On this, see also Anderson's book [88]. An account of the historical debate on the interpretation of space and motion is Julian Barbour's [89], a wonderful historical book. In the philosophy of science, the debate was reopened by a 1987 paper on the hole argument by John Earman and John Norton [90]. On the contemporary version of this debate, see [65, 91–93]. On the physical side of the discussion of what is "observable" in GR, see [71].

The discussion of the different notions of time follows [94]. A surprising and inspiring book on the subject is Fraser [95], a book that will convince the reader that the notion of time is far from being a monolithic concept. The literature on the problem of time in quantum gravity is vast. I list only a few pointers here, distinguishing various problems: origin of the "arrow of time" and the cosmological time asymmetry [96]; disappearance of the coordinate-time variable in canonical quantum gravity [97]; possibility of a consistent interpretation of quantum mechanics for systems without global time [26,98,99]; problems in choosing an "internal time" in general relativity, and the properties that such an internal time should have [66]; see also [100]. The presentation of the GPS observables follows [101]; see also [102,103].

3
Mechanics

In its conventional formulation, mechanics describes the evolution of states and observables in time. This evolution is governed by a hamiltonian. This is also true for special-relativistic theories, where evolution is governed by a representation of the Poincaré group, which includes a hamiltonian. This conventional formulation is not sufficiently broad because general-relativistic systems – in fact, the world in which we live – do not fit into this conceptual scheme. Therefore we need a more general formulation of mechanics than the conventional one. This formulation must be based on notions of "observable" and "state" that maintain a clear meaning in a general-relativistic context. A formulation of this kind is described in this chapter.

The conventional structure of conventional nonrelativistic mechanics already points rather directly to the relativistic formulation described here. Indeed, many aspects of this formulation are already utilized by many authors. For instance, Arnold [104] identifies the (presymplectic) space with coordinates (t, q^i, p_i) (time, lagrangian variables and their momenta) as the natural home for mechanics. Souriau has developed a beautiful and little-known relativistic formalism [105]. Probably the first to consider the point of view used here was Lagrange himself, in pointing out that the most convenient definition of "phase space" is the space of the physical motions [106]. Many of the tools used below are also used in hamiltonian treatments of generally covariant theories as constrained systems, although generally within a rather obscure interpretative cloud.

3.1 Nonrelativistic mechanics: *mechanics is about time evolution*

I begin with a brief review of conventional mechanics. This is useful to fix notations and introduce some notions that will play a role in the relativistic formalism. I give no derivations here: they are standard, and they can be obtained as a special case of the derivations in the next section.

Lagrangian. A dynamical system with m degrees of freedom describes the evolution in time t of m lagrangian variables q^i, where $i = 1, \ldots, m$. The space in which the variables q^i take value is the m-dimensional (nonrelativistic) configuration space $\mathcal{C}_0$. The dynamics of the system is determined by a single function of $2m$ variables $L(q^i, v^i)$, the lagrangian. Given two times t_1 and t_2 and two points q_1^i and q_2^i in $\mathcal{C}_0$, physical motions are such that the action

$$S[q] = \int_{t_1}^{t_2} dt \, L\left(q^i(t), \frac{dq^i(t)}{dt}\right) \tag{3.1}$$

is an extremum in the space of the motions $q^i(t)$ such that $q^i(t_1) = q_1^i$ and $q^i(t_2) = q_2^i$. A dynamical system is therefore specified by the couple $(\mathcal{C}_0, L)$. Physical motions satisfy the Lagrange equations

$$\frac{d}{dt} \, p_i\left(q^i(t), \frac{dq^i(t)}{dt}\right) = F_i\left(q^i(t), \frac{dq^i(t)}{dt}\right), \tag{3.2}$$

where momenta and forces are defined by

$$p_i(q^i, v^i) = \frac{\partial L(q^i, v^i)}{\partial v^i}, \qquad F_i(q^i, v^i) = \frac{\partial L(q^i, v^i)}{\partial q^i}. \tag{3.3}$$

Hamiltonian. The Lagrange equations can be cast in first-order form by using the lagrangian coordinates q^i and the momenta p_i as variables. Inverting the function $p_i(q^i, v^i)$ yields the function $v^i(q^i, p_i)$; inserting this in the function $F_i(q^i, v^i)$ defines the force $f_i(q^i, p_i) \equiv F_i(q^i, v^i(q^i, p_i))$ as functions of coordinates and momenta. The equations of motion (3.2) become

$$\frac{dq^i(t)}{dt} = v^i(q^i(t), p_i(t)), \qquad \frac{dp_i(t)}{dt} = f_i(q^i(t), p_i(t)). \tag{3.4}$$

These equations are determined by the function $H_0(q^i, p_i)$, the nonrelativistic hamiltonian defined by $H_0(q^i, p_i) = p_i v^i(q^i, p_i) - L(q^i, v^i(q^i, p_i))$. Indeed, (3.4) is equivalent to (3.2) with

$$v^i(q^i, p_i) = \frac{\partial H_0(q^i, p_i)}{\partial p_i}, \qquad f_i(q^i, p_i) = -\frac{\partial H_0(q^i, p_i)}{\partial q^i}. \tag{3.5}$$

Symplectic. The Hamilton equations (3.4)–(3.5) can be written in a useful and compact geometric language. The $2m$-dimensional space coordinatized by the coordinates q^i and the momenta p_i is the nonrelativistic phase space Γ_0. (The reason for the subscript $_0$ will be clear below.) Time evolution is a flow $(q^i(t), p_i(t))$ in this space; the vector field on Γ_0 tangent to this flow is

$$X_0 = v_i(q^i, p_i)\frac{\partial}{\partial q^i} + f_i(q^i, p_i)\frac{\partial}{\partial p_i}. \tag{3.6}$$

Therefore the dynamics is specified by assigning the vector field X_0 on Γ_0. Now, Γ_0 can be interpreted as the cotangent space T^*C_0. Any cotangent space carries a natural[1] one-form $\theta_0 = p_i dq^i$, where $d\theta_0$ is nondegenerate. A space equipped with such a one-form has the remarkable property that every function f determines a vector field X_f via the relation $(d\theta_0)(X_f) = -df$. A straightforward calculation shows that the flow defined by H_0 is precisely the time evolution vector field (3.6). Therefore the equations of motion (3.4)–(3.5) can be written simply[2] as

$$(d\theta_0)(X_0) = -dH_0. \tag{3.7}$$

The two-form $\omega_0 = d\theta_0$ entering (3.7) is symplectic.[3] A dynamical system is determined by a triple $(\Gamma_0, \omega_0, H_0)$, where Γ_0 is a manifold, ω_0 is a symplectic two-form and H_0 is a function on Γ_0.

Presymplectic. A very elegant formulation of mechanics, and a crucial step in the direction of the relativistic theory, is provided by the presymplectic formalism. This formalism is based on the idea of describing motions by using the graph of the function $(q^i(t), p_i(t))$ instead of the functions themselves. The graph of the function $(q^i(t), p_i(t))$ is an unparametrized curve $\tilde{\gamma}$ in the $(2m + 1)$-dimensional space $\Sigma = R \times \Gamma_0$, with coordinates (t, q^i, p_i); it is formed by all the points $(t, q^i(t), p_i(t))$ in this space. The vector field

$$X = \frac{\partial}{\partial t} + v_i(q^i, p_i)\frac{\partial}{\partial q^i} + f^i(q^i, p_i)\frac{\partial}{\partial p^i} \tag{3.8}$$

is tangent to all these curves. (So is any other vector field obtained by scaling X, namely any vector field $X' = fX$, where f is a scalar function on Σ.) Now, consider the Poincaré one-form

$$\theta = p_i dq^i - H_0(q^i, p_i)dt \tag{3.9}$$

on Σ. The two-form $\omega = d\theta$ is closed but it is degenerate (every two-form is degenerate in odd dimensions); that is, there is a vector field X (called the null vector field of ω) satisfying

$$(d\theta)(X) = 0. \tag{3.10}$$

---

[1] It is defined intrinsically by $\theta_0(X)(s) = s(\pi X)$ where X is a vector field on T^*C_0, s a point in T^*C_0, and π the bundle projection.

[2] The contraction between a two-form and a vector is defined by $(\alpha \wedge \beta)(X) = \alpha(X)\beta - \beta(X)\alpha$.

[3] That is, closed and nondegenerate. Closed means $d\omega_0 = 0$; nondegenerate means that $\omega_0(X) = 0$ implies $X = 0$.

The integral curves[4] of the null vector field of a two-form ω are called the "orbits" of ω. It is easy to see that X given in (3.8) satisfies (3.10). Therefore the graphs of the motions are simply the orbits of $d\theta$. In other words, (3.10) is a rewriting of the equations of motion.

A space Σ equipped with a closed degenerate two-form ω is called presymplectic. A dynamical system is thus completely defined by a presymplectic space (Σ, ω). We use also the notation (Σ, θ), where $\omega = d\theta$.

Notice that (3.10) is homogeneous and therefore it determines X only up to scaling. This is consistent with the fact that the vector field tangent to the motions is defined only up to scaling. That is, consistent with the fact that motions are represented by *unparametrized* curves in Σ.

Finally, it is easy to see that the action (3.1) is simply the line integral of the Poincaré one-form (3.9) along the orbits: if $\tilde{\gamma}$ is an orbit $(t, q^i(t), p_i(t))$ of ω, then the action of the motion $q^i(t)$ is

$$S[q] = \int_{\tilde{\gamma}} \theta. \tag{3.11}$$

Extended. Finally, let me come to a formulation of dynamics that extends naturally to general-relativistic systems. In light of the presymplectic formulation described above, it is natural to consider the relativistic configuration space

$$\mathcal{C} = R \times \mathcal{C}_0 \tag{3.12}$$

coordinatized by the $m + 1$ variables (t, q^i) and to describe motions with the graphs of the functions $q^i(t)$, which are unparametrized curves in $\mathcal{C}$. Consider the cotangent space $T^*\mathcal{C}$, with coordinates (t, q^i, p_t, p_i) and the function

$$H(t, q^i, p_t, p_i) = p_t + H_0(q^i, p_i) \tag{3.13}$$

on this space. Let Σ be the surface in $T^*\mathcal{C}$ defined by

$$H(q^i, t, p_i, p_t) = 0. \tag{3.14}$$

We can coordinatize Σ with the coordinates (t, q^i, p_i). Since it is a cotangent space, $T^*\mathcal{C}$ carries a natural one-form, which is

$$\tilde{\theta} = p_i dq^i + p_t dt. \tag{3.15}$$

The restriction of this one-form to the surface (3.14) is precisely (3.9). Therefore the surface (3.14) is the presymplectic space that defines the dynamics.

---

[4] An integral curve of a vector field is a curve everywhere tangent to the field.

In other words, the dynamics is completely defined by the couple $(\mathcal{C}, H)$: a relativistic configuration space $\mathcal{C}$ and a function H on $T^*\mathcal{C}$. The graphs of the motions are simply the orbits of $d\tilde{\theta}$ on the surface (3.14).[5] I call H the relativistic hamiltonian.

Remarkably, the dynamics can be directly expressed in terms of a variational principle based on $(\mathcal{C}, H)$: An unparametrized curve γ in $\mathcal{C}$ describes a physical motion if $\tilde{\gamma}$ extremizes the integral

$$S[\tilde{\gamma}] = \int_{\tilde{\gamma}} \tilde{\theta} \tag{3.16}$$

in the class of the curves $\tilde{\gamma}$ in $T^*\mathcal{C}$ satisfying (3.14) whose restriction γ to $\mathcal{C}$ connects two given points (t_1, q_1^i) and (t_2, q_2^i).

The relativistic configuration space $\mathcal{C}$ has the structure (3.12) and the relativistic hamiltonian H has the form (3.13). As we shall see, the structure (3.12)–(3.13) does not survive in the relativistic formulation of mechanics.

Relativistic phase space. Denote Γ the space of the orbits of $d\theta$ in Σ. There is a natural projection $\pi : \Sigma \to \Gamma$ that sends each point of Σ to the curve to which it belongs. It is not hard to show that there is one and only one symplectic two-form ω_{ph} on Γ such that its pull-back to Σ is $d\theta$, namely $\pi^*\omega_{\mathrm{ph}} = d\theta$. Therefore Γ is a symplectic space. Γ is the space of the physical motions; I shall call it the relativistic phase space.

The relation between the relativistic phase space Γ and the nonrelativistic phase space $\Gamma_0 = T^*\mathcal{C}_0$ is the following. Γ_0 is the space of the instantaneous states: the states that the system can have at a fixed time $t = t_0$. On the other hand, Γ is the space of all solutions of the equations of motion. Now, fix a time, say $t = t_0$. If at $t = t_0$ the system is in an initial state in Γ_0 it will then evolve in a well-defined motion. The other way around, each motion determines an instantaneous state at $t = t_0$. Therefore there is a one-to-one mapping between Γ and Γ_0. The identification between Γ and Γ_0 depends on the t_0 chosen.

Hamilton–Jacobi. The Hamilton–Jacobi equation is

$$\frac{\partial S(q^i, t)}{\partial t} + H_0\left(q^i, \frac{\partial S(q^i, t)}{\partial q^i}\right) = 0. \tag{3.17}$$

If a family of solutions $S(q^i, Q^i, t)$ depending on m parameters Q_i is found, then we can compute the function

$$P_i(q^i, Q^i, t) = -\frac{\partial S(q^i, Q^i, t)}{\partial Q^i} \tag{3.18}$$

---

[5]More precisely, the projections of these orbits on $\mathcal{C}$.

by simple derivation. Inverting this function we obtain

$$q^i(t) = q^i(Q^i, P_i, t), \tag{3.19}$$

which are physical motions, namely the general solution of the equations of motion, where the quantities (Q^i, P_i) are the $2m$ integration constants.

Solutions of (3.17) can be found in the form $S(q^i, Q^i, t) = Et - W(q^i, Q^i)$ where E is a constant, and W satisfies

$$H_0\left(q^i, \frac{\partial W(q^i, Q^i)}{\partial q^i}\right) = E. \tag{3.20}$$

S is called the principal Hamilton–Jacobi function, W is called the characteristic Hamilton–Jacobi function.

The Hamilton–Jacobi equation (3.17) can be obtained from the classical limit of the Schrödinger equation.

The Hamilton function. Consider two points (t_1, q_1^i) and (t_2, q_2^i) in $\mathcal{C}$. The function on $\mathcal{G} = \mathcal{C} \times \mathcal{C}$

$$S(t_1, q_1^i, t_2, q_2^i) = \int_{t_1}^{t_2} dt \, L(q^i(t), \dot{q}^i(t)), \tag{3.21}$$

where $q^i(t)$ is the physical motion from $q_1^i(t_1)$ to $q_2^i(t_2)$ (that minimizes the action), is called the Hamilton function. Equivalently,

$$S(t_1, q_1^i, t_2, q_2^i) = \int_{\tilde{\gamma}} \theta, \tag{3.22}$$

where $\tilde{\gamma}$ is the orbit into Σ that projects to $q^i(t)$. Notice the difference between the action (3.1) and the Hamilton function (3.21): the first is a functional of the motion; the second is a function of the end points. It is not hard to see that the Hamilton function solves the Hamilton–Jacobi equation (in both sets of variables). The Hamilton function is therefore a preferred solution of the Hamilton–Jacobi equation. If we know the Hamilton function, we have solved the equations of motion because we obtain the general solution of the equations of motion in the form $q^i = q^i(t, Q^i, P_i, T)$ by simply inverting the function

$$P_i(t, q^i, T, Q^i) = \frac{\partial S(t, q^i, T, Q^i)}{\partial Q^i} \tag{3.23}$$

with respect to q^i. The resulting function $q^i(t, T, Q^i, P_i)$ is the general solution of the equations of motion where the integration constants are the initial coordinate and momenta Q^i, P_i at time T.

Thus, the action defines a dynamical system; the Hamilton function directly gives all the motions.[6] The Hamilton function (3.21) is the classical limit of the quantum mechanical propagator.

Example: a pendulum. Let α be the lagrangian variable describing the elongation of a simple harmonic oscillator, which I call "pendulum" for simplicity. The lagrangian is $L(\alpha, v) = (mv^2/2) - (m\omega^2\alpha^2/2)$; the nonrelativistic hamiltonian is $H_0(\alpha, p) = (p^2/2m) + (m\omega^2\alpha^2/2)$. The extended configuration space has coordinates (t, α) and the relativistic hamiltonian is

$$H(t, \alpha, p_t, p) = p_t + \frac{p^2}{2m} + \frac{m\omega^2\alpha^2}{2}. \tag{3.24}$$

Choose coordinates (t, α, p) on the constraint surface $H = 0$, which is therefore defined by $p_t = -H_0(\alpha, p)$. The restriction of the one-form $\tilde{\theta} = p_t\,dt + p\,d\alpha$ to this surface is

$$\theta = p\,d\alpha - \left(\frac{p^2}{2m} + \frac{m\omega^2\alpha^2}{2}\right) dt. \tag{3.25}$$

The presymplectic two-form is therefore

$$\omega = d\theta = dp \wedge d\alpha - \frac{p}{m} dp \wedge dt - m\omega^2\alpha\,d\alpha \wedge dt. \tag{3.26}$$

The orbits are obtained by integrating the vector field

$$X = X_t\frac{\partial}{\partial t} + X_\alpha\frac{\partial}{\partial \alpha} + X_p\frac{\partial}{\partial p} \tag{3.27}$$

satisfying $\omega(X) = 0$. Inserting (3.26) and (3.27) in $\omega(X) = 0$ we get

$$\omega(X) = X_t\left(-\frac{p}{m}dp - m\omega^2\alpha\,d\alpha\right) + X_\alpha\left(dp + m\omega^2\alpha\,dt\right) + X_p\left(-d\alpha + \frac{p}{m}dt\right)$$

$$= \left(-\frac{p}{m}X_t + X_\alpha\right)dp + \left(-m\omega^2\alpha X_t - X_p\right)d\alpha + \left(m\omega^2\alpha X_\alpha + \frac{p}{m}X_p\right)dt$$

$$= 0. \tag{3.28}$$

Writing $dt(\tau)/d\tau = X_t, d\alpha(\tau)/d\tau = X_\alpha, dp(\tau)/d\tau = X_p$, equation (3.28) reads

$$\frac{d\alpha(\tau)}{d\tau} - \frac{p}{m}\frac{dt(\tau)}{d\tau} = 0, \qquad -\frac{dp(\tau)}{d\tau} - m\omega^2\alpha\frac{dt(\tau)}{d\tau} = 0, \tag{3.29}$$

together with a third equation dependent on the first two. Equation (3.29) can be written as

$$\frac{d\alpha(t)}{dt} = \frac{p}{m}, \qquad \frac{dp(t)}{dt} = -m\omega^2\alpha, \tag{3.30}$$

which are the Hamilton equations of the pendulum. We can write its general solution in the form

$$\alpha(t) = a\,e^{i\omega t} + \bar{a}\,e^{-i\omega t}. \tag{3.31}$$

The Hamilton function $S(\alpha_1, t_1, \alpha_2, t_2)$ is the preferred solution of the Hamilton–Jacobi equation

$$\frac{\partial S(\alpha, t)}{\partial t} + \frac{1}{2m}\left(\frac{\partial S(\alpha, t)}{\partial \alpha}\right)^2 + \frac{m\omega^2\alpha^2}{2} = 0, \tag{3.32}$$

---

[6] Hamilton (talking about himself in the third person): "Mr. Lagrange's function *states* the problem, Mr. Hamilton's function *solves* it" [107].

obtained by computing the action of the physical motion $\alpha(t)$ that goes from $\alpha(t_1) = \alpha_1$ to $\alpha(t_2) = \alpha_2$. This motion is given by (3.31) with

$$a = \frac{\alpha_1 e^{-i\omega t_2} + \alpha_2 e^{-i\omega t_1}}{2i \sin[(\omega(t_1 - t_2))]}. \tag{3.33}$$

Inserting this in the action and integrating we obtain the Hamilton function

$$S(\alpha_1, t_1, \alpha_2, t_2) = m\omega \frac{2\alpha_1\alpha_2 - (\alpha_1^2 + \alpha_2^2) \cos[(\omega(t_1 - t_2))]}{2 \sin[(\omega(t_1 - t_2))]}. \tag{3.34}$$

This concludes the short review of nonrelativistic mechanics. I now consider the generalization of this formalism to relativistic systems.

3.2 Relativistic mechanics

3.2.1 Structure of relativistic systems: partial observables, relativistic states

Is there a version of the notions of "state" and "observable" broad enough to apply naturally to relativistic systems? I begin by introducing the main notions and tools of covariant mechanics in the context of a simple system.

The pendulum revisited. Say we want to describe the small oscillations of a pendulum. To this aim, we need *two* measuring devices: a clock and a device that reads the elongation of the pendulum. Let t be the reading of the clock (in seconds) and α the reading of the device measuring the elongation of the pendulum (in centimeters). Call the variables t and α the *partial observables* of the pendulum. (I use also *relativistic observables* or simply *observables* if there is no risk of confusion with the nonrelativistic notion of observable, which is different.)

A useful observation is a reading of the time t and the elongation α, *together*. Thus, an observation yields a pair (t, α). Call a pair obtained in this manner an *event*.

Let $\mathcal{C}$ be the two-dimensional space with coordinates t and α. Call $\mathcal{C}$ the *event space* of the pendulum. (I use also *relativistic configuration space* or simply *configuration space* if there is no risk of confusion with the nonrelativistic configuration space $\mathcal{C}_0$, which is different.)

Experience shows we can find mathematical laws characterizing *sequences* of events. This is the reason we can do science. These laws have the following form. Call an unparametrized curve γ in $\mathcal{C}$ a *motion* of the system. Perform a sequence of measurements of pairs (t, α), and find that the points representing the measured pairs sit on a motion γ. Then we say that γ is a *physical motion*. We express a motion as a relation in $\mathcal{C}$

$$f(\alpha, t) = 0. \tag{3.35}$$

Thus a motion γ is a relation, or a *correlation*, between partial observables.

Then, disturb the pendulum (push it with a finger) and repeat the entire experiment over. At each repetition of the experiment, a different motion γ is found. That is, a different mathematical relation of the form (3.35) is found. Experience shows that the space of the physical motions is very limited: it is just a two-dimensional space. Only a two-dimensional space of curves γ is realized in Nature.

In the case of the small oscillations of a frictionless pendulum, we can coordinatize the physical motions by the two real numbers $A \geq 0$ and $0 \leq \phi < 2\pi$, and (3.35) is given by

$$f(\alpha, t; A, \phi) = \alpha - A\sin(\omega t + \phi) = 0. \tag{3.36}$$

This equation gives a curve γ in $\mathcal{C}$ for each couple (A, ϕ).

Let Γ be the two-dimensional space of the physical motions, coordinatized by A and ϕ. Γ is the *relativistic phase space* of the pendulum (or the *space of the motions*). A point in Γ is also called a *relativistic state*. (Or a *Heisenberg state*, or simply a *state* if there is no risk of confusion with the nonrelativistic notion of state, which is different.)

Equation (3.36) is the mathematical law that captures the empirical information we have on the pendulum. This equation is the *evolution equation* of the system. The function f is the *evolution function* of the system.

A relativistic state is determined by a couple (A, ϕ). It determines a curve γ in the (t, α) plane. That is, it determines a correlation between the two partial observables t and α, via (3.36). If we disturb the pendulum by interacting with it or if we start a new experiment over, we have a new state. The state remains the same if we observe the pendulum and the clock without disturbing them (here we disregard quantum theory, of course).

Summarizing: *each state in the phase space Γ determines a correlation between the observables in the configuration space $\mathcal{C}$.* The set of these relations is captured by the evolution equation (3.36), namely by the vanishing of a function

$$f : \Gamma \times \mathcal{C} \to R. \tag{3.37}$$

The evolution equation $f = 0$ expresses all predictions that can be made using the theory. Equivalently, these predictions are captured by the surface $f = 0$ in the cartesian product of the phase space with the configuration space.

General structure of the dynamical systems. The $(\mathcal{C}, \Gamma, f)$ language described above is general. It is sufficient to describe all predictions of conventional mechanics. On the other hand, it is broad enough to describe general-relativistic systems. All fundamental systems can be described (to the accuracy at which quantum effects can be disregarded) by making use of these concepts:

(i) The relativistic *configuration space* C of the partial *observables*.

(ii) The relativistic *phase space* Γ of the relativistic *states*.

(iii) The *evolution equation* $f = 0$, where $f : \Gamma \times C \to V$.

Here V is a linear space. The state in the phase space Γ is fixed until the system is disturbed. Each state in Γ determines (via $f = 0$) a motion γ of the system, namely it describes a relation, or a set of relations, between the observables in C.

A motion is not necessarily a one-dimensional curve in C: it can be a surface in C of any dimension k. If $k > 1$, we say that there is gauge invariance. For a system with gauge invariance we call "motion" the motion itself and any curve within it. In this chapter we shall not deal much with systems with gauge invariance, but we shall mention them where relevant.

Predictions are obtained as follows. We first perform enough measurements to determine the state. (In reality the state of a large system is often "guessed" on the basis of incomplete observations and reasonable assumptions, justified inductively.) Once the state is so determined or guessed, the evolution equation predicts all the possible events, namely all the allowed correlations between the observables in any subsequent measurement.

In the example of the pendulum, for instance, the equation predicts the value of α that can be measured together with any given t, or the values of t that can be measured together with any given α. These predictions are valid until the system is disturbed.

The definitions of observable, state, configuration space and phase space given here are different from the conventional definitions. In particular, notions of instantaneous state, evolution in time, observable at a fixed time, play no role here. These notions make no sense in a general-relativistic context. For nonrelativistic systems, the usual notions can be recovered from the definitions given. The relation between the relativistic definitions considered here and the conventional nonrelativistic notions is discussed in Section 3.2.4.

The task of mechanics is to find the (C, Γ, f) description for all physical systems. The first step, kinematics, consists in the specification of the observables that characterize the system. Namely, it consists in the specification of the configuration space C and its physical interpretation. Physical interpretation means the association of coordinates on C with measuring devices. The second step, dynamics, consists in finding the phase space Γ and the function f that describe the physical motions of the system.

In the next section, I describe a relativistic hamiltonian formalism for mechanics based on the relativistic notions of state and observable defined here.

3.2.2 Hamiltonian mechanics

Elementary physical systems can be described by hamiltonian mechanics.[7] Once the kinematics – that is, the space $\mathcal{C}$ of the partial observables q^a – is known, the dynamics – that is, Γ and f – is fully determined by giving a surface Σ in the space Ω of the observables q^a and their momenta p_a. The surface Σ can be specified by giving a function $H : \Omega \to R^k$. Σ is then defined by $H = 0$.[8] Denote $\tilde{\gamma}$ a curve in Ω (observables and momenta) and γ its restriction to $\mathcal{C}$ (observables alone). H determines the physical motions via the following

> **Variational principle.** *A curve γ connecting the events q_1^a and q_2^a is a physical motion if $\tilde{\gamma}$ extremizes the action*
>
> $$S[\tilde{\gamma}] = \int_{\tilde{\gamma}} p_a \, dq^a \qquad (3.38)$$
>
> *in the class of the curves $\tilde{\gamma}$ satisfying*
>
> $$H(q^a, p_a) = 0 \qquad (3.39)$$
>
> *whose restriction γ to $\mathcal{C}$ connects q_1^a and q_2^a.*

All (relativistic and nonrelativistic) hamiltonian systems can be formulated in this manner.

If $k = 1$, H is a scalar function, and is sometimes called the hamiltonian constraint. The case $k > 1$ is the case in which there is gauge invariance. In this case, the system (3.39) is sometimes called the system of the "constraint equations." I call H the *relativistic hamiltonian*, or, if there is no ambiguity, simply the *hamiltonian*. I denote the pair $(\mathcal{C}, H)$ as a *relativistic dynamical system*. The generalization to field theory is discussed in Section 3.3.

The relativistic hamiltonian H is related to, but should not be confused with, the usual nonrelativistic hamiltonian, denoted H_0 in this book. H always exists, while H_0 exists only for nonrelativistic systems.

---

[7]Perhaps because they are the classical limit of a quantum system.
[8]Different Hs that vanish on the same surface Σ define the same physical system.

Indeed, notice that this formulation of mechanics is similar to the extended formulation of nonrelativistic mechanics defined in Section 3.1. The novelty is that $\mathcal{C}$ and H do not have the structure (3.12)–(3.13). The discussion above shows that this structure is *not* necessary in order to have a well-defined physical interpretation of the formalism. A nonrelativistic system is characterized by the fact that one of its partial observables q^a is singled out by having the special role of an independent variable t. This does not happen in a relativistic system. The following simple example shows that the relativistic formulation of mechanics is a proper generalization of standard mechanics.

Timeless double pendulum. I now introduce a genuinely timeless system, which I will repeatedly use as a simple model to illustrate the theory. Consider a mechanical model with two partial observables, say a and b, whose dynamics is defined by the relativistic hamiltonian

$$H(a, b, p_a, p_b) = -\frac{1}{2} \left(p_a^2 + p_b^2 + a^2 + b^2 - 2E \right), \tag{3.40}$$

where E is a constant. The extended configuration space is $\mathcal{C} = R^2$. The constraint surface has dimension 3; it is the sphere of radius $\sqrt{2E}$ in $T^*\mathcal{C}$. The phase space has dimension 2. The motions are curves in the (a, b) space. For each state, the theory predicts the correlation between a and b.

A straightforward calculation (see below) shows that the evolution equation determined by H is an ellipse in the (a, b) space

$$f(a, b; \alpha, \beta) = \left(\frac{a}{\sin \alpha} \right)^2 + \left(\frac{b}{\cos \alpha} \right)^2 + 2 \frac{a}{\sin \alpha} \frac{b}{\cos \alpha} \cos \beta - 2E^2 \sin^2 \beta = 0, \tag{3.41}$$

where α and β parametrize Γ. Therefore motions are closed curves and in fact ellipses in $\mathcal{C}$. The system does not admit a conventional hamiltonian formulation, because for a nonrelativistic hamiltonian system motions in $\mathcal{C} = R \times \mathcal{C}_0$ are monotonic in $t \in R$ and therefore cannot be closed curves.

The example is not artificial. There exist cosmological models that have precisely this structure. For instance, we can identify a with the radius of a maximally symmetric universe and b with the spatially constant value of a field representing the matter content of that universe, and adopt the approximation in which these are the only two variables that govern the large-scale evolution of the universe. Then the dynamics of general relativity reduces to a system with the structure (3.40).

The associated nonrelativistic system. The system (3.40) can also be viewed as follows. Consider a physical system, which we denote the "associated nonrelativistic system," formed by two noninteracting harmonic oscillators. Let me stress that the associated nonrelativistic system is a *different* physical system than the timeless double pendulum considered above. The timeless double pendulum has one degree of freedom, its associated nonrelativistic system has two degrees of freedom. The partial observables of the associated nonrelativistic system are the two elongations a and b, *and* the time t. The *nonrelativistic* hamiltonian that governs the evolution in t is

$$H_0(a, b, p_a, p_b) = \frac{1}{2} \left(p_a^2 + p_b^2 + a^2 + b^2 - 2E \right). \tag{3.42}$$

It it has the same form as the *relativistic* hamiltonian (3.40) of the timeless double pendulum.[9] The constant term $2E$, of course, has no effect on the equations of motion; it only redefines the energy. Physically, we can view the relation between the two systems as follows. Imagine that we take the associated nonrelativistic system but we decide to ignore the clock that measures t: we consider just measurements of the two observables a and b. Furthermore, assume that the energy of the double pendulum is constrained to vanish, namely,

$$\frac{1}{2} \left(p_a^2 + p_b^2 + a^2 + b^2 \right) = E. \tag{3.44}$$

Then the observed relation between the measurements of a and b is described by the relativistic system (3.40).

Geometric formalism. As for nonrelativistic hamiltonian mechanics, the equations of motion can be expressed in an elegant geometric form. The variables (q^a, p_a) are coordinates on the cotangent space $\Omega = T^*\mathcal{C}$. Equation (3.39) defines a surface Σ in this space. The cotangent space carries the natural one-form

$$\tilde{\theta} = p_a \mathrm{d}q^a. \tag{3.45}$$

Denote θ the restriction of $\tilde{\theta}$ to the surface Σ. The two-form $\omega = \mathrm{d}\theta$ on Σ is degenerate: it has null directions. The integral surfaces of these null directions are the *orbits* of ω on Σ. Each such orbit projects from $T^*\mathcal{C}$ to $\mathcal{C}$ to give a surface in $\mathcal{C}$. These surfaces are the motions.

Consider the case $k = 1$. In this case Σ has dimension $2n-1$, the kernel of ω is generically one-dimensional, and the motions are generically one-dimensional. Let $\tilde{\gamma}$ be a motion on Σ, and X be a vector tangent to the motion; then

$$\omega(X) = 0. \tag{3.46}$$

To find the motions, we have just to integrate this equation. Equation (3.46) is the equation of motion. X is defined by the homogeneous equation (3.46) only up to a multiplicative factor. Therefore the tangent of the orbit is defined only up to a multiplicative factor, and so the parametrization of the orbit is not determined by (3.46).

The case $k > 1$ is analogous. In this case Σ has dimension $2n - k$, the kernel of ω is, generically, k-dimensional and the motions are, generically, k-dimensional. X is then a k-dimensional multi-tangent, and it still satisfies (3.46).

Let $\pi : \Sigma \to \Gamma$ be the projection map that associates with each point of the constraint surface the motion to which the point belongs. The

---

[9]The *relativistic* hamiltonian of the associated nonrelativistic system is

$$H(a, b, t, p_a, p_b, p_t) = p_t + \frac{1}{2} \left(p_a^2 + p_b^2 + a^2 + b^2 - 2E \right). \tag{3.43}$$

projection π equips the phase space Γ with a symplectic two-form ω_{ph} defined to be the two-form whose pull-back to Σ under π is ω. Locally it exists and it is unique precisely because ω is degenerate along the orbits.

Relation with the variational principle. Let $\tilde{\gamma}$ be an orbit of ω on Σ, such that its restriction γ in $\mathcal{C}$ is bounded by the initial and final events q_1 and q_2. Let $\tilde{\gamma}'$ be a curve in Σ infinitesimally close to $\tilde{\gamma}$, such that its restriction γ' is also bounded by q_1 and q_2. Let δs_1 (and δs_2) be the difference between the initial (and final) points of $\tilde{\gamma}$ and $\tilde{\gamma}'$. The four curves $\tilde{\gamma}$, δs_1, $-\tilde{\gamma}'$ and $-\delta s_2$ form a closed curve in Σ. Consider the integral of ω over the infinitesimal surface bounded by this curve. This integral vanishes because at every point of the surface one of the tangents is (to first order) a null direction of ω (the surface is a strip parallel to the motion $\tilde{\gamma}$). But $\omega = d\theta$ and therefore, by Stokes theorem, the integral of θ along the closed curve vanishes as well. The integral of $\theta = p_a dq^a$ along δs_1 and δs_2 is zero because q^a is constant along these segments. Therefore

$$\int_{\tilde{\gamma}} \theta + \int_{-\tilde{\gamma}'} \theta = 0, \tag{3.47}$$

or

$$\delta \int_{\tilde{\gamma}} \theta = 0, \tag{3.48}$$

for any variation in the class considered. This is precisely the variational principle stated in Section 3.2.

Hamilton equations. Consider first the case $k = 1$. Motions are one-dimensional. Parametrize the curve with an arbitrary parameter τ. That is, describe a motion (in Ω) with the functions $(q^a(\tau), p_a(\tau))$. These functions satisfy the Hamilton system

$$H(q^a, p_a) = 0, \tag{3.49}$$

$$\frac{dq^a(\tau)}{d\tau} = N(\tau)\ v^a(q^a(\tau), p_a(\tau)),$$

$$\frac{dp_a(\tau)}{d\tau} = N(\tau)\ f_a(q^a(\tau), p_a(\tau)), \tag{3.50}$$

where

$$v^a(q^a, p_a) = \frac{\partial H(q^a, p_a)}{\partial p_a}, \qquad f_a(q^a, p_a) = -\frac{\partial H(q^a, p_a)}{\partial q^a}. \tag{3.51}$$

The function $N(\tau)$ is called the "Lapse function." It is arbitrary. Different choices of $N(\tau)$ determine different parameters τ along the motion. To obtain a monotonic parametrization we need $N(\tau) > 0$. A preferred parametrization can be obtained by taking $N(\tau) = 1$, that is, replacing (3.50)–(3.51) by the equations (written in the usual compact form)

$$\dot{q}^a = \frac{\partial H}{\partial p_a}, \qquad \dot{p}_a = -\frac{\partial H}{\partial q^a}, \tag{3.52}$$

where the dot indicates derivative with respect to τ. This choice is called the Lapse $= 1$ gauge. It is not preferred in a physical sense. In particular, different but physically equivalent hamiltonians H defining the same surface Σ determine different preferred parametrizations. Nevertheless, it is often the easiest gauge to compute with.

If $k > 1$, the function H has components H^j, with $j = 1, \ldots, k$ and motions are k-dimensional surfaces. We can parametrize a motion with k arbitrary parameters $\vec{\tau} = \{\tau_j\}$. Namely, we can represent it using the $2n$ functions $q^a(\vec{\tau}), p_a(\vec{\tau})$ of k parameters τ_j. These equations satisfy the system given by (3.49) and

$$\frac{\partial q^a(\vec{\tau})}{\partial \tau_j} = N_j(\vec{\tau}) \, \frac{\partial H^j(q^a, p_a)}{\partial p_a}, \qquad \frac{\partial p_a(\vec{\tau})}{\partial \tau_j} = -N_j(\vec{\tau}) \, \frac{\partial H^j(q^a, p_a)}{\partial q^a}. \tag{3.53}$$

A motion is determined by the full k-dimensional surface in $\mathcal{C}$, we can choose a particular curve $\vec{\tau}(\tau)$ on this surface, where τ is an arbitrary parameter, and represent the motion by the one-dimensional curve $q^a(\tau) = q^a(\vec{\tau}(\tau))$ in $\mathcal{C}$. This satisfies the system formed by (3.49) and

$$\frac{dq^a(\tau)}{d\tau} = N_j(\tau) \, \frac{\partial H^j(q^a, p_a)}{\partial p_a}, \qquad \frac{dp_a(\tau)}{d\tau} = -N_j(\tau) \, \frac{\partial H^j(q^a, p_a)}{\partial q^a} \tag{3.54}$$

for k arbitrary functions of one variable $N_j(\tau)$. Different choices of the functions $N_j(\tau)$ determine different curves on the single surface that defines a motion. These are gauge-equivalent representations of the same motion.

It is important to stress that the parameters τ or τ_j are an artifact of this technique. They have no physical significance. They are absent in the geometric formalism as well as in the Hamilton–Jacobi formalism, as we shall see below. The physical content of the theory is in the motion in $\mathcal{C}$, not in the way the motion is parametrized. That is, the physical information is not in the functions $q^a(\tau)$: it is in the image of these functions in $\mathcal{C}$.

Relation with the variational principle. Parametrize the curve $\tilde{\gamma}$ with a parameter τ. The action (3.38) reads

$$S = \int d\tau \, p_a(\tau) \, \frac{dq^a(\tau)}{d\tau}. \tag{3.55}$$

The constraint (3.39) can be implemented in the action with lagrange multipliers $N_i(\tau)$. This defines the action

$$S = \int d\tau \left(p_a \frac{dq^a}{d\tau} - N_i \, H^i(p_a, q^a) \right). \tag{3.56}$$

Varying this action with respect to $N_i(\tau)$, $q^a(\tau)$ and $p_a(\tau)$ gives the Hamilton equation (3.49), (3.54).

Example: double pendulum. Consider the system defined by the hamiltonian (3.40). The Hamilton equations (3.49), (3.52) in the Lapse $= 1$ gauge give

$$\dot{a} = p_a, \qquad \dot{b} = p_b, \qquad \dot{p}_a = -a, \qquad \dot{p}_b = -b, \qquad a^2 + b^2 + p_a^2 + p_b^2 = 2E. \tag{3.57}$$

The general solution is

$$a(\tau) = A_a \sin(\tau), \qquad b(\tau) = A_b \sin(\tau + \beta), \tag{3.58}$$

where $A_a = \sqrt{2E} \sin\alpha$ and $A_b = \sqrt{2E} \cos\alpha$. The motions are given by the image in $\mathcal{C}$ of these curves. These are the ellipses (3.41). The parametrization of the curves (3.58) has no physical significance. The physics is in the unparametrized ellipses in $\mathcal{C}$ and in the relation between a and b they determine.

Hamilton–Jacobi. Hamilton–Jacobi formalism is elegant, general and powerful; it has a direct connection with quantum theory, and is conceptually clear. The *relativistic* formulation of Hamilton–Jacobi theory is simpler than the conventional nonrelativistic version, indicating that the relativistic formulation unveils a natural and general structure of mechanical systems.

The relativistic Hamilton–Jacobi formalism is given by the system of k partial differential equations

$$H\left(q^a, \frac{\partial S(q^a)}{\partial q^a}\right) = 0 \tag{3.59}$$

for the function $S(q^a)$ defined on the extended configuration space $\mathcal{C}$. Let $S(q^a, Q^i)$ be a family of solutions, parametrized by the $n - k$ constants of integration Q^i. Pose

$$f^i(q^a, P_i, Q^i) \equiv \frac{\partial S(q^a, Q^i)}{\partial Q^i} + P_i = 0 \tag{3.60}$$

for $n - k$ arbitrary constants P_i. This is the evolution equation. The constants Q^i, P_i coordinatize a $2(n - k)$-dimensional space Γ. This is the phase space.

The form of the relativistic Hamilton–Jacobi equation (3.59) is simpler than the usual nonrelativistic Hamilton–Jacobi equation (3.17). Furthermore, there is no equation to invert, as in the nonrelativistic formalism. Notice also that the function $S(q^a, Q^i)$ can be identified with the *principal* Hamilton–Jacobi function $S(t, q^i, Q^i) = Et + W(q^i, Q^i)$ of the nonrelativistic formalism as well as with the *characteristic* Hamilton–Jacobi function $W(q^i, Q^i)$, since (3.59) is formally like (3.20) with vanishing energy. The two functions are in fact identified in the relativistic formalism.

Example: double pendulum. The Hamilton–Jacobi equation of the timeless system (3.40) is

$$\left(\frac{\partial S(a, b)}{\partial a}\right)^2 + \left(\frac{\partial S(a, b)}{\partial b}\right)^2 + a^2 + b^2 - 2E = 0. \tag{3.61}$$

A one-parameter family of solutions is given by

$$S(a, b, A) = \frac{a}{2}\sqrt{A^2 - a^2} + \frac{A^2}{2}\arctan\left(\frac{a}{\sqrt{A^2 - a^2}}\right)$$
$$+ \frac{b}{2}\sqrt{2E - A^2 - b^2} + \frac{2E - A^2}{2}\arctan\left(\frac{b}{\sqrt{2E - A^2 - b^2}}\right). \quad (3.62)$$

The general solution (3.41) of the system is directly obtained by writing

$$\frac{\partial S(a, b, A)}{\partial A} - \phi = 0, \quad (3.63)$$

where ϕ is an integration constant.

Derivation of the Hamilton–Jacobi formalism. Since the phase space Γ is a symplectic space, we can locally choose canonical coordinates (Q^i, P_i) over it. These coordinates can be pulled back to Σ, where they are constant along the orbits. In fact, they label the orbits. Let $\theta_{\mathrm{ph}} = P_i dQ^i$; therefore $d\theta_{\mathrm{ph}} = \omega$ But $\omega = d\theta = d(p_a dq^a)$ so on Σ we have

$$d(\theta_{\mathrm{ph}} - \theta) = d(P_i dQ^i - p_a dq^a) = 0. \quad (3.64)$$

This implies that there should locally exist a function S on Σ such that

$$P_i dQ^i - p_a dq^a = -dS. \quad (3.65)$$

Let us choose q^a and Q^i as independent coordinates on Σ. Then (3.65) reads

$$dS(q^a, Q^i) = p_a(q^a, Q^i)dq^a - P_i(q^a, Q^i)dQ^i, \quad (3.66)$$

that is,

$$\frac{\partial S(q^a, Q^i)}{\partial q^a} = p_a(q^a, Q^i), \quad (3.67)$$

$$\frac{\partial S(q^a, Q^i)}{\partial Q^i} = -P_i(q^a, Q^i). \quad (3.68)$$

By the definition of Σ, we have $H(q^a, p_a) = 0$, which, using (3.67), gives the Hamilton–Jacobi equation (3.59). Equation (3.68) is then immediately the evolution equation (3.60).

In other words, $S(q^a, Q^i)$ is the generating function of a canonical transformation that relates the observables and their momenta (q^a, p_a) to new canonical variables (Q^i, P_i) satisfying $\dot{Q}^i = 0, \dot{P}_i = 0$. These new variables are constants of motion and therefore define Γ. The relation between $\mathcal{C}$ and Γ given by the canonical transformation equations (3.67)–(3.68) is the evolution equation.

3.2.3 Nonrelativistic systems as a special case

Here I discuss in more detail how the notions and the structures of conventional mechanics described in Section 3.1 are recovered from the relativistic formalism. A nonrelativistic system is simply a relativistic dynamical system in which one of the partial observables q^a is denoted t and called "time," and the hamiltonian H has the form

$$H = p_t + H_0, \quad (3.69)$$

where H_0 is independent from p_t and is called the nonrelativistic hamiltonian. The quantity $E = -p_t$ is called energy. The device that measures the partial observable t is called a clock.

The relativistic configuration space therefore has the structure

$$C = R \times C_0, \qquad (3.70)$$

with coordinates $q^a = (t, q^i)$, where $i = 1, \dots, n$–1. The space C_0 is the usual nonrelativistic configuration space. Accordingly, the cotangent space $\Omega = T^*C$ has coordinates $(q^a, p_a) = (t, q^i, p_t, p_i)$.

If H has the form (3.69), the relativistic Hamilton–Jacobi equation (3.59) becomes the conventional nonrelativistic Hamilton–Jacobi equation (3.17).

Given a state and a value t of the clock observable, we can ask what are the possible values of the observables q^i such that (q^i, t) is a possible event. That is, we can ask what is the value of q^i "when" the time is t. The solution is obtained by solving the evolution function $f^i(q^i, t; Q^i, P_i) = 0$ for the q^i. This gives

$$q^i = q^i(t; Q^i, P_i), \qquad (3.71)$$

which is interpreted as the evolution equation of the variables q^i in the time t. The form (3.69) of the hamiltonian guarantees that we can solve f with respect to the q^i, because the Hamilton equation for t (in the gauge Lapse = 1) is simply $t = \tau$, which can be inverted.

In the parametrized hamiltonian formalism, the evolution equation for $t(\tau)$ is trivial and gives, taking advantage of the freedom in rescaling τ, just $t = \tau$. Using this, equations (3.53) become the conventional Hamilton equations and (3.49) simply fixes the value of p_t, namely the energy.

In the presymplectic formalism, the surface Σ turns out to be

$$\Sigma = R \times \Gamma_0, \qquad (3.72)$$

where the coordinate on R is the time t and $\Gamma_0 = T^*C_0$ is the nonrelativistic phase space. The restriction of $\tilde{\theta}$ to this surface has the Cartan form

$$\theta = p_i dq^i - H_0 dt = \theta_0 - H_0 dt. \qquad (3.73)$$

We can take the vector field X to have the form

$$X = \frac{\partial}{\partial t} + X_0, \qquad (3.74)$$

where X_0 is a vector field on Γ_0. Then the equation of motion (3.46) reduces to the equation

$$(\mathrm{d}\theta_0)(X_0) = -\mathrm{d}H_0, \qquad (3.75)$$

which is the geometric form of the conventional Hamilton equations. Thus, H determines how the variables in Γ_0 are correlated to the variable t. That is, "how the variables in Γ_0 evolve in time." In this sense, the nonrelativistic hamiltonian H_0 generates "evolution in the time t." This evolution is generated in Γ_0 by the hamiltonian flow X_0 of H_0. A point $s = (q^i, p_i)$ in Γ_0 is taken to the point $s(t) = (q^i(t), p_i(t))$ where

$$\frac{\mathrm{d}s(t)}{\mathrm{d}t} = X_0(s(t)). \qquad (3.76)$$

The evolution of an observable (not depending explicitly on time) defined by $A_t(s) = A(s(t)) = A(s, t)$ can be written, introducing the Poisson bracket notation

$$\{A, B\} = -X_A(B) = X_B(A) = \sum_i \left(\frac{\partial A}{\partial q^i} \frac{\partial B}{\partial p_i} - \frac{\partial A}{\partial p_i} \frac{\partial B}{\partial q^i} \right), \quad (3.77)$$

as

$$\frac{\mathrm{d}A_t}{\mathrm{d}t} = \{A_t, H_0\}. \qquad (3.78)$$

Instantaneous states and relativistic states. The nonrelativistic definition of *state* refers to the properties of a system *at a certain moment of time.* Denote this conventional notion of state as the "instantaneous state." The space of the instantaneous states is the conventional nonrelativistic phase space Γ_0. Let's fix the value $t = t_0$ of the time variable, and characterize the instantaneous state in terms of the initial data. For the pendulum these are position and momentum, (α_0, p_0), at $t = t_0$. Thus (α_0, p_0) are coordinates on Γ_0.

On the other hand, a relativistic state is a solution of the equations of motion. (If there is gauge invariance, a state is a gauge equivalence class of solutions of the equations of motion.) The relativistic phase space Γ is the space of the solutions of the equations of motion.

Given a value t_0 of the time, there is a one-to-one correspondence between initial data and solutions of the equations of motion: each solution of the equation of motion determines initial data at $t = t_0$; and each choice of initial data at t_0 determines uniquely a solution of the equations of motion. Therefore there is a one-to-one correspondence between instantaneous states and relativistic states. Therefore the relativistic phase space Γ is isomorphic to the nonrelativistic phase space: $\Gamma \sim \Gamma_0$. However, the isomorphism depends on the time t_0 chosen, and the physical interpretation of the two spaces is quite different. One is a space of states at a given time, the other a space of motions.

In the case of the pendulum, the nonrelativistic phase space Γ_0 can be coordinatized with (α_0, p_0); the relativistic phase space Γ can be coordinatized with (A, ϕ). The identification map $(A, \phi) \mapsto (\alpha_0, p_0)$ is given by

$$\alpha_0(A, \phi) = A \sin(\omega t_0 + \phi), \tag{3.79}$$
$$p_0(A, \phi) = \omega m A \cos(\omega t_0 + \phi). \tag{3.80}$$

The nonrelativistic phase space Γ_0 plays a double role in nonrelativistic hamiltonian mechanics: it is the space of the instantaneous states, but it is also the arena of nonrelativistic hamiltonian mechanics, over which H_0 is defined. In the relativistic context, this double role is lost: one must distinguish the cotangent space $\Omega = T^*C$ over which H is defined from the phase space Γ which is the space of the motions. This distinction will become important in field theory where Ω is finite-dimensional while Γ is infinite-dimensional.

In a nonrelativistic system, X_0 generates a one-parameter group of transformations in Γ_0, the hamiltonian flow of H_0 on Γ_0. Instead of having the observables in C_0 depending on t, one can shift perspective and view the observables in C_0 as time-independent objects and the states in Γ_0 as time-dependent objects. This is a classical analog of the shift from the Heisenberg to the Schrödinger picture in quantum theory, and can be called the "classical Schrödinger picture."

In the relativistic theory there is no special "time" variable, C does not split naturally as $C = R \times C_0$, the constraints do not have the form $H = p_t + H_0$ and the description of the correlations in terms of "how the variables in C_0 evolve in time" is not available in general. In a system that does not admit a nonrelativistic formulation, the classical Schrödinger picture in which states evolve in time is not available: only the relativistic notions of state and observable make sense.

Special-relativistic systems. There are relativistic systems that do not admit a nonrelativistic formulation, such as the example of the double pendulum discussed above. There are also systems that can be given a nonrelativistic formulation, but their structure is far more clean in the relativistic formalism. Lorentz-invariant systems are typical examples. They can be formulated in the conventional hamiltonian picture only at the price of breaking Lorentz invariance. The choice of a preferred Lorentz frame specifies a preferred Lorentz time variable $t = x^0$. The predictions of the theory are Lorentz invariant, but the formalism is not. This way of dealing with the mechanics of special-relativistic systems hides the simplicity and symmetry of its hamiltonian structure. The relativistic hamiltonian formalism, exemplified below for the case of a free particle is manifestly Lorentz invariant.

Example: relativistic particle. The configuration space $\mathcal{C}$ is a Minkowski space $\mathcal{M}$, with coordinates x^μ. The dynamics is given by the hamiltonian $H = p^\mu p_\mu + m^2$, which defines the mass-m Lorentz hyperboloid $\mathcal{K}_m$. The constraint surface Σ is therefore given by $\Sigma = T^*\mathcal{M}|_{H=0} = \mathcal{M} \times \mathcal{K}_m$. The null vectors of the restriction of $d\theta = dp_\mu \wedge dx^\mu$ to Σ are

$$X = p_\mu \frac{\partial}{\partial x^\mu}, \tag{3.81}$$

because $\omega(X) = p^\mu dp_\mu = 2d(p^2) = 0$ on $p^\mu p_\mu = -m^2$. The integral lines of X, namely the lines whose tangent is X, are

$$x^\mu(\tau) = P^\mu \tau + X^\mu, \qquad p^\mu(\tau) = P^\mu, \tag{3.82}$$

which give the physical motions of the particle. The space of these lines is six-dimensional (it is coordinatized by the eight numbers (X^μ, P^μ) but $P^\mu P_\mu = -m^2$ and (P^μ, X^μ) defines the same line as $(P^\mu, X^\mu + P^\mu a)$ for any a), and represents the phase space. The motions are thus the timelike straight lines in $\mathcal{M}$.

Notice that all notions used are completely Lorentz invariant. A state is a time-like geodesic; an observable is any Minkowski coordinate; a correlation is a point in Minkowski space. The theory is about correlations between Minkowski coordinates, that is, observations of the particle at certain spacetime points. On the other hand, the split $\mathcal{M} = R \times R^3$ necessary to define the usual hamiltonian formalism, is observer dependent.

The relativistic formulation of mechanics is not only more general, but also more simple and elegant and better operationally founded, than the conventional nonrelativistic formulation. This is true whether one uses the Hamilton equations, the geometric language, or the Hamilton–Jacobi formalism.

3.2.4 *Discussion:* mechanics is about relations between observables

The key difference between the relativistic formulation of mechanics discussed in this chapter and the conventional one – and, in particular, between the relativistic definitions of state and observable and the conventional ones – is the role played by time. In the nonrelativistic context, time is a primary concept. Mechanics is defined as the theory of the evolution in time. In the definition considered here, on the other hand, no special partial observable is singled out as the independent variable. Mechanics is defined as the theory of the correlations between partial observables.

Technically, $\mathcal{C}$ does not split naturally as $\mathcal{C} = R \times \mathcal{C}_0$, the constraints do not have the form $H = p_t + H_0$ and the Schrödinger-like description of correlations in terms of "how states and observables evolve in time" is not available in general.

It is important to understand clearly the meaning of this shift of perspective.

The first point is that *it is possible* to formulate conventional mechanics in this time-independent language. In fact, the formalism of mechanics becomes even more clean and symmetric (for instance, Lorentz covariant) in this language. This is a remarkable fact by itself. What is remarkable is that the formal structure of mechanics doesn't really treat the time variable on a different footing than the other variables. The structure of mechanics is the formalization of what we have understood about the physical structure of the world. Therefore, we can say that the physical (more precisely, mechanical) structure of the world is quite blind to the fact that there is anything "special" about the variable t.

Historically, the idea that in a relativistic context we need the time-independent notion of state has been advocated particularly by Dirac (see [148] in Chapter 5) and by Souriau [105]. The advantages of the relativistic notion of state are multi-fold. In special relativity, for instance, time transforms with other variables, and there is no covariant definition of instantaneous state. In a Lorentz-invariant field theory, in particular, the notion of instantaneous state breaks explicit Lorentz covariance: the instantaneous state is the value of the field on a simultaneity surface, which is such for a certain observer only. The relativistic notion of state, on the other hand, is Lorentz invariant.

The second point is that this shift in perspective is *forced* in general relativity where the notion of a special spacelike surface over which initial data are fixed conflicts with diffeomorphism invariance. A generally covariant notion of instantaneous state, or a generally covariant notion of observable "at a given time," makes little physical sense. Indeed, none of the various notions of time that appear in general relativity (coordinate time, proper time, clock time) play the role that t plays in nonrelativistic mechanics. A consistent definition of state and observable in a generally covariant context cannot explicitly involve time.

The physical reason for this difference is discussed in Chapter 2. In nonrelativistic physics, time and position are defined with respect to a system of reference bodies and clocks that are implicitly assumed to exist and not to interact with the physical system studied. In gravitational physics, one discovers that no body or clock exists which does not interact with the gravitational field: the gravitational field affects directly the motion and the rate of any reference body or clock. Therefore, one cannot separate reference bodies and clocks from the dynamical variables of the system. General relativity – in fact, any generally covariant theory – is always a theory of interacting variables that necessarily include the physical bodies and clocks used as references to characterize spacetime points.

In the example of the pendulum discussed in Section 3.2.1, for instance, we can assume that the pendulum and the clock do not interact. In a general-relativistic context the two always interact and $\mathcal{C}$ does not split into $\mathcal{C}_0$ and R.

Summarizing, it is only in the nonrelativistic limit that mechanics can be seen as the theory of the evolution of the physical variables in time. In a fully relativistic context, *mechanics is a theory of correlations between partial observables*.

3.2.5 Space of boundary data $\mathcal{G}$ and Hamilton function S

I describe here the relativistic version of a structure that plays an important role in the quantum theory.

Hamilton function. Notice that the Hamilton function defined in (3.21) is naturally a function on (two copies of) the relativistic configuration space $\mathcal{C}$. In fact, its definition extends to the relativistic context: given two events q^a and q_0^a in $\mathcal{C}$, the Hamilton function is defined as

$$S(q^a, q_0^a) = \int_{\tilde{\gamma}} \theta, \tag{3.83}$$

where $\tilde{\gamma}$ is the orbit in Σ of the motion that goes from q_0^a to q^a. This is also the value of the action along this motion. For instance, for a nonrelativistic system we can write

$$S(q^a, q_0^a) = \int_{\gamma} \theta = \int_{\gamma} p_a \mathrm{d}q^a \tag{3.84}$$

$$= \int_0^1 p_a(\tau)\dot{q}^a(\tau)\mathrm{d}\tau = \int_0^1 \left(p_i(\tau)\dot{q}^i(\tau) + p_t(\tau)\dot{t}(\tau) \right) \mathrm{d}\tau$$

$$= \int_0^1 \left(p_i(\tau)\dot{q}^i(\tau) - H_0(\tau)\dot{t}(\tau) \right) \mathrm{d}\tau$$

$$= \int_{t_0}^t \left(p_i(t)\frac{\mathrm{d}q^i(t)}{\mathrm{d}t} - H_0(t) \right) \mathrm{d}t$$

$$= \int_{t_0}^t L\left(q^i, \frac{\mathrm{d}q^i(t)}{\mathrm{d}t} \right) \mathrm{d}t, \tag{3.85}$$

where L is the lagrangian. From the definition, we have

$$\frac{\partial S(q^a, q_0^a)}{\partial q^a} = p_a(q^a, q_0^a), \tag{3.86}$$

where $p_a(q^a, q_0^a)$ is the value of the momentum at the final event. Notice that this value depends on q^a *as well as on* q_0^a. The derivation of this equation is less obvious than appears at first sight: I leave the details to the acute reader.

It follows from (3.86) that $S(q^a, q_0^a)$ satisfies the Hamilton–Jacobi equation (3.59). The quantities q_0^a can be seen as the Hamilton–Jacobi integration constants. Notice that they are n, not $n-1$. Equations (3.60) now read

$$f^a(q^a; q_0^a, p_{a0}) = \frac{\partial S(q^a, q_0^a)}{\partial q_0^a} + p_{a0} = 0. \tag{3.87}$$

Therefore, the phase space is directly (over-)coordinatized by initial coordinates and momenta (q_0^a, p_{a0}). These are not independent for two reasons. First, they satisfy the equation $H = 0$. Second, different sets $(q_0^a(\tau), p_{a0}(\tau))$ along the same motion determine the same motion. Furthermore, one of the equations (3.87) turns out to be dependent on the others.

$S(q^a, q_0^a)$ satisfies the Hamilton–Jacobi equation in both sets of variables, namely it satisfies also

$$H\left(q_0^a, -\frac{\partial S(q^a, q_0^a)}{\partial q_0^a}\right) = 0, \tag{3.88}$$

where the minus sign comes from the fact that the second set of variables is in the lower integration boundary in (3.83).

If there is more than one physical motion γ connecting the boundary data, the Hamilton function is multivalued. If $\gamma_1, \ldots, \gamma_n$ are distinct solutions with the same boundary values, we denote its different branches as

$$S_i(q_1^a, q_2^a) = \int_{\tilde{\gamma}_i} \theta. \tag{3.89}$$

The Hamilton function is strictly related to the quantum theory. It is the phase of the propagator $W(q^a, q_0^a)$, which, as we shall see in Chapter 5, is the main object of the quantum theory. If S is single valued, we have

$$W(q^a, q_0^a) \sim A(q^a, q_0^a) \, e^{\frac{i}{\hbar} S(q^a, q_0^a)} \tag{3.90}$$

up to higher terms in $\hbar$. If S is multivalued,

$$W(q^a, q_0^a) \sim \sum_i A_i(q^a, q_0^a) \, e^{\frac{i}{\hbar} S_i(q^a, q_0^a)}. \tag{3.91}$$

Example: free particle. In the case of the free particle, the value of the classical action, along the motion is

$$S(x, t, x_0, t_0) = \int_0^1 (p_t \dot{t} + p \dot{x}) \mathrm{d}t = p_t \int_{t_0}^t \mathrm{d}t + p \int_{x_0}^x \mathrm{d}x$$

$$= -\frac{m(x - x_0)^2}{2(t - t_0)} + m\frac{(x - x_0)^2}{t - t_0}$$

$$= \frac{m(x - x_0)^2}{2(t - t_0)}. \tag{3.92}$$

It is easy to check that S solves the Hamilton–Jacobi equation of the free particle. The first of the two equations (3.87) gives the evolution equation

$$\frac{\partial S(x, t, x_0, t_0)}{\partial x_0} + p_0 = -m\frac{x - x_0}{t - t_0} + p_0 = 0. \tag{3.93}$$

The second equation constrains the p_t integration constant

$$\frac{\partial S(x, t, x_0, t_0)}{\partial t_0} + p_{t0} = -\frac{1}{2m}p_0^2 + p_{t0} = 0. \tag{3.94}$$

Recall that the propagator of the Schrödinger equation of the free particle is

$$W(x, t, x_0, t_0) = \frac{1}{\sqrt{i\hbar(t - t_0)}}\, e^{\frac{i}{\hbar}\frac{m(x - x_0)^2}{2(t - t_0)}} = \frac{1}{\sqrt{i\hbar(t - t_0)}}\, e^{\frac{i}{\hbar}S(x, t, x_0, t_0)}. \tag{3.95}$$

Example: double pendulum. The Hamilton function of the timeless system (3.40) can be computed directly from its definition. This gives

$$S(a, b, a', b') = S\left(a, b, a', b';\ A(a, b, a', b')\right), \tag{3.96}$$

where

$$S(a, b, a', b';\ A) = S(a, b, A) - S(a', b', A). \tag{3.97}$$

$S(a, b, A)$ is given in (3.62) and $A(a, b, a', b')$ is the value of A of the ellipse (3.41) that crosses (a, b) and (a', b'). This value can be obtained by noticing that (3.58) imply, with little algebra, that

$$A^2 = \frac{a^2 + a'^2 - 2aa' \cos \tau}{\sin^2 \tau} \tag{3.98}$$

and

$$E = \frac{(a^2 + b^2 + a'^2 + b'^2) - 2(aa' + bb') \cos \tau}{\sin^2 \tau}. \tag{3.99}$$

The second equation can be solved for $\tau(a, b, a', b')$, and inserting this in the first gives $A(a, b, a', b')$. It is not complicated to check that the derivative of $\partial S(a, b, a', b';\ A)/\partial A$ vanishes when $A = A(a, b, a', b')$. Using this it is easy to see that (3.96) solves the Hamilton–Jacobi equation in both sets of variables.

Notice that, for given (a, b, a', b'), equation (3.98) gives A as a function of τ. We can therefore consider also the function

$$S(a, b, a', b';\ \tau) = S(a, b, A(\tau)) - S(a', b', A(\tau)), \tag{3.100}$$

which is the value of the action of the nonrelativistic system formed by two harmonic oscillators evolving in a physical time τ, with a nonrelativistic hamiltonian H, that is, it is the Hamilton function of this system. With some algebra, this can be written also as

$$S(a, b, a', b'; \tau) = M\tau + \frac{(a^2 + b^2 + a'^2 + b'^2)\cos\tau - 2(aa' + bb')}{\sin\tau}. \tag{3.101}$$

As for A, we have immediately

$$\left.\frac{\partial S(a, b, a', b'; \tau)}{\partial \tau}\right|_{\tau = \tau(a, b, a', b')} = 0. \tag{3.102}$$

This means that the Hamilton function of the timeless system is numerically equal to the Hamilton function of the two oscillators for the "correct" time τ needed to go from (a', b') to (a, b) staying on a motion of total energy E. And that this "correct" time $\tau = \tau(a, b, a', b')$ is the one that minimizes the Hamilton function of the two oscillators.

More precisely, for given (a, b, a', b') there are two paths connecting (a', b') with (a, b): these are the two paths in which the ellipse that goes through (a', b') and (a, b) is cut by these two points. Denote S_1 and S_2 the two values of the action along these paths. Their relation is easily obtained by noticing that the action along the entire ellipse is easily computed as

$$S_1 + S_2 = 2\pi E. \tag{3.103}$$

The space of the boundary data $\mathcal{G}$. The Hamilton function is a function on the space $\mathcal{G} = \mathcal{C} \times \mathcal{C}$. An element $\alpha \in \mathcal{G}$ is an ordered pair of elements of the extended configuration space $\mathcal{C}$: $\alpha = (q^a, q_0^a)$. Notice that α is the ensemble of the boundary conditions for a physical motion. For a nonrelativistic system, $\alpha = (t, q^i, t_0, q_0^i)$; the motion begins at q_0^i at time t_0 and ends at q^i at time t.

The space $\mathcal{G}$ carries a natural symplectic structure. In fact, let $i : \mathcal{G} \to \Gamma$ be the map that sends each pair to the orbit that the pair defines. Then we can define the two-form $\omega_{\mathcal{G}} = i^*\omega_{\mathrm{ph}}$, where ω_{ph} is the symplectic form of the phase space, defined in Section 3.2.2. In other words, $\alpha = (q^a, q_0^a)$ can be taken as a natural over-coordinatization of the phase space. Instead of coordinatizing a motion with initial positions and momenta, we coordinatize it with initial and final positions. In these coordinates, the symplectic form is given by $\omega_{\mathcal{G}}$.

The two-form $\omega_{\mathcal{G}}$ can be computed without having first to compute Γ and ω_{ph}. Denote $\tilde{\gamma}_\alpha$ the orbit in Σ with boundary data α, and γ_α its projection to $\mathcal{C}$. Then α is the boundary of γ_α. We write $\alpha = \partial\gamma_\alpha$. Denote s and s_0 the initial and final points of $\tilde{\gamma}_\alpha$ in Σ. That is, $s = (q^a, p_a)$ and $s_0 = (q_0^a, p_{0a})$, where, in general, both p_a and p_{0a} depend on q^a and on q_0^a. Let $\delta\alpha = (\delta q^a, \delta q_0^a)$ be a vector (an infinitesimal displacement) at α. Then the following is true:

$$\omega_{\mathcal{G}}(\alpha)(\delta_1\alpha, \delta_2\alpha) = \omega_{\mathcal{G}}(q^a, q_0^a)((\delta_1 q^a, \delta_1 q_0^a), (\delta_2 q^a, \delta_2 q_0^a))$$
$$= \omega(s)(\delta_1 s, \delta_2 s) - \omega(s_0)(\delta_1 s_0, \delta_2 s_0). \tag{3.104}$$

Notice that $\delta_1 s$, the variation of s, is determined by $\delta_1 q$ as well as by $\delta_1 q_0$, and so on. This equation expresses $w_{\mathcal{G}}$ directly in terms of w. As we shall see, this equation admits an immediate generalization in the field theoretical framework, where w will be a five-form and $w_{\mathcal{G}}$ is a two-form.

Now fix a pair $\alpha = (q^a, q_0^a)$ and consider a small variation of only one of its elements, say

$$\delta\alpha = (\delta q^a, 0). \tag{3.105}$$

This defines a vector $\delta\alpha$ at α on $\mathcal{G}$, which can be pushed forward to Γ. If the variation is along the direction of the motion, then the push forward vanishes, that is $i_*\delta\alpha = 0$ because α and $\alpha + \delta\alpha$ define the same motion. It follows that if the variation is along the direction of the motion, $w_{\mathcal{G}}(\delta\alpha) = 0$. Therefore the equation

$$w_{\mathcal{G}}(X) = 0 \tag{3.106}$$

gives the solutions of the equations of motion.

Thus, the pair $(\mathcal{G}, w_{\mathcal{G}})$ contains all the relevant information of the system. The null directions of $w_{\mathcal{G}}$ define the physical motions, and if we divide $\mathcal{G}$ by these null directions, the factor space is the physical phase space, equipped with the physical symplectic structure.

Example: free particle. The space $\mathcal{G}$ has coordinates $\alpha = (t, x, t_0, x_0)$. Given this point in $\mathcal{G}$, there is one motion that goes from (t_0, x_0) to (t, x), which is

$$t(\tau) = t_0 + (t - t_0)\tau, \tag{3.107}$$
$$x(\tau) = x_0 + (x - x_0)\tau. \tag{3.108}$$

Along this motion,

$$p = m\frac{x - x_0}{t - t_0}, \tag{3.109}$$

$$p_t = -\frac{m(x - x_0)^2}{2(t - t_0)^2}. \tag{3.110}$$

The map $i : \mathcal{G} \to \Gamma$ is thus given by

$$P = p = m\frac{x - x_0}{t - t_0}, \tag{3.111}$$

$$Q = x - \frac{p}{m}t = x - \frac{x - x_0}{t - t_0}t, \tag{3.112}$$

and therefore the two-form $w_{\mathcal{G}}$ is

$$w_{\mathcal{G}} = i^*w_\Gamma = dP(t, x, t_0, x_0) \wedge dQ(t, x, t_0, x_0)$$

$$= m\, d\frac{x - x_0}{t - t_0} \wedge d\left(x - \frac{x - x_0}{t - t_0}t\right)$$

$$= \frac{m}{t - t_0}\left(dx - \frac{x - x_0}{t - t_0}dt\right) \wedge \left(dx_0 - \frac{x - x_0}{t - t_0}dt_0\right). \tag{3.113}$$

Immediately, we see that a variation $\delta\alpha = (\delta t, \delta x, 0, 0)$ (at constant (x_0, t_0)) such that $\omega_{\mathcal{G}}(\delta\alpha) = 0$ must satisfy

$$\delta x = \frac{x - x_0}{t - t_0} \delta t. \tag{3.114}$$

This is precisely a variation of x and t along the physical motion (determined by (x_0, t_0)). Therefore $\omega_{\mathcal{G}}(\delta\alpha) = 0$ gives again the equations of motion. The two null directions of $\omega_{\mathcal{G}}$ are thus given by the two vector fields

$$X = \frac{x - x_0}{t - t_0} \partial_x + \partial_t, \tag{3.115}$$

$$X_0 = \frac{x - x_0}{t - t_0} \partial_{x_0} + \partial_{t_0}, \tag{3.116}$$

which are in involution (their Lie bracket vanishes), and therefore define a foliation of $\mathcal{G}$ with two-dimensional surfaces. These surfaces are parametrized by P and Q, given in (3.111), (3.112) and, in fact,

$$X(P) = X(Q) = X_0(P) = X_0(Q) = 0. \tag{3.117}$$

We have simply recovered in this way the physical phase space: the space of these surfaces is the phase space Γ and the restriction of $\omega_{\mathcal{G}}$ to it is the physical symplectic form ω_{ph}.

Physical predictions from S. There are several different ways of deriving physical predictions from the Hamilton function $S(q^a, q_0^a)$.

- Equation (3.87) gives the evolution function f in terms of the Hamilton function.

- If we can measure the partial observables q^a as well as their momenta p_a, then the Hamilton function can be used for making predictions as follows. Let

$$p_a^1(q_1^a, q_2^a) = \frac{\partial S(q_1^a, q_2^a)}{\partial q_1^a},$$

$$p_a^2(q_1^a, q_2^a) = \frac{\partial S(q_1^a, q_2^a)}{\partial q_2^a}. \tag{3.118}$$

The two equations

$$p_a^1 = p_a^1(q_1^a, q_2^a),$$
$$p_a^2 = p_a^2(q_1^a, q_2^a) \tag{3.119}$$

relate the four partial observables of the quadruplet $(q_1^a, p_a^1, q_2^a, p_a^2)$. The theory predicts that it is possible to observe the quadruplet $(q_1^a, p_a^1, q_2^a, p_a^2)$ only if this satisfies (3.119). In this way the classical theory determines which combinations of values of partial observables can be observed.

- Alternatively, we can fix two points q_i^a and q_f^a in $\mathcal{C}$ and ask whether a third point q^a is on the motion determined by q_i^a and q_f^a. That is, ask whether or not we could observe the correlation q^a, given that the correlations q_i^a and q_f^a are observed. A moment of reflection will convince the reader that if the answer to this question is positive then

$$S(q_f^a, q^a) + S(q^a, q_i^a) = S(q_f^a, q_i^a). \tag{3.120}$$

because the action is additive along the motion. Furthermore, the incoming momentum at q^a and the outgoing one must be equal, therefore

$$\frac{\partial S(q_f^a, q^a)}{\partial q^a} = -\frac{\partial S(q^a, q_f^a)}{\partial q^a}. \tag{3.121}$$

3.2.6 Evolution parameters

A physical system is often defined by an action which is the integral of a lagrangian in an evolution parameter. But there are two different physical meanings that the evolution parameter may have.

We have seen that the variational principle governing any hamiltonian system can be written in the form (here $k = 1$)

$$S = \int d\tau \left(p_a \frac{dq^a}{d\tau} - NH(p_a, q^a) \right). \tag{3.122}$$

The action is invariant under reparametrizations of the evolution parameter τ. The evolution parameter τ has no physical meaning: there is no measuring device associated with it.

On the other hand, consider a nonrelativistic system, where $q^a = (t, q^i)$ and $H = p_t + H_0$. The action (3.122) becomes

$$S = \int d\tau \left(p_t \frac{dt}{d\tau} + p_i \frac{dq^i}{d\tau} - N(p_t + H_0(p_i, q^i)) \right). \tag{3.123}$$

Varying N we obtain the equation of motion

$$p_t = -H_0. \tag{3.124}$$

Inserting this relation back into the action, we obtain

$$S = \int d\tau \left(-H_0 \frac{dt}{d\tau} + p_i \frac{dq^i}{d\tau} \right). \tag{3.125}$$

We can now change the integration variable from τ to $t(\tau)$. Defining (in bad physicists' notation) $q^i(t) \equiv q^i(\tau(t))$ and so on, we can write

$$S = \int d\tau \, \frac{dt}{d\tau} \left(-H_0 + p_i \frac{dq^i}{dt} \right) = \int dt \left(p_i \frac{dq^i(t)}{dt} - H_0 \right). \quad (3.126)$$

The evolution parameter in the action is no longer an arbitrary unphysical parameter τ. It is one of the partial observables: the time observable t.

If we are given an action, we must understand whether the evolution parameter in the action is a partial observable, such as t, or an unphysical parameter, such as τ. If the action is invariant under reparametrizations of its evolution parameter, then the evolution parameter is unphysical. If it is not, then the evolution parameter is a partial observable.

The same is true if the action is given in lagrangian form. In performing the Legendre transform from the lagrangian to the hamiltonian formalism, the consequence of the invariance of the action under reparametrizations is doublefold. First, the relation between velocities and momenta cannot be inverted. The map from the space of the coordinates and velocities $(q^a, \dot{q}^a)$ to the space of coordinates and momenta (q^a, p_a) is not invertible. The image of this map is a subspace Σ of Ω, and we can characterize Σ by means of an equation $H = 0$ for a suitable hamiltonian H. Second, the canonical hamiltonian computed via the Legendre transform vanishes on Σ. In the language of constrained system theory, this is because the canonical hamiltonian generates evolution in the parameter of the action; since this is unphysical, this evolution is gauge; the generator of a gauge is a constraint, and therefore vanishes on Σ.

The evolution parameter in the action is often denoted t whether it is a partial observable or an unphysical parameter. One should not confuse the t in the first case with the t in the second case. They have very different physical interpretations. The time coordinate t in Maxwell theory is a partial observable. The time coordinate t in GR is an unphysical parameter. The fact that the two are generally denoted with the same letter and with the same name is a very unfortunate historical accident.

Example: relativistic particle As we have seen, the hamiltonian dynamics of a relativistic particle is defined by the relativistic hamiltonian $H = p_\mu p^\mu + m^2$, namely by the action principle

$$S = \int d\tau \left(p_\mu \dot{x}^\mu - \frac{N}{2} (p_\mu p^\mu + m^2) \right). \quad (3.127)$$

The relation between velocities and momenta, obtained by varying p_μ, is $\dot{x}^\mu = N p^\mu$. The inverse Legendre transform therefore gives

$$S = \frac{1}{2} \int d\tau \left(\frac{\dot{x}_\mu \dot{x}^\mu}{N} - N m^2 \right). \quad (3.128)$$

We can also get rid of the Lagrange multiplier N from this action by writing its equation of motion

$$-\frac{\dot{x}_\mu \dot{x}^\mu}{N^2} - m^2 = 0, \qquad (3.129)$$

which is solved by

$$N = \frac{\sqrt{-\dot{x}_\mu \dot{x}^\mu}}{m}, \qquad (3.130)$$

and inserting this relation back into the action. This gives

$$S = m \int d\tau \sqrt{-\dot{x}_\mu \dot{x}^\mu}, \qquad (3.131)$$

which is the best known reparametrization invariant action for the relativistic particle.

3.2.7 * Complex variables and reality conditions

In GR it is often convenient to use complex dynamical variables, since these simplify the form of the dynamical equations. A particularly convenient choice is a mixture of complex and real variables, where one canonical variable is complex, while the conjugate one is real. As we shall see, the selfdual connection (2.19), which is complex, naturally leads to canonical variables of this type. To exemplify how the use of such variables affects dynamics, consider a free particle with coordinate x, momentum p, and hamiltonian $H_0(x, p) = p^2/2m$, and assume we want to describe its dynamics in terms of the variables (x, z) where

$$z = x - ip. \qquad (3.132)$$

In terms of these variables the nonrelativistic hamiltonian reads

$$H_0(x, z) = -\frac{1}{2m}(x - z)^2. \qquad (3.133)$$

Consider z as a configuration variable and ix as its momentum variable. The Hamilton–Jacobi equation becomes

$$\frac{\partial S(z, t)}{\partial t} = -H_0\left(-i\frac{\partial S(z, t)}{\partial z}, z\right) = \frac{1}{2m}\left(i\frac{\partial S(z, t)}{\partial z} + z\right)^2. \qquad (3.134)$$

This is solved by

$$S(z, t; k) = kz + \frac{i}{2}z^2 - \frac{k^2}{2m}t. \qquad (3.135)$$

Equating the derivative of S with respect to the parameter k to a constant we obtain the solution:

$$C = \frac{\partial S(z, t; k)}{\partial k} = z - \frac{k}{m}t, \qquad (3.136)$$

that is,

$$z(t) = \frac{k}{m} t + C. \tag{3.137}$$

This is not the end of the story, since so far k and C can be arbitrary *complex* constants. To find the good solutions, corresponding to real x and p, we have to remind ourselves that z and x are not truly independent, since x is the real part of z:

$$z + \bar{z} = 2x, \tag{3.138}$$

that is,

$$z + \bar{z} = -2\mathrm{i}\, \frac{\partial S}{\partial z}. \tag{3.139}$$

Inserting the solutions (3.137) in the l.h.s., we get

$$\mathrm{Im}\,[k]\, \frac{t}{m} + \mathrm{Im}[C] = -k. \tag{3.140}$$

Therefore k is real and the imaginary part of C is $-k$. This immediately gives the correct solution.

Equation (3.138) is called the *reality condition*. The example illustrates that in the Hamilton–Jacobi formalism the reality condition restricts the values of the Hamilton–Jacobi constants once the solutions of the evolution equations are inserted.

3.3 Field theory

There are several ways in which a field theory can be cast in hamiltonian form. One possibility is to take the space of the fields at fixed time as the nonrelativistic configuration space Q. This strategy badly breaks special- and general-relativistic invariance. Lorentz covariance is broken by the fact that one has to choose a Lorentz frame for the t variable. Far more disturbing is the conflict with general covariance. The very foundation of generally covariant physics is the idea that the notion of a simultaneity surface over all the Universe is devoid of physical meaning. It is better to found hamiltonian mechanics on a notion not devoid of physical significance.

A second alternative is to formulate mechanics on the space of the solutions of the equations of motion. The idea goes back to Lagrange. In the generally covariant context, a symplectic structure can be defined over this space using a spacelike surface, but one can show that the definition is surface independent and therefore it is well defined. This strategy has been explored by several authors [108]. The structure is viable in principle and has the merit of showing that the hamiltonian formalism is intrinsically covariant. In practice, it is difficult to work with the space of solutions to the field equations in the case of an interacting theory. Therefore we must either work over a space that we can't even coordinatize, or coordinatize the space with initial data on some instantaneity surface and therefore effectively go back to the conventional fixed-time formulation.

The third possibility, which I consider here, is to use a covariant *finite*-dimensional space for formulating hamiltonian mechanics. I noted above that in the relativistic context the double role of the phase space, as the arena of mechanics and the space of the states, is lost. The space of the states, namely the phase space Γ, is infinite-dimensional in field theory, essentially by definition of field theory. But this does not imply that the arena of hamiltonian mechanics has to be infinite-dimensional as well. The natural arena for relativistic mechanics is the extended configuration space $\mathcal{C}$ of the partial observables. Is the space of the partial observables of a field theory finite- or infinite-dimensional?

3.3.1 Partial observables in field theory

Consider a field theory for a field $\phi(x)$ with N components. The field is defined over spacetime M with coordinates x and takes values in an N-dimensional target space T

$$\phi\colon M \longrightarrow T$$
$$x \longmapsto \phi(x). \tag{3.141}$$

For instance, this could be Maxwell theory for the electric and magnetic fields $\phi = (\vec{E}, \vec{B})$, where $N = 6$. In order to make physical measurements on the field described by this theory we need N measuring devices to measure the components of the field ϕ, and four devices (one clock and three devices giving us the distance from three reference objects) to determine the spacetime position x. Field values ϕ and positions x are therefore the partial observables of a field theory. Therefore the operationally motivated relativistic configuration space for a field theory is the *finite-dimensional* space

$$\mathcal{C} = M \times T, \tag{3.142}$$

which has dimension $4 + N$. A correlation is a point (x, ϕ) in $\mathcal{C}$. It represents a certain value (ϕ) of the fields at a certain spacetime point (x). This is the obvious generalization of the (t, α) correlations of the pendulum of the example in Section 3.2.1.

A physical motion γ is a physically realizable ensemble of correlations. A motion is determined by a solution $\phi(x)$ of the field equations. Such a solution determines a 4-dimensional surface in the $((4 + N)$-dimensional) space $\mathcal{C}$: the surface is the graph of the function (3.141). Namely, the ensemble of the points $(x, \phi(x))$. The space of the solutions of the field equations, namely the phase space Γ, is therefore an (infinite-dimensional) space of 4d surfaces γ in the $(4 + N)$-dimensional configuration space $\mathcal{C}$. Each state in Γ determines a surface γ in $\mathcal{C}$.

Hamiltonian formulations of field theory defined directly on $\mathcal{C} = M \times T$ are possible and have been studied. The main reason is that in a local field theory the equations of motion are local, and therefore what happens at a point depends only on the neighborhood of that point. There is no need, therefore, to consider full spacetime to find the hamiltonian structure of the field equations. I refer the reader to the beautiful and detailed paper [109] and the ample references therein, for a discussion of this kind of approach. I give a simple and self-contained illustration of the formalism below, with the emphasis on its general covariance.

3.3.2 * Relativistic hamiltonian mechanics

Consider a field theory on Minkowski space M. Call $\phi^A(x^\mu)$ the field, where $A = 1, \ldots, N$. The field is a function $\phi : M \to T$, where $T = R^N$ is the target space, namely it is the space in which the field takes values. The extended configuration space of this theory is the finite-dimensional space $\mathcal{C} = M \times T$, with coordinates $q^a = (x^\mu, \phi^A)$. The coordinates q^a are the $(4 + N)$ partial observables whose relations are described by the theory. A solution of the equations of motion defines a four-dimensional surface γ in $\mathcal{C}$. If we coordinatize this surface using the coordinates x^μ, then this surface is given by $[x^\mu, \phi^A(x^\mu)]$, where $\phi^A(x^\mu)$ is a solution of the field equations. If, alternatively, we use an arbitrary parametrization with parameters τ^ρ, $\rho = 0, 1, 2, 3$, then the surface is given by $[x^\mu(\tau^\rho), \phi^A(\tau^\rho)]$, where $\phi^A(x^\mu(\tau^\rho)) = \phi^A(\tau^\rho)$.

In the case of a finite number of degrees of freedom (and no gauges), motions are given by one-dimensional curves. At each point of the curve, there is one tangent vector, and momenta coordinatize the one-forms. In field theory, motions are four-dimensional surfaces, and have four independent tangents X_μ, or a "quadritangent" $X = \epsilon^{\mu\nu\rho\sigma} X_\mu \otimes X_\nu \otimes X_\rho \otimes X_\sigma$ at each point. Accordingly, momenta coordinatize the four-forms. Let $\Omega = \Lambda^4 T^* \mathcal{C}$ be the bundle of the four-forms $p_{abcd} dq^a \wedge dq^b \wedge dq^c \wedge dq^d$ over $\mathcal{C}$. A point in Ω is thus a pair (q^a, p_{abcd}). The space Ω carries the canonical four-form

$$\tilde{\theta} = p_{abcd} \, dq^a \wedge dq^b \wedge dq^c \wedge dq^d. \tag{3.143}$$

In general, given the finite-dimensional space $\mathcal{C}$ of the partial observables q^a, dynamics is defined by a relativistic hamiltonian $H : \Omega \to V$, where $\Omega = \Lambda^4 T^* \mathcal{C}$ and V is a vector space. Denote $\tilde{\gamma}$ a four-dimensional surface in Ω and γ the projection of this surface on $\mathcal{C}$. The physical motions γ are determined by the following.

Variational principle. *A surface γ with a boundary α is a physical motion if $\tilde{\gamma}$ extremizes the integral*

$$S[\tilde{\gamma}] = \int_{\tilde{\gamma}} p_{abcd} \, dq^a \wedge dq^b \wedge dq^c \wedge dq^d \tag{3.144}$$

in the class of the surfaces $\tilde\gamma$ satisfying

$$H(q^a, p_{abcd}) = 0, \tag{3.145}$$

and whose restriction γ to $\mathcal{C}$ is bounded by α.

This is a completely straightforward generalization of the variational principle of Section 3.2. Equation (3.145) defines a surface Σ in Ω. As before, we denote θ the restriction of $\tilde\theta$ to Σ, and $\omega = \mathrm{d}\theta$.

For a field theory on Minkowski space without gauges, the system (3.145) is given by

$$p_{ABCD} = p_{ABC\mu} = p_{AB\mu\nu} = 0, \tag{3.146}$$
$$H = \pi + H_0(x^\mu, \phi^A, p_A^\mu) = 0, \tag{3.147}$$

where H_0 is DeDonder's covariant hamiltonian [110] (see below for an example). It is convenient to use the notation $p_{\mu\nu\rho\sigma} = \pi\epsilon_{\mu\nu\rho\sigma}$ and $p_{A\nu\rho\sigma} = p_A^\mu\epsilon_{\mu\nu\rho\sigma}$ for the nonvanishing momenta and to use coordinates (x^μ, ϕ^A, p_A^μ) on Σ. On the surface defined by (3.146),

$$\tilde\theta = \pi\, \mathrm{d}^4x + p_A^\mu\, \mathrm{d}\phi^A \wedge \mathrm{d}^3x_\mu, \tag{3.148}$$

where we have introduced the notation $\mathrm{d}^4x = \mathrm{d}x^0 \wedge \mathrm{d}x^1 \wedge \mathrm{d}x^2 \wedge \mathrm{d}x^3$ and $\mathrm{d}^3x_\mu = \mathrm{d}^4x(\partial_\mu) = \frac{1}{3!}\epsilon_{\mu\nu\rho\sigma}\mathrm{d}x^\nu \wedge \mathrm{d}x^\rho \wedge \mathrm{d}x^\sigma$. On Σ, defined by (3.146) and (3.147),

$$\theta = \tilde\theta|_\Sigma = -H_0(x^\mu, \phi^A, p_A^\mu)\mathrm{d}^4x + p_A^\mu\, \mathrm{d}\phi^A \wedge \mathrm{d}^3x_\mu, \tag{3.149}$$

and ω is the five-form

$$\omega = -\mathrm{d}H_0(x^\mu, \phi^A, p_A^\mu) \wedge \mathrm{d}^4x + \mathrm{d}p_A^\mu \wedge \mathrm{d}\phi^A \wedge \mathrm{d}^3x_\mu. \tag{3.150}$$

An orbit of ω is a four-dimensional surface m immersed in Σ, such that at each of its points a quadruplet X of tangents to the surface satisfies

$$\omega(X) = 0. \tag{3.151}$$

I leave to the reader the exercise of showing that the projection of an orbit on $\mathcal{C}$ is a physical motion.

In more detail, let $(\partial_\mu, \partial_A, \partial_\mu^A)$ be the basis in the tangent space of Σ determined by the coordinates (x^μ, ϕ^A, p_A^μ). Parametrize the surface with arbitrary parameters τ^ρ. The surface is then given by the points $[x^\mu(\tau^\rho), \phi^A(\tau^\rho), p_A^\mu(\tau^\rho)]$. Let $\partial_\rho = \partial/\partial\tau^\rho$. Then let

$$X_\rho = \partial_\rho x^\mu(\tau^\rho)\, \partial_\mu + \partial_\rho\phi^A(\tau^\rho)\, \partial_A + \partial_\rho p_A^\mu(\tau^\rho)\, \partial_\mu^A. \tag{3.152}$$

Then $X = X_0 \otimes X_1 \otimes X_2 \otimes X_3$ is a rank four tensor on Σ. If $\omega(X) = 0$, then $\phi^A(x^\mu)$ determined by $\phi^A(x^\mu(\tau^\rho)) = \phi^A(\tau^\rho)$ is a physical motion.

Summarizing, the canonical formalism of field theory is completely defined by the couple $(\mathcal{C}, H)$, where $\mathcal{C}$ is the *finite*-dimensional space of the partial observables (field values and spacetime coordinates) and H a hamiltonian on the finite-dimensional space $\Omega = \Lambda^4 T^* \mathcal{C}$. Equivalently, it is completely defined by the finite-dimensional presymplectic space (Σ, θ). The formalism as well as its interpretation make sense even in the case in which the coordinates of $\mathcal{C}$ do not split into x^μ and ϕ^A and the relativistic hamiltonian does not have the particular form (3.146)–(3.147).

Example: scalar field. As an example, consider a scalar field $\phi(x^\mu)$ on Minkowski space, satisfying the field equations

$$\partial_\mu \partial^\mu \phi(x^\mu) + m^2 \phi(x^\mu) + V'(\phi(x^\mu)) = 0. \tag{3.153}$$

Here the Minkowski metric has signature $[+, -, -, -]$ and $V'(\phi) = dV(\phi)/d\phi$. The field is a function $\phi : M \to T$, where here $T = R$. The relativistic configuration space of this theory is the five-dimensional space $\mathcal{C} = \mathcal{M} \times T$ with coordinates (x^μ, ϕ). The space Ω has coordinates $(x^\mu, \phi, \pi, p^\mu)$ (equation (3.146) is trivially satisfied) and carries the canonical four-form

$$\tilde{\theta} = \pi \, d^4 x + p^\mu \, d\phi \wedge d^3 x_\mu. \tag{3.154}$$

The dynamics is defined on this space by the DeDonder relativistic hamiltonian

$$H = \pi + H_0 = 0, \tag{3.155}$$

$$H_0 = \frac{1}{2} \left(p^\mu p_\mu + m^2 \phi^2 + 2V(\phi) \right). \tag{3.156}$$

Therefore, we can use coordinates (x^μ, ϕ, p^μ) on the surface Σ defined by these equations and (3.149) gives

$$\theta = -\frac{1}{2} \left(p^\mu p_\mu + m^2 \phi^2 + 2V(\phi) \right) d^4 x + p^\mu \, d\phi \wedge d^3 x_\mu. \tag{3.157}$$

The couple (Σ, θ) defines the presymplectic formulation of the system; ω is the five-form

$$\omega = d\theta = - \left(p^\mu dp_\mu + m^2 \phi d\phi + V'(\phi)d\phi \right) \wedge d^4 x + dp^\mu \wedge d\phi \wedge d^3 x_\mu. \tag{3.158}$$

A tangent vector has the form

$$V = X^\mu \partial_{x^\mu} + X^\phi \partial_\phi + Y^\mu \partial_{p^\mu}. \tag{3.159}$$

If we coordinatize the orbits of ω with the coordinates x^μ, at every point we have the four independent tangent vectors

$$X_\mu = \partial_{x^\mu} + (\partial_\mu \phi)\partial_\phi + (\partial_\mu p^\rho)\partial_{p^\rho} \tag{3.160}$$

and the quadritangent $X = \epsilon^{\mu\nu\rho\sigma} X_\mu \otimes X_\nu \otimes X_\rho \otimes X_\sigma$. Inserting (3.160) and (3.158) in $\omega(X) = 0$, a straightforward calculation yields

$$\partial_\mu \phi(x) = p_\mu(x), \tag{3.161}$$

$$\partial_\mu p^\mu(x) = -m^2 \phi(x) - V'(\phi(x)), \tag{3.162}$$

and therefore precisely the field equations (3.153). Notice that the canonical formalism is manifestly Lorentz covariant, and no equal-time initial data surface has to be chosen.

A state is a 4d surface $(x, \phi(x))$ in the extended configurations space $\mathcal{C}$. It represents a set of combinations of measurements of partial observables that can be realized in Nature. The phase space Γ is the infinite-dimensional space of these states. A state determines whether or not a certain correlation (x, ϕ), or a certain set of correlations $(x_1, \phi_1) \ldots (x_n, \phi_n)$ can be observed. They can be observed if the points (x_i, ϕ_i) lie on the 4d surface that represents the state. Conversely, the observation of a certain set of correlations gives us information on the state: the surface has to pass through the observed points.

3.3.3 *The space of boundary data $\mathcal{G}$ and the Hamilton function S*

The space of boundary data $\mathcal{G}$ described in Section 3.2.5 plays a key role in quantum theory. In the finite-dimensional case $\mathcal{G}$ is the cartesian product of the extended configuration space with itself, but the same is not true in the field theoretical context, where we need an infinite number of boundary data to characterize solutions. Recall that in the finite-dimensional case $\mathcal{G}$ is the space of the possible *boundaries of a motion* in $\mathcal{C}$. In field theory, a motion is a 4d surface in $\mathcal{C}$. Its boundary is a three-dimensional surface α without boundaries, in $\mathcal{C}$. Let us therefore define $\mathcal{G}$ in field theory as a space of oriented three-dimensional surfaces α without boundaries in $\mathcal{C}$. As $\mathcal{C} = M \times T$, the boundary data α includes a 3d boundary surface σ in spacetime as well as the value φ of the field on this surface.

More precisely, let x^μ be spacetime coordinates in M, and ϕ^A coordinates in the target space. Coordinatize the 3d surface α with 3d coordinates $\vec{\tau} = (\tau^1, \tau^2, \tau^3)$. Then α is given by the functions

$$\alpha = [\sigma, \varphi], \tag{3.163}$$

$$\sigma: \quad \vec{\tau} \mapsto x^\mu(\vec{\tau}), \tag{3.164}$$

$$\varphi: \quad \vec{\tau} \mapsto \varphi^A(\vec{\tau}). \tag{3.165}$$

The functions $x^\mu(\vec{\tau})$ define the 3d surface σ without boundaries in spacetime. The functions $\varphi^A(\vec{\tau})$ define the value of the field $\phi(x)$ on this surface:

$$\phi^A(x(\vec{\tau})) \equiv \varphi^A(\vec{\tau}). \tag{3.166}$$

Say σ is the boundary of a connected region $\mathcal{R}$ of M. Then, generically, φ determines a solution $\phi(x)$ of the equations of motion in the interior $\mathcal{R}$, such that $\phi|_\sigma = \varphi$. Imagine that σ is a cylinder in Minkowski space. To determine a solution in the interior, we need the initial value of the field on the bottom of the cylinder, its final value on the top of the cylinder, as well as spatial boundary conditions on the side of the cylinder. The data α determine all these field values as well as the spacetime location of the cylinder itself. These data form the field theoretical generalization of the set $(t, q^i; t', q'^i)$ which form the argument of the Hamilton function and of the quantum propagator in finite-dimensional mechanics. Alternatively, the surface α need not be connected. For instance, it can be formed by two components, which we can view as initial and final configurations.

The Hamilton function $S[\alpha] = S[\sigma, \varphi]$ is defined as the action of the solution of the equations of motion $\phi(x)$, such that $\phi|_\sigma = \varphi$, in $\mathcal{R}$. We shall see below that $S[\alpha]$ satisfies a functional Hamilton–Jacobi equation, and can be seen as the classical limit of a quantum mechanical propagator.[10]

We can give a more formal definition of $S[\alpha]$, analogous to the definition (3.83). Let γ be the motion in $\mathcal{C}$ bounded by α. Let $\tilde{\gamma}$ be the lift of γ to Σ. That is, let $\tilde{\gamma}$ be the orbit of ω that projects down to γ. Then

$$S[\alpha] = \int_{\tilde{\gamma}} \theta. \tag{3.167}$$

Example: scalar field. For a scalar field, for instance,

$$S[\alpha] = \int_{\tilde{\gamma}} \theta = \int_{\tilde{\gamma}} (\pi d^4 x + p^\mu d\phi \wedge d^3 x_\mu) = \int_{\mathcal{R}} (\pi + p^\mu \partial_\mu \phi) \, d^4 x$$

$$= \int_{\mathcal{R}} \left(-\frac{1}{2} p^\mu p_\mu - \frac{1}{2} m^2 \phi^2 - V(\phi) + p^\mu \partial_\mu \phi \right) d^4 x \tag{3.168}$$

$$= \int_{\mathcal{R}} \left(\frac{1}{2} \partial_\mu \phi \partial^\mu \phi - \frac{1}{2} m^2 \phi^2 - V(\phi) \right) d^4 x$$

$$= \int_{\mathcal{R}} L(\phi, \partial_\mu \phi) \, d^4 x, \tag{3.169}$$

where L is the lagrangian density, and we have used the equation of motion $p_\mu = \partial_\mu \phi$.

It is not hard to compute the Hamilton function for a free scalar field, in the special case in which α is formed by the two spacelike parallel hypersurfaces $x^\mu(\vec{\tau}) = (t_1, \vec{\tau})$ and $x^\mu(\vec{\tau}) = (t_2, \vec{\tau})$ and by the values $\phi_1(\vec{x})$ and $\phi_2(\vec{x})$ of the field on these surfaces. The calculation is simplified by the fact that a free field is essentially a collection of oscillator with modes of wavelength $\vec{k}$ and frequency $\omega(\vec{k}) = \sqrt{|\vec{k}|^2 + m^2}$. Using this fact and (3.34) it is straightforward to compute the field for given boundary values and its action. This gives

$$S(\phi_1, t_1, \phi_2, t_2) = \int d^3 k \, \omega(\vec{k}) \frac{2\overline{\phi_1(\vec{k})}\phi_2(\vec{k}) - (|\phi_1|^2(\vec{k}) + |\phi_2|^2(\vec{k})) \, \cos[\omega(\vec{k})(t_1 - t_2)]}{2\sin[\omega(t_1 - t_2)]}, \tag{3.170}$$

where $\phi(\vec{k})$ are the Fourier components of $\phi(\vec{x})$.

* *The symplectic structure on $\mathcal{G}$.* As in the finite-dimensional case, we can define a symplectic structure on $\mathcal{G}$. Let s be the 3d surface in Σ that bounds $\tilde{\gamma}$. That is, $s = [x^\mu(\vec{\tau}), \varphi^A(\vec{\tau}), p_A^\mu(\vec{\tau})]$, where the momenta $p_A^\mu(\vec{\tau})$ are determined by the solution of the field equations determined by the entire α.

---

[10] $S[\alpha]$ is only defined on the regions of $\mathcal{G}$ where this solution exists, and it is multivalued where there is more than one solution.

Define a two-form on $\mathcal{G}$ as follows

$$\omega_{\mathcal{G}}[\alpha] = \int_s \omega. \tag{3.171}$$

The form $\omega_{\mathcal{G}}$ is a two-form: it is the integral of a five-form over a 3d surface. More precisely, let $\delta\alpha$ be a small variation of α. This variation can be seen as a vector field $\delta\alpha(\vec{\tau})$ defined on α. This variation determines a corresponding small variation δs, which, in turn, is a vector field $\delta s(\vec{\tau})$ over s. Then

$$\omega_{\mathcal{G}}[\alpha](\delta_1\alpha, \delta_2\alpha) = \int_{s_\alpha} \omega(\delta_1 s, \delta_2 s). \tag{3.172}$$

Thus, the five-form ω on the finite-dimensional space Σ defines the two-form $\omega_{\mathcal{G}}$ on the infinite-dimensional space $\mathcal{G}$.

Consider a small local variation $\delta\alpha$ of α. This means varying the surface α_M in Minkowski space, as well as varying the value of the field over it. Assume that this variation satisfies the field equations: that is, the variation of the field is the correct one, for the solution of the field equations determined by α. We have

$$\omega_{\mathcal{G}}[\alpha](\delta\alpha) = \int_{s_\alpha} \omega(\delta s). \tag{3.173}$$

But the variation δs is by construction along the orbit, namely in the null direction of ω and therefore the right-hand side of this equation vanishes. It follows that if $\delta\alpha$ is an infinitesimal physical motion, then

$$\omega_{\mathcal{G}}(\delta\alpha) = 0. \tag{3.174}$$

The pair $(\mathcal{G}, \omega_{\mathcal{G}})$ contains all the relevant information on the system. The null directions of $\omega_{\mathcal{G}}$ determine the variations of the 3-surface α along the physical motions. The space $\mathcal{G}$ divided by these null directions, namely the space of the orbits of these variations is the physical phase space Γ, and the $\omega_{\mathcal{G}}$, restricted to this space, is the physical symplectic two-form of the system.

Example: scalar field. Let's compute $\omega_{\mathcal{G}}$ in a slightly more explicit form for the example of the scalar field. From the definition (3.171),

$$\omega_{\mathcal{G}}[\alpha] = \int_s \omega = \int_s \mathrm{d}\pi \wedge \mathrm{d}^4 x + \mathrm{d}p^\mu \wedge \mathrm{d}\phi \wedge \mathrm{d}^3 x_\mu$$

$$= \int_s (p^\nu \mathrm{d}p_\nu + m^2 \phi \mathrm{d}\phi + V' \mathrm{d}\phi) \wedge \mathrm{d}^4 x + \mathrm{d}p^\mu \wedge \mathrm{d}\phi \wedge \mathrm{d}^3 x_\mu$$

$$= \int_{\alpha_M} \mathrm{d}^3 x_\nu \left[(p_\mu - \partial_\mu \phi)\mathrm{d}p^\mu \wedge \mathrm{d}x^\nu + (m^2\phi + V' + \partial_\mu p^\mu)\mathrm{d}\phi \wedge \mathrm{d}x^\nu + \mathrm{d}p^\nu \wedge \mathrm{d}\phi \right]$$

$$= \int_{\alpha_M} \mathrm{d}^3 x_\nu \, \mathrm{d}p^\nu \wedge \mathrm{d}\phi, \tag{3.175}$$

where we have used the x^μ coordinates themselves as integration variables, and there-fore the integrand fields are the functions of the x^μ. Notice that since the integral is on s, the p^μ in the integrand is the one given by the solution of the field equation deter-mined by the data on α. Therefore it satisfies the equations of motion (3.161)–(3.162), which we have used above. Using (3.161) again, we have

$$\omega_\mathcal{G}[\alpha] = \int_{\alpha_M} d^3x \, n_\nu \, d(\nabla^\nu \phi) \wedge d\phi. \tag{3.176}$$

In particular, if we consider variations $\delta\alpha$ that do not move the surface and such that the change of the field on the surface is $\delta\phi(x)$, we have

$$\omega_\mathcal{G}[\alpha](\delta_1\alpha, \delta_2\alpha) = \int_{\alpha_M} d^3x \, n_\nu \left(\delta_1\phi\nabla^\nu\delta_2\phi - \delta_2\phi\nabla^\nu\delta_1\phi\right). \tag{3.177}$$

This formula can be directly compared with the expression of the symplectic two-form given on the space of the solutions of the field equations in [108]. The expression is the same, but with a nuance in the interpretation: $\omega_\mathcal{G}$ is not defined on the space of the solutions of the field equations; it is defined on the space of the lagrangian data $\mathcal{G}$, and the normal derivative $n_\nu\nabla^\nu\phi$ of these data is determined by the data themselves via the field equations.

3.3.4 Hamilton–Jacobi

A Hamilton–Jacobi equation for the field theory can be written as a local equation on the boundary satisfied by the Hamilton function. I illustrate here the derivation of the Hamilton–Jacobi equation in the case of the scalar field, leaving the generalization to the interested reader. From the definition

$$S[\alpha] = \int_{\tilde\gamma} \theta = \int_{\tilde\gamma} (\pi d^4x + p^\mu d\phi \wedge d^3x_\mu) \tag{3.178}$$

we can write

$$\frac{\delta S[\alpha]}{\delta x^\mu(\vec\tau)} = \pi(\vec\tau) \, n_\mu(\vec\tau) + \epsilon_{\mu\nu\rho\sigma} \, p^\nu(\vec\tau) \, \partial_i\phi(\vec\tau) \, \partial_j x^\rho(\vec\tau) \, \partial_k x^\sigma(\vec\tau) \, \epsilon^{ijk}, \tag{3.179}$$

where

$$n_\mu(\vec\tau) = \frac{1}{3!}\epsilon_{\mu\nu\rho\sigma}\partial_1 x^\nu(\vec\tau)\partial_2 x^\rho(\vec\tau)\partial_3 x^\sigma(\vec\tau) \tag{3.180}$$

is the normal to the 3-surface σ. The momentum π depends on the full α. Contracting this equation with n^μ we obtain

$$\pi(\vec\tau) = n^\mu(\vec\tau)\frac{\delta S[\alpha]}{\delta x^\mu(\vec\tau)} + p^i(\vec\tau) \, \partial_i\phi(\vec\tau). \tag{3.181}$$

Using the equation of motion $p_\mu = \partial_\mu\phi$ this becomes

$$\pi(\vec\tau) = n^\mu(\vec\tau)\frac{\delta S[\alpha]}{\delta x^\mu(\vec\tau)} + \partial^i\phi(\vec\tau)\partial_i\phi(\vec\tau). \tag{3.182}$$

Also

$$\frac{\delta S[\alpha]}{\delta \varphi(\vec{\tau})} = p^{\mu}(\vec{\tau})n_{\mu}(\vec{\tau}).$$

(3.183)

The derivation of these two equations requires steps analogous to the ones we used to derive (3.86).

Now, from (3.155) and (3.156) we have that the scalar field dynamics is governed by the equation

$$\pi + \frac{1}{2}\left(p^{\mu}p_{\mu} + m^2\phi^2 + 2V(\phi)\right) = 0.$$

(3.184)

We split p_{μ} into its normal ($p = p^{\mu}n_{\mu}$) and tangential (p^i) components (so that $p^{\mu} = p^i\partial_i x^{\mu} + pn^{\mu}$), obtaining

$$\pi + \frac{1}{2}\left(p^2 - p^i p_i + m^2\phi^2 + 2V(\phi)\right) = 0.$$

(3.185)

Inserting (3.182) and (3.183), we obtain

$$\frac{\delta S[\alpha]}{\delta x^{\mu}(\vec{\tau})}n^{\mu}(\vec{\tau}) + \frac{1}{2}\left[\left(\frac{\delta S[\alpha]}{\delta\varphi(\vec{\tau})}\right)^2 + \partial_j\varphi(\vec{\tau})\partial^j\varphi(\vec{\tau}) + m^2\varphi^2(\vec{\tau}) + 2V(\varphi(\vec{\tau}))\right] = 0.$$

(3.186)

This is the Hamilton–Jacobi equation. Notice that the function

$$S[x^{\mu}(\vec{\tau}), \varphi(\vec{\tau})] = S[\sigma, \varphi] = S[\alpha]$$

(3.187)

is a function of the surface, not the way the surface is parametrized. Therefore it is invariant under a change of parametrization. It follows that

$$\frac{\delta S[\alpha]}{\delta x^{\mu}(\vec{\tau})}\partial_j x^{\mu}(\vec{\tau}) + \frac{\delta S[\alpha]}{\delta\varphi(\vec{\tau})}\partial_j\varphi(\vec{\tau}) = 0.$$

(3.188)

(This equation can be obtained also from the tangential component of (3.179).) The two equations (3.186) and (3.188) govern the Hamilton–Jacobi function $S[\alpha]$.

The connection with the nonrelativistic field theoretical Hamilton–Jacobi formalism is the following. We can restrict the formalism to a preferred choice of parameters $\vec{\tau}$. Choosing $\tau^j = x^j$, we obtain S in the form $S[t(\vec{x}), \phi(\vec{x})]$ and the Hamilton–Jacobi equation (3.186) becomes

$$\frac{\delta S}{\delta t(\vec{x})} + \frac{1}{2}\left[\left(\frac{\delta S[\alpha]}{\delta\phi(\vec{x})}\right)^2 + \partial_j\phi\partial^j\phi + m^2\phi^2 + 2V(\phi)\right] = 0. \quad (3.189)$$

Further restricting the surfaces to the ones of constant t gives the functional $S[t, \phi(\vec{x})]$, satisfying the Hamilton–Jacobi equation

$$\frac{\partial S}{\partial t} + \frac{1}{2} \int d^3\vec{x} \left[\left(\frac{\delta S}{\delta \phi(\vec{x})} \right)^2 + |\vec{\nabla}\phi|^2 + m^2\phi^2 + 2V(\phi) \right] = 0, \quad (3.190)$$

which is the usual nonrelativistic Hamilton–Jacobi equation

$$\frac{\partial S}{\partial t} + \mathcal{H} \left(\phi, \vec{\nabla}\phi, \frac{\delta S[\alpha]}{\delta \phi(\vec{x})} \right) = 0, \quad (3.191)$$

where $\mathcal{H}(\phi, \vec{\nabla}\phi, \partial_t\phi)$ is the nonrelativistic hamiltonian.

Canonical formulation on $\mathcal{G}$. We can write a hamiltonian density function $H(\vec{\tau})$ directly for the infinite-dimensional space $\mathcal{G}$. $H(\vec{\tau})$ is a function on the cotangent space $T^*\mathcal{G}$. We coordinatize this cotangent space with the functions $(x^\mu(\vec{\tau}), \varphi(\vec{\tau}))$ and their momenta $(\pi_\mu(\vec{\tau}), p(\vec{\tau}))$. The hamiltonian is then

$$H[x^\mu, \varphi, \pi_\mu, p](\vec{\tau}) = \pi_\mu(\vec{\tau})n^\mu(\vec{\tau}) + \frac{1}{2} \left[p^2(\vec{\tau}) + \partial_j\varphi(\vec{\tau})\partial^j\varphi(\vec{\tau}) \right.$$
$$\left. + m^2\varphi^2(\vec{\tau}) + 2V(\varphi(\vec{\tau})) \right], \quad (3.192)$$

and the Hamilton–Jacobi equation (3.186) reads

$$H \left[x^\mu, \varphi, \frac{\delta S[\alpha]}{\delta x^\mu}, \frac{\delta S[\alpha]}{\delta \varphi} \right] (\vec{\tau}) = 0. \quad (3.193)$$

If we restrict the surface $x^\mu(\vec{\tau})$ to the case $x^\mu(\vec{\tau}) = (t, \vec{\tau})$, then $H(\vec{\tau})$ becomes

$$H[x^\mu, \varphi, \pi_\mu, p](\vec{x}) = \pi_0(\vec{x}) + H_0(\vec{x}), \quad (3.194)$$

where $H_0(\vec{x})$ is the conventional nonrelativistic hamiltonian density

$$H_0[\phi, p] = \frac{1}{2} \left[p^2 + \partial_j\varphi\partial^j\varphi + m^2\varphi^2 + 2V(\varphi) \right]. \quad (3.195)$$

Physical predictions from S. The complete physical predictions of the theory can be obtained directly from the Hamilton function $S[\alpha] = S[\sigma, \varphi]$ as follows. Let $p(\vec{\tau})$ be a function on the surface σ. Define

$$F[\sigma, \varphi, p](\vec{\tau}) = \frac{\delta S[\sigma, \varphi]}{\delta \varphi(\vec{\tau})} - p(\vec{\tau}). \quad (3.196)$$

Given a closed surface σ in spacetime, we can observe field boundary values $\phi(x(\vec{\tau})) = \varphi(\vec{\tau})$ together with momenta $n^\mu \partial_\mu \phi(x(\vec{\tau})) = p(\vec{\tau})$ if and only if

$$F[\sigma, \varphi, p](\vec{\tau}) = 0. \tag{3.197}$$

This equation is equivalent to the equations of motion and expresses directly the physical content of the theory as a restriction on the partial observables that can be observed on a boundary surface.

As in the case of finite-dimensional systems, the general solution of the equations of motion can be obtained by derivations. For instance, let α be formed by two connected components that we denote $\alpha = [\sigma, \varphi]$ and $\alpha_0 = [\sigma_0, \varphi_0]$, parametrized by $\vec{\tau}$ and $\vec{\tau}_0$ respectively. Consider the equation for α

$$f[\alpha](\vec{\tau}) = \frac{\delta S[\alpha \cup \alpha_0]}{\delta \varphi_0(\vec{\tau}_0)} - p_0(\vec{\tau}_0) = 0, \tag{3.198}$$

where $p_0(\vec{\tau})$ is an arbitrary initial value momentum. This is the evolution equation that determines all surfaces α compatible with the initial data φ_0, p_0 on σ_0.

3.4 * Thermal time hypothesis

Earth lay with Sky, and after them was born Time
The wily, youngest and most terrible of her children.

Hesiod, *Theogony* [111]

In the macroscopic world, the physical variable t measured by a clock has peculiar properties. It is not easy to pinpoint these properties with precision without referring to a presupposed notion of time, but it is also difficult to deny that they exist. From the point of view developed in this book, at the fundamental level the variable t measured by a clock is on the same footing as any other partial observable. If we accept this idea, we have then to reconcile the fact that time is not a special variable at the fundamental level, with its peculiar properties at the macroscopic level. What is so special about time? An interesting possibility is that it is statistical mechanics, and therefore thermodynamics, that singles out t and gives it its special properties. I briefly illustrate this idea in this section.

The world around us is made up of systems with a large number of degrees of freedom, such as fields. We never measure the totality of these degrees of freedom. Rather, we measure certain macroscopic parameters, and make predictions on the basis of assumptions on the state of the other degrees of freedom. The viability of our choice of macroscopic parameters, and our assumptions about the state of the others, is justified a posteriori if the system of prediction works. We represent our incomplete knowledge and assumptions in terms of a statistical state ρ. The state ρ can be represented as a normalized positive function on the phase space Γ

$$\rho: \quad \Gamma \to R^+, \tag{3.199}$$

$$\int_\Gamma ds \, \rho(s) = 1. \tag{3.200}$$

$\rho(s)$ represents the assumed probability density of the state s in Γ. Then the expectation value of any observable $A : \Gamma \to R$ in the state ρ is

$$\rho[A] = \int_\Gamma ds \, A(s) \, \rho(s). \tag{3.201}$$

The fundamental postulate of statistical mechanics is that a system left free to thermalize reaches a time-independent equilibrium state that can be represented by means of the Gibbs statistical state

$$\rho_0(s) = Ne^{-\beta H_0(s)}, \tag{3.202}$$

where $\beta = 1/T$ is a constant – the inverse temperature – and H_0 is the nonrelativistic hamiltonian. Classical thermodynamics follows from this postulate. Time evolution $A_t = \alpha_t(A)$ of A is determined by (3.78). Equivalently, $A_t(s) = A(t(s))$ where $s(t)$ is the hamiltonian flow of H_0 on Γ. The correlation probability between A_t and B is given by

$$W_{AB}(t) = \rho_0[\alpha_t(A)B] = \int_\Sigma ds \, A(s(t)) \, B(s) \, e^{-\beta H_0(s)}. \tag{3.203}$$

In this chapter, we have seen that the formulas of mechanics do not single out a preferred variable, because all mechanical predictions can be obtained using the relativistic hamiltonian H which treats all variables on an equal footing instead of using the nonrelativistic hamiltonian H_0 which singles t out. Is this true also for statistical mechanics and thermodynamics? Equations (3.200)–(3.201) are meaningful also in the relativistic context, where Γ is the space of the solutions of the equations of motion. But this is not true for (3.202) and (3.203). These depend on the nonrelativistic hamiltonian. They depend on the fact that t is a variable different from the others. Equations (3.202) and (3.203) definitely single out t as a special variable. This observation indicates that the peculiar properties of the t variable have to do with statistical mechanics and thermodynamics, rather than with mechanics. With purely mechanical measurements we cannot recognize the time variable. With statistical or thermal measurements, we can.

Indeed, notice that if we try to pinpoint what is special about the variable t, we generally find features connected to thermodynamics: irreversibility, convergence to equilibrium, memory, feeling of "flow," and so on.

Indeed, there is an intriguing fact about (3.202) and (3.203). Imagine that we study a system which is in equilibrium at inverse temperature β and we do not know its nonrelativistic hamiltonian H_0. In principle, we can figure out H_0 simply by repeated microscopic measurements on copies of the system, without any need of observing time evolution. Indeed, if we find out the distribution of microstates ρ_0, then, up to an irrelevant additive constant, we have

$$H_0 = -\frac{1}{\beta} \ln \rho_0. \tag{3.204}$$

Therefore, in a statistical context we have, in principle, an operational procedure for determining which one is the time variable. First, measure ρ_0; second, compute H_0 from (3.204); third, compute the hamiltonian flow $s(t)$ of H_0 on Σ. The time variable t is the parameter of this flow. A "clock" is any measuring apparatus whose reading grows linearly with this flow. The multiplicative constant in front of H_0 just sets the unit in which time is measured. Up to this unit, we can find out which one is the time variable just by measuring ρ_0. This is in strident contrast with the purely mechanical context, where no operational procedure for singling out the time variable is available.

Now, let me come to the main observation. Imagine that we have a truly relativistic system where no partial observable is singled out as the time variable. Imagine that we make measurements on many copies of the system and find that the statistical state describing the system is given by a certain *arbitrary*[11] state ρ. *Define* the quantity

$$H_\rho = -\ln \rho. \tag{3.205}$$

Let $s(t_\rho)$ be the hamiltonian flow of H_ρ. Call t_ρ "thermal time." Call "thermal clock" any measuring device whose reading grows linearly with this flow. Given an observable A, consider the one-parameter family of observables A_{t_ρ} defined by $A_{t_\rho}(s) = A(t_\rho(s))$. Then it follows that the correlation probability between the observables A_{t_ρ} and B is given by

$$W_{AB}(t_\rho) = \int_\Sigma \mathrm{d}s \; A(t_\rho(s)) \, B(s) \, \mathrm{e}^{-H_\rho(s)}. \tag{3.206}$$

What is the difference between the physics described by (3.202)–(3.203) and that described by (3.205)–(3.206)? None! That is: *whatever* the

---

[11]For (3.205) to make sense, assume that ρ nowhere vanishes on Σ.

statistical state ρ, there exists always a variable t_ρ, measured by the thermal clock, with respect to which the system is in equilibrium and whose physics is the same as in the conventional nonrelativistic statistical case! This key observation naturally leads us to the following hypothesis.

> **The thermal time hypothesis.** In Nature, there is no preferred physical time variable t. There are no equilibrium states ρ_0 preferred a priori. Rather, all variables are equivalent: we can find the system in an arbitrary state ρ; if the system is in a state ρ, then a preferred variable is singled out by the state of the system. This variable is what we call time.

In other words, it is the statistical state that determines which variable is physical time, and not any a priori hypothetical "flow" that drives the system to a preferred statistical state. All variables are physically equivalent at the mechanical level. But if we restrict our observations to macroscopic parameters, and assume the other dynamical variables are distributed according to a statistical state ρ, then a preferred variable is singled out by this procedure. This variable has the property that correlations with respect to it are described precisely by ordinary statistical mechanics. In other words, it has precisely the properties that characterize our macroscopic time parameter.

In other words, when we say that a certain variable is "the time," we are not making a statement concerning the fundamental mechanical structure of reality.[12] Rather, we are making a statement about the statistical distribution we use to describe the macroscopic properties of the system that we describe macroscopically.

The hamiltonian H_ρ determined by a state ρ is called the thermal hamiltonian. The "thermal time hypothesis" is the idea that what we call "time" is simply the thermal time of the statistical state in which the world happens to be when described in terms of the macroscopic parameters we have chosen.

Let the system be in the mechanical microstate s. Describe it with macroscopic observables A_i. In general (but not always) there exists a statistical state ρ whose mean values give the correct predictions for the A_i, that is, $A_i(s) \sim \rho[A_i]$. Assuming it exits, ρ codes, in a sense, our ignorance of the microscopic details of the state. Intuitively, we can therefore say that the existence of time is the result of this ignorance of ours. Time is the expression of our ignorance of the microstate.

The thermal time hypothesis works surprisingly well in a number of cases. For example, if we start from a radiation-filled covariant cosmological model having no preferred time variable and write a statistical

---

[12]Time, $K\rho\acute{o}\nu o\varsigma$ comes *after* matter (Earth, $\Gamma\alpha\hat{\iota}\alpha$ and Sky, $O\grave{\upsilon}\rho\alpha\nu\acute{o}\varsigma$) also in Greek mythology. See Hesiod's quote [111] at the beginning of this section.

state representing the cosmological background radiation, then the thermal time of this state turns out to be precisely the Friedmann time [112]. Furthermore, we will see in Section 5.5.1 that this hypothesis extends in an extremely natural way to the quantum context, and even more naturally to the quantum field theoretical context where it leads also to a general abstract state-independent notion of time flow.

———

Bibliographical notes

The hamiltonian theory of systems with constraints is one of Dirac's many masterpieces. The theory is not just a technical complication of standard hamiltonian mechanics: it is a powerful generalization of mechanics which remains valid in the general-relativistic context. The title of Dirac's initial work on the subject was "Generalized Hamiltonian dynamics" [113]. The theory is synthesized in [114]. For modern accounts and developments, see [115]. For the notion of partial observable I have followed [116]. On the general structure of mechanics, I have followed [117, 118]. For a nontrivial example of relational evolution treated in detail, see [119].

The canonical treatment of field theory on finite-dimensional spaces derives from the Weyl and DeDonder's calculus of variations [110, 120]. A beautiful, comprehensive and mathematically precise discussion of covariant hamiltonian field theory is in [109], which contains complete references to the literature on the subject. See also [121].

The idea of the thermal origin of time was introduced in [112, 122] in the context of classical field theory and was independently suggested by Alain Connes. It is developed in quantum field theory (see Section 5.5.1) in [125], see also [124]. For a related Boltzmann-like approach, see [123].

4

Hamiltonian general relativity

I begin this chapter by presenting the Hamilton–Jacobi formulation of GR. This is the basis of the quantum theory.

In the remainder of the chapter I present formulations of hamiltonian GR on a finite-dimensional configuration space, along the lines illustrated at the end of the previous chapter.

This order of presentation is inverse to the logical order, which should start from the finite-dimensional configuration space of the partial observables. But I do not want to force the hurried reader to navigate through the entire chapter before finding the few simple equations that are the basis of the quantum theory.

I take the cosmological constant to be zero and ignore matter fields, leaving to the reader the generally easy exercise of adding the cosmological and matter terms to the relevant equations.

4.1 Einstein–Hamilton–Jacobi

GR can be expressed in terms of a complex field $A_a^i(\vec{\tau})$ and a 3d real momentum field $E_i^a(\vec{\tau})$, defined on a three-dimensional space σ without boundaries, satisfying the reality conditions

$$A_a^i + \overline{A_a^i} = \Gamma_a^i[E],\tag{4.1}$$

where Γ is defined below in (4.23)–(4.24). The theory is defined by the hamiltonian system

$$D_a E_i^a = 0,\tag{4.2}$$

$$E_i^a F_{ab}^i = 0,\tag{4.3}$$

$$F_{ab}^{ij} E_i^a E_j^b = 0,\tag{4.4}$$

where $F_{ab}^{ij} = \epsilon^{ij}{}_k F_{ab}^k$ (see pg. xxii) and D_a and F_{ab}^i are the covariant derivative and the curvature of A_a^i defined by

$$D_a v_i = \partial_a v_i + \epsilon_{ijk} A_a^j v^k, \tag{4.5}$$

$$F_{ab}^i = \partial_a A_b^i - \partial_b A_a^i + \epsilon^i{}_{jk} A_a^j A_b^k. \tag{4.6}$$

I sketch the derivation of these equations from the lagrangian formalism below. An indirect derivation via a finite-dimensional canonical formulation is given at the end of this chapter.

The Hamilton–Jacobi system is given in terms of the functional $S[A]$ by writing

$$E_i^a(\vec{\tau}) = \frac{\delta S[A]}{\delta A_a^i(\vec{\tau})} \tag{4.7}$$

in the hamiltonian system. The first two equations that we obtain

$$D_a \frac{\delta S[A]}{\delta A_a^i(\vec{\tau})} = 0, \qquad F_{ab}^i(\vec{\tau}) \frac{\delta S[A]}{\delta A_a^i(\vec{\tau})} = 0, \tag{4.8}$$

require that $S[A]$ is invariant under 3d diffeomorphisms (diffs) and local $SO(3)$ transformations, as I will show in a moment. The last reads

$$F_{ab}^{ij}(\vec{\tau}) \frac{\delta S[A]}{\delta A_a^i(\vec{\tau})} \frac{\delta S[A]}{\delta A_b^j(\vec{\tau})} = 0. \tag{4.9}$$

This is the Hamilton–Jacobi equation of GR. It defines the dynamics of GR.

Smeared form. Equivalently, we can integrate equations (4.2)–(4.4) against suitable "test" functions, and demand the integral to vanish for any such function. For the first two, we get

$$G[\lambda] = -\int d^3\tau \lambda^i \, D_a E_i^a = \int d^3\tau D_a \lambda^i E_i^a = 0, \tag{4.10}$$

$$C[f] = -\int d^3\tau f^a F_{ab}^i E_i^b = 0. \tag{4.11}$$

The quantities $D_a \lambda^i$ and $f^a F_{ab}^i$ that appear in these equations are the infinitesimal transformations of the connection under an internal gauge transformation with generator $\lambda^i(\vec{\tau})$ and under (the combination of an internal gauge transformation and) an infinitesimal diffeomorphism generated by the vector field $f^a(\vec{\tau})$

$$\delta_\lambda A_a^i = D_a \lambda^i, \qquad \delta_f A_a^i = f^b F_{ab}^i. \tag{4.12}$$

Therefore the smeared form of (4.8) reads

$$\int d^3\tau \; \delta_\lambda A^i_a(\vec{\tau}) \; \frac{\delta S[A]}{\delta A^i_a(\vec{\tau})} = 0, \qquad \int d^3\tau \; \delta_f A^i_a(\vec{\tau}) \; \frac{\delta S[A]}{\delta A^i_a(\vec{\tau})} = 0, \quad (4.13)$$

which is the requirement that $S[A]$ is invariant under gauge and diffeo-morphisms.

The quantity (4.4), on the other hand, is a density of weight two. To be able to integrate it against a scalar quantity and get a well-defined result, we need a density of weight one. This can be obtained by dividing the hamiltonian by the square roote of the determinant of E, exploiting the freedom in the definition of the hamiltonian. The Poisson bracket, derived below in (4.25), between the volume

$$\mathbf{V} = \int d^3x \; \sqrt{|\det E(x)|} \qquad (4.14)$$

and the connection is

$$\{\mathbf{V}, A^i_a(x)\} = (8\pi i G) \; \frac{E^b_j(x) E^c_k(x) \epsilon_{abc} \epsilon^{ijk}}{4\sqrt{|\det E(x)|}}. \qquad (4.15)$$

Using this, we can write (4.4) in the form

$$H[N] = \int N \; \mathrm{tr}(F \wedge \{\mathbf{V}, A\}) = 0. \qquad (4.16)$$

This form of the hamiltonian will prove convenient in the quantum theory. Equations (4.10), (4.11), and (4.16) define GR.

4.1.1 *3d fields.* "The length of the electric field is the area"

What is the relation between the 4d fields used in Chapter 2 and the 3d fields used above? Consider a solution $(e^I_\mu(x), A^i_\mu(x))$ of the Einstein equations (2.21). Choose a 3d surface $\sigma : \vec{\tau} = (\tau^a) \mapsto x^\mu(\vec{\tau})$ without boundaries in the coordinate space. The four-dimensional forms A^i (the selfdual connection, defined in (2.19)), Σ^i (the 4d Plebanski two-form, defined in (2.23)) and e^I (the gravitational field, introduced in (2.1)) induce the three-dimensional forms

$$A^i(\vec{\tau}) = A^i_a(\vec{\tau}) \; d\tau^a, \qquad (4.17)$$

$$\Sigma^i(\vec{\tau}) = \Sigma^i_{ab}(\vec{\tau}) \; d\tau^a \wedge d\tau^b, \qquad (4.18)$$

$$e^I(\vec{\tau}) = e^I_a(\vec{\tau}) \; d\tau^a \qquad (4.19)$$

on σ. The 3d field E is defined as the vector density associated to Σ^i, that is

$$E^{ai}(\vec{\tau}) = \epsilon^{abc} \; \Sigma^i_{bc}(\vec{\tau}). \qquad (4.20)$$

Let's write $e^I(\vec{\tau}) = (e^0(\vec{\tau}), e^i(\vec{\tau}))$. Choose a gauge in which

$$e^0(\vec{\tau}) = 0. \tag{4.21}$$

(The extension of the formalism to a more general gauge deserves to be investigated. See [127].) It is easy to see that in this gauge $E_i^a(\vec{\tau})$ is real and

$$E_i^a(\vec{\tau}) = \frac{1}{2}\, \epsilon_{ijk}\, \epsilon^{abc}\, e_b^j(\vec{\tau})\, e_c^k(\vec{\tau}). \tag{4.22}$$

The connection $\Gamma^i[E](\vec{\tau}) = \epsilon^i{}_j{}^k \Gamma^j{}_k[E](\vec{\tau})$ used in (4.1) is defined by

$$\mathrm{d}e^i + \Gamma^i{}_j[E] \wedge e^j = 0 \tag{4.23}$$

(this is the first Cartan structure equation for σ) which is solved by

$$\Gamma_{ak}^j = \frac{1}{2}\, e_k^b(\partial_a e_b^j - \partial_b e_a^j + e^{cj} e_{al}\partial_b e_c^l). \tag{4.24}$$

That is, it is the spin connection of the triad e_a^i. It is also easy to verify that in this gauge the two quantities $A_a^i(\vec{\tau})$ and $E_i^a(\vec{\tau})$, defined by (4.17) and (4.20), satisfy the "reality condition" (4.1).

The quantity $E_i^a(\vec{\tau})$ is ($8\pi i G$ times) the momentum conjugate to $A_a^i(\vec{\tau})$. Hence, we can write immediately the Poisson brackets

$$\{A_a^i(\vec{\tau}), E_j^b(\vec{\tau}')\} = (8\pi i G)\, \delta_a^b \delta_j^i \delta^3(\vec{\tau}, \vec{\tau}'). \tag{4.25}$$

In Maxwell and Yang–Mills theories, the momentum conjugate to the three-dimensional connection A is called electric field. The field E is therefore called the gravitational electric field. In the gauge (4.21) we are considering, E is determined just by $e_a^i(\tau)$, the triad field of σ. Equation (4.22) shows that E is the inverse matrix of the triad $e_a^i(\tau)$ multiplied by its determinant

$$E^{ai} = (\det e)e^{ai}. \tag{4.26}$$

I sketch here the derivation of the basic equations of the hamiltonian formalism, namely the Poisson brackets (4.25) and the constraint system (4.2)–(4.4). For a detailed discussion of this derivation, see, for instance, I [2, 9, 20, 126]. An indirect derivation via a finite-dimensional canonical formulation is given at the end of this chapter. We can start, for instance, from the action (2.27), without the cosmological constant, and write it as follows:

$$S[\Sigma, A] = \frac{-i}{16\pi G} \int \Sigma_i \wedge F^i = \frac{-i}{16\pi G} \int \Sigma_{i\mu\nu} F^i_{\rho\sigma} \epsilon^{\mu\nu\rho\sigma} \mathrm{d}^4 x$$

$$= \frac{-i}{8\pi G} \int \left(\Sigma_{iab} F^i_{c0} + \Sigma_{i0a} F^i_{bc} \right) \epsilon^{abc} \mathrm{d}^4 x$$

$$= \frac{-i}{8\pi G} \int \left(E^c_i (\partial_0 A^i_c - \partial_c A^i_0 + \epsilon^i_{jk} A^j_0 A^k_c) + P_{iIJ} e^J_a e^J_0 F^i_{bc} \epsilon^{abc} \right) \mathrm{d}^4 x$$

$$= \frac{-i}{8\pi G} \int \left(E^c_i \dot{A}^i_c + A^i_0 \mathrm{D}_c E^c_i + \frac{1}{2} (\epsilon^i_{jk} e^j_a e^k_0 + i e^0_0 e^i_a) F_{ibc} \epsilon^{abc} \right) \mathrm{d}^4 x$$

$$= \frac{-i}{8\pi G} \int \left(E^c_i \dot{A}^i_c + \lambda^i_0 (\mathrm{D}_c E^c_i) + \lambda^b (E^a_i F^i_{ab}) + \lambda (E^a_j E^b_k F^{jk}_{ab}) \right) \mathrm{d}^4 x. \tag{4.27}$$

The dot over A indicates time derivative. I have used the gauge condition $e^0_i = 0$, and the Lagrange multipliers are multiples of the nondynamical variables A^i_0, e^0_0, e^i_0. The first term shows that $E^c_i / 8\pi G$ is the momentum conjugate to A^i_c; varying with respect to the Lagrange multipliers yields the constraint system (4.2)–(4.4).

The geometry of the three surface. In Section 2.1.4, we saw that the gravitational field has a metric interpretation. The metric structure inherited by σ depends on the gravitational electric field E. In particular, consider a two-dimensional surface $S : \sigma = (\sigma^1, \sigma^2) \mapsto \vec{\tau}(\sigma^i)$ embedded in the three-dimensional surface σ. What is the area of S? From the definition of the area, equation (2.70), we have, in a few steps,

$$\mathbf{A}(S) = \int_S \mathrm{d}^2 \sigma \, |E|. \tag{4.28}$$

Here the norm is defined by $|v| = \sqrt{\delta^{ij} v_i v_j}$ and

$$E_i(\sigma) = E^a_i(\vec{\tau}(\sigma)) \, n_a(\sigma), \tag{4.29}$$

the normal to the surface being defined by

$$n_a(\sigma) = \epsilon_{abc} \frac{\partial \tau^b(\sigma)}{\partial \sigma^1} \frac{\partial \tau^c(\sigma)}{\partial \sigma^2}. \tag{4.30}$$

Equation (4.28) can be interpreted as the surface integral of the norm of the two-form

$$E_i = E^a_i \, \epsilon_{abc} \, \mathrm{d}x^b \wedge \mathrm{d}x^c \tag{4.31}$$

and written

$$\mathbf{A}(S) = \int_S |E|. \tag{4.32}$$

Thus E_i^a, or, more precisely its norm, or "length" $|E|$, defines the area element. We could therefore say that in gravity *"the length of the electric field is the area,"* or, more precisely, the area of a surface is the flux of (the norm of) the gravitational electric field across the surface.

Using (2.73), a similar calculation gives the volume of a 3d region R

$$\mathbf{V}(R) = \int_R \mathrm{d}^3\tau \ \sqrt{|\det E|}. \tag{4.33}$$

If we know the area of any surface and the volume of any region, we know the geometry.

These expressions for area and volume in terms of the gravitational electric field E play a major role in quantum gravity. The corresponding quantum operators have a discrete spectrum, their eigenstates are known and determine a convenient basis in the quantum state space.

For later use, notice that $\det E = \det\left((\det e)e^{-1}\right) = (\det e)^3 (\det e)^{-1} = (\det e)^2$. Hence $\sqrt{n \cdot n} = |\det e| = \sqrt{|\det E|}$, where n is defined in (4.30).

Phase space, states, and relation with the Einstein equations. A *state* of GR is an equivalence class of 4d field configurations $e_\mu^I(x)$ solving the Einstein equations, under the two gauge transformations (2.123) and (2.124). The space Γ of these equivalence classes is the phase space of GR.

Given one solution of the Einstein equations, consider a 3d surface σ without boundaries in coordinate space. Let $A_a^i(\vec{\tau})$ be the connection on σ induced by the 4d selfdual connection $A_\mu^i(x)$. A state determines a family of possible 3d fields $A_a^i(\vec{\tau})$, called compatible with the state, obtained by changing the representative in the equivalence class of solutions, or, equivalently, changing σ. A family of 3d fields $A_a^i(\vec{\tau})$ compatible with a state can be obtained, in principle, from a solution of the Hamilton–Jacobi system, as follows.

In general, to solve the system we need a solution $S[A, \alpha]$ of the Hamilton–Jacobi system depending on a sufficiently large number α_n of parameters. A state is then determined by constants α_n and β_n as follows. The equation

$$F[A, \alpha] = \frac{\partial S[A, \alpha]}{\partial \alpha_n} - \beta_n = 0 \tag{4.34}$$

determines the $A_a^i(\vec{\tau})$ compatible with the state. In this sense, a solution $S[A, \alpha]$ of the HJ equation (4.9) contains the solution of the Einstein equations. In what follows I focus on the particular solution of the Hamilton–Jacobi system provided by the Hamilton function.

4.1.2 Hamilton function of GR and its physical meaning

A preferred solution of the Hamilton–Jacobi equation is the Hamilton function $S[A]$. This is defined as the value of the action of the region $\mathcal{R}$ bounded by $\sigma = \partial\mathcal{R}$, computed on a solution of the field equations determined by the boundary value A.

A boundary value A on σ determines a solution $(e^I_\mu(x), A^i_\mu(x))$ of the Einstein equations in the region $\mathcal{R}$. In turn, this solution induces on σ the 3d field $E[A]$. The Hamilton function satisfies

$$\frac{\delta S[A]}{\delta A^i_a(\vec{\tau})} = E^a_i(\vec{\tau})[A]. \tag{4.35}$$

Notice that $E^a_i(\vec{\tau})[A]$ is the value of $E^a_i(\vec{\tau})$ which is determined via the Einstein equations, by the value of A on the *entire* surface σ. Define the functional

$$F[A, E](\vec{\tau}) = \frac{\delta S[A]}{\delta A^i_a(\vec{\tau})} - E^a_i(\vec{\tau}); \tag{4.36}$$

then the equation

$$F[A, E] = 0 \tag{4.37}$$

is equivalent to the Einstein equations. It expresses the conditions that the Einstein equations put on the possibility of having fields A and E on a 3d surface.

This can be viewed as an evolution problem in the special case in which we take σ to be formed by two connected components, σ_{in} and σ_{out}, bounding a single connected region $\mathcal{R}$. For instance, σ_{in} and σ_{out} could be two spacelike surfaces of a spatially closed universe. In this case a solution is determined by the components $[A_{\text{in}}, A_{\text{out}}]$ of the connection on σ_{in} and σ_{out}, and also by the components $[A_{\text{in}}, E_{\text{in}}]$ of the connection and electric field on σ_{in} alone. We can write

$$F[A_{\text{out}}, A_{\text{in}}, E_{\text{in}}](\vec{\tau}) = \frac{\delta S[A_{\text{in}}, A_{\text{out}}]}{\delta A_{\text{in}}{}^i_a(\vec{\tau})} - E_{\text{in}}{}^a_i(\vec{\tau}) = 0. \tag{4.38}$$

Taking A_{out} as the unknown and A_{in} and E_{in} as data, this equation gives the general solution of the Einstein equations: for fixed A_{in} and E_{in}, it is solved by all 3d connections A_{out} on σ that are compatible with a solution bounded by a 3d surface with "initial conditions" A_{in} and E_{in}. That is, (4.38) determines all fields A_{out} that can "evolve" from the "initial conditions" A_{in} and E_{in}. Therefore the Hamilton function $S[A]$ contains the full solution of the Einstein equations. $S[A]$ expresses the full dynamics of GR.

As we shall see, the full dynamics of quantum GR is contained in the corresponding quantum propagator $W[A]$. To the first relevant order in $\hbar$, $W[A]$ will be related to $e^{-\frac{i}{\hbar}S[A]}$.

I have made no request that the 3d surface σ be spacelike. In particular, I have avoided the issue of whether arbitrary boundary values A admit or determine an interpolating solution. In general, the function $S[A]$ will be defined only on a region, and can be multivalued in some other region. These issues are important and I refer the interested reader to [109] and references therein for literature on the topic. On the other hand, I think that the insistence on spacelike surfaces might be more tied to our prerelativistic thinking habits than to their relations with Cauchy problems. In view of the construction of the quantum theory, these problems can perhaps be postponed. If needed, the requirement that the 3d surface is spacelike can be implemented as a restriction on the momentum E.

Experiments. Suppose we knew explicitly the Hamilton function $S[A]$. How could we compare the theory with experience? The answer is simple. We should measure the 3d fields A and E on a closed 3-surface σ. The theory predicts that the only fields (A, E) we could measure are the ones that satisfy (4.36) and (4.37). Therefore the theory determines which 3d fields could be measured and which could not. In turn, this determines restrictions (namely predictions) on any other quantity depending on these fields. Several important observations are in order.

First, the prediction is local in the sense that it regards a finite region of spacetime. Observables that require the full spacetime or the full space to be observed are not realistic.

Second and most important: where is the surface σ located? Which surface σ should we consider? The remarkable answer is: it doesn't matter. This is a key point in the interpretation of GR and should be understood in detail.

Consider a concrete experimental situation. Consider for instance a scattering experiment in a particle accelerator, or the propagation and reception of waves (electromagnetic or gravitational). In a nonrelativistic situation, say on Minkowski spacetime, we can view the situation as follows. We have a certain number of objects and detectors located in certain known positions of spacetime. We measure the initial, or incoming, data. We measure the final, or outgoing, data. Furthermore, we specify spatial boundary values (that forbid, for instance, spurious incoming radiation). The initial, final and boundary values of the fields can be represented by the value of the fields on a compact 3d surface σ. These data, however, are not sufficient to make theoretical predictions: we also need to know the location of σ in spacetime. To fix the ideas, say that σ is a cylinder in Minkowski space. The height of the cylinder, for instance, is the time lapse between the beginning and the end of the experiment.

Notice that the only relevant aspects of the location of objects, apparatus and detectors are their relative distances and time lapses. Therefore, the only relevant aspect of the location of σ is the value of the metric on the surface and in its interior. Indeed, if we displace the surface (that is, the full experiment) in such a way that the geometry of the experiment remains the same, we expect that the outcome will not change. Since the geometry of the interior is dictated (on Minkowski) by the geometry of σ, we actually need to know only the geometry of σ. It is this geometry that determines the relative distances and time lapses between emissions and detections. Thus, the full data that we need in a prerelativistic situation are:

(i) the value of the dynamical fields on σ; and

(ii) the geometry of σ.

Consider now the general-relativistic situation. The same data as above are needed, but now the geometry of σ is determined by the value of the dynamical fields on σ, because the geometry is determined by the gravitational field! Therefore the data that we need is

(i) the value of the dynamical fields on σ,

and nothing else!

The "location" of σ in the coordinate manifold is irrelevant, because it only reflects the arbitrary choice of coordinatization of spacetime. In other words, the distances and the time lapses among the detectors are precisely part of the boundary data (A, E) on σ. For instance, if σ is a cylinder, the time lapse between the initial and final measurement is precisely coded in the value of the gravitational field on the vertical (timelike) side of the cylinder. Asking what happens after a longer time means nothing but asking what happens for larger values of E on the side of the cylinder.

4.2 Euclidean GR and real connection

4.2.1 Euclidean GR

In this section I describe a field theory different from GR, but which plays an important role in quantum gravity. This is often called "euclidean GR." Usual physical GR is then denoted "lorentzian," to emphasize its distinction from euclidean GR. Euclidean GR can be defined by the same equations as GR, for instance the action (2.13), with the only difference that indices $I, J, \ldots$ in the internal space are raised and lowered with the euclidean metric δ_{IJ} instead of the Minkowski metric η_{IJ}. Accordingly, the euclidean spin connection ω is an $SO(4)$ connection, instead of an $SO(3,1)$ connection.

It is still convenient to define the selfdual connection A as in (2.19), but the appropriate selfdual projector P is now defined without the imaginary factor, that is

$$A^i = \omega^i + \omega^{0i}. \tag{4.39}$$

Therefore the selfdual connection A is *real* in the euclidean case. The absence of the imaginary factor gives immediately the Poisson brackets

$$\{A_a^i(\vec{\tau}), E_j^b(\vec{\tau}')\} = (8\pi G)\ \delta_a^b \delta_j^i \delta^3(\vec{\tau}, \vec{\tau}'), \tag{4.40}$$

instead of (4.25).

There is an important difference between the lorentzian and euclidean cases. In the euclidean case, the connection lives in the $so(4)$ algebra. This algebra decomposes as $so(4) = so(3) \oplus so(3)$. The real connection (4.39) is simply one of the two components. Therefore (4.39) has *half* the information of ω.

In the lorentzian case, on the other hand, the Lorentz algebra $so(3,1)$ does not decompose at all. However, its complexification $so(3,1;C)$ decomposes as $so(3,1;C) = so(3;C) \oplus so(3;C)$. A real ω determines two complex components which are complex conjugate to each other, and each component contains the same information as ω itself. In this case, indeed, the connection (2.19) has three complex components, which is precisely the *same* information as the six real components of ω.

Remarkably, the canonical formalism for the euclidean theory parallels completely the one for the lorentzian theory. The theory is defined by the same Hamilton–Jacobi equations (4.2)–(4.4), with the only difference that (in the gauge (4.21)), the reality conditions (4.1) are replaced by

$$E^{ai} - \overline{E^{ai}} = 0, \qquad A_a^i - \overline{A_a^i} = 0. \tag{4.41}$$

The world is described by *lorentzian* GR, not by euclidean GR. Why then is *euclidean* GR useful at all? Because euclidean GR plays a role in the search for a physical quantum theory of gravity in several ways. These will be discussed in more detail in the second part of the book, but it is appropriate to anticipate some of these reasons here.

First, the key difficulty of quantum gravity is to understand how to formulate a nontrivial generally covariant quantum field theory. Euclidean GR is an example of a nontrivial generally covariant field theory which is simpler than lorentzian GR, because the reality conditions are simpler. Therefore, a complete and consistent formulation of euclidean quantum GR is not yet a quantum theory of gravity, but is probably a major step in that direction. Euclidean GR is a highly nontrivial model of the true theory.

Second, it is well known that the euclidean version of flat-space quantum field theories is strictly connected to the physical lorentzian version. Under wide assumptions, one can prove that physical n-point functions are analytical continuations of the ones of the euclidean theory. Naively, one can simply Wick-rotate the time coordinate in the imaginary plane. More precisely, solid theorems from axiomatic quantum field theory assure us that Wightman distributions are indeed the analytic continuation of the moments of an euclidean process (the Schwinger functions), under very general hypotheses. Defining the euclidean quantum field theory is therefore equivalent to defining the physical theory. In fact, calculations are routinely performed in the euclidean region in standard quantum field theory. We cannot assume naively that the same remains true in quantum gravity. There is no Wick rotation to consider (recall the coordinate t is irrelevant for observable amplitudes anyway!), and we are outside the hypotheses of the axiomatic approach. Therefore we cannot, as we do on flat space, content ourself to define the euclidean quantum field theory and lazily be sure that a consistent physical theory will follow.

Still, the very strict connection between the euclidean and the lorentzian theory that exists on flat space strongly suggests that some connection between euclidean and lorentzian quantum GR is likely to exist. Stephen Hawking, in particular, has explored the hypothesis that physical quantum gravity could be directly defined in terms of the quantization of the euclidean theory. There are various indications for that. First, the formal functional path integral of the euclidean theory solves the Wheeler–DeWitt equation for the lorentzian theory as well. Second, there is a standard technique for obtaining the vacuum of a quantum theory by propagating for an infinite euclidean time; the adaptation of this idea to gravity led Jim Hartle and Stephen Hawking to the idea that a quantum gravitational "vacuum" is obtained from propagation in imaginary time, or, equivalently, from the quantum euclidean theory.

Finally, as I show in the next section, the lorentzian theory admits a formulation that has the same kinematics as the euclidean theory. It is therefore reasonable to expect that the kinematical features of the two theories are the same, and therefore kinematical aspects of the physical theory can be studied in the euclidean context.

4.2.2 Lorentzian GR with a real connection

Let us return to lorentzian GR. In this context, define the quantity

$$A^i = \omega^i + \omega^{0i} \qquad (4.42)$$

precisely as in (4.39). This quantity does not transform as a connection under a local Lorentz transformation (as it does in the euclidean case), but it is still a well-defined field. If we fix the gauge (4.21), then the reduced local internal gauge invariance is $SO(3)$ and A defined in (4.42) transforms as a connection under $SO(3)$ transformations. For this reason, it is denoted the "real connection" of the lorentzian theory.

Remarkably, we can take the real connection, or more precisely, its three-dimensional restriction to the boundary surface, as a canonical coordinate. Lorentzian GR, in other words, can be expressed in terms of a real $SO(3)$ connection. The reality conditions are trivial. The only difference with respect to the euclidean theory is the form of the hamiltonian, which acquires another, more complicated term with respect to (4.4):

$$H = (F_{ab}^{ij} + 2K_{[a}^i K_{b]}^j)\, E_i^a E_j^b, \qquad (4.43)$$

where $K_a^i = A_a^i - \Gamma_a^i[E]$. (See for instance [20].) The connection (4.42) and the hamiltonian (4.43) provide a second hamiltonian formalism for GR, alternative to the one described at the beginning of this chapter.

4.2.3 Barbero connection and Immirzi parameter

Finally, there is a third possible formalism for lorentzian GR. It consists in using the connection

$$A^i = \omega^i + \gamma\omega^{0i}, \tag{4.44}$$

where γ is an arbitrary complex parameter. This is called the Barbero connection: it derives naturally from the use of the Holst action (see Sec. 2.1.1.) The case $\gamma = i$ gives the selfdual connection. When γ is real, it is called the Immirzi parameter. In this case the reality conditions are still trivial (that is, $\overline{A} = A$) and the hamiltonian is a small modification of (4.43) (see [20]),

$$H = (F_{ab}^{ij} + (\gamma^2 + 1)K_{[a}^i K_{b]}^j)\, E_i^a E_j^b. \tag{4.45}$$

We will use this formalism in the quantum theory.

Since γ scales the term ω^{0i}, which is the one that has nonvanishing Poisson brackets with E, it is easy to see that the Poisson brackets between the Barbero connection and the electric field are

$$\{A_a^i(\vec{x}), E_j^b(\vec{y})\} = (8\pi\gamma G)\, \delta_a^b \delta_j^i \delta^3(x, y). \tag{4.46}$$

The fact that γ can be arbitrary is important because, as we shall see, the quantum theories obtained starting with different values of γ lead to *different* physical predictions. That is, in pure gravity γ has no effect in the classical theory but has an effect in the quantum theory. (In the presence of minimally coupled fermions, γ appears in the equation of motion [128].) Presumably, the presence of this parameter reflects a one-parameter quantization ambiguity of the theory: γ is a parameter of the quantum theory that is absent in the classical theory, such as, for instance, the θ parameter of the QCD θ-vacua. In fact, γ can also be introduced as the constant in front of a topological term added to the action, precisely as the θ parameter in QCD. Such terms do not affect the classical equations of motion but affect the quantum theory.

As we shall see in Chapter 8, γ enters in several key predictions of the quantum theory. In particular, it enters in the computation of the black-hole entropy. Comparing the black-hole entropy with the one determined thermodynamically then determines γ. A calculation along these lines, sketched in Chapter 8, suggests the value

$$\gamma \approx 0.2375. \tag{4.47}$$

It has also been repeatedly suggested that γ may determine the relation between the bare and renormalized Newton constant. Nevertheless, the physical interpretation of this parameter is not yet clear.

4.3 * Hamiltonian GR

I give here a formulation of canonical GR on a finite-dimensional configuration space, along the lines described in Section 3.3.2.

4.3.1 Version 1: real SO(3,1) connection

Let T be the space on which the fields e and ω take value. This is a $(16 + 24)$-dimensional space with coordinates $(e_\mu^I, \omega_\mu^{IJ})$. Let $\Sigma = M \times T$ be the $(4 + 16 + 24)$-dimensional space with coordinates $(x^\mu, e_\mu^I, \omega_\mu^{IJ})$. Consider the four-form

$$\theta = \epsilon_{IJKL} \, e_\mu^I \, e_\nu^J \, \mathrm{D}\omega_\rho^{KL} \wedge \mathrm{d}x^\mu \wedge \mathrm{d}x^\nu \wedge \mathrm{d}x^\rho \qquad (4.48)$$

defined on this space. Here, the covariant differential D is defined by

$$\mathrm{D}\omega_\rho^{KL} = \mathrm{d}\omega_\rho^{KL} + \omega_{\sigma I}^{K} \, \omega_\rho^{IL} \, \mathrm{d}x^\sigma. \qquad (4.49)$$

This structure defines GR, as follows. Consider a four-dimensional surface $\tilde\gamma$ in Σ. Recall from Section 3.3.2 that we say that $\tilde\gamma$ is an orbit of ω if the quadritangent X to the orbit is in the kernel of the five-form $\omega = \mathrm{d}\theta$:

$$\mathrm{d}\theta(X) = 0. \qquad (4.50)$$

The orbits of ω are the solutions of the Einstein equations. If we use the x as coordinates on the $\tilde\gamma$, then $\tilde\gamma$ is represented by

$$\tilde\gamma = (x^\mu, e_\mu^I(x), \omega_\mu^{IJ}(x)). \qquad (4.51)$$

If $\tilde\gamma$ is an orbit of ω then the functions $e_\mu^I(x), \omega_\mu^{IJ}(x)$ solve the Einstein equations. The demonstration is a straightforward calculation along the lines sketched for the scalar field example in Section 3.3.2.

4.3.2 Version 2: complex SO(3) connection

Consider the space Σ with coordinates $(x^\mu, A_\mu^i, e_\mu^I)$ where A_μ^i is complex and e_μ^I is real. Define the gauge-covariant differential acting on all quantities with internal i indices as

$$\mathrm{D}v^i = \mathrm{d}v^i + \epsilon_{jk}^i A_\mu^j v^k \mathrm{d}x^\mu; \qquad (4.52)$$

and
$$DA^i_\mu = dA^i_\mu + \epsilon^i_{jk}A^j_\nu A^k_\mu dx^\nu. \tag{4.53}$$

GR is defined by the four-form

$$\theta = P_{IJi}\, e^I \wedge e^J \wedge DA^i, \tag{4.54}$$

where P^i_{IJ} is the selfdual projector defined in (2.17). Indeed, the orbits $(x^\mu, A^i_\mu(x^\mu), e^I_\mu(x^\mu))$ of $\omega = d\theta$ satisfy the Einstein equations, in the form

$$e^I \wedge (de_J + P_{JKi}\, A^i \wedge e^K) = 0, \tag{4.55}$$

$$P_{IJi} \wedge e^I \wedge F^i = 0, \tag{4.56}$$

where F^i is the curvature of A^i. The calculation is straightforward.

4.3.3 Configuration space and hamiltonian

Above I have defined canonical GR directly as a presymplectic (Σ, θ) system. This form can be derived from a configuration space and a hamiltonian, namely from the $(\mathcal{C}, H)$ formalism described in Section 3.3.2 as follows.

Consider the finite-dimensional space $\tilde{\mathcal{C}}$ with coordinates (x^μ, A^i_μ). Here, A^i_μ is a complex matrix. Assuming immediately (3.146), the corresponding space Ω has coordinates $(x^\mu, A^i_\mu, \pi, p^{\mu\nu}_i)$ and carries the canonical four-form

$$\tilde{\theta} = \pi d^4 x + p^{\mu\nu}_i dA^i_\nu \wedge d^3 x_\mu. \tag{4.57}$$

Using D, the canonical form (4.57) reads

$$\tilde{\theta} = p d^4 x + p^{\mu\nu}_i DA^i_\mu \wedge d^3 x_\nu, \tag{4.58}$$

where $p = \pi - p^{\mu\nu}_i A^j_\nu A^k_\mu \epsilon^i_{jk}$. Also, define

$$E^i_{\mu\nu} = \epsilon_{\mu\nu\rho\sigma}\, \delta^{ij} p^{\rho\sigma}_j \tag{4.59}$$

and the forms $A^i = A^i_\mu dx^\mu, DA^i = dA^i_\mu \wedge dx^\mu + A^j_\nu A^k_\mu \epsilon^i_{jk} dx^\nu \wedge dx^\nu, E^i = E^i_{\mu\nu} dx^\mu \wedge dx^\nu$, and so on, on Ω.

GR is defined by the hamiltonian system

$$p = 0, \tag{4.60}$$

$$p^{\mu\nu}_i + p^{\nu\mu}_i = 0, \tag{4.61}$$

$$\bar{E}^i \wedge E^j = 0, \tag{4.62}$$

$$(\delta_{ik}\delta_{jl} - \frac{1}{3}\delta_{ij}\delta_{kl})E^i \wedge E^j = 0. \tag{4.63}$$

The key point is that the constraints (4.62), (4.63) imply that there exists a real four by four matrix e^I_μ, where $I = 0, 1, 2, 3$, such that $E^i_{\mu\nu}$ is the selfdual part of $e^I_\mu e^J_\nu$. In fact, it is easy to check that (4.62) and (4.63) are solved by

$$E^i = P^i_{IJ} \, e^I \wedge e^J, \qquad (4.64)$$

and the counting of degrees of freedom indicates that this is the unique solution. Therefore, we can use the coordinates $(x^\mu, A^i_\mu, e^I_\mu)$ on the constraint surface Σ (where A^i_μ is complex and e^I_μ is real) and the induced canonical four-form is (4.54). Thus, we recover the above (Σ, θ) structure.

4.3.4 Derivation of the Hamilton–Jacobi formalism

Let α be a three-dimensional surface in $\tilde{\mathcal{C}}$. Thus $\alpha = [x^\mu(\vec{\tau}), A^i_\mu(\vec{\tau})]$, where $\vec{\tau} = (\tau^1, \tau^2, \tau^3) = (\tau^a)$. Define the functional

$$S[\alpha] = \int_{\tilde{\gamma}} \theta, \qquad (4.65)$$

as in (3.167). That is, $\tilde{\gamma}$ is the four-dimensional surface in Σ which is an orbit of $d\theta$, and therefore a solution of the field equations, and is such that the projection of its boundary to $\tilde{\mathcal{C}}$ is α. From the definition (4.54),

$$\frac{\delta S[\alpha]}{\delta A^i_\mu(\vec{\tau})} = P_{iIJ} \, \epsilon^{\mu\nu\rho\sigma} e^J_\rho(\vec{\tau}) e^I_\sigma(\vec{\tau}) n_\nu(\vec{\tau}), \qquad (4.66)$$

where n_ν is defined in (3.180). Since from this equation we have immediately

$$n_\mu(\vec{\tau}) \frac{\delta S[\alpha]}{\delta A^i_\mu(\vec{\tau})} = 0, \qquad (4.67)$$

it follows that the dependence of $S[\alpha]$ on $A^i_\mu(\vec{\tau})$ is only through the restriction of $A^i(\vec{\tau})$ to the 3-surface α_M, that is, only through the components

$$A^i_a(\vec{\tau}) = \partial_a x^\mu(\vec{\tau}) A^i_\mu(\vec{\tau}). \qquad (4.68)$$

Thus $S[\alpha] = S[x^\mu(\vec{\tau}), A^i_a(\vec{\tau})]$ and

$$\frac{\delta S[\alpha]}{\delta A^i_a(\vec{\tau})} = P_{iJK} \, \epsilon^{abc} \, \partial_b x^\rho(\vec{\tau}) \partial_c x^\sigma(\vec{\tau}) e^J_\rho(\vec{\tau}) e^K_\sigma(\vec{\tau}) n_\nu(\vec{\tau}) \equiv E^a_i(\vec{\tau}). \quad (4.69)$$

Therefore E^a_i is the conjugate momentum to the connection A^i_a. Notice that E^a_i is the dual of the restriction to the boundary surface σ of the Plebanski two-form $\Sigma^i = \Sigma^i_{\mu\nu} dx^\mu \wedge dx^\nu$ defined in (2.23). Assume for simplicity that the boundary surface is given by $x^0 = 0$ and coordinatized by $\vec{x}(\vec{\tau}) = \vec{\tau}$, and that we have chosen the gauge $e_b^0(\vec{\tau}) = 0$. Then $n_\mu = (1, 0, 0, 0)$ and

$$E^{ai} = \epsilon^{abc} \Sigma_{bc}^i. \tag{4.70}$$

Its real part is the densitized inverse triad:

$$\operatorname{Re} E_i^a = -\epsilon_{ijk} \, \epsilon^{abc} \, e_b^j e_c^k = \det(e) \, e_i^a, \tag{4.71}$$

where e_i^a is the matrix inverse to the "triad" one-form e_a^i. Its imaginary part is

$$\operatorname{Im} E_i^a = \epsilon^{abc} \, e_b^i e_c^0. \tag{4.72}$$

The projection of the field equations (4.56) on σ, written in terms of E_i^a, read $\mathrm{D}_a E_i^a = 0$, $F_{ab}^i E^{ai} = 0$ and $F_{ab}^i E^{ai} E^{bk} \epsilon_{ijk} = 0$, where D_a and F_{ab}^i are the covariant derivative and the curvature of A_a^i. Using (4.69) these give the three Hamilton–Jacobi equations of GR

$$\mathrm{D}_a \frac{\delta S[\alpha]}{\delta A_a^i(\vec{\tau})} = 0, \tag{4.73}$$

$$\frac{\delta S[\alpha]}{\delta A_a^i(\vec{\tau})} F_{ab}^i = 0, \tag{4.74}$$

$$F_{ab}^{ij}(\vec{\tau}) \frac{\delta S[\alpha]}{\delta A_a^i(\vec{\tau})} \frac{\delta S[\alpha]}{\delta A_b^j(\vec{\tau})} = 0. \tag{4.75}$$

Kinematical gauges. Equation (4.73) could have been obtained by simply observing that $S[\alpha]$ is invariant under local $SU(2)$ gauge transformations on the 3-surface. Under one such transformation generated by a function $f^i(\vec{\tau})$ the variation of the connection is $\delta_f A_a^i = \mathrm{D}_a f^i$. Therefore, S satisfies

$$0 = \delta_f S = \int \mathrm{d}^3 \vec{\tau} \, \delta_f A_a^i(\vec{\tau}) \frac{\delta S[\alpha]}{\delta A_a^i(\vec{\tau})} = \int \mathrm{d}^3 \vec{\tau} \, \mathrm{D}_a f^i(\vec{\tau}) \frac{\delta S[\alpha]}{\delta A_a^i(\vec{\tau})}$$

$$= -\int \mathrm{d}^3 \vec{\tau} \, f^i(\vec{\tau}) \, \mathrm{D}_a \frac{\delta S[\alpha]}{\delta A_a^i(\vec{\tau})}. \tag{4.76}$$

This gives (4.73). Next, the action is invariant under a change of coordinates on the 3-surface α_M. Under one such transformation generated by a function $f^a(\vec{\tau})$, the variation of the connection is $\delta_f A_a^i = f^b \partial_b A_a^i + A_b^i \partial_a f^b$. Integrating by parts as in (4.76), this gives

$$\partial_b A_a^i \frac{\delta S[\alpha]}{\delta A_a^i(\vec{\tau})} + (\partial_b A_a^i) \frac{\delta S[\alpha]}{\delta A_a^i(\vec{\tau})} = 0, \tag{4.77}$$

which, combined with (4.73), gives (4.74). Thus, (4.73) and (4.74) are simply the requirement that $S[\alpha]$ is invariant under internal gauge and changes of coordinates on the 3-surface. The three equations (4.73), (4.74) and (4.75) govern the dependence of S on $A_a^i(\vec{\tau})$.

Dropping the coordinates. It is easy to see that S is independent from $x^\mu(\vec{\tau})$. A change of coordinates $x^\mu(\vec{\tau})$ tangential to the surface cannot affect the action, which is independent of the coordinates used. More formally, the invariance under change of parameter $\vec{\tau}$ implies

$$\frac{\delta S[\alpha]}{\delta x^\mu(\vec{\tau})} \partial_j x^\mu(\vec{\tau}) = \frac{\delta S[\alpha]}{\delta A_a^i(\vec{\tau})} \delta_j A_a^i(\vec{\tau}), \tag{4.78}$$

and we have already seen that the right-hand side vanishes. The variation of S under a change of $x^\mu(\vec{\tau})$ normal to the surface is governed by the Hamilton–Jacobi equation proper, equation (3.186). In the present case, following the same steps as for the scalar field, we obtain

$$\frac{\delta S[\alpha]}{\delta x^\mu(\vec{\tau})} n_\mu(\vec{\tau}) + \epsilon_{ijk} F_{ab}^i \frac{\delta S[\alpha]}{\delta A_a^j(\vec{\tau})} \frac{\delta S[\alpha]}{\delta A_b^k(\vec{\tau})} = 0. \tag{4.79}$$

But the second term vanishes because of (4.75). Therefore $S[\alpha]$ is independent of tangential as well as normal parts of $x^\mu(\vec{\tau})$: S depends only on $[A_a^i(\vec{\tau})]$.

We can thus drop altogether the spacetime coordinates x^μ from the extended configuration space. Define a smaller extended configuration space $\mathcal{C}$ as the 9d complex space of the variables A_a^i. Geometrically, this can be viewed as the space of the linear mappings $A : D \to sl(2; C)$, where $D = R^3$ is a "space of directions" and we have chosen the complex selfdual basis in the $sl(2; C)$ algebra. We then identify the space $\mathcal{G}$ as a space of parametrized 3d surfaces A, with components $[A_a^i(\vec{\tau})]$ and without boundaries in $\mathcal{C}$. GR is defined on this space by the Hamilton–Jacobi system

$$D_a \frac{\delta S[A]}{\delta A_a^i(\vec{\tau})} = 0, \tag{4.80}$$

$$\frac{\delta S[A]}{\delta A_a^i(\vec{\tau})} F_{ab}^i = 0, \tag{4.81}$$

$$F_{ab}^{ij}(\vec{\tau}) \frac{\delta S[A]}{\delta A_a^i(\vec{\tau})} \frac{\delta S[\alpha]}{\delta A_b^j(\vec{\tau})} = 0. \tag{4.82}$$

These are the equations presented at the beginning of this chapter, on which we will base quantum gravity.

Equivalently, we can solve immediately (4.80) and (4.81) by defining the space $\mathcal{G}_0$ of the equivalence classes of 3d $SU(2)$ connections A under gauge and 3d diffeomorphisms $(A_a{}^i(\vec{\tau}) = \frac{\partial \tau'^b}{\partial \tau^a} A_b'^i(\vec{\tau}'(\vec{\tau})))$ transformation. Then GR is defined by the sole equation (4.82) on this space (where functions $S[A_a^i(\vec{\tau})]$ overcoordinatize $\mathcal{G}_0$). Accordingly, we can interpret GR as the dynamical system defined by the extended configuration space

$\mathcal{G}_0$ and the relativistic hamiltonian

$$H(\vec{\tau}) = F_{ab}^{ij}(\vec{\tau})\ E_i^a(\vec{\tau})\ E_j^b(\vec{\tau}). \tag{4.83}$$

4.3.5 Reality conditions

The two variables on which we have based the canonical formulation of GR described above are a complex 3d connection A_a^i and its complex conjugate momentum E^{ai}. They have 9 complex components each. On the other hand, the degrees of freedom of GR have $(9+9)$ *real* components, of which $(2+2)$ are physical degrees of freedom, 7 are constrained and 7 are gauges. The explanation of the apparent doubling of the components is that A and E are like the coordinates $z = x + ip$ and $\bar{z} = x - ip$ over the phase space of a one-dimensional system. That is, they are not independent of each other.

To find out these relations, let us write the real and imaginary parts of A and E. From their definition, we have

$$\operatorname{Re} A_a^i = \omega_a^i, \quad (4.84) \qquad\qquad \operatorname{Re} E_i^a = \det(e)\ e_i^a, \quad (4.86)$$
$$\operatorname{Im} A_a^i = \omega_a^{0i}, \quad (4.85) \qquad\qquad \operatorname{Im} E_i^a = \epsilon^{abc}\ e_b^0\ e_{ic}. \quad (4.87)$$

We have chosen a gauge in which $e_a^0 = 0$. Then (4.87) implies that E is real. Recall that the tetrad and the connection ω are related by the equation $de^I = \omega_J^I \wedge e^J$. Projecting this equation on the 3-surface, we obtain

$$de^i = \omega_j^i \wedge e^j + \omega_0^i \wedge e^0. \tag{4.88}$$

In the gauge chosen, the last term vanishes and ω_j^i is the spin connection of the triad e^i. Hence, (4.84) implies that the real part of A satisfies (4.1). Without fixing the gauge $e_a^0 = 0$ the reality conditions are a bit more cumbersome.

Bibliographical notes

The hamiltonian formulation of GR was developed independently by Peter Bergmann and his group [129] and by Dirac [130]. The long-term goal of both was quantum gravity. The main tool for this, the hamiltonian theory of constrained systems, was developed for this purpose. The great algebraic complexity of the hamiltonian formalism was dramatically reduced by the introduction of the ADM variables by Arnowitt, Deser and

Misner [131], and then by the selfdual connection variables systematized by Ashtekar [132].

The conventional derivation of the fundamental equations (4.2–4.4) from the lagrangian formalism can be found in many books and articles; see, for instance, I [2, 9, 20, 126]. See also the original articles [132]. The expression (4.16) of the hamiltonian, which plays an important role in the quantum theory, was introduced by Thomas Thiemann [133]. The usefulness of the Barbero connection was pointed out in [134]; on its geometrical interpretation see [136]. The importance of the Immirzi parameter for the quantum theory in [135]. An (inconclusive) discussion on the Immirzi parameter and its physical interpretation is in [137].

For the finite-dimensional formulation, I have followed here [138, 139]. On other versions of this formalism, see [140]. For the covariant Hamilton–Jacobi formalism for GR, see also [141].

5

Quantum mechanics

Quantum mechanics (QM) is not just a theory of micro-objects: it is our current fundamental theory of motion. It expresses a deeper understanding of Nature than classical mechanics. Precisely as classical mechanics, the conventional formulation of QM describes evolution of states and observables in time. Precisely as classical mechanics, this is not sufficient to deal with general relativistic systems, because these systems do not describe evolution in time; they describe correlations between observables. Therefore, a formulation of QM slightly more general than the conventional one – or a quantum version of the relativistic classical mechanics discussed in the previous chapter – is needed. In this chapter I discuss the possibility of such a formulation. In the last section I discuss the general physical interpretation of QM.

QM can be formulated in a number of more or less equivalent formalisms: canonical (Hilbert spaces and self-adjoint operators), covariant (Feynman's sum-over-histories), algebraic (states as linear functionals over an abstract algebra of observables) and others. Generally, but not always, we are able to translate these formalisms into one another, but often what is easy in one formulation is difficult in another. A general-relativistic sum-over-histories formalism has been developed by Jim Hartle [26]. Here I focus on the canonical formalism, because the canonical formalism has provided the mathematical completeness and precision needed to explicitly construct the mathematical apparatus of quantum gravity. Later I will consider alternative formalisms.

5.1 Nonrelativistic QM

Conventional QM can be formulated as follows.

States. The states of a system are represented by vectors ψ in a complex separable Hilbert space $\mathcal{H}_0$.

Observables. Each observable quantity $\mathcal{A}$ is represented by a self-adjoint operator A on $\mathcal{H}_0$. The possible values that $\mathcal{A}$ can take are the numbers in the spectrum of A.

Probability. The average of the values that $\mathcal{A}$ takes over many equal states represented by ψ is $\bar{a} = \langle \psi | A | \psi \rangle / \langle \psi | \psi \rangle$.

Projection. If the observable $\mathcal{A}$ takes values in the spectral interval I, the state ψ becomes then the state $P_I \psi$, where P_I is the spectral projector on the interval I.

Evolution. States evolve in time according to the Schrödinger equation

$$i\hbar \partial_t \psi(t) = H_0 \psi(t) \tag{5.1}$$

where H_0 is the hamiltonian operator corresponding to the energy. Equivalently, states do not evolve in time but observables do and their evolution is governed by the Heisenberg equation

$$\frac{d}{dt} A(t) = -\frac{i}{\hbar} [A(t), H_0]. \tag{5.2}$$

A given quantum system is defined by a family (generally an algebra) of operators A_i, including H_0, defined over an Hilbert space $\mathcal{H}_0$.

This scheme for describing Nature differs substantially from the newtonian one. Here are the main features of the physical content of the above scheme.

Probability. Predictions are only probabilistic.

Quantization. Some physical quantities can take certain discrete values only (are "quantized").

Superposition principle. If a system can be in a state A, where a physical quantity q has value a, as well as in state B, where q has value b, then the system can also be in states (denoted $\psi = c_a A + c_b B$, with $|c_a|^2 + |c_b|^2 = 1$) where q has value a with probability $|c_a|^2$, and value b with probability $|c_b|^2$.

Uncertainty principle. There are couples of (conjugate) variables that cannot have determined values at the same time.

Effect of observations on predictions. The properties we expect the system to have at some time t_2 are determined not only by the properties we know the system had at time t_0, but also by the properties we know the system has at the time t_1, where $t_0 < t_1 < t_2$.[1]

---

[1] Bohr expressed this fact by saying that observation affects the observed system. But formulations such as Bohm's or consistent histories force us to express this physical fact using more careful wording.

In Section 5.6, I discuss the physical content of QM in more depth.

In general, a quantum system $(\mathcal{H}_0, A_i, H_0)$ has a classical limit which is a mechanical system describing the results of observations made on the system at scales and with accuracy larger than the Planck constant. In the classical limit, Heisenberg uncertainty can be neglected and the observables A_i can be taken as coordinates of a commutative phase space Γ_0. Quantum commutators define classical Poisson brackets and (5.2) reduces to Hamilton equation (3.78).

If the classical limit is known, the search for a quantum system from which this limit may derive is called the *quantization* problem. There is no reason for the quantization problem to have a unique solution. The existence of distinct solutions is denoted "quantization ambiguity." Experience shows that the simplest quantization of a given classical system is very often the physically correct one. If we are given a classical system defined by a nonrelativistic configuration space $\mathcal{C}_0$ with coordinates q^i and by a nonrelativistic hamiltonian $H_0(q^i, p_i)$, then a solution of the quantization problem can be obtained by interpreting the Hamilton–Jacobi equation (3.17) as the eikonal approximation of the wave function (5.1) that governs the quantum dynamics [142]. This can be achieved by defining multiplicative operators q^i, derivative operators $p_i = -i\hbar\frac{\partial}{\partial q^i}$ and the hamiltonian operator

$$H_0 = H_0\left(q^i, -i\hbar\frac{\partial}{\partial q^i}\right) \tag{5.3}$$

on the Hilbert space $\mathcal{H}_0 = L_2[\mathcal{C}_0]$, the space of the square integrable functions on the nonrelativistic configuration space [143].

In a special-relativistic context this structure remains the same, but the *Evolution* postulate above is extended to the requirement that $\mathcal{H}_0$ carries a unitary representation of the Poincaré group and H_0 is the generator of the time translations of this representation.

This structure is not generally relativistic. In particular, the notions of "state" and "observable" used above are the *nonrelativistic* ones. Can the structure of QM be extended to the relativistic framework? In Section 5.2, I discuss such an extension. As a preliminary step, however, in the rest of this section I introduce and illustrate some tools needed for this reformulation, in the context of a very simple system – as I did for classical mechanics.

5.1.1 Propagator and spacetime states

Nonrelativistic formulation. The quantum theory of the pendulum can be written on the Hilbert space $\mathcal{H}_0 = L_2[R]$ of wave functions $\psi_0(\alpha)$, in terms of the multiplicative position operator α, the momentum operator $p_\alpha = -i\hbar \frac{\partial}{\partial \alpha}$ and the hamiltonian

$$H_0 = -\frac{\hbar^2}{2m} \frac{\partial^2}{\partial \alpha^2} + \frac{m\omega^2}{2} \alpha^2. \tag{5.4}$$

More precisely, the theory is defined on a *rigged* Hilbert space, or Gelfand triple. A Gelfand triple $\mathcal{S} \subset \mathcal{H} \subset \mathcal{S}'$ is formed by a Hilbert space $\mathcal{H}$, a proper subset $\mathcal{S}$ dense in $\mathcal{H}$ and equipped with a weak topology, and the dual $\mathcal{S}'$ of $\mathcal{S}$, with their natural identifications. A manifold M with a measure dx determines a rigged Hilbert space $\mathcal{S}_M \subset \mathcal{H}_M \subset \mathcal{S}'_M$ where $\mathcal{S}_M$ is the space of smooth functions on M with fast decrease (Schwarz space), $\mathcal{H}_M = L_2[M, dx]$ and $\mathcal{S}'_M$ is the space of the tempered distributions on M. This setting allows us in particular to deal with eigenstates of observables with continuous spectrum and Fourier transforms.

The operators (here $\hbar = 1$)

$$\alpha(t) = e^{itH_0} \alpha e^{-itH_0}, \tag{5.5}$$

which solve (5.2) are the Heisenberg position operators that give the position at any time t. Denote $|\alpha; t\rangle$ the generalized eigenstate of the operator $\alpha(t)$ with eigenvalue α (which are in $\mathcal{S}'$)

$$\alpha(t)|\alpha; t\rangle = \alpha|\alpha; t\rangle \tag{5.6}$$

and $|\alpha\rangle = |\alpha; 0\rangle$. Clearly $|\alpha; t\rangle = e^{itH_0}|\alpha\rangle$. Given a state $|\psi\rangle$, the Schrödinger wave function

$$\psi(\alpha, t) = \langle \alpha; t|\psi\rangle = \langle \alpha|e^{-itH_0}|\psi\rangle \tag{5.7}$$

satisfies the Schrödinger equation (5.1). Conversely, each solution of the Schrödinger equation, restricted to $t = 0$, defines a state in $\mathcal{H}_0$. Therefore, there is a one-to-one correspondence between states at fixed time $\psi_0(\alpha)$ and solutions of the Schrödinger equation $\psi(\alpha, t)$. I call $\mathcal{H}$ the space of the solutions of the Schrödinger equation. Thanks to the identification just mentioned, $\mathcal{H}$ is a Hilbert space, isomorphic to the Hilbert space $\mathcal{H}_0$ of the states at fixed time. I call

$$R_0 : \qquad \mathcal{H} \to \mathcal{H}_0 \tag{5.8}$$
$$\psi(\alpha, t) \mapsto \psi_0(\alpha) = \psi(\alpha, 0) \tag{5.9}$$

the identification map. The relation between $\mathcal{H}$ and $\mathcal{H}_0$ is analogous to the relation between the spaces Γ and Γ_0 in classical mechanics, discussed in Chapter 3.

The propagator is defined as

$$W(\alpha, t, \alpha', t') = \langle \alpha; t|\alpha'; t'\rangle = \langle \alpha|e^{-i(t-t')H_0}|\alpha'\rangle$$
$$= \sum_n H_n(\alpha) \, e^{-iE_n(t-t')} \overline{H_n(\alpha')}, \tag{5.10}$$

where $H_n(\alpha)$ is the eigenfunction of H_0 with eigenvalue E_n. Explicitly, a straightforward calculation that can be found in many books gives

$$W(\alpha, t, \alpha', t') = \sqrt{\frac{m\omega}{ih\sin[\omega(t-t')]}}\; e^{\frac{i\omega m}{2\hbar}\left[\frac{(\alpha^2+\alpha'^2)\cos[\omega(t-t')]-2\alpha\alpha'}{\sin^2[\omega(t-t')]}\right]}, \quad (5.11)$$

where $h = 2\pi\hbar$. The propagator satisfies the Schrödinger equation in the variables (α, t) (and the conjugate equation in the variables (α', t')). ·

Spacetime states. It is convenient to consider the following states. Given any compact support complex function $f(\alpha, t)$, the state

$$|f\rangle = \int d\alpha\, dt\; f(\alpha, t)\; |\alpha; t\rangle \qquad (5.12)$$

is in $\mathcal{H}_0$ and is called the "spacetime smeared state," or simply the "spacetime state" of the function $f(\alpha, t)$. Since standard normalizable states are dimensionless (for $\langle\psi|\psi\rangle = 1$ to make sense) and the states $|\alpha; t\rangle$ have dimension $L^{-1/2}$, the function f must have dimensions $T^{-1}L^{-1/2}$. These states generalize the conventional wave packets for which $f(\alpha, t) = f(\alpha)\delta(t)$. Conventional wave packets can be thought of as being associated with results of *instantaneous* position measurements with finite resolution in space; as I will illustrate later on, spacetime states can be associated with realistic measurements, where the measuring apparatus has finite resolution in space as well as in time. The Schrödinger wave function of $|f\rangle$ is

$$\psi_f(\alpha, t) = \langle\alpha; t|f\rangle$$
$$= \langle\alpha; t|\int d\alpha' dt'\; f(\alpha', t')\; |\alpha'; t'\rangle$$
$$= \int d\alpha' dt'\; W(\alpha, t, \alpha', t')\; f(\alpha', t') \qquad (5.13)$$

and satisfies the Schrödinger equation. The scalar product of two spacetime states is

$$\langle f|f'\rangle = \int d\alpha\, dt\, d\alpha' dt'\; \overline{f(\alpha, t)}\; W(\alpha, t, \alpha', t')\; f'(\alpha', t'). \quad (5.14)$$

In particular, we can associate a normalized state $|R\rangle$ to each spacetime region R

$$|R\rangle = C_R \int_R d\alpha\, dt\; |\alpha; t\rangle, \qquad (5.15)$$

where the factor

$$C_R^{-2} = \int d\alpha \, dt \, d\alpha' dt' \, W(\alpha, t, \alpha', t') \tag{5.16}$$

fixes the normalization $\langle R|R \rangle = 1$, as well as giving the state the right dimensions.

5.1.2 Kinematical state space $\mathcal{K}$ and "projector" P

As discussed in Chapter 3, the kinematics of a pendulum is described by two partial observables: time t and elongation α. These coordinatize the relativistic configuration space $\mathcal{C}$. The classical relativistic formalism treats α and t on an equal footing. The quantum relativistic formalism, as well, treats α and t on an equal footing, and therefore it is based on functions $f(\alpha, t)$ on $\mathcal{C}$.

To be precise, let $\mathcal{S} \subset \mathcal{K} \subset \mathcal{S}'$ be the Gelfand triple defined by $\mathcal{C}$ and the measure $d\alpha dt$. That is, $\mathcal{S}$ is the space of the smooth functions $f(\alpha, t)$ on $\mathcal{C}$ with fast decrease, $\mathcal{K} = L^2[\mathcal{C}, d\alpha dt]$ and $\mathcal{S}'$ is formed by the tempered distributions over $\mathcal{C}$.

I call $\mathcal{S}$ the "kinematical state space" and its elements $f(\alpha, t)$ "kinematical states."

In the relativistic formalism, the dynamics of the system is defined by the relativistic hamiltonian $H(\alpha, t, p, p_t)$, given in (3.24). The quantum dynamics is defined by the "Wheeler–DeWitt" (WdW) equation

$$H \, \psi(\alpha, t) = 0, \tag{5.17}$$

where

$$
\begin{aligned}
H &= H\left(\alpha, t, -i\hbar \frac{\partial}{\partial \alpha}, -i\hbar \frac{\partial}{\partial t}\right) \\
&= -i\hbar \frac{\partial}{\partial t} + H_0 \\
&= -i\hbar \frac{\partial}{\partial t} - \frac{\hbar^2}{2m} \frac{\partial^2}{\partial \alpha^2} + \frac{m\omega^2}{2} \alpha^2,
\end{aligned}
\tag{5.18}
$$

and H_0 is given in (5.4). In the case of the pendulum, (5.17) reduces to the Schrödinger equation (5.1), but (5.17) is more general than the Schrödinger equation, because in general H does not have the nonrelativistic form $H = p_t + H_0$. Solutions $\psi(\alpha, t)$ of this equation form a linear space $\mathcal{H}$, which carries a natural scalar product that I will construct in a moment. The key object for the relativistic quantum theory is the operator

$$P = \int d\tau \, e^{-i\tau H} \tag{5.19}$$

defined on $\mathcal{S}'$. This operator maps arbitrary functions $f(\alpha, t)$ into solutions of the WdW equation (5.17), namely into $\mathcal{H}$.

To see this, expand a function $f(\alpha, t)$ as

$$f(\alpha, t) = \sum_n \int dE \; f_n(E) \; H_n(\alpha) \; e^{-iEt}. \tag{5.20}$$

Acting with P on this function we obtain

$$\begin{aligned}
[Pf](\alpha, t) &= \int d\tau \; e^{-i\tau H} \sum_n \int dE \; f_n(E) \; H_n(\alpha) \; e^{-iEt} \\
&= \int d\tau \sum_n \int dE \; e^{-i\tau(-E+E_n)} \; f_n(E) \; H_n(\alpha) \; e^{-iEt} \\
&= \sum_n \int dE \; \delta(E - E_n) \; f_n(E) \; H_n(\alpha) \; e^{-iEt} \\
&= \sum_n \psi_n \; H_n(\alpha) \; e^{-iE_n t},
\end{aligned} \tag{5.21}\cdot$$

where $\psi_n = f_n(E_n)$, which is the general solution of (5.17). Therefore P sends arbitrary functions into solutions of the WdW equation. Intuitively: $P \sim \delta(H)$.

The integral kernel of P is the propagator (5.10). Indeed, the inverse of (5.20) gives

$$\psi_n = f_n(E_n) = \int d\alpha dt \; \overline{H_n(\alpha)} \; e^{iE_n t} \; f(\alpha, t). \tag{5.22}$$

Inserting this in (5.21), we have

$$\begin{aligned}
[Pf](\alpha, t) &= \sum_n \int d\alpha' dt' \; \overline{H_n(\alpha')} \; e^{iE_n t'} H_n(\alpha) \; e^{-iE_n t} \; f(\alpha', t') \\
&= \int d\alpha' dt' \; W(\alpha, t, \alpha', t') \; f(\alpha', t').
\end{aligned} \tag{5.23}$$

P is often called "the projector," although improperly so. Intuitively, it "projects" on the space of the solutions of the WdW equation. In some systems (when 0 is an eigenvalue in the discrete spectrum of H) P is indeed a projector. But generically, and in particular for the nonrelativistic systems (where 0 is in the continuum spectrum of H), P is not a projector because its domain is smaller than the full $\mathcal{S}'$. In particular, it does not contain the solutions of the WdW equation, namely P's codomain. The domain of P contains, on the other hand, $\mathcal{S}$.

The matrix elements of P,

$$\langle f|P|f'\rangle_\kappa = \int d\alpha \, dt \, d\alpha' dt' \; \overline{f(\alpha, t)} \; W(\alpha, t, \alpha', t') \; f'(\alpha', t'), \tag{5.24}$$

define a degenerate inner product in $\mathcal{S}$. Dividing $\mathcal{S}$ by the kernel of this inner product, that is, identifying f and f' if $Pf = Pf'$, and completing in norm, we obtain a Hilbert space. But if $Pf = Pf'$, then f and f' define the same solution of the WdW equation. In fact, they define the solution that corresponds to the spacetime state $|f\rangle$ defined above. Therefore, an element of this Hilbert space corresponds to a solution of the WdW equation: the Hilbert space can be identified with the space of the solutions of the WdW equation $\mathcal{H}$. Therefore,

$$P : \; \mathcal{S} \to \mathcal{H}$$
$$f \mapsto |f\rangle. \tag{5.25}$$

It follows that P directly equips the space $\mathcal{H}$ of the solutions with a Hilbert space structure: if $\psi = Pf$ and $\psi' = Pf'$ are two solutions of the WdW equation (5.17), their scalar product is *defined* by

$$\langle \psi | \psi' \rangle \equiv \langle f | P | f' \rangle_{\mathcal{K}}, \tag{5.26}$$

where the right-hand side is the scalar product in $\mathcal{K}$, and is explicitly given in (5.24).

Notice that the scalar product on the space of the solutions of the WdW equation can be defined just by using the relativistic operator P, without any need of picking out t as a preferred variable.

For all nonrelativistic systems, the configuration space has the structure $\mathcal{C} = \mathcal{C}_0 \times R$, where $t \in R$ and a function $\psi(\alpha, t)$ in $\mathcal{H}$ is uniquely determined by its restriction $\psi_t = R_t \psi$ on $\mathcal{C}_0$, for a fixed t

$$\psi_t(\alpha) \equiv \psi(\alpha, t). \tag{5.27}$$

For each t, denote $\mathcal{H}_t$ the space of the $L_2[\mathcal{C}_0]$ functions $\psi_t(\alpha)$, so that $R_t : \mathcal{H} \to \mathcal{H}_t$. The spaces $\mathcal{H}$ and $\mathcal{H}_t$ are in one-to-one correspondence: the inverse map R_t^{-1} is the evolution determined by equation (5.17). In particular, $\mathcal{H}_0$ is the Hilbert space used in the nonrelativistic formulation of the quantum theory. Under the identification between $\mathcal{H}$ and $\mathcal{H}_0$ given by R_0, the scalar product defined above is precisely the usual scalar product of the nonrelativistic Hilbert space.

This can be directly seen by noticing that the right-hand side of (5.24) is precisely (5.14). More explicitly, let $\psi(\alpha, t) = \sum_n \psi_n H_n(\alpha) e^{-iE_n t}$ be a function in $\mathcal{H}$, namely a solution of the WdW equation. Its restriction to $t = 0$ is $\psi_0(\alpha) = \sum_n \psi_n H_n(\alpha)$ and its norm in $\mathcal{H}_0$ is $||\psi_0||^2 = \int d\alpha \, |\psi_0(\alpha)|^2 = \sum_n |\psi_n|^2$. A function f such that $Pf = \psi$ is, for instance, simply $f(\alpha, t) = \psi_0(\alpha)\delta(t) = \sum_n \psi_n H_n(\alpha) \int dE e^{-iEt}$. (This is actually not in $\mathcal{S}_0$, but we could take a sequence of functions in $\mathcal{S}_0$ converging to f. But f is in the domain of P and such a procedure would not give anything new.) The norm of ψ is

$$
\begin{aligned}
||\psi||^2 &= \langle f|f\rangle_{\mathcal{H}} = \langle f|P|f\rangle_{\mathcal{K}} \\
&= \int d\tau \int d\alpha \int dt\, \overline{f(\alpha,t)}\, e^{-i\tau H}\, f(\alpha,t) \\
&= \int d\tau \int d\alpha \int dt\, \sum_n \overline{\psi_n}\, \overline{H_n(\alpha)} \int dE e^{iEt} e^{-i\tau H} \sum_m \psi_m\, H_m(\alpha) \int dE' e^{-iE't} \\
&= \sum_n \int d\tau \int dt \int dE \int dE'\, |\psi_n|^2 e^{iEt} e^{-i\tau(E'-E_n)}\, e^{-iE't} \\
&= \sum_n |\psi_n|^2 = ||\psi_0||^2.
\end{aligned}
\tag{5.28}
$$

5.1.3 Partial observables and probabilities

Consider two events (α,t) and (α',t') in the extended configuration space. Suppose we have observed the event (α',t'). What is the probability of observing the event (α,t)?

To measure this probability, we need measuring apparata for α and for t. In general, these apparata will have a certain resolution, say $\Delta\alpha$ and Δt. The proper question is therefore what is the probability of observing an event included in the region $R = (\alpha \pm \Delta\alpha, t \pm \Delta t)$? It is important to remark that no realistic measuring device, or detector, can have $\Delta\alpha = 0$, nor $\Delta t = 0$. Most QM textbooks put much emphasis on the fact that $\Delta\alpha > 0$ and completely ignore the fact that $\Delta t > 0$. Consider thus two regions R and R'. If a detector at R' has detected the pendulum, what is the probability $\mathcal{P}_{R,R'}$ that a detector at R detects the pendulum?

If the regions R and R' are much smaller than any other physical quantity in the problem, including the spatial and temporal separation of R and R', a direct application of perturbation theory shows that

$$
\mathcal{P}_{R,R'} = \gamma^2\, |\langle R|R'\rangle|^2,
\tag{5.29}
$$

where γ^2 is a dimensionless constant related to the efficiency of the detector. (We may assume that a "perfect" detector is defined by $\gamma = 1$.) The reader can repeat the calculation himself, or find it for instance in [144]. Explicitly, we can write this probability as the modulus square of the amplitude $\mathcal{P}_{R,R'} = |\mathcal{A}_{R,R'}|^2$

$$
\mathcal{A}_{R,R'} = \gamma\, \frac{\langle R|R'\rangle}{\sqrt{\langle R|R\rangle}\ \sqrt{\langle R'|R'\rangle}},
\tag{5.30}
$$

$$
\langle R|R'\rangle = \int_R d\alpha dt \int_{R'} d\alpha' dt'\, W(\alpha,t,\alpha',t').
\tag{5.31}
$$

Therefore the propagator has all the information about transition probabilities.

Assume that R is sufficiently small, so that the wave function $\psi(\alpha, t) = \langle \alpha; t | R' \rangle$ is constant within R and has the value $\psi(\alpha, t)$. Then we can write the probability of the pendulum being detected in R as

$$\mathcal{P}_R = \gamma \, (V_R C_R)^2 \, |\psi(\alpha, t)|^2, \tag{5.32}$$

where V_R is the volume of the region R. Now, assume the region R has sides $\Delta\alpha, \Delta t$. A direct calculation (see [144]) shows that if $\Delta t \ll m\Delta\alpha^2/\hbar$, then $(V_R C_R)^2$ is proportional to $\Delta\alpha$; therefore,

$$\mathcal{P}_R \sim \Delta\alpha \, |\psi(\alpha, t)|^2. \tag{5.33}$$

So, for small regions we have the two important results that (i) the temporal resolution of the detector drops out from the detection probability, and (ii) the probability is proportional to the spacial resolution of the detector. Because of (i), we can forget the temporal resolution of the detector and take the idealized limit of an *instantaneous* detector. Because of (ii), we can associate a probability *density* in α to each infinitesimal interval $d\alpha$ in α. Fixing the overall normalization by requiring that an idealized perfect detector covering all values of α detects with certainty, this yields the results that $|\psi(\alpha, t)|^2$ is the probability density in α to detect the system at (α, t) with an instantaneous detector. That is, we recover the conventional probabilistic interpretation of the wave function from (5.29).

In the opposite limit, when $\Delta t \gg m\Delta\alpha^2/\hbar$, $(V_R C_R)^2$ is proportional to $(\Delta t)^{-1/2}$. Therefore,

$$\mathcal{P}_R \sim (\Delta t)^{-1/2} \, |\psi(\alpha, t)|^2, \tag{5.34}$$

and we cannot associate a probability density in t with this detector, because the detection probability does not scale linearly with Δt. The different behavior of the probability in α and t is a consequence of the specific form of the dynamics.

Partial observables in quantum theory. Recall that α and t are partial observables. They determine commuting self-adjoint operators in $\mathcal{K}$. These act simply by multiplication. Their common generalized eigenstates $|\alpha, t\rangle$ are in $\mathcal{S}$. The states $|\alpha, t\rangle$ satisfy

$$\langle \alpha, t | P | \alpha', t' \rangle = W(\alpha, t, \alpha', t'). \tag{5.35}$$

We can view the states $|\alpha, t\rangle$ as "kinematical states" that do not know anything about dynamics. They correspond to a single quantum event. The "kinematical" scalar product of these states in $\mathcal{K}$, given below in (5.36), expresses only their independence; while the "physical" scalar product of these states in $\mathcal{H}$ given in (5.35), expresses the physical relation between the two events: it determines the probability that one event happens given that the other happened.

Do not confuse $|\alpha, t\rangle$ with $|\alpha; t\rangle$. The first is an eigenstate of α and t, the second is an eigenstate of $\alpha(t)$. They both determine (generalized) functions on $\mathcal{C}$. The state $|\alpha, t\rangle$ determines a delta distribution at the point (α, t)

$$\langle \alpha', t' | \alpha, t \rangle = \delta(\alpha', \alpha)\delta(t', t); \tag{5.36}$$

while the state $|\alpha; t\rangle$ determines a solution of the Schrödinger equation. This solution has support all over $\mathcal{C}$ and is such that, on the line $t = constant$, it is a delta function in α

$$\langle \alpha'; t | \alpha; t \rangle = \delta(\alpha', \alpha), \tag{5.37}$$

while for different t's

$$\langle \alpha; t | \alpha'; t' \rangle = W(\alpha, t, \alpha', t'). \tag{5.38}$$

The relation between the two is simply

$$|\alpha; t\rangle = P|\alpha, t\rangle. \tag{5.39}$$

Notice that (5.38) and (5.39) give

$$W(\alpha, t, \alpha', t') = \langle \alpha, t | P^\dagger P | \alpha, t \rangle_{\mathcal{H}}, \tag{5.40}$$

which is consistent with (5.35) because the definition of the scalar product in $\mathcal{H}$ (indicated in (5.40) by $\langle \cdot | \cdot \rangle_{\mathcal{H}}$) is (5.26).

5.1.4 Boundary state space K and covariant vacuum $|0\rangle$

In this subsection I introduce some notions that play an important role in the field theoretical context. Fix two times $t = 0$ and t. Let $\mathcal{H}_0 = L_2[R, d\alpha]$ be the space of the instantaneous quantum states ψ_0 at $t = 0$. Let $\mathcal{H}_t \sim \mathcal{H}_0$ be the space of the instantaneous states ψ_t at t. The probability amplitude of measuring a state ψ_t at t if the state ψ_0 was measured at $t = 0$ is

$$A = \langle \psi_t | e^{-iH_0 t} | \psi_0 \rangle. \tag{5.41}$$

Consider the boundary state space

$$\mathcal{K}_t = \mathcal{H}_t^* \otimes \mathcal{H}_0 = L_2[R^2, d\alpha d\alpha']. \tag{5.42}$$

The linear functional ρ_t defined by

$$\rho_t(\psi_t \otimes \psi_0) = \langle \psi_t | e^{-iH_0 t} | \psi_0 \rangle \tag{5.43}$$

is well defined on $\mathcal{K}_t$. This functional captures the entire dynamical information about the system. A linear functional on a Hilbert space defines a state. I denote $|0_t\rangle$ the state defined by ρ_t

$$\rho_t(\psi) = \langle 0_t | \psi \rangle_{\mathcal{K}_t} \tag{5.44}$$

and call it the "dynamical vacuum" state in boundary state space $\mathcal{K}_t$.

These definitions can be given the following physical interpretation. We make a measurement on the system at $t = 0$ and a measurement at t. We can measure the positions (α, α'), or the momenta, or other combinations. The outcomes of the two measurements are not independent, because of the dynamics but, to start with, let's ignore the dynamics. All possible outcomes of measurements at $t = 0$ (with their kinematical relations) are described by instantaneous states at $t = 0$, namely by the nonrelativistic Hilbert space $\mathcal{H}_0$. Similarly for t. If we ignore the dynamical correlations, we can view the two measurements as if they were done on two independent systems, and therefore we can describe the outcomes of the two measurements using the Hilbert space $\mathcal{K}_t$. Dynamics is a correlation between the two measurements. These correlations are described by a probability amplitude associated with any given couple of states. Namely, to any state in $\mathcal{K}_t$.

It is a simple exercise, that I leave to the reader, to show that in the representation $\mathcal{K}_t = L_2[R^2, d\alpha d\alpha']$ the state $|0_t\rangle$ is precisely the propagator

$$\langle 0_t \,|\, \alpha, \alpha'\rangle = W(\alpha, t, \alpha', 0). \tag{5.45}$$

Dynamical vacuum versus Minkowski vacuum. Denote $|0_M\rangle$ the lowest eigenstate of H_0 in $\mathcal{H}_0$,

$$\langle \alpha | 0_M \rangle = H_0(\alpha) = \frac{1}{\sqrt{2\pi}} \, e^{-\frac{1}{2}\alpha^2} \tag{5.46}$$

and call it the "Minkowski" vacuum, because of its analogy with the vacuum state of the quantum field theories on Minkowski space. Consider the analytic continuation in imaginary time of the propagator (5.10)

$$W(\alpha, -it, \alpha', 0) = \langle \alpha | e^{-H_0 t} | \alpha' \rangle = \sum_n H_n(\alpha) \, e^{-E_n t} \, \overline{H_n(\alpha')}. \tag{5.47}$$

For large t, only the lowest-energy state survives in the sum, and we have

$$W(\alpha, -it, \alpha', 0) \xrightarrow{\;t\to\infty\;} H_0(\alpha) \, e^{-E_0 t} \, \overline{H_0(\alpha')}. \tag{5.48}$$

Using the definitions of the previous section, this can be written as

$$\lim_{t\to\infty} e^{E_0 t} \, |0_{-it}\rangle = |0_M\rangle \otimes \langle 0_M|. \tag{5.49}$$

(The ket and bra in the right-hand side are in $\mathcal{H}_0$, while the ket in the left-hand side is in $\mathcal{K} = \mathcal{H}_0^* \otimes \mathcal{H}_0$.) This expression relates the dynamical vacuum $|0_t\rangle$ and the Minkowski vacuum $|0_M\rangle$. We will use this equation to find the quantum states corresponding to Minkowski spacetime from the spinfoam formulation of quantum gravity.

The boundary state space K *and covariant vacuum* $|0\rangle$. The construction above can be given a more covariant formulation as follows. Consider the Hilbert space

$$\mathsf{K} = \mathcal{K}^* \otimes \mathcal{K} = L_2[R^4, \mathrm{d}\alpha \, \mathrm{d}t \, \mathrm{d}\alpha' \mathrm{d}t'] = L_2[\mathcal{G}]. \tag{5.50}$$

I call this space the "total" quantum space. The propagator defines a preferred state $|0\rangle$ in K

$$\langle \alpha, t, \alpha', t' | 0 \rangle = W(\alpha, t, \alpha', t'). \tag{5.51}$$

I call this state the covariant vacuum state.

To run a complete experiment in a one-dimensional quantum system we need to measure two events: a "preparation" and a "measurement." The space K describes all possible (a priori equal) outcomes of the measurements of these two events. Any couple of measurements is represented by operators on K and any outcome is represented by a state $\psi \in \mathsf{K}$ which is an eigenstate of these operators. The dynamics is given by the bra $\langle 0 |$. The probability amplitude of the given outcome is determined by

$$A = \langle 0 | \psi \rangle. \tag{5.52}$$

This is a compact and fully covariant formulation of quantum dynamics.

5.1.5 * Evolving constants of motion

The interpretation of the theory is already entirely contained in (5.29). Still, to make the connection with the nonrelativistic formalism more direct, we can also consider operators related to observable quantities whose probability distribution can be predicted by the theory.

In the classical theory, if we know the (relativistic) state of the pendulum, we can predict the value of α when t has value, say, $t = T$. In the quantum theory, there is an operator that corresponds to this physical prediction. It is, of course, the Heisenberg position operator (5.5) for $t = T$, that is, $\alpha(T)$. (For clarity, it is convenient to distinguish the particular numerical value T from the argument of the wave function t.) I now define and characterize this operator in a relativistic language.

First of all, notice that the operator $\alpha(T)$ defined on $\mathcal{H}_0$ in (5.5) is in fact well defined on $\mathcal{H}$, as

$$\alpha(T) = R_0^{-1}\alpha(T)R_0 = R_0^{-1}\mathrm{e}^{iTH_0} \, \alpha \, \mathrm{e}^{-iTH_0} R_0 = R_T^{-1}\alpha \, R_T. \tag{5.53}$$

The operator $\alpha(T)$ can be directly defined on $\mathcal{H}$, without referring to $\mathcal{H}_0$, as follows. Consider the operator

$$a(T) = \mathrm{e}^{-i\omega(T-t)} \left(\alpha + i\frac{p_\alpha}{m\omega} \right) \tag{5.54}$$

and its real part

$$\alpha(T) = \mathrm{Re}\,[a(T)] = \frac{a(T) + a^\dagger(T)}{2}, \tag{5.55}$$

defined on S. These operators commute with H for any T. Therefore they are well defined on the space of the solutions of (5.17), namely on $\mathcal{H}$. The restriction of the operator (5.55) to $\mathcal{H}$ is precisely the operator (5.53).

The operator $\alpha(T)$ is characterized by two properties. First, the fact that it commutes with the hamiltonian

$$[\alpha(T), H] = 0. \tag{5.56}$$

Second, if we put $T = t$ in the expressions (5.54), (5.55), we obtain α. That is, $\alpha(T)$ is defined as an operator function $\alpha(T)(\alpha, p_\alpha, t)$ such that

$$\alpha(T)(\alpha, p_\alpha, T) = \alpha. \tag{5.57}$$

Intuitively, these two equations determine $\alpha(T)$ since the second fixes it at $t = T$ and the first evolves it for all t. Operators of this kind are called "evolving constants of motion." They are "evolving" because they describe the evolution (here the evolution of α with respect to t); they are "constants of motion" because they commute with the hamiltonian. In GR, the operators of this kind are independent from the temporal coordinate.

5.2 Relativistic QM

In the previous section, I used the example of a pendulum to introduce a certain number of notions on which a relativistic hamiltonian formulation of QM can be based. It is now time to attempt a general theory of relativistic QM.

5.2.1 General structure

Kinematical states. Kinematical states form a space S in a rigged Hilbert space $S \subset \mathcal{K} \subset S'$.

Partial observables. A partial observable is represented by a self-adjoint operator in $\mathcal{K}$. Common eigenstates $|s\rangle$ of a complete set of commuting partial observables are denoted quantum events.

Dynamics. Dynamics is defined by a self-adjoint operator H in $\mathcal{K}$, the (relativistic) hamiltonian. The operator from S to S'

$$P = \int d\tau\, e^{-i\tau H} \tag{5.58}$$

is (sometimes improperly) called the projector. (The integration range in this integral depends on the system.) Its matrix elements

$$W(s, s') = \langle s|P|s' \rangle \tag{5.59}$$

are called transition amplitudes.

Probability. Discrete spectrum: the probability of the quantum event s given the quantum event s' is

$$P_{ss'} = |W(s, s')|^2, \tag{5.60}$$

where $|s\rangle$ is normalized by $\langle s|P|s \rangle = 1$. Continuous spectrum: the probability of a quantum event in a small spectral region R given a quantum event in a small spectral region R' is

$$P_{RR'} = \left| \frac{W(R, R')}{\sqrt{W(R, R)} \, \sqrt{W(R', R')}} \right|^2, \tag{5.61}$$

where

$$W(R, R') = \int_R ds \int_{R'} ds' \, W(s, s'). \tag{5.62}$$

To this we may add:

Boundary quantum space and covariant vacuum. For a finite number of degrees of freedom, the boundary Hilbert space $\mathsf{K} = \mathcal{K}^* \otimes \mathcal{K}$ represents any observations of pairs of quantum events. The covariant vacuum state $|0\rangle \in \mathsf{K}$ defined by

$$\langle 0|(\psi \otimes \psi')\rangle_\mathsf{K} = \langle \psi|P|\psi' \rangle_\mathcal{K} \tag{5.63}$$

expresses the dynamics. It determines the correlation probability amplitude of any such observation. The extension to QFT is considered in Section 5.3.5.

States. A physical state is a solution of the Wheeler–DeWitt equation

$$H\psi = 0. \tag{5.64}$$

Equivalently, it is an element of the Hilbert space $\mathcal{H}$ defined by the quadratic form $\langle \cdot |P| \cdot \rangle$ on $\mathcal{S}$. (Elements of $\mathcal{K}$ are called kinematical states and elements of K are called boundary states.)

Complete observables. A complete observable $\mathcal{A}$ is represented by a self-adjoint operator on $\mathcal{H}$. A self-adjoint operator A in $\mathcal{K}$ defines a complete observable if

$$[A, H] = 0. \tag{5.65}$$

Projection. If the value of the observable $\mathcal{A}$ is restricted to the spectral interval I, the state ψ becomes the state $P_I \psi$, where P_I is the spectral projector on the interval I. If an event corresponding to a sufficiently small region R is detected, the state becomes $|R\rangle$.

A relativistic quantum system is defined by a rigged Hilbert space of kinematical states $\mathcal{K}$ and a set of partial observables A_i including a relativistic hamiltonian operator H. Alternatively, it is defined by giving the projector P.

Axiomatizations are meant to be clarifying, not prescriptive. The structure defined above is still tentative and perhaps incomplete. There are aspects of this structure that deserve to be better understood, clarified, and specified. Among these is the precise meaning of the "smallness" of the region R in the case of the continuum spectrum, and the correct treatment of repeated measurements. On the other hand, the conventional structure of QM is *certainly* physically incomplete, in the light of GR. The above is an attempt to complete it, making it general relativistic.

5.2.2 Quantization and classical limit

In general, a quantum system $(\mathcal{K}, A_i, H)$ has a classical limit which is a relativistic mechanical system $(\mathcal{C}, H)$ describing the results of observations on the system at scales and with accuracy larger than the Planck constant. In the classical limit, Heisenberg uncertainty can be neglected and a commuting set of partial observables A_i can be taken as coordinates of a commutative relativistic configuration space $\mathcal{C}$.

If we are given a classical system defined by a nonrelativistic configuration space $\mathcal{C}$ with coordinates q^a and by a relativistic hamiltonian $H(q^a, p_a)$, a solution of the quantization problem is provided by the multiplicative operators q^a, the derivative operators

$$p_a = -i\hbar \frac{\partial}{\partial q^a}, \qquad (5.66)$$

and the hamiltonian operator

$$H = H\left(q^a, -i\hbar \frac{\partial}{\partial q^a}\right) \qquad (5.67)$$

on the Hilbert space $\mathcal{K} = L_2[\mathcal{C}, dq^a]$, or, more precisely, the Gelfand triple determined by $\mathcal{C}$ and the measure dq^a. The physics is entirely contained in the transition amplitudes

$$W(q^a, q'^a) = \langle q^a | P | q'^a \rangle, \qquad (5.68)$$

where the states $|q^a\rangle$ are the eigenstates of the multiplicative operators q^a.

In turn, the space K has the structure

$$\mathsf{K} = L_2[\mathcal{G}]. \tag{5.69}$$

As we shall see, this remains true in field theory and in quantum gravity. The space $\mathcal{G}$ was defined in Section 3.2.5 for finite-dimensional systems, in Section 3.3.3 for field theories, and in Section 4.3.4 in the case of gravity.

In the limit $\hbar \to 0$, the Wheeler–DeWitt equation becomes the relativistic Hamilton–Jacobi equation (3.59) and the propagator has the form (writing $q \equiv (q^a)$)

$$W(q, q') \sim \sum_i A_i(q, q') \; e^{\frac{i}{\hbar}S_i(q, q')}, \tag{5.70}$$

where $S_i(q, q')$ are the different branches of the Hamilton function, as in (3.89). Now, the reverse of each path is still a path. The Hamilton function and the amplitude of a reversed path acquires a minus, giving

$$W(q, q') \sim \sum_i A_i(q, q') \sin\left[\tfrac{1}{\hbar} S_i(q, q')\right], \tag{5.71}$$

and W is real. Assuming only one path matters,

$$W(q, q') \sim A(q, q') \sin\left[\tfrac{1}{\hbar} S(q, q')\right] \tag{5.72}$$

and we can write, for instance,

$$\lim_{\hbar \to 0} \frac{1}{W} \; i\hbar\frac{\partial}{\partial q^a} \; i\hbar\frac{\partial}{\partial q^b} \; W(q, q') = \frac{\partial S(q, q')}{\partial q^a} \frac{\partial S(q, q')}{\partial q^b}. \tag{5.73}$$

This equation provides a precise relation between a quantum theory (entirely defined by the propagator $W(q, q')$) and a classical theory (entirely defined by the Hamilton function $S(q, q')$). Using (3.86) and (5.66) this equation can be written in the suggestive form

$$\lim_{\hbar \to 0} \frac{1}{W} \; p_a p_b W(q, q') = p_a(q, q') \; p_b(q, q'). \tag{5.74}$$

5.2.3 Examples: pendulum and timeless double pendulum

Pendulum. An example of relativistic formalism is provided by the quantization of the pendulum described in the previous section: the kinematical state space is $\mathcal{K} = L_2[R^2, d\alpha dt]$. The partial observable operators are the multiplicative operators α and t acting on the functions $\psi(\alpha, t)$ in $\mathcal{K}$. Dynamics is defined by the operator H given in (5.18). The Wheeler–DeWitt equation is therefore

$$\left(-i\hbar \frac{\partial}{\partial t} - \frac{\hbar^2}{2m} \frac{\partial^2}{\partial \alpha^2} + \frac{m\omega^2}{2} \alpha^2\right) \Psi(\alpha, t) = 0. \tag{5.75}$$

$\mathcal{H}$ is a space of solutions of this equation. The "projector" operator P : $\mathcal{K} \to \mathcal{H}$ defined by H is given in (5.23), and defines the scalar product in $\mathcal{H}$. Its matrix elements $W(\alpha, t, \alpha', t')$ between the common eigenstates of α and t are given by the propagator (5.11). They express all predictions of the theory. Because of the specific form of H, these define a probability density in α but not in t, as explained in Section 5.1.3.

Equivalently, the quantum theory can be defined by the boundary state space $\mathsf{K} = L_2[\mathcal{G}]$, where $\mathcal{G}$ is the boundary space of the classical theory, with coordinates (α, t, α', t'), and the covariant vacuum state $\langle \alpha, t, \alpha', t'|0\rangle = W(\alpha, t, \alpha', t')$, which determines the amplitude $A = \langle 0|\psi\rangle$ of any possible outcome $\psi \in \mathsf{K}$ of a preparation/measurement experiment.

Timeless double pendulum. An example of a relativistic quantum system which cannot be expressed in terms of conventional relativistic quantum mechanics is provided by the quantum theory of the timeless system (3.40). The kinematical Hilbert space $\mathcal{K}$ is $L_2[R^2, dadb]$, and the Wheeler–DeWitt equation is

$$\frac{1}{2}\left(-\hbar^2 \frac{\partial^2}{\partial a^2} - \hbar^2 \frac{\partial^2}{\partial b^2} + a^2 + b^2 - 2E\right) \Psi(a, b) = 0. \tag{5.76}$$

Below I describe this system in some detail.

States. Since $H = H_a + H_b - E$, where H_a (resp. H_b) is the harmonic oscillator hamiltonian in the variable a (resp. b), this equation is easy to solve by using the basis that diagonalizes the harmonic oscillator. Let

$$\psi_n(a) = \langle a|n\rangle = \frac{1}{\sqrt{n!}} H_n(a)\, e^{-a^2/2\hbar} \tag{5.77}$$

be the normalized nth eigenfunction of the harmonic oscillator, with eigenvalue $E_n = \hbar(n + 1/2)$. Here $H_n(a)$ is the nth Hermite polynomial. Then, clearly,

$$\Psi_{n_a n_b}(a, b) = \psi_{n_a}(a)\psi_{n_b}(b) \equiv \langle a, b|n_a, n_b\rangle \tag{5.78}$$

solves (5.76) if

$$\hbar(n_a + n_b + 1) = E. \tag{5.79}$$

Therefore the quantum theory exists (with this ordering) only if $E/\hbar = N + 1$ is an integer, which we assume from now on. The general solution of (5.76) is

$$\Psi(a, b) = \sum_{n=0,N} c_n \, \psi_n(a) \, \psi_{N-n}(b). \tag{5.80}$$

Therefore $\mathcal{H}$ is an $(N+1)$-dimensional proper subspace of $\mathcal{K}$. An orthonormal basis is formed by the $N + 1$ states $|n, N - n\rangle$, with $n = 0, \ldots, N$.

Projector. The projector $P : \mathcal{S} \to \mathcal{H}$ is in fact a true projector, and can be written explicitly as

$$P = \sum_{n=0,N} |n, N - n\rangle\langle n, N - n|. \tag{5.81}$$

This can be obtained from (5.58) by taking the integration range to be 2π, determined by the range of τ in the classical hamiltonian evolution, or by the fact that H is the generator of an $U(1)$ unitary action on $\mathcal{K}$, with period 2π. Indeed,

$$\int_0^{2\pi} d\tau \, e^{-\frac{i}{\hbar}\tau H} = \int_0^{2\pi} d\tau \sum_{n_a, n_b} |n_a, n_b\rangle e^{-\frac{i}{\hbar}\tau(h(n_a + n_b + 1) - E)} \langle n_a, n_b|$$

$$= \sum_{n_a, n_b} |n_a, n_b\rangle \delta(n_a + n_b + 1 - E/h)\langle n_a, n_b|$$

$$= P. \tag{5.82}$$

Transition amplitudes. The transition amplitudes are the matrix elements of P. In the basis that diagonalizes a and b

$$W(a, b, a', b') = \langle a, b|P|a', b'\rangle = \sum_{n=0,N} \langle a, b|n, N - n\rangle\langle n, N - n|a', b'\rangle. \tag{5.83}$$

Explicitly, this is

$$W(a, b, a', b') = \sum_{n=0,N} \frac{1}{\sqrt{n!(N-n)!}} H_n(a) H_{N-n}(b)$$

$$\times H_n(a') H_{N-n}(b') \, e^{-(a^2 + b^2 + a'^2 + b'^2)/2\hbar}. \tag{5.84}$$

This function codes all the properties of the quantum system. Roughly, it determines the probability density of measuring (a, b) if (a', b') was measured. Let us study its properties.

Semiclassical limit of the projector. Notice that by inserting (5.82) into (5.83) we can write the projector as

$$W(a, b, a', b') = \int_0^{2\pi} d\tau \; \langle a, b | e^{-\frac{i}{\hbar}H\tau} | a', b' \rangle$$

$$= \int_0^{2\pi} d\tau \; e^{\frac{i}{\hbar}E\tau} \langle a | e^{-\frac{i}{\hbar}H_a\tau} | a' \rangle \langle b | e^{-\frac{i}{\hbar}H_b\tau} | b' \rangle. \quad (5.85)$$

$$W(a, b, a', b') = \int_0^{2\pi} d\tau \; e^{\frac{i}{\hbar}E\tau} W(a, a', \tau) \; W(b, b', \tau), \quad (5.86)$$

where $W(a, a', \tau)$ is the propagator of the harmonic oscillator in a physical time τ, given in (5.11). Inserting (5.11) in (5.86) we obtain

$$W(a, b, a', b') = \int_0^{2\pi} d\tau \; \frac{1}{\sin \tau} \; e^{-\frac{i}{\hbar}S(a,b,a',b',\tau)}, \quad (5.87)$$

where $S(a, b, a', b', \tau)$ is given in (3.101). We can evaluate this integral in a saddle-point approximation. This gives

$$W(a, b, a', b') \sim \sum_i \frac{1}{\sin \tau_i} \; e^{-\frac{i}{\hbar}S(a,a',b,b'\tau_i)}, \quad (5.88)$$

where the τ_i are determined by

$$\frac{\partial S(a, b, a', b'; \tau)}{\partial \tau}\bigg|_{\tau=\tau_i(a,b,a',b')} = 0. \quad (5.89)$$

But this is precisely (3.102), that defines the value of τ giving the Hamilton function of the timeless system. This equation has two solutions, corresponding to the two portions into which the ellipse is cut. The relation between the two actions is given in (3.103). Recalling that $E/\hbar$ is an integer, this gives

$$W(a, b, a', b') \sim \frac{1}{\sin \tau(a, b, a', b')} \left(e^{-\frac{i}{\hbar}S(a,a',b,b')} - e^{\frac{i}{\hbar}S(a,a',b,b')} \right), \quad (5.90)$$

that is,

$$W(a, b, a', b') \sim \frac{1}{\sin \tau(a, b, a', b')} \; \sin\left[\frac{1}{\hbar}S(a, a', b, b')\right], \quad (5.91)$$

as in (5.72). Here $\sim$ indicates equality in the lowest order in $\hbar$. This equation expresses the precise relation between the quantum theory and the classical theory.

Propagation "forward and backward in time." Notice that the two terms in (5.90) have two natural interpretations. One is that they represent the two classical paths going from (a', b') to (a, b) in $\mathcal{C}$. The other, more interesting, interpretation is that they correspond to a trajectory going from (a', b') to (a, b) and a "time reversed" trajectory going from (a, b) to (a', b'). In fact, the projector (which, recall, is real) can be naturally interpreted as the sum of two propagators: one going forward and one going backward in the parameter time τ.

The distinction between forward and backward in the parameter time τ has no physical significance in the classical theory, because the physics is only in the ellipses in $\mathcal{C}$, not in the *orientation* of the ellipses.

However, in the quantum theory we can identify in $\mathcal{H}$ "clockwise-moving" and "anticlockwise-moving" components. These components are the eigenspaces of the positive and negative eigenvalues of the angular momentum operator $L = a\partial_b - b\partial_a$ (or $L = \partial_\phi$ where $a = r\sin\phi, b = r\cos\phi$). Thus, we can write wave packets "traveling along the ellipses purely forward, or purely backward in the parameter time." If we consider only a local evolution in a small region of $\mathcal{C}$, and we interpret, say, b as the independent time variable, and a as the dynamical variable, then these two components have, respectively, positive and negative energy. In a sense, they can be viewed as particles and antiparticles.

5.3 Quantum field theory

I assume the reader is familiar with standard quantum field theory (QFT). Here I illustrate the connection between QFT and the relativistic formalism developed above, and I recall a few techniques that will be used in Part II and are not widely known. Of particular importance are the distinction between Minkowski vacuum and covariant vacuum, the functional representation of a field theory, and the construction of the physical Hilbert space of lattice Yang–Mills theory.

In Chapter 3 we have seen that a classical field theory can be defined covariantly by the boundary space $\mathcal{G}$ of closed surfaces α in a finite-dimensional space $\mathcal{C}$ and a relativistic hamiltonian H on $T^*\mathcal{G}$. For instance, in a scalar field theory $\mathcal{C} = M \times R$ has coordinates (x^μ, ϕ), where x^μ is a point in Minkowski space and ϕ a field value. A surface α is determined by the two functions

$$\alpha = [x^\mu(\vec{\tau}), \varphi(\vec{\tau})], \tag{5.92}$$

and determines a boundary 3-surface $x^\mu(\vec{\tau})$ in Minkowski space M and boundary values $\phi(x(\tau)) = \varphi(\tau)$ of the field on this surface.

A quantization of the theory can be obtained, precisely as in the finite-dimensional case, in terms of a boundary state space K of functionals $\Psi[\alpha]$ on $\mathcal{G}$. Notice however that the difference between the kinematical state space $\mathcal{K}$ and the boundary state space K is far less significant in field theory than for finite-dimensional systems. In the finite-dimensional case, the states $\psi(q^a)$ in $\mathcal{K}$ are functions on the extended configuration space $\mathcal{C}$, while the states $\psi(q^a, q^{a\prime})$ in K are functions on the boundary space $\mathcal{G} = \mathcal{C} \times \mathcal{C}$. In the field theoretical case, both states have the form $\Psi[\alpha]$. The difference is that the states in $\mathcal{K}$ are functions of an "initial" surface α, where $x^\mu(\vec{\tau})$ can be, for instance, the spacelike surface $x^0 = 0$; in this case α contains only one-half of the data needed to determine a solution of the field equations. On the other hand, the states $\Psi[\alpha]$ in K are functions of a *closed* surface α. In fact, the only difference between $\mathcal{K}$ and K is in the global topology of α. If we disregard this, and consider local equations, we can confuse $\mathcal{K}$ and K (see Section 5.3.5).

The relativistic hamiltonian is given in (3.192). The complete solution of the classical dynamics is known if we know the Hamilton function $S[\alpha]$, which is the value of the action

$$S[\alpha] = S[\mathcal{R}, \phi] = \int_{\mathcal{R}} \mathcal{L}(\phi(x), \partial_\mu \phi(x)) \mathrm{d}^4 x, \tag{5.93}$$

where $\mathcal{R}$ is the four-dimensional region bounded by $x(\vec{\tau})$ and $\phi(x)$ is the solution of the equations of motion in this region, determined by the boundary data $\phi(x(\vec{\tau})) = \varphi(\vec{\tau})$. If there is more than one of these solutions, we write them as $\phi_i(x)$ and the Hamilton function is multivalued

$$S_i[\alpha] = S[\mathcal{R}, \phi_i] = \int_{\mathcal{R}} \mathcal{L}(\phi_i(x), \partial_\mu \phi_i(x)) \mathrm{d}^4 x. \tag{5.94}$$

The relativistic Hamiltonian gives rise to the Wheeler–DeWitt equation

$$H\left[x^\mu, \phi, -\mathrm{i}\hbar \frac{\delta}{\delta x^\mu}, -\mathrm{i}\hbar \frac{\delta}{\delta \varphi}\right](\vec{\tau}) \, \Psi[\alpha] = 0, \tag{5.95}$$

precisely as in the finite-dimensional case. The Hamilton–Jacobi equation (3.193) can be interpreted as the eikonal approximation for this wave equation.

The complete solution of the dynamics is known if we know the propagator $W[\alpha]$, which is a solution of this equation. Formally, the field propagator can be written as a functional integral

$$W[\alpha] = \int_{\phi(x(\vec{\tau}))=\varphi(\vec{\tau})} [\mathrm{D}\phi] \, \mathrm{e}^{-\frac{\mathrm{i}}{\hbar} S[\mathcal{R},\phi]}. \tag{5.96}$$

Of course, one should not confuse the field propagator $W[\alpha]$ with the Feynman propagator. The first propagates field, the second the particles of a QFT. The first is a functional of a surface and the value of the field on this surface, the second is a function of two spacetime points. To the lowest order in $\hbar$, the saddle-point approximation gives

$$W[\alpha] \sim \sum_i A_i[\alpha] \, e^{-\frac{i}{\hbar} S_i[\alpha]}. \tag{5.97}$$

There are two characteristic difficulties in the field theoretical context that are absent in finite dimensions: the definition of the scalar product, and the need to regularize operator products.

First, in finite dimensions, a measure dq^a on $\mathcal{C}$ is sufficient to define an associated L_2 Hilbert space of wave functions. In the field theoretical case, we have to define the scalar product in some other way. The scalar product must respect the invariances of the theory and must be such that real classical variables be represented by self-adjoint operators. This is because self-adjoint operators have a real spectrum, and the spectrum determines the values that a quantity can take in a measurement. Given a set of linear operators on a linear space, the requirement that they are self-adjoint puts stringent conditions on the scalar product. As we shall see, in all cases of interest these requirements are sufficient to determine the scalar product.

Second, local operators are in general distributions and their products are ill defined. Operator products arise in physical observable quantities as well as in the dynamical equation, namely in the Wheeler–DeWitt equation. In particular, functional derivatives are distributions. In the classical Hamilton–Jacobi equation we have products of functional derivatives of the Hamilton–Jacobi functional, which are well-defined products of functions. In the corresponding quantum Wheeler–DeWitt equation, these become products of functional-derivative operators which are ill defined without an appropriate renormalization procedure. The definition of generally covariant regularization techniques will be a major concern in the second part of the book.

5.3.1 Functional representation

Consider a simple free scalar theory, where $V = 0$. I describe this well-known QFT in some detail in order to illustrate certain techniques that play a role in quantum gravity. In particular, I illustrate the *functional* representation of quantum field theory, a simple form of the Wheeler–DeWitt equation, the general form of $W[\alpha]$, and its physical interpretation. The functional representation is the representation in which the field operator is diagonal. The quantum states will be represented as functionals $\Psi[\phi] = \langle \phi | \Psi \rangle$, where $|\phi\rangle$ is the (generalized) eigenstate of the field operator with eigenvalue $\phi(\vec{x})$. The relation between this representation and the conventional one on the Fock basis $|\vec{k}_1, \ldots, \vec{k}_1\rangle$ is precisely the same as the relation between the Schrödinger representation $\psi(x)$ and the one on the energy basis $|n\rangle$, for a simple harmonic oscillator. I also illustrate the way in which the scalar product on the space of the solutions of the Wheeler–DeWitt equation is determined by the reality properties of the field operators.

To start with, and to connect the generally covariant formalism described above with conventional QFT, let's restrict the surface $x(\vec{\tau})$ in α to a spacelike surface $x^{\mu}(\vec{\tau}) = (t, \vec{\tau})$ in Minkowski space. Then $\alpha = [t, \phi(\vec{x})]$ and $\Psi[\alpha] = \Psi[t, \phi(\vec{x})]$. The Hamilton–Jacobi equation (3.186) reduces to (3.190). The corresponding quantum Wheeler–DeWitt equation becomes

$$i\hbar \frac{\partial}{\partial t} \Psi = H_0 \Psi, \tag{5.98}$$

where the nonrelativistic hamiltonian operator H_0 is

$$H_0 = \int d^3x \, H_0 \left[\phi, -i\hbar \frac{\delta}{\delta \phi}\right](\vec{x}), \tag{5.99}$$

and $H_0[\phi, p](\vec{x})$ is given in (3.195). The factor ordering of this operator can be chosen in order to avoid the divergence that would result from the naive factor ordering

$$H_{0 \text{ naive}} = \frac{1}{2} \int d^3x \left[-\hbar^2 \frac{\delta}{\delta\phi(\vec{x})} \frac{\delta}{\delta\phi(\vec{x})} + |\vec{\nabla}\phi|^2(\vec{x}) + m^2\phi^2(\vec{x})\right]. \tag{5.100}$$

The Fourier modes

$$\phi(\vec{k}) = (2\pi)^{-3/2} \int d^3x \, e^{+i\vec{k}\cdot\vec{x}} \phi(\vec{x}) \tag{5.101} \cdot$$

decouple

$$H_0 = \frac{1}{2} \int d^3\vec{k} \left[p^2(\vec{k}) + \omega^2(\vec{k})\phi^2(\vec{k})\right], \tag{5.102}$$

where $\omega = \sqrt{|\vec{k}|^2 + m^2}$. The dangerous divergence is produced by the vacuum energy of the quantum oscillators associated with each mode $\vec{k}$, and can be avoided by normal ordering. In terms of the positive and negative frequency fields

$$a(\vec{k}) = \frac{i}{\sqrt{2\omega}} \, p(\vec{k}) + \sqrt{\frac{\omega}{2}} \phi(\vec{k}), \tag{5.103}$$

$$a^{\dagger}(\vec{k}) = -\frac{i}{\sqrt{2\omega}} \, p(-\vec{k}) + \sqrt{\frac{\omega}{2}} \phi(-\vec{k}), \tag{5.104}$$

the hamiltonian reads

$$H_0 = \int d^3\vec{k} \, \omega(\vec{k}) \, a^{\dagger}(\vec{k}) \, a(\vec{k}). \tag{5.105}$$

We define the quantum hamiltonian by this equation, where

$$a(\vec{k}) = \frac{\hbar}{\sqrt{2\omega}} \frac{\delta}{\delta\phi(\vec{k})} + \sqrt{\frac{\omega}{2}}\phi(\vec{k}), \tag{5.106}$$

$$a^{\dagger}(\vec{k}) = -\frac{\hbar}{\sqrt{2\omega}} \frac{\delta}{\delta\phi(-\vec{k})} + \sqrt{\frac{\omega}{2}}\phi(-\vec{k}). \tag{5.107}$$

The lowest-energy eigenvector of the hamiltonian has vanishing eigenvalue and is called the Minkowski vacuum state. This state is usually denoted $|0\rangle$; I denote it here as $|0_{\mathrm{M}}\rangle$, where M stands for Minkowski, in order to distinguish it from other vacuum states that will be introduced later on. The Minkowski vacuum state is determined by $a(\vec{k})|0_{\mathrm{M}}\rangle = 0$. In the functional representation, this state reads

$$\Psi_{0_{\mathrm{M}}}[\phi] \equiv \langle\phi|0_{\mathrm{M}}\rangle \tag{5.108}$$

and is determined by

$$a(\vec{k})\Psi_{0_{\mathrm{M}}}[\phi] = \frac{\hbar}{\sqrt{2\omega}} \frac{\delta}{\delta\phi(\vec{k})}\Psi_{0_{\mathrm{M}}}[\phi] + \sqrt{\frac{\omega}{2}}\phi(\vec{k})\Psi_{0_{\mathrm{M}}}[\phi] = 0. \tag{5.109}$$

The solution of this equation gives the functional form of the vacuum state

$$\Psi_{0_M}[\phi] = Ne^{-\frac{1}{2\hbar}\int d^3k \ \omega(\vec{k})\phi(\vec{k})\phi(\vec{k})}. \tag{5.110}$$

The one-particle state with momentum $\vec{k}$ is created by $a^{\dagger}(\vec{k})$:

$$\Psi_{\vec{k}}[\phi] \equiv \langle\phi|\vec{k}\rangle = a^{\dagger}(\vec{k})\Psi_{0_{\mathrm{M}}}[\phi] = \sqrt{2\omega} \ \phi(\vec{k}) \ \Psi_{0_{\mathrm{M}}}[\phi]. \tag{5.111}$$

It has energy $\hbar\omega(\vec{k})$. Therefore, the time-dependent state

$$\Psi_{\vec{k}}[t, \phi] \equiv \sqrt{2\omega} \ e^{-i\omega(\vec{k})t} \ \phi(\vec{k}) \ \Psi_{0_{\mathrm{M}}}[\phi] \tag{5.112}$$

is a solution of the Wheeler–DeWitt equation (5.98).

A generic one-particle state with wave function $f(\vec{k})$ is defined by

$$|f\rangle \equiv \int \frac{d^3k}{\sqrt{2\omega}} \ f(\vec{k}) \ |\vec{k}\rangle, \tag{5.113}$$

and its functional representation is therefore

$$\Psi_f[\phi] \equiv \langle\phi|f\rangle = \int d^3k \ f(\vec{k}) \ \phi(\vec{k}) \ \Psi_0[\phi] \tag{5.114}$$

or

$$\Psi_f[\phi] = \phi[f] \ \Psi_0[\phi], \tag{5.115}$$

where

$$\phi[f] = \int d^3k \ f(\vec{k}) \ \phi(\vec{k}).$$ (5.116)

The corresponding solution of the Wheeler–DeWitt equation (5.98) is

$$\Psi_f[t, \phi] = \int d^3k \ f(\vec{k}) \ e^{-i\omega(\vec{k})t} \ \phi(\vec{k}) \ \Psi_0[\phi]$$ (5.117)

or, in Fourier transform,

$$\Psi_f[t, \phi] = \int d^3x \ F(t, \vec{x}) \ \phi(\vec{x}) \ \Psi_0[\phi],$$ (5.118)

where

$$F(x) = F(t, \vec{x}) = (2\pi)^{-3/2} \int d^3k \ e^{i(\vec{k}\cdot\vec{x} - \omega(\vec{k})t)} \ f(\vec{k})$$ (5.119)

is a *positive-energy* solution of the Klein–Gordon equation.

The n-particle states $|\vec{k}_1, \dots, \vec{k}_n\rangle$ can be obtained using again the creation operator $a^\dagger(\vec{k})$ in the well-known way. They have energy $\hbar(\omega_1 + \dots + \omega_n)$ where $\omega_i = \omega(\vec{k}_i)$. The general solution of the Wheeler–DeWitt equation is therefore

$$\Psi[t, \phi] = \sum_n \int \frac{d^3k_1 \dots d^3k_n}{\sqrt{2\omega_1 \dots 2\omega_n}} \ f(\vec{k}_1, \dots, \vec{k}_n) \ e^{-i(\omega_1 + \dots + \omega_n)t}$$
$$\times a^\dagger(k_1) \dots a^\dagger(k_n)\Psi_0[\phi].$$ (5.120)

The space $\mathcal{F}$ of these solutions, labeled by the functions $f(\vec{k}_1, \dots, \vec{k}_n)$ is the physical state space $\mathcal{H}$ of the theory. Since $\Psi[t, \phi]$ is determined by $\Psi[\phi] = \Psi[0, \phi]$, we can also represent the quantum states by their value on the $t = 0$ surface, namely as functionals $\Psi[\phi]$.

Scalar product. The scalar product can be determined on the space of the solutions of the Wheeler–DeWitt equation from the requirement that real quantities are represented by self-adjoint operators. The scalar field $\phi(\vec{x})$ and its momentum $p(\vec{x})$ are real. Therefore, we must demand that the corresponding operators are self-adjoint. It follows that the operator $a^\dagger(\vec{k})$ is the adjoint of the operator $a(\vec{k})$. Using this, we obtain easily

$$\langle \vec{k}|\vec{k}'\rangle = \langle a^\dagger(\vec{k})0|a^\dagger(\vec{k}')0\rangle = \langle 0|a(\vec{k})a^\dagger(\vec{k}')0\rangle = \hbar\delta(\vec{k} - \vec{k}').$$ (5.121)

It follows from (5.113) that

$$\langle f|f'\rangle = \hbar \int \frac{d^3k}{2\omega} \ \overline{f(\vec{k})} \ f'(\vec{k}).$$ (5.122)

(Recall that $d^3k/2\omega$ is the Lorentz-invariant measure.) Therefore, the one-particle state space is $\mathcal{H}_1 = L_2[R^3, d^3k/2\omega]$. Let us write

$$f(\vec{x}) = (2\pi)^{-3/2} \int \frac{d^3k}{2\omega} \, e^{i\vec{k}\vec{x}} \, f(\vec{k}). \qquad (5.123)$$

The spacetime function

$$f(x) = f(t, \vec{x}) = (2\pi)^{-3/2} \int \frac{d^3k}{2\omega} \, e^{i(\vec{k}\cdot\vec{x} - \omega(\vec{k})t)} \, f(\vec{k}) \qquad (5.124)$$

is a *positive-frequency* solution of the Klein–Gordon equation with initial value $f(0, \vec{x}) = f(\vec{x})$ (not to be confused with the one defined in (5.119), which is $F = i\partial_t f$). Then, easily,

$$\langle f | f' \rangle = i\hbar \int d^3x \left[\overline{f(x)} \partial_0 f'(x) - f'(x) \partial_0 \overline{f(x)} \right]_{t=0}. \qquad (5.125)$$

This is the well-known Klein–Gordon scalar product, which is positive definite on the positive-frequency solutions.

Notice that the one-particle Hilbert space can be represented in various equivalent manners. It is

- the space of the positive-frequency solutions $f(x)$ of the Klein–Gordon equation, with the scalar product (5.125);
- the space $H = L_2[R^3, d^3k/2\omega]$ of the functions $f(\vec{k})$;
- the space $H = L_2[R^4, \delta(k^2 + m^2)\theta(k^0)d^4k]$ of the functions $f(k)$;
- the space $H = L_2[R^3, d^3x]$ of the functions

$$f(\vec{x}) = \int \frac{d^3k}{\sqrt{2\omega}} e^{i\vec{k}\cdot\vec{x}} \, f(\vec{k}) \qquad (5.126)$$

  (the position operator $\vec{x}$ in this representation is obviously self-adjoint: it is the well-known Newton–Wigner operator, which has a far more complicated form in other representations);
- and so on.

Using the same technique the entire space $\mathcal{F}$ can be equipped with a scalar product. The resulting Hilbert space is of course the well-known Fock space over this one-particle Hilbert space.

5.3.2 Field propagator between parallel boundary surfaces

Consider now a surface Σ_t formed by two parallel spacelike planes in Minkowski space: say $x_1^\mu(\vec{\tau}) = (t_1, \vec{\tau})$ and $x_2^\mu(\vec{\tau}) = (t_2, \vec{\tau})$. Consider two scalar fields $\varphi_1(\vec{\tau})$, $\varphi_2(\vec{\tau})$ on these planes. Let α be the union of these two surfaces, with their fields; that is, α is formed by two disconnected components $\alpha = \alpha_1 \cup \alpha_2 = [x_1^\mu(\vec{\tau}), \varphi_1(\vec{\tau})] \cup [x_2^\mu(\vec{\tau}), \varphi_2(\vec{\tau})]$. Consider the field propagator (5.96) for this value of α. Thus, $W[\alpha] = W[t_1, \varphi_1, t_2, \varphi_2]$. In this case, we can simply write

$$W[t_1, \varphi_1, t_2, \varphi_2] = \langle t_1, \varphi_1 | t_2, \varphi_2 \rangle = \langle \varphi_1 | e^{-\frac{i}{\hbar} H_0(t_1 - t_2)} | \varphi_2 \rangle. \qquad (5.127)$$

The calculation of the propagator is simplified by the fact that the quantum field theory is essentially a collection of one harmonic oscillator for each mode k. Using the propagator of the harmonic oscillator given in (5.11), one obtains with some algebra

$$W[t_1, \varphi_1, t_2, \varphi_2]$$
$$= \mathcal{N} \exp\left\{ -\frac{i}{2\hbar} \int \frac{d^3k}{(2\pi)^3} \omega \left[\frac{(|\varphi_1|^2 + |\varphi_2|^2)\cos[\omega(t_1 - t_2)] - 2\overline{\varphi_1}\varphi_2}{\sin[\omega(t_1 - t_2)]} \right] \right\},$$
$$(5.128)$$

where $\mathcal{N}$ is the formal divergent normalization factor

$$\mathcal{N} \sim \prod_{\vec{k}} \sqrt{\frac{m\omega(\vec{k})}{h}} \exp\left\{ -\frac{V}{2} \int \frac{d^3k}{(2\pi)^3} \ln\left[\sin[\omega(\vec{k})(t_1 - t_2)] \right] \right\}. \quad (5.129)$$

This has the form (5.97): see the classical Hamilton function given in (3.170).

Minkowski vacuum from the euclidean field propagator. The state space at time zero, $\mathcal{H}_{t=0}$, is Fock space, where the field operators $\varphi(\vec{x}) = \phi(\vec{x}, t)$ and the hamiltonian H_0 are defined. Fock space is separable and therefore admits countable bases. Choose a basis $|n\rangle$ of eigenstates of H_0 with eigenvalues E_n, and consider the operator

$$W(T) = \sum_n e^{-\frac{T}{\hbar} E_n} |n\rangle\langle n|. \qquad (5.130)$$

In the large-T limit, this becomes the projection on the only eigenstate with vanishing energy, namely the Minkowski vacuum

$$\lim_{T \to \infty} W(T) = |0_M\rangle\langle 0_M|. \qquad (5.131)$$

In the functional Schrödinger representation, the operator (5.130) reads

$$W[\varphi_1, \varphi_2, T] = \langle \varphi_1 | e^{-\frac{i}{\hbar} H_0(-iT)} | \varphi_2 \rangle = W[0, \varphi_1, iT, \varphi_2]. \qquad (5.132)$$

Therefore, it is the analytical continuation of the field propagator (5.127), and satisfies the euclidean Schrödinger equation

$$-\hbar \frac{\partial}{\partial T} W[\varphi_1, \varphi_2,, T] = H_{\varphi_1} W[\varphi_1, \varphi_2,, T]. \qquad (5.133)$$

We can obtain the vacuum (up to normalization) as

$$\Psi_{0_{\mathrm{M}}}[\varphi] = \langle \varphi | 0_{\mathrm{M}} \rangle = \lim_{T \to \infty} W[\varphi, 0, T]. \qquad (5.134)$$

We can derive all particle scattering amplitudes from the functional $W[\varphi_1, \varphi_2, T]$. For instance, the 2-point function can be obtained as the analytic continuation of the Schwinger function

$$S(x_1, x_2) = \lim_{T \to \infty} \int D\varphi_1 D\varphi_2 \; W[0, \varphi_1, T] \varphi_1(\vec{x}_1)$$
$$\times W[\varphi_1, \varphi_2, (t_1 - t_2)] \varphi_2(\vec{x}_2) \; W[\varphi_2, 0, T]. \qquad (5.135)$$

This can be generalized to any n-point function where the times $t_1, \ldots, t_n$ are on the $t = 0$ and the $t = T$ surfaces; these, in turn, are sufficient to compute all scattering amplitudes, since time dependence of asymptotic states is trivial.

$W[\varphi_1, \varphi_2, T]$ admits the well-defined functional integral representation

$$W[\varphi_1, \varphi_2, T] = \int_{\substack{\phi|_{t=T}=\varphi_1 \\ \phi|_{t=0}=\varphi_2}} D\phi \; e^{-\frac{1}{\hbar} S_T^E[\phi]}. \qquad (5.136)$$

Here the integral is over all fields ϕ on the strip $\mathcal{R}$ bounded by the two surfaces $t = 0$ and $t = T$, with fixed boundary value. The action $S_T^E[\phi]$ is the euclidean action. Notice that using this functional integral representation, the expression (5.135) for the Schwinger function becomes the well-known expression

$$S(x_1, x_2) = \int D\phi \; \phi(x_1) \; \phi(x_2) \; e^{-\frac{1}{\hbar} S^E[\phi]}, \qquad (5.137)$$

obtained by joining at the two boundaries the three functional integrals in the regions $t < t_2$, $t_2 < t < t_1$ and $t_1 < t$. The functional $W[\varphi_1, \varphi_2, T]$ can be computed explicitly in the free field theory. Its expression in terms of the Fourier transform $\tilde{\varphi}$ of φ is the analytic continuation of (5.128)

$$W[\varphi_1, \varphi_2, T] = \mathcal{N} \exp \left\{ -\frac{1}{2\hbar} \int \frac{d^3 k}{(2\pi)^3} \; \omega \left(\frac{|\tilde{\varphi}_1|^2 + |\tilde{\varphi}_2|^2}{\tanh(\omega T)} - \frac{2\tilde{\varphi}_1 \overline{\tilde{\varphi}_2}}{\sinh(\omega T)} \right) \right\}.$$

$$(5.138)$$

The dynamical vacuum $|0_{\Sigma_T}\rangle$. Consider the boundary state space $\mathcal{K}_{\Sigma_t}$, associated with the *entire* surface Σ_t, as in Section 5.1.4. That is, define $\mathcal{K}_{\Sigma_t} = \mathcal{H}_t \otimes \mathcal{H}_0^*$. Denote $\varphi = (\varphi_1, \varphi_2)$ a field on Σ_t. The field basis of the Fock space induces the basis $|\varphi\rangle = |\varphi_1, \varphi_2\rangle \equiv |\varphi_1\rangle_t \otimes \langle\varphi_2|_0$ in $\mathcal{K}_{\Sigma_t}$; the vectors $|\Psi\rangle$ of K_{Σ_t} are written in this basis as functionals $\Psi[\varphi] = \Psi[\varphi_1, \varphi_2] \equiv \langle\varphi_1, \varphi_2|\Psi\rangle$. This is the field theoretical generalization of the boundary state space defined in (5.42).

The functional W defines a preferred state in this Hilbert space, as in (5.43)–(5.44). Denote this state $|0_{\Sigma_t}\rangle$, and call it the dynamical vacuum. It is defined by $\langle\varphi|0_{\Sigma_t}\rangle \equiv W[t, \varphi_1, 0, \varphi_2]$. This state expresses the dynamics from $t = 0$ to t. A state in the tensor product of two Hilbert spaces defines a linear mapping between the two spaces. The linear mapping from $\mathcal{H}_{t=0}$ to $\mathcal{H}_{t=T}$ defined by $|0_{\Sigma_T}\rangle$ is precisely the time evolution e^{-iHt}.

The interpretation of this state is the same as in the finite-dimensional case. The tensor product of two quantum state spaces describes the ensemble of the measurements described by the two factors. Therefore, $\mathcal{K}_{\Sigma_t}$ is the space of the possible results of all measurements performed at time 0 *and* at time t. Observations at two different times are correlated by the dynamics. Hence $\mathcal{K}_{\Sigma_t}$ is a "kinematical" state space, in the sense that it describes more outcomes than the physically realizable ones. Dynamics is then a restriction on the possible outcomes of observations. It expresses the fact that measurement outcomes are correlated. The linear functional $\langle0_{\Sigma_t}|$ on $\mathcal{K}_{\Sigma_t}$ assigns an amplitude to any outcome of observations. This amplitude gives us the correlation between outcomes at time 0 and outcomes at time t.

Therefore, the theory can be represented as follows. The Hilbert space $\mathcal{K}_{\Sigma_t}$ describes all possible outcomes of measurements made on Σ_t. The dynamics is given by a single bra state $\langle0_{\Sigma_t}| : \mathcal{K}_t \to C$. For a given collection of measurement outcomes described by a state $|\Psi\rangle$, the quantity $\langle0_{\Sigma_t}|\Psi\rangle$ gives the correlation probability amplitude between these measurements.

Using (5.131), we have then the relation between the dynamical vacuum and the Minkowski vacuum (the bra/ket mismatch is apparent only, as the three states are in different spaces)

$$\lim_{t\to\infty} |0_{\Sigma_{-it}}\rangle = |0_M\rangle \otimes \langle0_M|. \tag{5.139}$$

5.3.3 Arbitrary boundary surfaces

So far, I have considered only boundary surfaces formed by two parallel spacelike planes. This restriction is sufficient and convenient in ordinary QFT on Minkowski space, but it has no meaning in a generally covariant context. It is therefore necessary to consider arbitrary boundary surfaces, so let us study the extension of the formalism to the case where the surface

Σ, instead of being formed by two parallel planes, is the boundary of a (sufficiently regular) arbitrary *finite* region of spacetime $\mathcal{R}$.

Let Σ be a closed, connected 3d surface in Minkowski spacetime, with the topology (but, in general, not the geometry) of a 3-sphere, and $\Sigma = \partial\mathcal{R}$. Let φ be a scalar field on Σ and consider the functional

$$W[\varphi, \Sigma] = \int_{\phi|_\Sigma = \varphi} D\phi \; e^{-S_\mathcal{R}^E[\phi]}. \tag{5.140}$$

The integral is over all 4d fields on $\mathcal{R}$ that take the value φ on Σ, and the action in the exponent is the euclidean action where the 4d integral is over $\mathcal{R}$. In the free theory the integral is a well-defined gaussian integral and can be evaluated. The classical equations of motion with boundary value φ on Σ form an elliptic system which in general has a solution $\phi_{cl}[\varphi]$, that can be obtained by integration from the Green function for the shape $\mathcal{R}$. A change of variable in the integral reduces it to a trivial gaussian integration times $e^{-S_\mathcal{R}^E[\varphi_{cl}]}$. Here $S_\mathcal{R}^E[\varphi]$ is the field theoretical Hamilton function: the action of the bulk field determined by the boundary condition φ.

$W[\varphi, \Sigma]$ can be defined in the Minkowski regime as well. If Σ is a rectangular box in Minkowski space, let $\varphi = (\varphi_{out}, \varphi_{in}, \varphi_{side})$ be the components of the field on the spacelike bases and timelike side. Consider the field theory defined in the box, with time-dependent boundary conditions φ_{side}, and let $U[\varphi_{side}]$ be the evolution operator from $t = 0$ to $t = T$ generated by the (time-dependent) hamiltonian of the theory. Then we can write

$$W[\varphi, \Sigma] \equiv \langle \varphi_{out} | U[\varphi_{side}] | \varphi_{in} \rangle. \tag{5.141}$$

In particular, if φ_{side} is constant in time, W can be obtained by analytic continuation from the euclidean functional. More generally, we can write the formal definition

$$W[\varphi, \Sigma] = \int_{\phi|_\Sigma = \varphi} D\phi \, e^{iS_\mathcal{R}[\phi]}. \tag{5.142}$$

Notice that $W[\varphi, \Sigma]$ is a function on the space $\mathcal{G}$ defined in Section 3.3.3. This space represents all possible ensembles of classical field measurements on a closed surface, namely the minimal data for a local experiment. Formally, functions on $\mathcal{G}$ define the quantum state space K, and $W[\varphi, \Sigma]$ defines the preferred covariant vacuum state $|0\rangle$ in K.

Local Schrödinger equation. $W[\varphi, \Sigma]$ satisfies a local functional equation that governs its dependence on Σ. Let $\vec{\tau}$ be arbitrary coordinates on Σ. Represent the surface and the boundary fields as $\Sigma : \vec{\tau} \mapsto x^\mu(\vec{\tau})$ and $\varphi : \vec{\tau} \mapsto \varphi(\vec{\tau})$. Let $n^\mu(\vec{\tau})$ be the unit length normal to Σ. Then

$$n^\mu(\vec{\tau}) \frac{\delta}{\delta x^\mu(\vec{\tau})} W[\varphi, \Sigma] = H(\vec{\tau}) \, W[\varphi, \Sigma], \tag{5.143}$$

where $H(\vec{x})$ is an operator obtained by replacing $\pi(\vec{x})$ by $-i\delta/\delta\varphi(x)$ in the hamiltonian density

$$H(\vec{x}) = g^{-\frac{1}{2}}\pi^2(\vec{x}) + g^{\frac{1}{2}}\left(|\vec{\nabla}\varphi|^2 + m^2\varphi^2\right). \qquad (5.144)$$

Here, g is the determinant of the induced metric on Σ and the norm is taken in this metric (see [145, 146]). The local Hamilton–Jacobi equation (3.186) can be viewed as the eikonal approximation of this equation. Since W is independent from the parametrization, we have

$$\frac{\partial x^\mu(\vec{\tau})}{\partial \vec{\tau}} \frac{\delta}{\delta x^\mu(\vec{\tau})} W[\varphi, \Sigma] = \vec{P}(\vec{\tau})\, W[\varphi, \Sigma], \qquad (5.145)$$

where the linear momentum is $\vec{P}(\vec{\tau}) = \vec{\nabla}\phi(\vec{\tau})\,\delta/\delta\varphi(\vec{\tau})$. If Σ is spacelike, (5.143) is the (euclidean) Tomonaga–Schwinger equation.

We expect a local equation like (5.143) to hold in any field theory. If the theory is generally covariant, the functional W will be independent from Σ, and therefore the left-hand side of the equation will vanish, leaving only the hamiltonian operator acting on the field variables, namely a Wheeler–DeWitt equation.

5.3.4 What is a particle?

Choose Σ to be a cylinder Σ_{RT}, with radius R and height T, with the two bases on the surfaces $t = 0$ and $t = T$. Given two compact support functions φ_1 and φ_2, defined on $t = 0$ and $t = T$ respectively, we can always choose R large enough for the two compact supports to be included in the bases of the cylinder. Then

$$\lim_{R\to\infty} W[\varphi_1, \varphi_2, \Sigma_{RT}] = W[\varphi_1, \varphi_2, T] \qquad (5.146)$$

because the euclidean Green function decays rapidly and the effect of having the side of the cylinder at finite distance goes rapidly to zero as R increases. Equation (5.135) illustrates how scattering amplitudes can be computed from $W[\varphi_1, \varphi_2, T]$. In turn, (5.146) indicates how $W[\varphi_1, \varphi_2, T]$ can be obtained from $W[\varphi, \Sigma]$, where Σ is the boundary of a finite region. Therefore, knowledge of $W[\varphi, \Sigma]$ allows us to compute particle scattering amplitudes. We expect this to remain true in the perturbative expansion of an interacting field theory as well, where $\mathcal{R}$ includes the interaction region.

The limits $T, R \to \infty$ seem to indicate that arbitrarily large surfaces Σ are needed to compute vacuum and particle scattering amplitudes. But notice that the convergence of $W[\varphi_1, \varphi_2, T]$ to the vacuum projector is dictated by (5.130) and is exponential in the mass gap, or the Compton frequency of the particle. Thus T at laboratory scales is largely sufficient to guarantee arbitrarily accurate convergence. In the euclidean regime, rotational symmetry suggests the same to hold for the $R \to \infty$ limit. Thus, the limits can be replaced by choosing R and T at laboratory scales. (At least for the vacuum, which does not require analytic continuation.)

The conventional notions of vacuum and particle states are global in nature. How is it possible that we can recover them from the local functional $W[\varphi, \Sigma]$? This is an important question that plays a role in QFT on curved spacetime and in quantum gravity. To answer this question, notice that realistic particle detectors are *finitely* extended. How can a finitely extended detector detect particles, if particles are globally defined objects?

The answer is that there exist two distinct notions of particle. Fock particle states are "global," while the physical states detected by a localized detector (eigenstates of local operators describing detection) are "local" particle states. Local particle states are close to (in a suitable topology), but distinct from, the global particle states. In conventional QFT, we use a global particle state in order to conveniently approximate the local particle state detected by a detector. Global particle states, indeed, are far easier to deal with.

Therefore, the global nature of the conventional definition of vacuum and particles is not dictated by the physical properties of particles: it is an approximation adopted for convenience. Replacing the limits $R \to \infty$ and $T \to \infty$ with finite macroscopic R and T we miss the *exact* global vacuum or n-particle state, but we can nevertheless describe local experiments. The restriction of QFT to a finite region of spacetime must describe completely experiments confined to this region.

Global and local particles in a simple finite system. The distinction between global particles and local particles can be illustrated in a very simple system. Consider two weakly coupled harmonic oscillators. Let the total hamiltonian of the system be

$$H = \frac{1}{2}(\dot{p}_1^2 + q_1^2) + \frac{1}{2}(\dot{p}_2^2 + q_2^2) - 2\lambda q_1 q_2 = H_1 + H_2 - \lambda V. \qquad (5.147)$$

Consider a measuring apparatus that interacts only with the first oscillator and measures the quantity H_1. The Hilbert space of the system is $\mathcal{H} = L_2[R^2, dq_1 dq_2]$. On this space, the quantity H_1 is represented by the operator $-\hbar^2 \partial^2/\partial q_1^2 + q_1^2$. The operator has a discrete spectrum $E = (n + 1/2)\hbar$. If the result of the measurement is the eigenvalue $(1 + 1/2)\hbar$, let us say that "there is one local particle in the first oscillator." In particular, a one-local-particle state is the common eigenstate of H_1 and H_2

$$\psi_{\text{local}}(q_1, q_2) = q_1 e^{-(q_1^2 + q_2^2)/2\hbar}, \qquad (5.148)$$

in which there is one local particle in the first oscillator and no local particles in the second.

Next, let us diagonalize the full hamiltonian H. This can easily be done by finding the normal modes of the system, which are $q_\pm = (q_1 \pm q_2)/\sqrt{2}$, and have frequencies

$\omega_{\pm}^2 = 1 \pm \lambda$. The eigenvalues of H are therefore $E = \hbar(n_+\omega_+ + n_-\omega_- + 1)$. We call $|n_+, n_-\rangle$ the corresponding eigenstates, and $N = n_+ + n_-$ the global-particle number. In particular, we call "one-global-particle state" all states with $N = 1$, namely any state of the form $|\psi\rangle = \alpha|1, 0\rangle + \beta|0, 1\rangle$. Notice that this is precisely the definition of one-particle states in QFT: a one-particle state is an arbitrary linear combination of states $|k\rangle$ where there is a single quantum in one of the modes. In particular, consider the one-global-particle state $|\psi\rangle = (|1, 0\rangle + |0, 1\rangle)/\sqrt{2}$. This is a global particle which is maximally localized on the first oscillator. A straightforward calculation gives, to first order in λ

$$\psi_{\text{global}}(q_1, q_2) = (q_1 + \frac{\lambda}{4}q_2)e^{-(q_1^2 + q_2^2 - 2\lambda q_1 q_2)/2\hbar}. \tag{5.149}$$

The two states ψ_{local} and ψ_{global} are different and have different physical meaning. The state ψ_{global} is the kind of state that is called a one-particle state in QFT. It is the one-particle state which is most localized on the first oscillator. On the other hand, if our measuring apparatus interacts only with the first oscillator then what we measure is not ψ_{global}, it is ψ_{local}, which is an eigenstate of an operator that acts only on the variable q_1.

In QFT we confuse the two kinds of states. In the formalism we use global-particle states such as ψ_{global}. However, particle detectors are localized in space. (A local measuring apparatus can only interact with the components of the field in a finite region, like the apparatus that interacts only with the variable q_1 in the example.) Therefore, they measure particle states such as ψ_{local}. Strictly speaking, therefore, the interpretation of the particle states measured by particle detectors as global-particle states is a mistake, because a global-particle state can never be an eigenstate of a local measuring apparatus, and therefore cannot be detected by a local apparatus.

The reason we can nevertheless use this interpretation successfully is that the states ψ_{local} and ψ_{global} are very similar. In the example, their distance in the Hilbert norm vanishes to first order in lambda:

$$(\psi_{\text{global}}, \psi_{\text{local}}) = 0(\lambda). \tag{5.150}$$

The error we make in using ψ_{global} to describe the physical state ψ_{local} is small if λV is small. In the field theoretical case, λV represents the interaction energy between the region inside the detector and the region outside the detector; this energy is very small compared to the energy of the state itself, for all the states of interest. We can effectively approximate the local-particle states that are detected by our measuring apparatus, by means of the global-particle states, which are easier to deal with.

On the other hand, the argument shows that global-particle states are not *required* for dealing with the realistic observed particles; they are just a convenient approximation. If we can define local-particle states by means of a local formalism, we are not making a mistake: rather, we are simply not using an approximation that was convenient on flat space, but may not be viable in a generally covariant context.

5.3.5 *Boundary state space* K *and covariant vacuum* $|0\rangle$

Finally, consider the space $\mathcal{G}$ of the variable $\alpha = (\Sigma, \varphi)$, where Σ is a closed 3d surface in spacetime. Call K the space of functions $\psi[\alpha] = \psi[\Sigma, \varphi]$. This space represents all possible outcomes of ensembles of measurements on the boundary of a finite region of spacetime. The measurements include spacetime localization measurements that determine the surface Σ as well as field (or particle) measurements that determine φ (or a function of φ). K is the boundary quantum space.

There is a preferred state $|0\rangle$ in K, given by

$$\langle \Sigma, \varphi | 0 \rangle = W[\Sigma, \varphi]. \tag{5.151}$$

If the functional integral can be defined, this is given by (5.142). The state $|0\rangle$ expresses the dynamics entirely. As we shall see, this formulation of QFT makes sense in quantum gravity.

In general, K is a space of functions over $\mathcal{G}$. Recall that $\mathcal{G}$ is the space of data needed to determine a classical solution: two events in the finite-dimensional case; a set of events forming a 3d closed surface in the field case.

In the case of a finite-dimensional theory, a classical solution in some interval is determined by *two* events in $\mathcal{C}$. In the quantum theory, a complete experiment consists of *two* events, a preparation and a quantum measurement. In this case, $\mathsf{K} = L_2[\mathcal{G}] = L_2[\mathcal{C} \times \mathcal{C}] \sim L_2[\mathcal{C}] \otimes L_2[\mathcal{C}] = \mathcal{K} \otimes \mathcal{K}$ is the space representing *two* quantum events, while $\mathcal{K} = L_2[\mathcal{C}]$ is the space representing a single quantum event.

In the field theoretical case, a classical solution in a region $\mathcal{R}$ is determined by infinite events in $\mathcal{C}$, forming a closed 3d surface, namely by a 3d surface $\Sigma = \partial R$ in spacetime and the field φ on it. In the quantum theory, a complete experiment requires measurements (or assumptions) on the entire Σ. In this case, $\mathsf{K} \sim L_2[\mathcal{G}]$ is the space describing the observation of the entire boundary surface Σ and the measurements on it.

The boundary of $\mathcal{R}$ can be formed by two (or even more) connected components Σ. In this case, we can decompose K into the tensor product of one factor $\mathcal{K}$ associated with each component. The space $\mathcal{K}$ is then a space of functionals of the connected surface Σ and the field on it. Since the Wheeler–DeWitt equation is local, it looks the same on $\mathcal{K}$ and on K. Therefore, the distinction between K and $\mathcal{K}$ is of much less importance in the field theoretical context than in the finite-dimensional case. The space K is associated with the idea of the full data characterizing an experiment on a closed surface Σ, while the space $\mathcal{K}$ is associated with the idea of an "initial data" surface Σ.

5.3.6 Lattice scalar product, intertwiners and spin network states

An interacting quantum field theory can be constructed as a perturbation expansion around a free theory. An alternative is to define a cut-off theory, with a large but finite number of degrees of freedom, using a lattice. One expects then to recover physical predictions as suitable limits as the lattice spacing is taken to zero. I illustrate here the definition of the scalar product in a lattice gauge theory, since the same technique is used in quantum gravity.

Consider a three-dimensional lattice Γ with L links l and N nodes n. To define a Yang–Mills theory for a compact Yang–Mills group G on this lattice, we associate a group element U_l to each link l and we consider the Hilbert space $\tilde{\mathcal{K}}_\Gamma = L_2[G^L, dU_l]$, where G^L is the product of L copies of G and $dU_l \equiv dU_l \ldots dU_l$ is the Haar measure on the group. Quantum states in $\tilde{\mathcal{K}}_\Gamma$ are functions $\Psi(U_l)$ of L group elements. The scalar product of two states is given by

$$\langle \Psi | \Phi \rangle \equiv \int dU_1 \ldots dU_L \, \overline{\Psi(U_1, \ldots, U_L)} \, \Phi(U_1, \ldots, U_L). \qquad (5.152)$$

An orthonormal basis of states in $\tilde{\mathcal{K}}_\Gamma$ can be obtained as follows. Let j label unitary irreducible representations of G, and let $(R^j(U))^\alpha_\beta$ be the matrix elements of the representation. The Peter–Weyl theorem tells us that the states $|j, \beta, \alpha\rangle$ defined by $\langle U|j, \beta, \alpha\rangle = (R^j(U))^\alpha_\beta$ form an orthonormal basis in $L_2[G, dU]$. A basis in $\tilde{\mathcal{K}}_\Gamma$ is therefore given by the states

$$|j_l, \beta_l, \alpha_l\rangle \equiv |j_1, \ldots, j_L, \beta_1, \ldots, \beta_L, \alpha_1, \ldots, \alpha_L\rangle \qquad (5.153)$$

defined by $\langle U_l | j_l, \beta_l, \alpha_l \rangle = \prod_l (R^{j_l}(U_l))^{\alpha_l}_{\beta_l}$.

The theory is invariant under local Yang–Mills transformations on the lattice. These depend on a group element λ_n for each node n. The variables U_l transform under a gauge transformation as $U_l \mapsto \lambda_{l_i}{}^{-1} U_l \lambda_{l_f}$ where the link l goes from the initial node l_i to the final node l_f. Hence the gauge-invariant states are the ones satisfying

$$\Psi(U_l) = \Psi(\lambda_{l_i}{}^{-1} U_l \lambda_{l_f}). \qquad (5.154)$$

These states form a linear subspace $\tilde{\mathcal{K}}^0_\Gamma$ of $\tilde{\mathcal{K}}_\Gamma$: the space $\tilde{\mathcal{K}}^0_\Gamma$ is the (fixed-time) Hilbert space of the gauge-invariant states of the theory. An orthonormal basis of states in $\tilde{\mathcal{K}}^0_\Gamma$ can be obtained using the notion of an intertwiner.

Intertwiners. Consider N irreducible representations $j_1, \ldots, j_N$. Consider the tensor product of their Hilbert spaces

$$\mathcal{H}_{j_1, \ldots, j_N} = \mathcal{H}_{j_1} \otimes \ldots \otimes \mathcal{H}_{j_N}. \qquad (5.155)$$

This space can be decomposed into a sum of irreducible components. In particular, let $\mathcal{H}^0_{j_1, \ldots, j_N}$ be the subspace formed by the invariant vectors, namely the subspace that transforms in the trivial representation. This space is k-dimensional, where k is the multiplicity with which the trivial representation appears in the decomposition. It is, of course, a Hilbert space, and therefore we can choose an orthonormal basis in it. We call the elements i of this basis "intertwiners" between the representations $j_1, \ldots, j_N$.

More explicitly, elements of $\mathcal{H}_{j_1, \ldots, j_N}$ are tensors $v^{\alpha_1 \cdots \alpha_N}$ with one index in each representation. Elements of $\mathcal{H}^0_{j_1, \ldots, j_N}$ are tensors $v^{\alpha_1 \cdots \alpha_N}$ that are invariant under the action of G on all their indices. That is, they satisfy

$$R^{(j_1)\alpha_1}{}_{\beta_1}(U) \ldots R^{(j_N)\alpha_N}{}_{\beta_N}(U) \, v^{\beta_1 \cdots \beta_N} = v^{\alpha_1 \cdots \alpha_N}. \tag{5.156}$$

The intertwiners $v_i^{\alpha_1 \cdots \alpha_N}$ are a set of k such invariant tensors, which are orthonormal in the scalar product of $\mathcal{H}^0_{j_1, \ldots, j_N}$. That is, they satisfy

$$\overline{v_i^{\alpha_1 \cdots \alpha_N}} \, v_{i' \, \alpha_1 \ldots \alpha_N} = \delta_{ii'}. \tag{5.157}$$

If the space $\mathcal{H}_j$ carries the representation j, its dual space $\mathcal{H}_j^*$ carries the dual representation j^*. An intertwiner i between n dual representations $j_1^*, \ldots, j_n^*$ and m representations $j_1, \ldots, j_m$ is an invariant tensor in the space $(\otimes_{i=1,n} \mathcal{H}_{j_i}^*) \otimes (\otimes_{k=1,m} \mathcal{H}_{j_k})$, that is, a covariant map

$$i : \; (\otimes_{i=1,n} \mathcal{H}_{j_i}) \longrightarrow (\otimes_{k=1,m} \mathcal{H}_{j_k}), \tag{5.158}$$

or an invariant tensor with n lower indices and m upper indices.

Now, associate a representation j_l to each link l, and an intertwiner i_n in each node n (in the tensor product of the representations associated with the links adjacent to the node) of the lattice. The set $s = (\Gamma, j_l, i_n)$ is called a "spin network." Each spin network s defines a state $|s\rangle$ by

$$\langle U_l | s \rangle = \psi_s(U_l) = \prod_l R^{j_l}(U_l) \cdot \prod_n i_n, \tag{5.159}$$

where the raised dot indicates index contraction. Notice that the indices (not indicated in the equation) match, as on each side of the dot there is one index for each couple node-link. The states $|s\rangle$ form a complete and orthonormal basis in $\tilde{\mathcal{K}}^0_\Gamma$

$$\langle s | s' \rangle = \delta_{ss'}. \tag{5.160}$$

This basis will play a major role in quantum gravity.

5.4 Quantum gravity

Finally, I sketch here the formal structure of quantum gravity. The actual mathematical definition of the quantities mentioned here is the task I undertake in the second part of the book.

5.4.1 Transition amplitudes in quantum gravity

In the presence of a background, QFT yields scattering amplitudes and cross sections for asymptotic particle states, and these are compared with data obtained in a lab. The conventional theoretical definition of these amplitudes involves infinitely extended spacetime regions and relies on symmetry properties of the background. In a background-independent context this procedure becomes problematic. For instance, background independence implies immediately that any 2-point function $W(x, y)$ is constant for $x \neq y$, as mentioned in Section 1.1.4. How can the formalism control the localization of the measuring apparatus?

We have seen above that in the context of a simple scalar field theory, local physics can be expressed in terms of a functional $W[\varphi, \Sigma]$ that depends on field boundary eigenstates φ and the geometry of the 3d surface Σ that bounds $\mathcal{R}$. Physical predictions concerning measurements performed in the finite region $\mathcal{R}$, including scattering amplitudes between particles detected in the lab, can be expressed in terms of $W[\varphi, \Sigma]$. The functional satisfies a local version of the Schrödinger equation. The geometry of Σ codes the relative spacetime localization of the particle detectors. $W[\varphi, \Sigma]$ can be expressed as a functional integral over a finite spacetime region $\mathcal{R}$ of spacetime. In the euclidean regime, the functional integral is well defined and can be used to determine the Minkowski vacuum state.

This technique can be extended to quantum gravity, namely to a diffeomorphism-invariant context. The effect of diffeomorphism invariance is that the functional W turns out to be independent of the location of Σ. At first sight, this seems to leave us in the characteristic interpretative obscurity of background-independent QFT: the independence of W from Σ is equivalent to the independence of $W(x, y)$ from x and y, mentioned above.

But a closer look reveals it is not so. The boundary field includes the gravitational field, which is the metric, and therefore the argument of W *does* describe the metric of the boundary surface, that is, the relative spacetime location of the detectors, as explained in Section 4.1.2. Therefore the relative location of the detectors, lost with Σ because of general covariance, comes back with φ, as this now includes the boundary value of the gravitational field. The boundary value of the gravitational field plays the double role previously played by φ and Σ. In fact, this is precisely the core of the conceptual novelty of general relativity: there is no a priori distinction between localization measurements and measurements of dynamical variables.

More formally, in a background-dependent theory the space $\mathcal{G}$ is a space of couples (Σ, φ), but in a general-relativistic theory the space $\mathcal{G}$ is just a space of fields on a closed differential surface. In pure GR, we can take $\mathcal{G}$ as the space of the gravitational connections A on a closed surface. Accordingly, the space K is a space of functionals of the field A on a closed surface. These functionals are invariant under 3d diffeomorphisms of the surface. In the second part of the book, the space K will be built explicitly. As explained in the previous section, the functional W determines a preferred state $|0\rangle$ in K. This is the covariant vacuum state, which contains the dynamical information of the theory.

A key result of the theory developed in the second part of the book is that the eigenstates of the gravitational field on a 3d surface are not smooth fields. They present a characteristic Planck-scale discreteness. These eigenstates determine a preferred basis $|s\rangle$ in K, labeled by the "spin networks" s that will be described in detail in the second part of the book. Each state $|s\rangle$ describes a "quantum geometry of space," namely the possible result of a complete measurement of the gravitational field on the 3d surface. We shall express W in this preferred basis

$$W(s) = \langle 0|s\rangle. \tag{5.161}$$

Therefore, because of the Planck-scale discreteness of space, in the gravitational context, the analog of $W[\varphi, \Sigma]$ is the functional $W(s)$. A definition of $W(s)$ in the canonical quantum theory will be given below in (7.37). As we shall see, the covariant vacuum state $|0\rangle$ will simply be related to the spin network state with no nodes and no links. A sum-over-histories definition of $W(s)$ will be given below in (9.21).

A case of particular interest is the one in which we can separate the boundary surface Σ into two components. For instance, these can be disconnected. Accordingly, we can write s as $(s_{\text{out}}, s_{\text{in}})$ and the associated amplitude as

$$W(s_{\text{out}}, s_{\text{in}}) = \langle 0|s_{\text{out}}, s_{\text{in}}\rangle = \langle s_{\text{out}}|P|s_{\text{in}}\rangle, \tag{5.162}$$

where P is the projector on the solutions of the Wheeler–DeWitt equation. A sum-over-histories expression of $W(s_{\text{out}}, s_{\text{in}})$ is given in terms of histories that go from s_{in} to s_{out}.

5.4.2 Much ado about nothing: the vacuum

The notion of "vacuum state" plays a central role in QFT on a background spacetime. The vacuum is the basis over which Fock space is built. In gravity, on the other hand, the notion of vacuum is very ambiguous. This fact contributes to make quantum gravity sharply different from conventional QFTs. However, this is not a difficulty: a preferred notion of vacuum is not needed for a quantum theory to be well defined. The quantum theory of a harmonic oscillator has a vacuum state, but the quantum theory of a free particle does not. In this respect, general relativity resembles more a free particle than a harmonic oscillator.

Notice that even the terminology of classical GR is confusing with respect to the notion of vacuum: in relativistic parlance, *all* solutions of the Einstein equations without a source term are called "vacuum solutions."

We use three distinct notions of vacuum in quantum gravity.

Covariant vacuum. The first is the nonperturbative, or covariant, vacuum state $|0\rangle$ defined in Sections 5.1.4 and 5.3.2. This is the state in the boundary state space that defines the dynamics. Intuitively, it is defined by the sum-over-histories on a region bounded by the given boundary data. If the metric boundary data are chosen to be spacelike, this is the Hartle–Hawking state. In the context we are considering, instead, the boundary surface bounds a finite 4d region of spacetime, and the state $|0\rangle$ is a background-independent way of coding quantum dynamics.

Empty state. The state $|\emptyset\rangle$ is the kinematical quantum state of the gravitational field in which the volume of space is zero, namely in which there is no physical space. As we shall see, it is related to the covariant vacuum state $|0\rangle$.

Minkowski vacuum. A different notion of vacuum is the Minkowski vacuum state $|0_M\rangle$. The quantum state $|0_M\rangle$ that describes the Minkowski vacuum is not singled out by the dynamics alone. Instead, it is singled out as the lowest eigenstate of an energy H_T which is the variable canonically conjugate to a nonlocal function of the gravitational field defined as the proper time T along a given worldline. This is analogous to the identification of the energy with a momentum p_0, under the choice of a specific Lorentz time x^0. To find this state in quantum gravity, we can use the procedure employed in (5.49) and (5.139). This will be briefly discussed at the end of Chapter 9. Alternatively, in an asymptotically flat context we expect $|0_M\rangle$ to be the lowest eigenstate of the ADM energy.

The notion of vacuum is strictly connected to the notion of energy. The vacuum can be defined as the state with lowest energy. In GR the notion of energy is ambiguous, and the ambiguity in the definition of energy is reflected in the ambiguity in defining the vacuum. Indeed, we can identify several notions of energy in GR.

Canonical energy. The canonical energy, namely the generator H of translations in coordinate time, vanishes identically in any general-relativistic theory. In this sense, all physical states of quantum gravity are vacuum states.

Matter energy. The energy-momentum tensor T^I_μ of the nongravitational fields is well defined, and therefore the energy $E_{\text{matter}} = T^0_0$ of the nongravitational fields is well defined. In classical GR, a vacuum solution is a solution with $E_{\text{matter}} = 0$. In this sense, vacuum states are all the pure gravity physical states without matter.

Gravitational energy. The energy of the gravitational field E_{gravity} is, strictly speaking, (minus) the left-hand side of the time–time component of the Einstein equations; so, the time–time component of the Einstein equations reads $E_{\text{gravity}} + E_{\text{matter}} = 0$. That is, the total energy vanishes; see, for instance, [147].

ADM energy. We can associate an energy E_{ADM} to an isolated system surrounded by a region where the gravitational field is approximately minkowskian. Such a system can be described by asymptotically flat solutions of the Einstein equations. For such a system, we can identify the energy with the generator E_{ADM} of time translations in the asymptotic Minkowski space. Given asymptotic flatness, E_{ADM} is minimized by the Minkowski solution. In this sense, the Minkowski solution is "the vacuum" of the asymptotic minkowskian theory.

The fact that the notions of energy and vacuum are so ambiguous in GR should not be disconcerting. There is nothing essential in these notions: a quantum theory and its predictions are meaningful also in the absence of them. The notions of energy and vacuum play an important role in non-general-relativistic physics just because of the accidental fact that we live in a region of the Universe which happens to have a peculiar symmetry: translation invariance in newtonian or special-relativistic time.

5.5 *Complements

5.5.1 Thermal time hypothesis and Tomita flow

The thermal time hypothesis discussed in Section 3.4 extends nicely to QM and very nicely to QFT.

QM. In QM, the time flow is given by

$$A_t = \alpha_t(A) = e^{itH_0} A e^{-itH_0}. \tag{5.163}$$

A statistical state is described by a density matrix ρ. It determines the expectation values of any observable A via

$$\rho[A] = \text{tr}[A\rho]. \tag{5.164}$$

This equation defines a positive functional ρ on the observables' algebra. The relation between a quantum Gibbs state ρ_0 and H_0 is the same as in (3.202). That is

$$\rho_0 = N e^{-\beta H_0}. \tag{5.165}$$

Correlation probabilities can be written as

$$W_{AB}(t) = \rho_0[\alpha_t(A)B] = \text{tr}[e^{itH_0} A\ e^{-itH_0} B\ e^{-\beta H_0}]. \qquad (5.166)$$

Notice that it follows immediately from the definition that

$$\rho_0[\alpha_t(A)B] = \rho_0[\alpha_{(-t-i\beta)}(B)A], \qquad (5.167)$$

namely

$$W_{AB}(t) = W_{BA}(-t - i\beta). \qquad (5.168)$$

A state ρ_0 over an algebra, satisfying the relation (5.167) is said to be KMS (Kubo–Martin–Schwinger) with respect to the flow α_t.

We can generalize easily the thermal time hypothesis. Given a generic state ρ the thermal hamiltonian is defined by

$$H_\rho = \ln\rho, \qquad (5.169)$$

and the thermal time flow is defined by

$$A_{t_\rho} = \alpha_{t_\rho}(A)\ = e^{it_\rho H_\rho} A e^{-it_\rho H_\rho}. \qquad (5.170)$$

ρ is a KMS state with respect to the thermal time flow.

QFT: Tomita flow. In QFT, finite-temperature states do not live in the same Hilbert space as the zero-temperature states. H_0 is a divergent operator on these finite-temperature states. This is to be expected, since in a thermal state there is a constant energy density and therefore a diverging total energy H_0. Therefore, (5.165) makes no sense in QFT. How, then, do we characterize the Gibbs states? The solution to this problem is well known: equation (5.167) can still be used to characterize a Gibbs state ρ_0 in the algebraic framework and can be taken as the basic postulate of statistical QFT: a Gibbs state ρ_0 over an algebra of observables is a KMS state with respect to the time flow $\alpha(t)$.

It follows that if we want to extend the thermal time hypothesis to field theory, we cannot use (5.169). Can we get around this problem? Is there a flow α_{t_ρ} which is KMS with respect to a *generic* thermal state ρ? Remarkably, the answer is yes. A celebrated theorem by Tomita states precisely that given any[2] state ρ over a von Neumann algebra,[3] there is always a flow α_t, called the Tomita flow of ρ, such that (5.167) holds.

---

[2] Any *separating* state ρ. A separating density matrix has no zero eigenvalues. This is the QFT equivalent of the condition stated in footnote 11 of Chapter 3.

[3] The observables' algebra is in general a C^* algebra. We obtain a von Neumann algebra by closing in the Hilbert norm of the quantum state space.

This theorem allows us to extend (3.205) to QFT: the thermal time flow α_{t_ρ} is defined in general as the Tomita flow of the statistical state ρ.

Thus the thermal time hypothesis can be readily extended to QFT: what we call the "flow of time" is simply the Tomita flow of the statistical state ρ in which the world happens to be, when it is described in terms of macroscopic parameters.

The flow α_{t_ρ} depends on the state ρ. However, a von Neumann algebra possesses also a more abstract notion of time flow, independent of ρ. This is given by the one-parameter group of outer automorphisms, formed by the equivalence classes of auto-morphisms under inner (unitary) automorphisms. Alain Connes has shown that this group is independent of ρ. It only depends on the algebra itself. Connes has stressed the fact that this group provides an abstract notion of time flow that depends only on the algebraic structure of the observables, and nothing else.

The thermal time hypothesis and the notion of thermal time have not yet been extensively investigated. They might provide the key by which to relate timeless fundamental mechanics with our experience of a world evolving in time.

5.5.2 The "choice" of the physical scalar product

The solutions of the Wheeler–DeWitt equation (5.64) form the linear space $\mathcal{H}$. This space is naturally equipped with a scalar product that makes it a Hilbert space. This scalar product is often denoted the "physical" scalar product, in order to distinguish it from the scalar product in $\mathcal{K}$, denoted the "kinematical" scalar product.

The relation between kinematical and physical scalar product depends on the hamil-tonian H. The space $\mathcal{H}$ is the eigenspace of H corresponding to the eigenvalue zero. In order for solutions to exist, the spectrum of H must therefore include zero. If zero is part of the *discrete* spectrum of H, then $\mathcal{H}$ is a proper subspace of $\mathcal{K}$: that is, the solutions of the Wheeler–DeWitt equation (5.64) are normalizable states in $\mathcal{K}$. In this case, the physical scalar product is the same as the kinematical scalar product, and there is no complication. But if zero is part of the *continuum* spectrum of H, then $\mathcal{H}$ is formed by generalized eigenvectors, which are in $\mathcal{S}'$ and not in $\mathcal{K}$. That is, the solutions of the Wheeler–DeWitt equation (5.64) are nonnormalizable states in $\mathcal{K}$. In this case, the physical scalar product is different from the kinematical scalar product. What is it?

In the quantum gravity (and quantum cosmology) literature there is a certain con-fusion regarding the issue of the definition of the physical scalar product. For instance, one often reads that this issue has to do with the notion of time. This is a conceptual mistake that derives from the observation that in a nonrelativistic theory there is a pre-ferred time variable and the problem of defining $\mathcal{H}$ starting from $\mathcal{K}$ does not appear. But the fact that the issue of defining the product in $\mathcal{H}$ appears in timeless systems doesn't imply that it cannot be resolved unless there is a time variable.

In fact, there are a large number of solutions to this issue, all essentially equivalent. Preferences vary; here are some of the solutions proposed.

(i) The scalar product can be defined on $\mathcal{H}$ using the matrix elements of the projector, as illustrated above.

(ii) Here is a general theorem on the issue. If H is a self-adjoint operator on a Hilbert space $\mathcal{K}$, then we can write

$$\mathcal{K} = \int_S ds \, \mathcal{H}_s. \tag{5.171}$$

Here S is the spectrum of H, ds is a measure on this spectrum, and $\mathcal{H}_s$ is a family of Hilbert spaces labeled by the eigenvalues s. The meaning of this integral over Hilbert spaces is the following: any vector $\psi \in \mathcal{K}$ can be written as a family ψ_s, where, for every s, $\psi_s \in \mathcal{H}_s$, and

$$(\psi, \phi)_\mathcal{K} = \int_S ds \, (\psi_s, \phi_s)_{\mathcal{H}_s}, \tag{5.172}$$

where $(\ ,\ )_\mathcal{H}$ is the scalar product in the Hilbert space $\mathcal{H}$ and, in this instance, the integral is a standard numerical integral. The relevance of this theorem is that it states that there is a *Hilbert* space $\mathcal{H}_0$. That is, a scalar product on the space of the solutions of $H\psi = 0$.

Here is a simple example of how the theorem works. Consider the space $\mathcal{K} = L_2[R^2, dxdy]$ and the self-adjoint operator $H = -id/dy$. The solutions of $H\psi = 0$ or

$$-i\frac{d}{dy}\psi(x,y) = 0 \tag{5.173}$$

are functions $\psi(x,y)$ constant in y and are nonnormalizable in $\mathcal{K}$. However, the decomposition (5.171), (5.172) is immediate:

$$\mathcal{K} = \int_R dy \, \mathcal{H}_y, \tag{5.174}$$

where $\mathcal{H}(y) = L_2[R, dx]$. In fact

$$(\psi, \phi)_\mathcal{K} = \int_{R^2} dx \, dy \, \overline{\psi(x,y)} \, \phi(x,y) = \int_R dy \, (\psi_y, \phi_y)_{\mathcal{H}_y}, \tag{5.175}$$

where $\psi_y(x) = \psi(x,y)$ and

$$(\psi_y, \phi_y)_{\mathcal{H}_y} = \int_R dx \, \overline{\psi_y(x)} \, \phi_y(x). \tag{5.176}$$

The space of the solutions of (5.173) is $\mathcal{H}(0)$ and has the natural Hilbert structure $\mathcal{H}(0) = L_2[R, dx]$.

(iii) Here is another solution. Pick a set of self-adjoint operators A_i in $\mathcal{K}$ that commute with H. These are well defined on the space $\mathcal{H}$, because if $H\psi = 0$, then $H(A_i\psi) = A_iH\psi = 0$. Now, *require* that the operators A_i be self-adjoint in the physical scalar product. For a sufficient number of operators, this requirement fixes the scalar product of $\mathcal{H}$.

In the example given in (ii) above, the obvious self-adjoint operators that commute with $H = -id/dy$ are x and $-id/dx$. These are well defined on the space of the functions of x alone. There is only one scalar product on this space of functions that makes x and $-id/dx$ self-adjoint: the one of $L_2[R, dx]$.

(iv) A convenient way of addressing the problem, especially in the case in which H is not a single operator but has many components, is given by the "group averaging" technique. Assume the Wheeler–DeWitt equation has the form $H_i\psi = 0$, where the self-adjoint operators $H_i\psi = 0$ are the generators of a unitary action of a group U on

$\mathcal{K}$. Assume also that $\mathcal{S}$ is invariant under this action and that we can find an invariant measure on the group, or, at least, on the orbit of the group in $\mathcal{K}$. Then we can generalize the operator $P : \mathcal{S} \to \mathcal{H}$ of (5.58):

$$P = \int_U d\tau \, U(\tau),$$ (5.177)

and write the physical scalar product as

$$(P\psi, P\phi)_{\mathcal{H}} \equiv (P\psi)(\phi) = \int_U d\tau \, (\psi|U(\tau)|\phi)_{\mathcal{K}}.$$ (5.178)

There certainly are other techniques as well. This is a field in which the same ideas have independently reappeared many times under different names (and with different levels of mathematical precision). All these techniques are generally equivalent. If there is a case in which they differ, we'll have to resort to physical arguments to find the physically correct choice.

5.5.3 Reality conditions and scalar product

Section 3.2.7 illustrated the possibility of using mixed complex and real dynamical variables, a strategy that will turn out to be useful in GR. Here, I illustrate what happens with the same choice in quantum theory. In particular, I illustrate the key role that the reality conditions play in quantum theory. Recall the simple example discussed in Section 3.2.7, a free particle described in the coordinates x and $z = x - ip$. We can write the quantum theory in terms of wave functionals $\psi(z)$ of the complex variable z. The Schrödinger equation gives immediately (see (3.134))

$$i\hbar \frac{\partial \psi(z,t)}{\partial t} = H_0 \left(\hbar \frac{\partial}{\partial z}, z \right) \psi(z,t) = -\frac{1}{2m} \left(\hbar \frac{\partial}{\partial z} - z \right)^2 \psi(z,t).$$ (5.179)

A complete family of solutions is given by

$$\psi_k(z,t) = e^{-\frac{1}{\hbar} S(z,t;k)},$$ (5.180)

where $S(z,t;k)$ is given in (3.135). Observe now that in the quantum theory the reality condition (3.138) becomes a relation between operators

$$z + z^\dagger = 2\hbar \frac{\partial}{\partial z}.$$ (5.181)

Notice that classical complex conjugation is translated into the adjoint operation: this is necessary in order for real quantities to be represented by self-adjoint operators. Now, (5.181) makes sense only after we have specified the scalar product, because the dagger operation is defined in terms of, and therefore depends on, the scalar product. Indeed, requiring the reality condition (5.181) to hold, amounts to posing a condition on the scalar product of the theory. Let us search for a scalar product of the form

$$(\psi, \phi) = \int dz d\bar{z} \, f(z, \bar{z}) \, \overline{\psi(z)} \, \phi(z),$$ (5.182)

where f is a function to specify. Imposing (5.181) gives the condition on f

$$(z + \bar{z}) f(z, \bar{z}) = -2\hbar \frac{\partial}{\partial z} f(z, \bar{z}).$$ (5.183)

This gives

$$f(z, \bar{z}) = e^{-(z+\bar{z})^2 / 4\hbar}.$$ (5.184)

Let us check whether the states (5.180) are well defined with respect to this product. Inserting (5.180) (at $t = 0$ for simplicity) and (5.184) in (5.182) gives

$$(\psi_k, \psi_{k'}) = \int dz d\bar{z} \; e^{-(z+\bar{z})^2/4\hbar} \; e^{\frac{i}{\hbar}(k\bar{z}-\frac{i}{2}\bar{z}^2)} \; e^{-\frac{i}{\hbar}(k'z+\frac{i}{2}z^2)}. \tag{5.185}$$

A simple change of variables shows that the integral in the imaginary part of z is finite and the integral in the real part of z is proportional to $\delta(k, k')$. Therefore the states ψ_k form a standard continuous orthogonal basis of generalized states. They are clearly eigenstates of the momentum, since

$$p\psi_k = i(x - z)\psi_k = i\left(\frac{\partial}{\partial z} - z\right)\psi_k = k\psi_k. \tag{5.186}$$

In fact, what we have developed is a simple rewriting of the standard Hilbert space of a quantum particle.

Notice that appearances can be misleading. For instance, for $k = 0$ the state ψ_k reads

$$\psi_0(z, t) = e^{+z^2/2\hbar}. \tag{5.187}$$

This looks like a badly nonnormalizable state, but it is not. It is a well-defined generalized state, since the negative exponential in the measure compensates for the positive exponential in the state.

5.6 *Relational interpretation of quantum theory

Quantum mechanics is one of the most successful scientific theories ever. However, its interpretation is controversial. What does the theory actually tell us about the physical world? This question sparked off a lively debate, which was intense during the 1930s, the early days of the theory, and is generating new interest today.

The possibility that the interpretation of an empirically successful theory could be debated should not surprise: examples abound in the history of science. For instance, the great scientific revolution was fueled by the debate on whether the efficacy of the copernican system should be taken as an indication that the Earth *really* moves. In more recent times, Einstein's celebrated contribution to special relativity consisted, to a large extent, just in understanding the physical interpretation (simultaneity is relative) of an already existing effective mathematical formalism (the Lorentz transformations). In these cases, as in the case of quantum mechanics, an overly strictly empiricist position could have circumvented the problem altogether, by reducing the content of the theory to a list of predicted numbers. But science would not then have progressed.

Quantum theory was first constructed for describing microscopic objects (atoms, electrons, photons) and the way these interact with macroscopic apparatuses built to measure their properties. Such interactions were called "measurements." The theory is formed by a mathematical formalism, which allows probabilities of alternative outcomes of such measurements to be calculated. If used just for this purpose, the theory raises no difficulty. But we expect the macroscopic apparatuses themselves – in fact, any physical system in the world – to obey quantum theory, and this seems to raise contradictions within the theory. Here I discuss these apparent contradictions, and a possible resolution. This resolution offers a precise answer to the question of what the quantum theory actually tells us about the physical world.

5.6.1 The observer observed

Measurements. A "measurement" of the variable A of a system S is an interaction between the system S and another system O, whose effect on O depends on the value that the variable A has at the time of the interaction. We say that the variable A is "measured" and that its value a is the "outcome of the measurement." For instance, let S be a particle that impacts on O, let the effect of this impact depend on the position of the particle and let q be the value of the position at the moment of the impact. Then we say that the position Q is measured and that the outcome of the measurement is q.

The term "measurement," and the common terminology used to describe measurement situations (S for "System," and O for "Observer") are very misleading because they evoke a human intentionally "observing" S and using an apparatus to gather data about it. There is nothing "human," or "intentional," in the definition of measurement given above. The system O does not need to be human, nor to be a special "apparatus," nor to be macroscopic. The measured value need not be stored. Any interaction between two physical systems is a measurement. The measured variable of the system S is the variable that determines the effect that the interaction has on O. This is true in classical as well as in quantum theory.

Classical states and quantum states. In classical mechanics, a system S is described by a certain number of physical variables $A, B, C, \ldots$ For instance, a particle is described by its position Q and velocity V. These variables change with time. They represent the contingent properties of the system. We say that the values of these variables determine, at every moment, the "state" of the system. If the value of the position Q of the particle is q and the value of its velocity V is v, we say that the state is (q, v). In classical mechanics a state is therefore a list of values of physical variables.

Quantum mechanics differs from classical mechanics because it assumes that the variables of the system do not have a determined value at all times. Werner Heisenberg introduced this key idea. According to quantum theory, an electron does not have a well-determined position at every time. When it is not interacting with an external system sensitive to its position, the electron can be "spread out" over different positions. It is in a "quantum superposition" of different positions.

It follows that in quantum mechanics the state of the system cannot be captured by giving the value of its variables. Instead, quantum theory introduces a novel notion of the "state" of a system, different from the classical list of variable values. The new notion of "quantum state" was introduced in the work of Erwin Schrödinger in the form of the "wave function" of the system. Paul Adrien Maurice Dirac gave it its general abstract formulation, in terms of a vector Ψ moving in an abstract vector space. From the knowledge of the state Ψ, we can compute the probability of the different measurement outcomes $a_1, a_2, \ldots$ of any variable A. That is, the probability of the different ways in which the system S can affect a system O in the course of an interaction.

The theory prescribes that at every such measurement we must update the value of Ψ, to take into account which of the different outcomes has been realized. This sudden change of the state Ψ depends on the outcome of the measurement and is therefore probabilistic. This is the "collapse of the wave function."

The notion of "state of the system" of classical mechanics is therefore split into two distinct notions in quantum theory: (i) the state Ψ that expresses the probability for the different ways the system S can interact with its surroundings; and (ii) the actual sequence of values $q_1, q_2, q_3, \ldots$ that the variables of S take in the course of the interactions. These are the called "measurement outcomes"; I prefer calling them "quantum events."

We can either think that Ψ is a "real" entity, or that it is nothing more than a theoretical bookkeeping for the quantum events, which are the "real" events. The choice of the relative ontological weight we attribute to the state Ψ or the quantum events $q_1, q_2, q_3, \ldots$ is a matter of convenience; empirical evidence alone does not uniquely determine what is "real." I think the second choice is cleaner, but in the following I refer to both.

The observer observed. The key problem of the interpretation of quantum mechanics is illustrated by the following situation. Consider a real physical situation, illustrated in Figure 5.1, in which at some time t, a system O interacts with a system S, and then, at a later time t', a third system O' interacts with the coupled system $[S + O]$, formed by S and O together. Let the effect on O of the first interaction depend on the variable A of the system S, and the effect on O' of the second interaction depend on the variable B of the coupled system $[S + O]$. (That is, we can say that O

Fig. 5.1   The observer observed.

measures the variable A of S at time t, and then O' measures the variable B of $[S + O]$ at time t'.) Before the first interaction, say S was in a state where A is a quantum superposition of two values a_1 and a_2. Say that at the first interaction O measures the value a_1 of the variable O. The puzzling question can be formulated in various equivalent manners.

- What is the state of S and O between the two interactions?

- Has the quantum event a_1 happened or not?

- Does the quantity A have a determined value after the first interaction or not?

Say that before the first interaction the state of S was $\Psi = c_1\Psi_1 + c_2\Psi_2$ where Ψ_1, Ψ_2 are states where A has values a_1, a_2 respectively. Then at time t we have

$$\begin{cases} c_1\Psi_1 + c_2\Psi_2 \;\rightarrow\; \Psi_1; \\ A \text{ takes the value } a_1. \end{cases} \tag{5.188}$$

However, the system O obeys the laws of quantum theory as well. Therefore we can also give a quantum description of the evolution of the coupled quantum system $(S + O)$ formed by S and O together. If we do so, no collapse happens. Instead, the effect of the interaction is the Schrödinger evolution

$$
\begin{cases}
\left(c_1 \Psi_1 + c_2 \Psi_2\right) \otimes \Phi \;\rightarrow\; \left(c_1 \Psi_1 \otimes \Phi_1 + c_2 \Psi_2 \otimes \Phi_2\right); \\
A \text{ is still in the superposition of the two values } a_1,\, a_2
\end{cases}
\tag{5.189}
$$

for suitable states Φ, Φ_1, Φ_2 of O.

What is real seems to depend on how we choose to describe the world. What is the real state of affairs of the world after the interaction between S and O: (5.188) or (5.189)? In either case we get a difficulty. If we say that after t the state has collapsed as in (5.188) and A has the value a_1, we get the wrong predictions about the second measurement at time t'. In fact, quantum theory allows us to predict the probability distribution of the possible outcomes of the second measurement, but, to compute this, we have to use the state (5.189), and not the state (5.188). Indeed, if B and A do not commute, this probability distribution can be affected by the interference between the two different "branches" in (5.189). In other words, we have to assume that the variable A was in a quantum superposition of the values a_1, a_2, and not determined.

But if we do so, and say that after the first measurement the state is (5.189), then we must say that A has no determined value at time t. But the situation is general: any measurement can be thought of as the first measurement of the example, and therefore we must conclude that no variable can take a definite value *ever*.

Thus we seem to get a contradiction in both cases: whether we think that the wave function has collapsed and a_1 was realized, or whether we think it hasn't. This is the core of the difficulty of the interpretation of quantum theory.

Real wave functions or real quantum events. Let us examine the above difficulty in a bit more detail, from the two points of view of the two possible ontologies of quantum theory.

If we think that Ψ is real but it never truly collapses, there is no simple and compelling reason why the world should appear as described by values of physical quantities that take determined values at each interaction, as it does. We experience particles in given positions, not particle wavefunctions. The relation between a noncollapsing wavefunction ontology and our experience of the world is very indirect and involuted. We need some complicated story to understand how specific observed values $q_1, q_2, q_3, \ldots$ can emerge from the sole Ψ. If this story is given (which is possible), we are then in a situation similar to the one of a quantum event ontology, to which I now turn.

I think it is preferable to take the quantum events as the actual elements of reality, and view Ψ just as a bookkeeping device, coding the events that happened in the past and their consequences. For instance, I prefer to say that the "reality" of a subatomic particle is expressed by the sequence of the positions of the particle revealed by the bubbles in a bubble chamber, not by the spherically symmetric wave function emerging from the interaction area. The reality of the electron is in the events where it reveals itself, interacting with its surrounding, not in the abstract probability amplitude for such events. From this perspective, the real events of the world are the "realizations" (the "coming to reality," the "actualization") of the values $q_1, q_2, q_3, \ldots$ in the course of the interactions between physical systems. These quantum events have an intrinsically discrete "quantized" granular structure.

This perspective, however, does not solve the above puzzle either. The key puzzle of quantum mechanics becomes the fact that the statement that the quantum event a_1 "has happened" can be at the same time true and not-true: has the quantum event a_1 happened or not? If we answer no, then we are forced to say that no quantum event *ever* happens, because the situation described above is completely general: any quantum event happening in the interaction of two systems S and O is "non-happening" as far as the effect of $(S + O)$ on a further system O' is concerned. If we say yes, then we contradict the predictions of quantum mechanics (about the second interaction).

The "second observer" puzzle captures the core conceptual difficulty of the interpretation of quantum mechanics: reconciling the possibility of quantum superposition with the fact that the world we observe and describe is characterized by determined values of physical quantities. More precisely, the puzzle shows that we cannot disentangle the two: according to the theory, a quantum event (a_1) can be at the same time realized and not realized.

A possible escape from the puzzle is to assume that there are "special" systems that produce the collapse and cause quantum events to happen. For instance, these could be "macroscopic" systems, or "sufficiently complex" systems, or "systems with memory," or the "gravitational field," or human "consciousness"... All these systems, and others, have been suggested as causing quantum collapse and generating quantum events. If this were correct, at some point we shall be able to measure violations of the predictions of QM. That is, QM as we know it would break down for those systems.

So far this breaking down of QM has never been observed. We can fancy a phenomenology that we have not yet observed that could bring back reality to the way we used to think it is. It is certainly worthwhile to investigate this possibility theoretically and experimentally. But we should not forget that reality might be truly different from what we thought, and might be simply demanding us to renounce some old prejudice. I think that the history of physics indicates that the productive attitude is not to resist the conceptual novelty of empirically successful theories but, rather, to make an effort to understand it. We should not force reality into our prejudices, but rather try to adapt our conceptual schemes to what we learn about the world.

5.6.2 Facts are interactions

I think that the key to the solution of the difficulty can be found in the observation that the two descriptions (5.188) and (5.189) refer to different systems: the first to O, the second to O'. More precisely, the first is relevant for describing the effects of interactions on O; the second for describing the effects of interactions on O'.

The solution of the puzzle can be found in the idea that quantum events are the elements of reality, but they are always relative to a physical system: the quantum event a_1 happens with respect to O, but it does not happen with respect to O'.

In other words, the way out from the puzzle is that the values of the variables of any physical system are relational. They do not express a property of the system S alone, but rather refer to the relation between this system and another system. The variable A has value a_1 with respect to O, but it has no determined values with respect to O'. This point of view is called the *relational interpretation of quantum mechanics*, or simply *relational quantum mechanics*.

The central idea of relational quantum mechanics is that there is no meaning in saying that a certain variable of the system S takes the value q. There is only meaning in saying that a variable has value q *with respect to* a system O. In the example discussed above, for instance, the fact that A takes the value a_1 with respect to O does not imply that A has the value a_1 also with respect to O'.

If we avoid all statements that are not referred to a physical system, we can get rid of all apparent contradictions of quantum theory. The apparent contradiction between the two statements that a variable has or hasn't a value is resolved by referring the statements to the different systems with which the system in question interacts. If I observe an electron at a certain position, I cannot conclude that the electron is there: I can only conclude that the electron *as seen by me* is there.

Indeed, quantum theory must be understood as an account of the way distinct physical systems *affect one another* when they interact, and not the way physical systems "are." This account exhausts all that can be said about the physical world. The physical world can be described as a network of interacting components, where there is no meaning to "the state of an isolated system." The state of a physical system is the network of its relationships with the surrounding systems. The physical structure of the world is identified with this network of relationships.

The unique account of the state of the world of the classical theory is thus shattered into a multiplicity of accounts, one for each possible "observing" physical system. Quantum mechanics is a theory about the physical description of physical systems relative to other systems, and this is a complete description of the world.

Of course we can pick a system O once and for all as "the observer system," and be concerned only with the effects of the rest of the world on this system. Each interaction between the rest of the world and O is correctly described by standard quantum mechanics. In this description, the quantum state Ψ collapses at each interaction with O. This description is completely self-consistent, but it treats O as if it were a special system: a classical, nonquantum system. If we want to describe O itself quantum mechanically, we can, but we have to pick a different system O' as the observer, and describe the way O interacts with O'. In this description, the quantum properties of O are taken into account, but not the ones of O', because this description describes the effects of the rest of the world on O'.

Consistency. This relativisation of actuality is viable thanks to a remarkable property of the formalism of quantum mechanics.

John von Neumann was the first to notice that the formalism of the theory treats the measured system (S) and the measuring system (O) differently, but the theory is surprisingly flexible on the choice of where to put the boundary between the two. Different choices give different accounts of the state of the world (for instance, the collapse of the wave function happens at different times); but this does not affect the predictions about the final observations. This flexibility reflects a general structural property of quantum theory which guarantees consistency among all the distinct "accounts of the world" of the different observing systems. The manner in which this consistency is realized, however, is subtle.

As a simple illustration of this phenomenon, consider the case in which a system O with two states Φ_1 and Φ_2 (say, a light-bulb which can be *on* or *off*) interacts with a system S with two states Ψ_1 and Ψ_2 (say, the spin of the electron, which can be *up* or *down*). Assume the interaction is such that if the spin is *up* (*down*) the light goes *on* (*off*). To start with, the electron can be in a superposition of its two states. In the account of the state of the electron that we can associate with the light, the wave function of the electron collapses to one of two states during the interaction, as in (5.188), and the light is then either *on* or *off*. But we can also consider the light/electron composite system as a quantum system and study the interactions of this composite system with another system O'. In the account associated to O', there is no collapse at the time of the interaction, and the composite system is still in the superposition of the two states [spin *up*/light *on*] and [spin *down*/light *off*] after the interaction, as in (5.189). As remarked above, it is necessary to assume this superposition because it accounts for measurable interference effects between the two states: if quantum mechanics is correct, these interference effects are truly observable by O'.

So, we have two discordant accounts of the same events: the one associated to O, where the spin has a determined value, and the one associated to O', where the spin is in a superposition. Now, can the two discordant accounts be compared and does the comparison lead to a contradiction?

They can be compared, because the information on the first account is stored in the state of the light and O' has access to this information. Therefore, O and O' can compare their accounts of the state of the world. However, the comparison does not lead to a contradiction *because the comparison is itself a <u>physical</u> process that must be understood in the context of quantum mechanics.*

Indeed, O' can physically interact with the electron and then with the light (or, equivalently, with the light and then with the electron). If, for instance, O' finds the spin of the electron *up*, quantum mechanics predicts that the observer will then consistently find the light on, because in the first measurement the state of the composite system collapses on its [spin *up*/light *on*] component, namely on the first term of the right-hand side of (5.189).

That is, the multiplicity of accounts does not lead to a contradiction precisely because the comparison between different accounts can only be a physical quantum interaction. Many common paradoxes of quantum mechanics follow from assuming that the communication between different observers violates quantum mechanics.[4] This internal self-consistency of the quantum formalism is general, and is perhaps its most remarkable

---

[4]The EPR (Einstein–Podolski–Rosen) apparent paradox might be among these. The two observers far from each other are physical systems. The standard account neglects the fact that each of the two is in a quantum superposition with respect to the other, until the moment they physically communicate. But this communication is a physical interaction and must be strictly consistent with causality.

aspect.[5] This self-consistency is a strong indication of the relational nature of the world.

5.6.3 Information

What appears with respect to O as a measurement of the variable A (with a specific outcome), appears with respect to O' simply as a dynamical process that establishes a *correlation* between S and O. As far as the observer O is concerned, the variable A of a system S has taken a certain value. As far as the second observer O' is concerned, the only relevant element of reality is that a correlation is established between S and O.

Concretely, this correlation appears in all further observations that O' would perform on the $[S + O]$ system. That is, the way the two systems S and O will interact with O' is characterized by the fact that there is a correlation: O' will find some properties of O correlated with some properties of S.

On the other hand, until it physically interacts with $[S+O]$, the system O' has no access to the actual outcomes of the measurements performed by O on S. These actual outcomes are real only with respect to O.

The existence of a correlation between the possible outcomes of a measurement performed by O' on S and the outcomes of a measurement performed by O' on O can be interpreted in terms of information. In fact, this correlation corresponds *precisely* to Shannon's definition of information. According to this definition, "O has information about S" *means* that we shall observe O and S in a subset of the set formed by the cartesian product of the possible states of O and the possible states of S. Thus, a measurement of S by O has the effect that "O has information about S." This statement has a precise technical meaning which refers to the possible outcomes of the observations by a third system O'.

On the other hand, if we interact a sufficient number of times with a physical system S, we can then predict (the distribution probability of the) future outcomes of our interactions with this system. In this sense, by interacting with S we can say we "have information" about S. (This

---

[5]In fact, one may conjecture that this peculiar consistency between the observations of different observers is the missing ingredient for a reconstruction theorem of the Hilbert space formalism of quantum theory. Such a reconstruction theorem is still unavailable. On the basis of reasonable physical assumptions, one is able to derive the structure of an orthomodular lattice containing subsets that form Boolean algebras which "almost," but not quite, implies the existence of a Hilbert space and its projectors' algebra. Perhaps an appropriate algebraic formulation of the condition of consistency between subsystems could provide the missing hypothesis to complete the reconstruction theorem.

information need not be stored or utilized, but its existence is the necessary physical condition for being able to store it or utilize it for predictions.)

Therefore, we have two distinct senses in which the physical theory is about information. But a moment of reflection shows that the two simply reflect the same physical reality, as it affects two different systems. On the one hand, O has information about S because it has interacted with S and the past interactions are sufficient to "give information," namely to determine (the probability distribution of) the result of future interactions. On the other hand, O has information about S in the sense that there are correlations in the outcomes of measurements that O' can make on the two.

There is a crucial subtle difference, that can be figuratively expressed as follows: O "knows" about S, while O' only knows *that* O knows about S, but does not know *what* O knows. As far as O' is concerned, a physical interaction between S and O establishes a correlation: it does not select an outcome.

These observations are sufficient to conclude that what precisely quantum mechanics is about is the information that physical systems have about one another.

The common unease with taking quantum mechanics as a fundamental description of Nature, referred to as the measurement problem, can be traced to the use of an incorrect notion, in the same way that unease with Lorentz transformations derived from the notion, shown by Einstein to be mistaken, of an observer-independent time. The incorrect notion that generates the unease with quantum mechanics is the notion of an observer-independent state of a system, or observer-independent values of physical quantities, or an observer-independent quantum event.

We can assume that all systems are equivalent, there is no a priori observer–observed distinction; the theory describes the information that systems have about one another. The theory is complete, because this description exhausts the physical world.

In physics, the move of deepening our insight into the physical world by relativizing notions previously treated as absolute has been applied repeatedly and very successfully. Here are a few examples.

The notion of the velocity of an object has been understood as meaningless, unless it is referred to a reference body with respect to which the object is moving. With special relativity, simultaneity of two distant events has been understood as meaningless, unless referred to a specific state of motion of something. (This something is usually denoted as "the observer" without, of course, any implication that the observer is human or has any other peculiar property besides having a state of motion. Similarly, the "observer system" O in quantum mechanics need not be human or have any other property beside the possibility of interacting with the "observed" system S.) With general relativity, the position in space and time of an object has been understood as meaningless, unless it is referred to the gravitational field, or to another dynamical physical entity.

The step proposed by the relational interpretation of quantum mechanics has strong analogies with these. In a sense it is a longer jump, since all the contingent (variable) properties of all physical systems are taken to be meaningful only as relative to a second physical system. This is not an arbitrary step. It is a conclusion which is difficult to escape, following from the observation – explained above in the example of the "second observer" – that a variable (of a system S) can have a well-determined value a_1 for one observer (O) and at the same time fail to have a determined value for another observer (O').

This way of thinking of the world has perhaps heavy philosophical implications. But it is Nature that is forcing us to this way of thinking. If we want to understand Nature, our task is not to frame Nature into our philosophical prejudices, but rather to learn how to adjust our philosophical prejudices to what we learn from Nature.

5.6.4 Spacetime relationalism versus quantum relationalism

I close with a very speculative suggestion. As discussed in Section 2.3, the main idea underlying GR is the relational interpretation of localization: objects are not located in spacetime. They are located with respect to one another. In the present section, I have observed that the lesson of QM is that quantum events and states of systems are relational: they make sense only with respect to another system. Thus, both GR and QM are characterized by a form of relationalism. Is there a connection between these two forms of relationalism?

Let us look closer at the two relations. In GR, the localization of an object S in spacetime is relative to another object (or field) O, to which S is *contiguous*. Contiguities, or, equivalently, Einstein's "spacetime coincidences" are the basic relations that construct spacetime. In QM, there are no absolute properties or facts: properties of a system S are relative to another system O with which S is *interacting*. Facts are interactions. Thus, interactions form the basic relations between systems.

But there is a strict connection between contiguity and interaction. On the one hand, S and O can interact only if they are contiguous; if they are nearby in spacetime: this is locality. Interaction requires contiguity. On the other hand, what does it mean that S and O are contiguous? What else does it mean besides the fact that they can interact?[6] Therefore,

---

[6]The very word "contiguous" derives from the Latin *cum-tangere*: to touch each other, that is, to inter-act.

contiguity is manifested by interacting. In a sense, the fact that interactions are local means that there is a sort of identity between being contiguous and interacting.

Thus, locality ties together very strictly the spacetime relationalism of GR with the relationalism underlying QM. It is tempting to try to develop a general conceptual scheme based on this observation. This could be a conceptual scheme in which contiguity is nothing else than a manifestation, or can be identified with, the existence of a quantum interaction. The spatiotemporal structure of the world would then be directly determined by who is interacting with whom. This is, of course, very vague, and might lead nowhere, but I find the idea intriguing.

———

Bibliographical notes

Textbooks on quantum theory are numerous. I think the best of all is the first of them: Dirac [148], because of Dirac's crystal-clear thinking. In the earlier editions, Dirac uses a relativistic notion of state (that does not evolve in time) as is done here. He calls these states "relativistic," as is done here. In later editions, he switches to Schrödinger states that evolve in time, explaining in a preface that it is easier to calculate with these, but it is a pity to give up relativistic states, which are more fundamental.

I have discussed the idea that QM remains consistent also in the absence of unitary time evolution in [98] and [149]. The same idea is developed by many authors, see [26], [150], and references therein.

In the past I have discussed relativistic systems only in terms of "evolving constants." The two-oscillators example used in the text was considered in these terms in [151, 152]. The probabilistic interpretation of the covariant formulation presented in this chapter is an evolution of this point of view, and derives from [144].

I have taken the discussion on the boundary formulation of QFT from [145]. The idea that quantum field theory must be formulated in terms of boundary data on a finite surface has been advocated by Robert Oeckl [153]. The derivation of the local Schrödinger equation is in [146] and [154]. The Tomonaga–Schwinger equation was introduced in [155]. On the difficulty of a direct interpretation of the n-point functions in quantum gravity, see for instance [156]. The Hartle–Hawking state was introduced in [157].

The possibility of defining the physical scalar product on the space of the solutions of the Wheeler–DeWitt equation even when these solutions are nonnormalizable in the kinematical Hilbert space has been discussed by many authors using a variety of techniques. A nice mathematical construction has been given by Don Marolf, see [158] and references therein.

The thermal time hypothesis was extended to QM and QFT in [125].

The relational interpretation presented here is discussed in [159, 160]; see also [161, 162]. An overview of similar points of view is in the online *Stanford Encyclopedia of Philosophy* [163]; on a possibly related point of view, see also [164]. The role of information in the foundations of quantum theory has been stressed by John Wheeler in [165, 166]. For a recent discussion on the role of information in the foundation of quantum theory, see for instance [167] and references therein. An original and fascinating point of view on the relational aspects of quantum and relativity is explored by David Finkelstein in [168].

Part II
Loop quantum gravity

– Now it's time to leave the capsule
if you dare . . .
– This is Major Tom to Ground Control
I'm stepping through the door
And I'm floating in a most peculiar way
And the stars look very different today. . .

David Bowie, *Space Oddity*

6
Quantum space

It is time to begin to put together the tools developed in the first part of the book, and build the quantum theory of spacetime. The strategy is simple. We "quantize" the canonical formulation of GR described at the beginning of Chapter 4, according to the relativistic QM formalism detailed in Chapter 5. This chapter deals with the kinematical part of the theory: states, partial observables and their eigenvalues. The next chapter deals with dynamics, namely with transition amplitudes.

6.1 Structure of quantum gravity

In Chapter 4, we have seen that GR can be formulated as the dynamical system defined by the Hamilton–Jacobi equation (4.9)

$$F_{ab}^{ij}(\vec{\tau}) \, \frac{\delta S[A]}{\delta A_a^i(\vec{\tau})} \, \frac{\delta S[A]}{\delta A_b^j(\vec{\tau})} = 0, \tag{6.1}$$

where the functional $S[A]$ is defined on the space $\mathcal{G}$ of the 3d $SU(2)$ connections $A_a^i(\vec{\tau})$, and is invariant under internal gauge transformations and 3d diffeomorphisms. That is,

$$\delta_f A_a^i(\vec{\tau}) \, \frac{\delta S[A]}{\delta A_a^i(\vec{\tau})} = 0, \qquad \delta_\lambda A_a^i(\vec{\tau}) \, \frac{\delta S[A]}{\delta A_a^i(\vec{\tau})} = 0, \tag{6.2}$$

where the variations $\delta_f A_a^i$ and $\delta_\lambda A$ are given in (4.12). Equivalently, the theory is defined by the hamiltonian $H[A, E] = F_{ab}^{ij} E_i^a E_j^b$ on $T^*\mathcal{G}$.

Following the prescription of Chapter 5, a quantization of the theory can be obtained in terms of complex-valued Schrödinger wave functionals $\Psi[A]$ on $\mathcal{G}$. The quantum dynamics is inferred from the classical dynamics by interpreting $S[A]$ as $\hbar$ times the phase of $\Psi[A]$. Namely, interpreting the classical Hamilton–Jacobi theory as the eikonal approximation of a quantum wave equation, as in [142]; semiclassical "wave packets" will then behave according to the classical theory. This can be obtained defining

225

the quantum dynamics by replacing derivatives of the Hamilton–Jacobi functional $S[A]$ with derivative operators. The two equations (6.2) remain unchanged: they simply force $\Psi[A]$ to be invariant under $SU(2)$ gauge transformations and 3d diffeomorphisms. Equation (6.1) gives

$$F_{ab}^{ij}(\vec{\tau})\,\frac{\delta}{\delta A_a^i(\vec{\tau})}\frac{\delta}{\delta A_b^j(\vec{\tau})}\,\Psi[A] = 0. \qquad (6.3)$$

This is the Wheeler–DeWitt equation, or Einstein–Schrödinger equation. It governs the quantum dynamics of spacetime. In other words, the dynamics is defined by the hamiltonian operator $H = H[A, -\mathrm{i}\hbar\,\delta/\delta A]$.

More precisely, we want a rigged Hilbert space $\mathcal{S} \subset \mathcal{K} \subset \mathcal{S}'$, where $\mathcal{S}$ is a suitable space of functionals $\Psi[A]$. Partial observables are represented by self-adjoint operators on $\mathcal{K}$. Their eigenvalues describe the quantization of physical quantities. The operator P, formally given by the field theoretical generalization of (5.58),

$$P \sim \int [DN]\, \mathrm{e}^{-\mathrm{i}\int \mathrm{d}^3\tau\; N(\vec{\tau})H(\vec{\tau})}, \qquad (6.4)$$

sends $\mathcal{S}$ to the space of the solutions of (6.3). Its matrix elements between eigenstates of partial observables define the transition amplitudes of quantum gravity. These determine all (probabilistic) dynamical relations between any measurement that we can perform.

A preferred state in $\mathcal{K}$ is $|\emptyset\rangle$, the eigenstate of the geometry with zero volume and zero area. The covariant vacuum is given by $|0\rangle = P|\emptyset\rangle$. If we assume that the surface Σ coordinatized by $\vec{\tau}$ is the entire boundary of a finite spacetime region, then we can identify $\mathcal{K}$ with the boundary space K. The correlation probability amplitude associated with a measurement of partial observables on the boundary surface is $A = W(s) = \langle 0|s\rangle$, where $|s\rangle$ is the eigenstate of the partial observables corresponding to the measured eigenvalues.

We must now build this structure concretely. As we shall see, in a precise technical sense, this structure is unique.

6.2 The kinematical state space $\mathcal{K}$

I construct here the quantum state space defined by the *real* connection (see Section 4.2). There are three reasons for this. First, the physical lorentzian theory can be formulated in terms of this connection: the only difference is that the hamiltonian operator is slightly more complicated than (6.3), as explained in Section 4.2.2. Second, things are far easier with the real connection, and it is better to do easy things first; in the construction of the quantum state space defined by the complex connection there are still some open technical complications [169]. Third, the real connection with the hamiltonian operator (6.3) defines the quantum euclidean theory, namely the quantum theory which has the theory defined in Section 4.2.1 as its classical limit; this is an interesting model by itself and is likely to be related to the physical theory, as was discussed in Section 4.2.

Cylindrical functions. Let $\mathcal{G}$ be the space of the smooth 3d real connections A defined everywhere on a 3d surface Σ, except, possibly, at isolated points.[1] Fix the topology of Σ, say to a 3-sphere. I now define a space $\mathcal{S}$ of functionals on $\mathcal{G}$.

We are now going to make use of the geometric interpretation of the field A as a connection. The $so(3)$ Lie algebra is the same as the $su(2)$ Lie algebra, and it is convenient to view A as an $su(2)$ connection. Let τ_i be a fixed basis in the $su(2)$ Lie algebra. I choose $\tau_i = -\frac{i}{2}\,\sigma_i$, where σ_i are the Pauli matrices (A.14). Write

$$A(\vec{\tau}) = A_a^i(\vec{\tau})\,\tau_i\,\mathrm{d}x^a. \qquad (6.5)$$

Recall from Section 2.1.5 that an oriented path γ in Σ and a connection A determine a group element $U(A,\gamma) = \mathcal{P}\exp\int_\gamma A$, called the holonomy of the connection along the path. For a given γ, the holonomy $U(A,\gamma)$ is a functional on $\mathcal{G}$. Consider an ordered collection Γ of smooth oriented paths γ_l with $l = 1,\ldots,L$ and a smooth function $f(U_1,\ldots,U_L)$ of L group elements. A couple (Γ, f) defines a functional of A

$$\Psi_{\Gamma,f}[A] = f(U(A,\gamma_1),\ldots,U(A,\gamma_L)). \qquad (6.6)$$

$\mathcal{S}$ is defined as the linear space of all functionals $\Psi_{\Gamma,f}[A]$, for all Γ and f. We call these functionals "cylindrical functions." In a suitable topology, which is not important to detail here, $\mathcal{S}$ is dense in the space of all continuous functionals of A.

I call Γ an "ordered oriented graph" embedded in Σ. I call simply "graph" an ordered oriented graph up to ordering and orientation, and denote it by the same letter Γ. Clearly, as far as cylindrical functions are concerned, changing the ordering or the orientation of a graph is just the same as changing the order of the arguments of the function f, or replacing arguments with their inverse.

Scalar product. I now define a scalar product on the space $\mathcal{S}$. If two functionals $\Psi_{\Gamma,f}[A]$ and $\Psi_{\Gamma,g}[A]$ are defined with the same ordered oriented graph Γ, define

$$\langle \Psi_{\Gamma,f} | \Psi_{\Gamma,g} \rangle \equiv \int \mathrm{d}U_1\ldots\mathrm{d}U_L\, \overline{f(U_1,\ldots,U_L)}\, g(U_1,\ldots,U_L)\,, \qquad (6.7)$$

where $\mathrm{d}U$ is the Haar measure on $SU(2)$. Notice the similarity with the lattice scalar product (5.152): equation (6.7) is the scalar product of a Yang–Mills theory on the lattice Γ.

---

[1]The reason for this technical choice will become clear below.

The extension of this scalar product to functionals defined on the same graph, but with different ordering or orientation, is obvious. But also the extension to functionals defined on different graphs Γ is simple. In fact, observe that different couples (Γ, f) and (Γ', f') may define the same functional. For instance, say Γ is the union of the L' curves in Γ' and L'' other curves, and that $f(U_1, \ldots, U_{L'}, U_{L'+1}, \ldots, U_{L'+L''}) = f'(U_1, \ldots, U_{L'})$; then, clearly, $\Psi_{\Gamma, f} = \Psi_{\Gamma', f'}$. Using this fact, it is clear that we can rewrite any two given functionals $\Psi_{\Gamma', f'}$ and $\Psi_{\Gamma'', g''}$ as functionals $\Psi_{\Gamma, f}$ and $\Psi_{\Gamma, g}$ having the same graph Γ, where Γ is the union of Γ' and Γ''. Using this fact, (6.7) becomes a definition valid for any two functionals in $\mathcal{S}$:

$$\langle \Psi_{\Gamma', f'} | \Psi_{\Gamma'', g''} \rangle \equiv \langle \Psi_{\Gamma, f} | \Psi_{\Gamma, g} \rangle. \tag{6.8}$$

Notice that even if (6.7) is similar to the scalar product of a lattice Yang–Mills theory, the difference is profound. Here we are dealing with a genuinely continuous theory, in which the states do not live on a single lattice Γ, but on *all* possible lattices in Σ. There is no cut-off on the degrees of freedom, as in lattice Yang–Mills theory.

Loop states and the loop transform. An important example of a finite norm state is provided by the case $(\Gamma, f) = (\alpha, \mathrm{tr})$. That is, Γ is formed by a single *closed* curve α, or a "loop," and f is the trace function on the group. We can write this state as Ψ_α, or simply in Dirac notation as $|\alpha\rangle$. That is,

$$\Psi_\alpha[A] = \Psi_{\alpha, \mathrm{tr}}[A] = \langle A | \alpha \rangle = \mathrm{tr}\, U(A, \alpha) = \mathrm{tr}\, \mathcal{P} e^{\oint_\alpha A}. \tag{6.9}$$

The very peculiar properties that these states have in quantum gravity, which will be illustrated later on, have motivated the entire LQG approach (and its name). The norm of Ψ_α is easily computed from (6.7)

$$|\Psi_\alpha|^2 = \int dU \, |\mathrm{tr}\, U|^2 = 1. \tag{6.10}$$

A "multiloop" is a collection $[\alpha] = (\alpha_1, \ldots, \alpha_n)$ of a finite number n of (possibly overlapping) loops. A "multiloop state" is defined as

$$\Psi_{[\alpha]}[A] = \Psi_{\alpha_1}[A] \ldots \Psi_{\alpha_n}[A]. = \mathrm{tr}\, U(A, \alpha_1) \ldots \mathrm{tr}\, U(A, \alpha_n). \tag{6.11}$$

The functional on loop space

$$\Psi[\alpha] = \langle \Psi_\alpha | \Psi \rangle \tag{6.12}$$

is called the "loop transform" of the state $\Psi[A]$. The functional $\Psi[\alpha]$ represents the quantum state as a functional on a space of loops. This formula, called the "loop transform," is the formula through which LQG was originally constructed (see for instance [170]). Using the measure $d\mu_0[A]$ mentioned below, this can be written as

$$\Psi[\alpha] = \int d\mu_0[A] \, \mathrm{tr}\, \mathcal{P} e^{\oint_\alpha A} \Psi[A]. \tag{6.13}$$

Intuitively, this is a sort of infinite-dimensional Fourier transform from the A space to the α space.

Kinematical Hilbert space. Define the kinematical Hilbert space $\mathcal{K}$ of quantum gravity as the completion of $\mathcal{S}$ in the norm defined by the scalar product (6.7),[2] and $\mathcal{S}'$ as the completion of $\mathcal{S}$ in the weak topology defined by (6.7).[3] This completes the definition of the kinematical rigged Hilbert space $\mathcal{S} \subset \mathcal{K} \subset \mathcal{S}'$.

Why this definition? The main reason is that the scalar product (6.7) is invariant under diffeomorphisms and local gauge transformations (Section 6.2.2 below) and it is such that real classical observables become self-adjoint operators (Section 6.5 below). These very strict conditions are the ones that the scalar product must satisfy in order to give a consistent theory with the correct classical limit. Furthermore, the main feature of this definition is that the loop states Ψ_α are normalizable. As we shall see later on, loop states are natural objects in quantum gravity. They diagonalize geometric observables and they are solutions of the Wheeler–DeWitt equation. Hence the kinematics as well as the dynamics select this space of states as the natural ones in gravity.

There are two objections that can be raised against the definition of $\mathcal{K}$ we have given. First, $\mathcal{K}$ is nonseparable. This objection would be fatal in the context of flat-space quantum field theory, but it turns out to be harmless in a general-relativistic context, because of diffeomorphism invariance. Indeed, the "excessive size" of the nonseparable Hilbert space will turn out to be just gauge: the physical Hilbert space $\mathcal{H}$ is separable. As we shall see in Section 6.4, it is sufficient to factor away the diffeomorphism gauge to obtain a separable Hilbert space $\mathcal{K}_{\text{diff}}$.

Second, loop states are normalizable in lattice Yang–Mills theory, but they are nonnormalizable in continuous Yang–Mills theory. By analogy, one might object that they should not be normalizable states in continuous quantum gravity either. As we shall see below, however, this analogy is misleading, again precisely because of the great structural difference between a diffeomorphism-invariant QFT and a QFT on a background. As we shall see, in continuous Yang–Mills theory a loop state describes an unphysical excitation that has infinitesimal transversal physical size. In gravity, on the other hand, a loop state describes a physical excitation that has *finite* (planckian) transversal physical size. This will become clear below in Section 6.6.2.

Boundary Hilbert space. There are two natural ways of defining the boundary space K. We can either define $\mathsf{K} = \mathcal{K}^* \otimes \mathcal{K}$ and describe the quantum geometry of a spacetime region bounded by an initial and a

---

[2] The space of the Cauchy sequences Ψ_n, where $||\Psi_m - \Psi_n||$ converges to zero.
[3] The space of the sequences Ψ_n such that $\langle \Psi_n | \Psi \rangle$ converges for all Ψ in $\mathcal{S}$.

final surface; or simply define $\mathsf{K} = \mathcal{K}$ interpreting the closed connected surface Σ as the boundary of a finite 4d spacetime region.

6.2.1 Structures in $\mathcal{K}$

The space $\mathcal{K}$ has a rich and beautiful structure. I mention here only a few aspects of this structure which are important for what follows, referring to the more mathematically oriented literature on this subject (see [20] and references therein) for more details.

Graph subspaces. The cylindrical functions with support on a given graph Γ form a subspace $\tilde{\mathcal{K}}_\Gamma$ of $\mathcal{K}$. By definition, $\tilde{\mathcal{K}}_\Gamma = L_2[SU(2)^L]$, where L is the number of paths in Γ. The space $\tilde{\mathcal{K}}_\Gamma$ is the (unconstrained) Hilbert space of a lattice gauge theory with spatial lattice Γ, as described in Section 5.3.6. If the graph Γ is contained in the graph Γ', the Hilbert space $\tilde{\mathcal{K}}_\Gamma$ is a proper subspace of the Hilbert space $\tilde{\mathcal{K}}_{\Gamma'}$. This nested structure of Hilbert spaces is called a projective family of Hilbert spaces. $\mathcal{K}$ is – and can be defined as – the projective limit of this family.

An orthonormal basis. The tool for finding a basis in $\mathcal{K}$ is the Peter–Weyl theorem, which states that a basis on the Hilbert space of L_2 functions on $SU(2)$ is given by the matrix elements of the irreducible representations of the group. Irreducible representations of $SU(2)$ are labeled by half-integer spin j. Call $\mathcal{H}_j$ the Hilbert space on which the representation j is defined and v^α its vectors. Write the matrix elements of the representation j as

$$R^{(j)\alpha}{}_\beta(U) = \langle U | j, \alpha, \beta \rangle. \tag{6.14}$$

For each graph Γ choose an ordering and an orientation. Then a basis

$$|\Gamma, j_l, \alpha_l, \beta_l\rangle \equiv |\Gamma, j_1, \ldots, j_L, \alpha_1, \ldots, \alpha_L, \beta_1, \ldots, \beta_L\rangle \tag{6.15}$$

in $\tilde{\mathcal{K}}_\Gamma$ is simply obtained by tensoring the basis (6.14). That is,

$$\langle A | \Gamma, j_l, \alpha_l, \beta_l \rangle = R^{(j_1)\alpha_1}{}_{\beta_1}(U(A, \gamma_1)) \ldots R^{(j_L)\alpha_L}{}_{\beta_L}(U(A, \gamma_L)). \tag{6.16}$$

This set of vectors in $\mathcal{K}$ is not a basis because the same vector appears in $\tilde{\mathcal{K}}_\Gamma$ and $\tilde{\mathcal{K}}_{\Gamma'}$ if Γ is contained in the graph Γ'. However it is very easy to get rid of the redundancy, because all $\tilde{\mathcal{K}}_\Gamma$ vectors belong to the trivial representation of the paths that are in Γ' but not in Γ. Therefore, an orthonormal basis of $\mathcal{K}$ is simply given by the states $|\Gamma, j_l, \alpha_l, \beta_l\rangle$ defined in (6.16) where the spins $j_l = \frac{1}{2}, 1, \frac{3}{2}, 2, \ldots$ never take the value zero. This fact justifies the following definition.

Proper graph subspaces. For each graph Γ, the *proper* graph subspace $\mathcal{K}_\Gamma$ is the subset of $\tilde{\mathcal{K}}_\Gamma$ spanned by the basis states with $j_j > 0$. It is easy to see that all proper subspaces $\mathcal{K}_\Gamma$ are orthogonal to each other, and they span $\mathcal{K}$; we can write this as

$$\mathcal{K} \sim \bigoplus_\Gamma \mathcal{K}_\Gamma. \tag{6.17}$$

The "null" graph $\Gamma = \emptyset$ is included in the sum; the corresponding Hilbert space is the one-dimensional space spanned by the state $\Psi[A] = 1$. This state is denoted $|\emptyset\rangle$; thus $\langle A|\emptyset\rangle = 1$.

* *$\mathcal{K}$ as an L_2 space.* I have defined $\mathcal{K}$ as the completion of $\mathcal{S}$ in the scalar product defined by the bilinear form (6.7). Can this space be viewed as a space of square integrable functionals in some measure? The answer is yes, and it involves a beautiful mathematical construction that I will not describe here, since it is not needed for what follows, and for which I refer the reader to [20]. Very briefly, $\mathcal{K} \sim L_2[\mathcal{A}, d\mu_0]$, where $\mathcal{A}$ is an *extension* of the space of the smooth connection. The extension includes distributional connections. The measure $d\mu_0$ is defined on this space and is called the Ashtekar–Lewandowski measure. The construction is analogous to the definition of the gaussian measure $d\mu_G[\phi] \sim$ "$e^{-\int dx dy \; \phi(x)G(x,y)\phi(y)}[d\phi]$," which, as is well known, needs to be defined on a space of distributions $\phi(x)$. The space $\mathcal{A}$ has the beautiful property of being the Gelfand spectrum of the abelian C^* algebra formed by the smooth holonomies of the connection A.

6.2.2 Invariances of the scalar product

The kinematical state space $\mathcal{S} \subset \mathcal{K} \subset \mathcal{S}'$ carries a natural representation of local $SU(2)$, and $Diff(\Sigma)$, simply realized by the transformations of the argument A. The scalar product defined above is invariant under these transformations. Therefore $\mathcal{K}$ carries a *unitary* representation of local $SU(2)$, and $Diff(\Sigma)$. Let us look at this in some detail.

Local gauge transformations. Under (smooth) local $SU(2)$ gauge transformations $\lambda : \Sigma \to SU(2)$ the connection A transforms inhomogeneously like a gauge potential, i.e.,

$$A \to A_\lambda = \lambda A \lambda^{-1} + \lambda d\lambda^{-1}. \tag{6.18}$$

This transformation of A induces a natural representation of local gauge transformations $\Psi(A) \to \Psi(A_{\lambda^{-1}})$ on $\mathcal{K}$. Despite the inhomogeneous transformation rule (6.18) of the connection, the holonomy transforms homogeneously (see Sec. 2.1.5)

$$U[A, \gamma] \to U[A_\lambda, \gamma] = \lambda(x_f^\gamma)\, U[A, \gamma]\, \lambda^{-1}(x_i^\gamma), \qquad (6.19)$$

where $x_i^\gamma, x_f^\gamma \in \Sigma$ are the initial and final points of the path γ. For a given (Γ, f), define

$$f_\lambda(U_1, \ldots, U_L) = f(\lambda(x_f^{\gamma_1})U_1\lambda^{-1}(x_i^{\gamma_1}), \ldots \lambda(x_f^{\gamma_L})U_L\lambda^{-1}(x_i^{\gamma_L})). \qquad (6.20)$$

It is then easy to see that the transformation of the quantum states is

$$\Psi_{\Gamma,f}(A) \to [U_\lambda \Psi_{\Gamma,f}](A) = \Psi_{\Gamma,f}(A_{\lambda^{-1}}) = \Psi_{\Gamma,f_{\lambda^{-1}}}(A). \qquad (6.21)$$

Since the Haar measure is invariant under right and left group transformations, it follows immediately that (6.7) is invariant. From (6.21) and from their definition it is easy to see that basis states $|\Gamma, j_l, \alpha_l, \beta_l\rangle$ transform as

$$U_\lambda |\Gamma, j_l, \alpha_l, \beta_l\rangle = R^{(j_1)\alpha_1}{}_{\alpha'_1}(\lambda^{-1}(x_{f_1}))\, R^{(j_1)\beta'_1}{}_{\beta_1}(\lambda(x_{i_1})) \cdots$$
$$R^{(j_L)\alpha_L}{}_{\alpha'_L}(\lambda^{-1}(x_{f_L}))\, R^{(j_L)\beta'_L}{}_{\beta_L}(\lambda(x_{i_L}))$$
$$|\Gamma, j_l, \alpha'_l, \beta'_l\rangle, \qquad (6.22)$$

where i_l and f_l are the points where the link l begins and ends.

(Extended) diffeomorphisms. Consider maps $\phi : \Sigma \to \Sigma$ that are continuous, invertible, and such that the map and its inverse are smooth everywhere, except, possibly, at a finite number of isolated points. Call these maps "extended diffeomorphisms" (or sometimes, loosely, just "diffeomorphisms"). Call the group formed by these maps $Diff^*$.

An example of an extended diffeomorphism which is not a proper diffeomorphism, in two dimensions, is the following. In polar coordinates, the map

$$r' = r, \qquad \phi' = \phi + \frac{1}{2}\sin\phi \qquad (6.23)$$

is continuous everywhere, while it is differentiable everywhere except at $r = 0$, where its jacobian is ill defined.

Under an extended diffeomorphism, the transformation of the connection is well defined (recall $A \in \mathcal{G}$ is defined everywhere on Σ except on a finite number of isolated points.) A transforms as a one-form,

$$A \to \phi^* A. \qquad (6.24)$$

Hence, $\mathcal{S}$ carries the representation U_ϕ of $Diff^*$ defined by $U_\phi \Psi(A) = \Psi((\phi^*)^{-1} A)$. The holonomy transforms as

$$U[A, \gamma] \to U[\phi^* A, \gamma] = U[A, \phi^{-1}\gamma], \qquad (6.25)$$

where $(\phi\gamma)(s) \equiv (\phi(\gamma(s)))$. That is, dragging A by a diffeomorphism ϕ is equivalent to dragging the curve γ. (Notice that if ϕ is not a proper diffeomorphism, the curve $\phi\gamma$ may fail to be smooth, at a finite number of points at most.) In turn, a cylindrical function $\Psi_{\Gamma,f}[A]$ is sent to a cylindrical function $\Psi_{\phi\Gamma,f}[A]$, namely one which is based on the shifted graph. Since the right-hand side of (6.7) does not depend explicitly on the graph, the diffeomorphism invariance of the inner product is immediate.

6.2.3 Gauge-invariant and diffeomorphism-invariant states

The kinematical state space $\mathcal{K}$ is a space of arbitrary wave functionals of the connection $\Psi[A]$, but recall that we need a space of states that are wave functionals invariant under local gauge transformations and diffeomorphisms. More formally, the two classical equations (4.80) and (4.81) must be implemented in the quantum theory, giving

$$D_a \frac{\delta}{\delta A_a^i(\vec{\tau})} \Psi[A] = 0, \tag{6.26}$$

$$F_{ab}^i(\vec{\tau}) \frac{\delta}{\delta A_a^i(\vec{\tau})} \Psi[A] = 0. \tag{6.27}$$

The same argument as used in Section 4.3.4 shows that these equations demand the invariance of Ψ under local $SU(2)$ transformations and diffeomorphisms. More precisely, the smearing functions $f^a(\vec{\tau})$ of Section 4.3.4 must be chosen in an appropriate class: here the relevant group is $Diff^*$ and therefore $f^a(\vec{\tau})$ can be any infinitesimal generator of $Diff^*$. Under this choice, (6.27) is equivalent to the requirement of invariance of the state under $Diff^*$.

Call $\mathcal{K}_0$ the space of the states invariant under local $SU(2)$, and $\mathcal{K}_{\text{diff}}$ the space of the states invariant under local $SU(2)$ and $Diff^*$. Recalling that we call $\mathcal{H}$ the space of solutions of the Wheeler–DeWitt equations, we have therefore the sequence of Hilbert spaces

$$\mathcal{K} \xrightarrow{SU(2)} \mathcal{K}_0 \xrightarrow{Diff^*} \mathcal{K}_{\text{diff}} \xrightarrow{H} \mathcal{H}, \tag{6.28}$$

where the three steps correspond to the implementation of the three equations that the wave functional must satisfy, namely (6.26), (6.27) and (6.3) respectively. I construct explicitly $\mathcal{K}_0$ in the next section and $\mathcal{K}_{\text{diff}}$ in the following one. (Except for the first one, the domain of the maps in (6.28) will turn out to be given by the first term of the corresponding rigged Hilbert space.)

6.3 Internal gauge invariance. The space $\mathcal{K}_0$

The space $\mathcal{K}_0$ is the space of the states in $\mathcal{K}$ invariant under local $SU(2)$ gauge transformations. I call $\mathcal{S}_0$ the gauge-invariant subspace of $\mathcal{S}$ and $\mathcal{S}_0'$ its dual. It is not difficult to see that $\mathcal{K}_0$ is a proper subspace of $\mathcal{K}$. Examples of finite norm $SU(2)$ invariant states are provided by the loop states defined in (6.9). In fact, a moment of reflection shows that the multiloop states are sufficient to span $\mathcal{K}_0$. In the first years of the development of LQG, multiloop states were used as a basis for $\mathcal{K}_0$; however, this basis is overcomplete, and this fact complicates the formalism. Nowadays we have a much better control of $\mathcal{K}_0$, thanks to the introduction of the spin network states, which can be seen as finite linear combinations of multiloop states forming a genuine orthonormal basis.

The spin network basis of quantum gravity is a simple extension of the spin network basis defined in Section 5.3.6 in the context of lattice gauge theory. As we shall see in the next section, however, diffeomorphism invariance will soon make the two cases very different, and connect the quantum gravity spin network basis to Penrose's old "spin network" idea that quantum states of the geometry can be described as abstract graphs carrying spins.

Spin networks. Denote "nodes" the end points of the oriented curves in Γ. Without loss of generality, assume that each set of curves Γ is formed by curves γ that, if they overlap at all, overlap only at nodes. Viewed in this way, Γ is in fact a graph immersed in the manifold, that is, a collection of nodes n, which are points of Σ, joined by links l, which are curves in Σ. The "outgoing multiplicity" m_{out} of a node is the number of links that begin at the node. The "ingoing multiplicity" m_{in} of a node is the number of links that end at the node. The multiplicity, or valence $m = m_{\text{in}} + m_{\text{out}}$ of a node is the sum of the two.

Given the graph Γ, for which an ordering and an orientation have been chosen, let j_l be an assignment of an irreducible representation, different from the trivial one, to each link l. Let i_n be an assignment of an intertwiner i_n to each node n. The notion of intertwiner was defined in Section 5.3.6. The intertwiner i_n associated with a node is between the representations associated with the links adjacent to the node. The triplet $S = (\Gamma, j_l, i_n)$ is called a "spin network embedded in Σ." A choice of j_l and i_n is called a "coloring" of the links, and the nodes, respectively.

6.3.1 Spin network states

Consider a spin network $S = (\Gamma, j_l, i_n)$, with L links and N nodes. The state $|\Gamma, j_l, \alpha_l, \beta_l\rangle$, defined above in (6.16), has L indices α_l and L indices

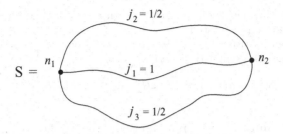

Fig. 6.1 A simple spin network with two trivalent nodes.

β_l. The N intertwiners i_n have, altogether, precisely a set of indices dual to these. The contraction of the two

$$|S\rangle \equiv \sum_{\alpha_l,\beta_l} v_{i_1}^{\beta_1\ldots\beta_{n_1}}{}_{\alpha_1\ldots\alpha_{n_1}} \, v_{i_2}^{\beta_{n_1+1}\ldots\beta_{n_2}}{}_{\alpha_{n_1+1}\ldots\alpha_{n_2}} \, \cdots$$
$$v_{i_N}^{\beta_{(n_{N-1}+1)}\ldots\beta_L}{}_{\alpha_{(n_{N-1}+1)}\ldots\alpha_L} \, |\Gamma, j_l, \alpha_l, \beta_l\rangle \qquad (6.29)$$

defines the spin network state $|S\rangle$. The pattern of the contraction of the indices is dictated by the topology of the graph itself: the index α_l (resp. β_l) of the link l is contracted with the corresponding index of the inter-twiner v_{i_n} of the node n where the link l starts (resp. ends). The gauge invariance of this state follows immediately from the transformation properties (6.22) of the basis states and the invariance of the intertwiners. As a functional of the connection, this state is

$$\Psi_S[A] = \langle A|S\rangle \equiv \left(\bigotimes_l R^{(j_l)}(H[A, \gamma_l]) \right) \cdot \left(\bigotimes_n i_n \right). \qquad (6.30)$$

The raised dot notation indicates the contraction between dual spaces: on the left, the tensor product of the matrices lives in the space $\otimes_l (\mathcal{H}^*_{j_l} \otimes \mathcal{H}_{j_l})$. On the right, the tensor product of all intertwiners lives precisely in the dual of this space.

Example. Let's say Γ has two nodes, n_1 and n_2, and three links l_1, l_2, l_3, each link begin-ning at n_1 and ending at n_2. Let the coloring of the links be $j_1 = 1, j_2 = 1/2, j_3 = 1/2$, see Figure 6.1. At each of the two nodes, we must therefore consider the tensor product of two fundamental and one adjoint representation of $SU(2)$. As is well known, the ten-sor product of these representations contains a single copy of the trivial representation; therefore there is only one possible intertwiner. A moment of reflection shows that this is given by the triple of Pauli matrices, $v^{i,AB} = \frac{1}{\sqrt{3}}\sigma^{i,AB}$ since these have precisely the invariance property

$$(R(U))^i{}_j \, U^A{}_C \, U^B{}_D \, \sigma^{j,CD} = \sigma^{i,AB}. \qquad (6.31)$$

Here $(R(U))^i{}_j$ is the adjoint representation, with $i, j = 1, 2, 3$ vector indices and $U^A{}_C$ is the fundamental representation, with $A, B = 0, 1$ spinor indices. The $\frac{1}{\sqrt{3}}$ is a normalization factor to satisfy (5.157). Therefore, there is only one possible coloring of the nodes in this case. The spin network state is then

$$\Psi_S[A] = \frac{1}{3} \, \sigma_{i,AB} \, (R(H[A, \gamma_1]))^i{}_j \, (H[A, \gamma_2])^A{}_C \, (H[A, \gamma_3])^B{}_D \, \sigma^{j,CD}. \tag{6.32}$$

Let me now enunciate the main fact concerning the spin network states: the ensemble of the spin network states $|S\rangle$ forms an orthonormal basis in $\mathcal{K}_0$. Orthonormality can be checked by a direct calculation. The basis is labeled by spin networks, namely graphs Γ and colorings (j_l, i_n).

Some comments. First, I have assumed the spins j_l to be all different from zero (a spin network containing a link l with $j_l = 0$ is identified with the spin network obtained by removing the link l). Second, this result is a simple consequence of the Peter–Weyl theorem, namely of the fact that the states $|\Gamma, j_l, \alpha_l, \beta_l\rangle$ form a basis in $\mathcal{K}$, and the very definition of the intertwiners. Third, the spin network basis is not unique, as it depends on the (arbitrary) choice of a basis in each space of intertwiners at each node. Notice also that, in the basis $|S\rangle = |\Gamma, j_l, i_n\rangle$, the label Γ runs over all nonoriented and nonordered graphs. However, for the definition of the coloring, an orientation and an ordering has to be chosen for each Γ.

The space $\mathcal{S}_0$ is the space of the *finite* linear combinations of spin network states, which is dense in $\mathcal{K}_0$, and $\mathcal{S}_0'$ is its dual.

6.3.2 * Details about spin networks

Orientation. There is an isomorphism ϵ between a representation j and its dual j^*. If j is the fundamental,

$$\begin{aligned} \epsilon : \quad & C^2 \to (C^2)^* \\ & \psi^A \mapsto \psi_A = \epsilon_{AB} \psi^B, \end{aligned} \tag{6.33}$$

where ϵ_{AB} is the antisymmetric tensor. This extends to all other representations, since they can be obtained from tensor products of the fundamental. Using this, we can raise and lower the indices of the intertwiners, and identify intertwiners with the same total number of indices. We can then ask what happens if we change the orientation of one of the links. Using (A.6), a straightforward calculation, that I leave to the reader, shows that the only change is to the overall sign, if the representation has half-integer spin. Thus, the only relevant orientation of the spin network is an overall *global* orientation.

Spin networks versus loop states. A spin network state can be decomposed into a finite linear combination of (multi-)loop states. The representation j can be written as the symmetrized tensor product of $2j$ fundamental representations. Therefore we can write the elements of $\mathcal{H}_j$ as completely symmetric complex tensors $\psi^{A_1, \ldots, A_{2j}}$ with $2j$ spinor indices $A_i = 0, 1$. In this basis, the representation matrices have the simple form

$$R^{(j)A_1, \ldots, A_{2j}}{}_{B_1, \ldots, B_{2j}}(U) = U^{(A_1}{}_{B_1} \ldots U^{A_{2j})}{}_{B_{2j})}, \tag{6.34}$$

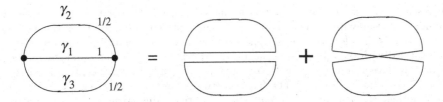

Fig. 6.2   Decomposition of a spin network state into loop states.

where the parentheses indicate complete symmetrization (see Appendix A1). In this basis, the intertwiners are simply combinations of the only two $SU(2)$ invariant tensors, namely ϵ_{AB} and δ_A^B. For instance a (nonnormalized) trivalent intertwiner between incoming representations j, j' and an outgoing representation j'' is

$$v_{A_1,\ldots,A_{2j},B_1,\ldots,B_{2j'}}{}^{C_1,\ldots,C_{2j''}} = \epsilon_{A_1 B_1}\cdots\epsilon_{A_a B_a}\,\delta_{B_{a+1}}^{C_1}\cdots\delta_{B_{2j'}}^{C_b}\,\delta_{A_{a+1}}^{C_{b+1}}\cdots\delta_{A_{2j}}^{C_{2j''}}, \quad (6.35)$$

where $j = a + c$, $j' = a + b$ and $j'' = b + c$. Now, when the two holonomy matrices of two contiguous links γ_1 and γ_2 are joined by δ_A^B, they give the holonomy of the curve obtained joining γ_1 and γ_2, which I denote $\gamma_1 \sharp \gamma_2$

$$H[A,\gamma_1]_B^A \delta_C^B H[A,\gamma_2]_D^C = H[A,\gamma_1\sharp\gamma_2]_D^A. \quad (6.36)$$

On the other hand, recall that $\epsilon_{AB}U^A{}_C\epsilon_{CD} = (U^{-1})^D{}_A$. Therefore,

$$\epsilon_{DB}H[A,\gamma_1]_B^A\epsilon_{AC}H[A,\gamma_2]_E^C = H[A,\gamma_1^{-1}\sharp\gamma_2]_E^D. \quad (6.37)$$

Therefore the tensors ϵ_{AB} and δ_A^B in the intertwiners simply join the segments in the arguments of the holonomies. Since the graph of the spin network is finite, a line of joining must close to a loop. A moment of reflection will convince the reader that a spin network state (6.29) is therefore equal to a linear combination of products of holonomies of closed lines that wrap along the graph. That is, it is a linear combination of multiloop states.

The decomposition of a spin network state in loop states can be obtained graphically as follows. Replace each link of the graph, colored with spin j, with $2j$ parallel strands. Symmetrize these strands along each link. The intertwiners at the nodes can be represented as collections of segments joining the strands of different links. By joining these segments with the strands one obtains a linear combination of multiloops. The spin network states can then be expanded in the corresponding loop states. Notice that this is analogous to the construction at the basis of the Kaufman–Lins recoupling theory illustrated in Appendix A1 (care should be taken with signs). For details on this construction, see [171].

Applying this rule to the state (6.32) illustrated in Figure 6.1, it is easy to see that

$$\Psi_S(A) = \frac{1}{2}\left[\Psi_{((\gamma_1\sharp\gamma_2^{-1}),(\gamma_1\sharp\gamma_3^{-1}))} + \Psi_{\gamma_1\sharp\gamma_2^{-1}\sharp\gamma_1\sharp\gamma_3^{-1}}\right]. \quad (6.38)$$

See Figure 6.2 for a graphical illustration of this decomposition.

Details on intertwiners. Given a graph and a coloring of its links, it may happen that there is no nonvanishing intertwiner at all associated with a node. For instance, a node with valence unity cannot exist in a spin network, because a single nontrivial irreducible representation does not contain any invariant subspace. If the node is bivalent, there is an intertwiner only if the incoming and outgoing representations are the same, and the intertwiner is the identity. The spin network with two links joined by a bivalent node is identified with one obtained replacing the two links with a single link. A trivalent node may have adjacent links with colorings j_1, j_2, j_3 only if these satisfy the Clebsch–Gordan conditions (A.10)–(A.11). The intertwiners are directly given by the Wigner $3j$-coefficients (A.16), up to normalization. Nontrivial intertwiner spaces begin only with nodes of valence four or higher. Intertwiners of an n-valent node can be labeled by $n - 2$ spins, as detailed in Appendix A1.

6.4 Diffeomorphism invariance. The space $\mathcal{K}_{\text{diff}}$

Let me now come to the second and far more crucial invariance: 3d diffeomorphism invariance. We have to find the diffeomorphism-invariant states.

Transformation properties of spin network states under diffeomorphisms. The spin network states $|S\rangle$ are not invariant under diffeomorphisms. A diffeomorphism moves the graph around on the manifold, and therefore changes the state. Notice, however, that a diffeomorphism may change more than the graph of a spin network, that is, the equation $U_\phi |\Gamma, j_l, i_n\rangle = |(\phi\Gamma, j_l, i_n)\rangle$ is not always correct. In particular, a diffeomorphism that leaves the graph Γ invariant may still affect a spin network state $|\Gamma, j_l, i_n\rangle$. This is because, for each graph, the definition of the spin network state requires the choice of an orientation and ordering of the links, and these can be changed by a diffeomorphism.

Here is an example. Let Γ be an "eyeglasses" graph formed by two loops α and β in the $j = 1/2$ representation connected by a path γ in the $j = 1$ representation. The space of the intertwiners at each node is one-dimensional, but this does not imply that there is no choice to be made for the basis, since if i is a normalized intertwiner, so is $-i$. With one choice, the state is

$$\Psi_S[A] = (U(A, \alpha))^A{}_B \, \sigma_i{}^B{}_A \, (R^{(1)}(U(A, \gamma)))^i{}_j \, \sigma^{jD}{}_C \, (U(A, \beta))^C{}_D. \tag{6.39}$$

Using elementary $SU(2)$ representation theory this can be rewritten (up to a normalization factor) as

$$\Psi_S[A] \sim \text{tr} H[A, \alpha\gamma\beta\gamma^{-1}] - \text{tr} H[A, \alpha\gamma\beta^{-1}\gamma^{-1}]. \tag{6.40}$$

Now consider a diffeomorphism ϕ that turns the loop β around, namely it reverses its orientation, while leaving α and γ as they are. Clearly this diffeomorphism will send the two terms of the last equation into each other, giving

$$U_\phi \Psi_S[A] = -\Psi_S[A], \tag{6.41}$$

while $\phi\Gamma = \Gamma$.

Given an oriented and ordered graph Γ, there is a finite discrete group G_Γ of maps g_k, such as the one of the example, that change its order or orientation and that can be obtained as a diffeomorphism. The elements g_k of this group act on $\mathcal{K}_\Gamma$.

A moment of reflection will convince the reader that the diff-invariant states are not in $\mathcal{K}_0$, they are in $\mathcal{S}'_0$. We are therefore in the common situation in which the solutions of a quantum equation must be searched in the extension of the Hilbert space, and the scalar product must be appropriately extended to the space of the solutions, as explained in Section 5.5.2.

The elements of $\mathcal{S}'_0$ are linear functionals Φ on the functionals $\Psi \in \mathcal{S}_0$. The requirement of diff invariance makes sense in $\mathcal{S}'_0$ because the action of the diffeomorphism group is well defined in $\mathcal{S}'_0$ by duality

$$(U_\phi \Phi)(\Psi) \equiv \Phi(U_{\phi^{-1}} \Psi). \tag{6.42}$$

Therefore a diff-invariant element Φ of $\mathcal{S}'_0$ is a linear functional such that

$$\Phi(U_\phi \Psi) \equiv \Phi(\Psi). \tag{6.43}$$

The space $\mathcal{K}_{\text{diff}}$ is the space of these diff-invariant elements of $\mathcal{S}'_0$. Remarkably, we have a quite good understanding of this space, whose elements can be viewed as the quantum states of physical space.

The space $\mathcal{K}_{\text{diff}}$. I now define a map $P_{\text{diff}} : \mathcal{S}_0 \to \mathcal{S}'_0$, and show that the (closure in norm of the) image of this map is precisely $\mathcal{K}_{\text{diff}}$. Let the state $P_{\text{diff}}\Psi$ be the element of $\mathcal{S}'_0$ defined by

$$(P_{\text{diff}}\,\Psi)(\Psi') = \sum_{\Psi'' = U_\phi \Psi} \langle \Psi'', \Psi' \rangle. \tag{6.44}$$

The sum is over all states Ψ'' in $\mathcal{S}_0$ for which there exist a $\phi \in \textit{Diff}^*$ such that $\Psi'' = U_\phi \Psi$. The key point is that this sum is always finite, and therefore well defined. To see this, notice that since Ψ and Ψ' are in $\mathcal{S}_0$, they can be expanded in a finite linear combination of spin network states. If a diffeomorphism changes the graph of a spin network state Ψ_S, then it takes it to a state orthogonal to itself. If it doesn't change the graph, then either it leaves the state invariant, so that no multiplicity appears in (6.44), or it changes the ordering or the orientation of the links; but these are discrete operations, giving at most a discrete multiplicity in the sum in (6.44). Therefore, the sum in (6.44) is always well defined. Clearly $P_{\text{diff}}\,\Psi$, is diff invariant, namely it satisfies (6.43). Furthermore, it is not difficult to convince oneself that the functionals of the form (6.44) span the space of the diff-invariant states. Therefore, the (closure in norm of the) image of P_{diff} is $\mathcal{K}_{\text{diff}}$. States related by a diffeomorphism are projected by P_{diff} to the same element of $\mathcal{K}_{\text{diff}}$:

$$P_{\text{diff}}\,\Psi_S = P_{\text{diff}}\,(U_\phi \Psi_S). \tag{6.45}$$

Finally, the scalar product on $\mathcal{K}_{\text{diff}}$ is naturally defined by

$$\langle P_{\text{diff}}\,\Psi_S, P_{\text{diff}}\,\Psi_{S'} \rangle_{\mathcal{K}_{\text{diff}}} \equiv (P_{\text{diff}}\,\Psi_S)(\Psi_{S'}) \tag{6.46}$$

(see section 5.5.2). This completely defines $\mathcal{K}_{\text{diff}}$. Equivalently, $\mathcal{K}_{\text{diff}}$ is defined by the bilinear form

$$\langle \Psi, \Psi' \rangle_{\mathcal{K}_{\text{diff}}} \equiv \langle \Psi | P_{\text{diff}} | \Psi' \rangle \equiv \sum_{\Psi''=\phi\Psi} \langle \Psi'', \Psi' \rangle \tag{6.47}$$

in $\mathcal{S}_0$.

To understand intuitively why the above definition works, consider the following formal argument. Imagine that we were able to define a measure $d\phi$ on $Diff^*$. We could then write diffeomorphism-invariant states by simply integrating an arbitrary state on the orbit of the diffeomorphism group:

$$P_{\text{diff}}\Psi \equiv \int_{Diff^*} [d\phi] \, U_\phi \Psi. \tag{6.48}$$

Therefore,

$$(P_{\text{diff}}\Psi')(\Psi) = \int_{Diff^*} [d\phi] \, (\Psi' U_\phi \Psi). \tag{6.49}$$

Let us assume for simplicity that $\Psi \in \mathcal{K}_\Gamma$ and $\Psi' \in \mathcal{K}_{\Gamma'}$ (the general case following by linearity). Then the right-hand side of (6.49) vanishes unless there is a ϕ that sends Γ in Γ'. If this is the case, the integral has support just on the subgroup of $Diff^*$ that leaves Γ invariant. Elements of this subgroup can either change the state Ψ or leave it invariant. Thus, we can rewrite (6.49) as

$$(P_{\text{diff}}\Psi')(\Psi) = \sum_{\Psi''=U_{\phi'}\Psi} \int_{D_{\Psi''}} [d\phi] \, (\Psi' U_\phi \Psi''), \tag{6.50}$$

where the integral is over the subgroup $D_{\Psi''}$ of $Diff^*$ that leaves Ψ'' invariant. But then we can take the scalar product out of the integral and write

$$(P_{\text{diff}}\Psi')(\Psi) = \sum_{\Psi''=U_\phi\Psi} (\Psi' U_\phi \Psi'') \left(\int_{D_{\Psi'}} [d\phi] \right). \tag{6.51}$$

If we now assume that the measure $d\phi$ is such that the volume of $D_{\Psi'}$ is unity, we recover the definition given above. Therefore, the definition (6.44) can be seen as a rigorous implementation of the intuitive "integration of the diffeomorphism group" of (6.48).

6.4.1 Knots and s-knot states

To understand the structure of $\mathcal{K}_{\text{diff}}$, consider the action of P_{diff} on the states of the spin network basis. To this aim, observe that a diffeomorphism sends a spin network state $|S\rangle$ to an orthogonal state, or to a state obtained by a change in the order of the orientation of the links. Denote $g_k|S\rangle$ the states that are obtained from $|S\rangle$ by changes of orientation or ordering, and that can be obtained via a diffeomorphism, as in the example above. The maps g_k form the finite discrete group G_Γ, therefore the range of the discrete index k is finite. Then it is easy to see that

$$\langle S | P_{\text{diff}} | S' \rangle = \begin{cases} 0 & \text{if } \Gamma \neq \phi\Gamma' \\ \sum_k \langle S | g_k | S' \rangle & \text{if } \Gamma = \phi\Gamma' \end{cases} . \tag{6.52}$$

An equivalence class K of unoriented graphs Γ under diffeomorphisms is called a "knot." Knots without nodes have been widely studied by the branch of mathematics called knot theory, with the aim of classifying them. Knots with nodes have also been studied in knot theory, but to a lesser extent. From the first line of (6.52), we see that two spin networks S and S' define orthogonal states in $\mathcal{K}_{\mathrm{diff}}$ unless they are knotted in the same way. That is, unless they are defined on graphs Γ and Γ' belonging to the same knot class K. Therefore the basis states in $\mathcal{K}_{\mathrm{diff}}$ are, first of all, labeled by knots K. We call $\mathcal{K}_K$ the subspace of $\mathcal{K}_{\mathrm{diff}}$ spanned by the basis states labeled by the knot K. That is,

$$\mathcal{K}_K = P_{\mathrm{diff}}\,\mathcal{K}_\Gamma \qquad\qquad (6.53)$$

for any $\Gamma \in K$.

The states in $\mathcal{K}_K$ are then distinguished only by the coloring of links and nodes. As observed before, the colorings are not necessarily orthonormal, due to the nontrivial action of the discrete symmetry group G_Γ. To find an orthonormal basis in $\mathcal{K}_K$ we have therefore to further diagonalize the quadratic form defined by the second line of (6.52). Denote $|s\rangle = |K, c\rangle$ the resulting states. The discrete label c is called the coloring of the knot K. Up to the complications due to the discrete symmetry G_Γ, it corresponds to the coloring of the links and the nodes of Γ. The states $|s\rangle = |K, c\rangle$ are called spin-knot states, or s-knot states.

6.4.2 The Hilbert space $\mathcal{K}_{\mathrm{diff}}$ is separable

The key property of knots is that they form a discrete set. Therefore, the label K is discrete. It follows that $\mathcal{K}_{\mathrm{diff}}$ admits a discrete orthonormal basis $|s\rangle = |K, c\rangle$. Thus, $\mathcal{K}_{\mathrm{diff}}$ is a separable Hilbert space. The "excessive size" of the kinematical Hilbert space $\mathcal{K}$ reflected in its nonseparability turns out to be just a gauge artifact.

The fact that knots without nodes form a discrete set is a classic result of knot theory. It is easy to understand it intuitively: first, if two loops without nodes can be continuously deformed into each other without crossing, then there is a diffeomorphism that sends one into the other; second, to change the node class we have to deform a link across another link, and this is a discrete operation. On the other hand, the fact that knots *with nodes* form a discrete set is nontrivial. Indeed, it depends on the fact that we have chosen the extension $Diff^*$ of the diffeomorphism group $Diff$. Had we chosen $Diff$ as the invariance group, the space of the knot classes would have been continuous.

To understand this, recall that the action of $Diff$ on the tangent space is linear. Consider a graph Γ that can be deformed continuously into a graph Γ'. Let p be the location of an n-valent node of Γ and p' the corresponding node on Γ'. Is there a diffeomorphism ϕ in $Diff$ sending Γ to Γ'? The answer in general is negative, for the following reason. The diffeomorphism must send p to p'. Hence $\phi(p) = p'$. The tangent space to p is sent into the tangent space to p' by the jacobian J_p of ϕ at p, which is a linear transformation in 3d. Let $\vec{v}_i$, $i = 1, \ldots, n$ be the tangents of the n links at p and $\vec{v}_i'$, $i = 1, \ldots, n$ be the tangents of the n links at p'. For ϕ to send the two nodes into each other, we must have

$$J_p \vec{v}_i = \vec{v}_i'. \tag{6.54}$$

But, in general, there is no linear transformation sending n given directions into n other given directions. In other words, (6.54) gives n linear conditions on the nine degrees of freedom of the jacobian matrix $(J_p)^a{}_b = \partial \phi^a(x)/\partial x^b|_p$. Therefore, in general, two graphs that can be transformed into each other continuously cannot be transformed into each other by a diffeomorphism. The equivalence classes are characterized by continuous parameters at the nodes.

On the other hand, maps in $Diff^*$ can freely transform these parameters. The reason is that thanks to the relaxation of the differentiability condition, an extended diffeomorphism $\phi \in Diff^*$ can act nonlinearly on the tangents. In Section 6.7, after the discussion of the physical interpretation of the knot states, I will discuss the physical reasons why $Diff^*$ is more appropriate than $Diff$ as gauge group.

This concludes the construction of the kinematical quantum state space of LQG. The physical meaning of the s-knot states in $\mathcal{K}_{\text{diff}}$ will become clear later on. It is now time to define the operators.

6.5 Operators

There are two basic field variables in the canonical theory, from which all measurable quantities can be constructed: the connection $A_a^i(\tau)$ and its momentum $E_i^a(\tau)$. I now define quantum operators corresponding to simple functions of these. Quantum states are functionals $\Psi[A]$ of the connection A. The momentum conjugate to the real connection A is $(1/8\pi G)E$ (see (4.40)). We can therefore define the two field operators

$$A_a^i(\tau)\Psi[A] = A_a^i(\tau)\ \Psi[A], \tag{6.55}$$

$$\frac{1}{8\pi G}E_i^a(\tau)\Psi[A] = -i\hbar \frac{\delta}{\delta A_a^i(\tau)}\ \Psi[A] \tag{6.56}$$

on functionals of $\Psi[A]$. In the following, I choose units in which $8\pi G = 1$; I will then restore physical units when needed. The first is a multiplicative operator; the second a functional derivative. However, both these operators send $\Psi[A]$ out of the state spaces that we have constructed. In particular, they are not well defined in $\mathcal{K}$. This can be easily cured by taking, instead of A and E, some simple function of these.

6.5.1 The connection A

The holonomy $U(A, \gamma)$ is well defined on $\mathcal{S}$. More precisely, let $U^A{}_B(A, \gamma)$ be the matrix elements of the group element $U(A, \gamma)$. Then

$$(U^A{}_B(A, \gamma)\Psi)[A] = U^A{}_B(A, \gamma)[A]\ \Psi[A]. \tag{6.57}$$

The right-hand side is clearly in $\mathcal{S}$ if $\Psi[A]$ is. In fact, any cylindrical function of the connection is immediately well defined as a multiplicative operator in $\mathcal{K}$.

For instance, consider a closed loop α and let $T_\alpha[A] = \mathrm{tr}U(A, \alpha)$. Consider the action of this operator on a spin network state $|S\rangle$ with a graph that does not intersect with α. Then clearly

$$T_\alpha|S\rangle = |S \cup \alpha\rangle \tag{6.58}$$

where S $\cup$ α is the spin network formed by S plus the loop α in the $j = 1/2$ representation.

Notice that all this is quite different from quantum field theory on a background spacetime, where field operators are operator-valued distributions, and therefore are well defined only when smeared in three dimensions. In (6.57), a well-defined operator is obtained by simply smearing (the path-ordered exponential of) the field in just one dimension: along the loop γ. This is a characteristic feature of diff-invariant quantum field theories.

6.5.2 The conjugate momentum E

To understand the action of E, we have to compute the functional derivative of the holonomy, the building block of the cylindrical functions. It is not hard to show that

$$\frac{\delta}{\delta A_a^i(x)} U(A, \gamma) = \int ds \; \dot{\gamma}^a(s) \; \delta^3(\gamma(s), x) \; [U(A, \gamma_1) \; \tau_i \; U(A, \gamma_2)]. \tag{6.59}$$

Here s is an arbitrary parametrization of the curve γ, $\gamma^a(s)$ are the coordinates of the curve, $\dot{\gamma}^a(s) \equiv d\gamma^a(s)/ds$ is the tangent to the curve in the point s, γ_1 and γ_2 are the two segments in which γ is separated by the point s. This is a crucial formula, that plays a major role in what follows. The diligent reader is therefore invited to derive it and understand it in detail. There are several possible derivations. The naive one is just to use a formal functional derivation of the expression (2.80). The rigorous one is to consider variations of the defining equation (2.78). See [172] for details.

Notice that the right-hand side of (6.59) is a distribution, but only a two-dimensional one, since one of the three deltas in δ^3 is in fact integrated over ds. It is therefore natural to search for an operator well defined on $\mathcal{K}$ by smearing E in two-dimensions. To this purpose, consider a two-dimensional surface $\mathcal{S}$ embedded in the 3d manifold.

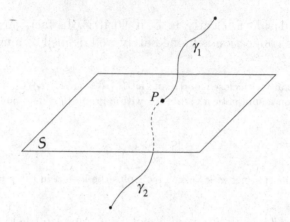

Fig. 6.3   A curve that intersects the surface at an individual point P.

Let $\vec{\sigma} = (\sigma^1, \sigma^2)$ be coordinates on the surface $\mathcal{S}$. The surface is defined by $\mathcal{S} : (\sigma^1, \sigma^2) \mapsto x^a(\sigma^1, \sigma^2)$. Consider the operator

$$E_i(\mathcal{S}) \equiv -\mathrm{i}\hbar \int_{\mathcal{S}} \mathrm{d}\sigma^1 \mathrm{d}\sigma^2 \, n_a(\vec{\sigma}) \frac{\delta}{\delta A_a^i(x(\vec{\sigma}))} \,, \qquad (6.60)$$

where

$$n_a(\vec{\sigma}) = \epsilon_{abc} \frac{\partial x^b(\vec{\sigma})}{\partial \sigma^1} \frac{\partial x^c(\vec{\sigma})}{\partial \sigma^2} \qquad (6.61)$$

is the normal one-form on $\mathcal{S}$ and ϵ_{abc} is the completely antisymmetric object (for the relativists, the Levi–Civita tensor of density weight (-1)).

The "grasp." Let us now compute the action of the operator $E_i(\mathcal{S})$ on the holonomy $U(A, \gamma)$. Assume for the moment that the end points of γ do not lie on the surface $\mathcal{S}$. For simplicity, let us also begin by assuming that the curve γ crosses the surface $\mathcal{S}$ at most once, and denote P the intersection point (if any), see Figure 6.3.

The curve is separated into two parts, $\gamma = \gamma_1 \cup \gamma_2$, by P. By using (6.59) and (6.60), we obtain

$$E_i(\mathcal{S}) \, U(A, \gamma)$$

$$= -\mathrm{i}\hbar \int_{\mathcal{S}} \mathrm{d}\sigma^1 \mathrm{d}\sigma^2 \, \epsilon_{abc} \frac{\partial x^a(\vec{\sigma})}{\partial \sigma^1} \frac{\partial x^b(\vec{\sigma})}{\partial \sigma^2} \frac{\delta}{\delta A_c^i(\vec{x}(\vec{\sigma}))} \, U(A, \gamma)$$

$$= -\mathrm{i}\hbar \int_{\mathcal{S}} \int_{\gamma} \mathrm{d}\sigma^1 \mathrm{d}\sigma^2 \mathrm{d}s \, \epsilon_{abc} \frac{\partial x^a}{\partial \sigma^1} \frac{\partial x^b}{\partial \sigma^2} \frac{\partial x^c}{\partial s} \, \delta^3\big(\vec{x}(\vec{\sigma}), \vec{x}(s)\big)$$

$$\times U(A, \gamma_1) \, \tau_i \, U(A, \gamma_2). \qquad (6.62)$$

A closer look at this result reveals a great simplification of the last integral. The integral vanishes unless the surface and the curve intersect. Assume that there is a single intersection point and further assume that it has coordinates $x^a = 0$. In the neighborhood of this point, consider the map $(\sigma^1, \sigma^2, s) \rightarrow (x^1, x^2, x^3)$ from the integration domain to coordinate space, defined by

$$x^a(\sigma^1, \sigma^2, s) = x^a(\sigma^1, \sigma^2) + x^a(s). \tag{6.63}$$

The jacobian of this map

$$J \equiv \frac{\partial(x^1, x^2, x^3)}{\partial(\sigma^1, \sigma^2, s^3)} = \epsilon_{abc} \frac{\partial x^a}{\partial \sigma^1} \frac{\partial x^b}{\partial \sigma^2} \frac{\partial x^c}{\partial s} \tag{6.64}$$

appears in the integral. We can therefore make the change of variables $(\sigma^1, \sigma^2, s) \rightarrow (x^1, x^2, x^3)$ in the integral.

The jacobian is nonvanishing, since I have required that there is only a single, non-degenerate point of intersection. The jacobian (6.64) and the integral (6.62) would vanish if the tangent vectors given by the partial derivatives in (6.64) were coplanar, i.e., if a tangent $\partial x^{a,b}(\vec{\sigma})/\partial \sigma^{1,2}$ to the surface were parallel to the tangent $\partial x^c(s)/\partial s$ of the curve. This happens, for instance, if the curve lies entirely in $\mathcal{S}$. Then there wouldn't be just a single intersection point. I consider these limiting cases later on.

With a change of variables in the integral we can easily perform the integration and get rid of the delta function. We obtain, remarkably,

$$\int_{\mathcal{S}} \int_{\gamma} d\sigma^1 d\sigma^2 ds \, \epsilon_{abc} \frac{\partial x^a(\vec{\sigma})}{\partial \sigma^1} \frac{\partial x^b(\vec{\sigma})}{\partial \sigma^2} \frac{\partial x^c(s)}{\partial s} \, \delta^3\big(\vec{x}(\vec{\sigma}), \vec{x}(s)\big) = \pm 1 \,. \tag{6.65}$$

In fact, this integral is a well-known analytic coordinate-independent expression for the intersection number between the surface $\mathcal{S}$ and the curve γ. It vanishes if there is no intersection. The sign is dictated by the relative orientation of the surface and the curve. Hence, we obtain the simple result

$$E_i(\mathcal{S}) \, U(A, \gamma) = \pm i\hbar \, U(A, \gamma_1) \, \tau_i \, U(A, \gamma_2). \tag{6.66}$$

The action of the operator $\hat{E}^i(\mathcal{S})$ on holonomies consists of just inserting the matrix $(\pm i\hbar \, \tau_i)$ at the point of intersection. We say that the operator $E^i(\mathcal{S})$ "grasps" γ.

The generalization to multiple intersections is immediate. Using P to label different intersection points, we have:

$$E_i(\mathcal{S}) \, U(A, \gamma) = \sum_{P \in (\mathcal{S} \cap \gamma)} \pm i\hbar \, U(A, \gamma_1^P) \, \tau_i \, U(A, \gamma_2^P). \tag{6.67}$$

For later use, I give here also the action of the operator $E_i(\mathcal{S})$ on the holonomy in an arbitrary representation j

$$E_i(\mathcal{S}) \, R^j\big(U(A, \gamma)\big) = \pm i\hbar \, R^j\big(U(A, \gamma_1)\big) \, ^{(j)}\tau_i \, R^j\big(U(A, \gamma_2)\big), \tag{6.68}$$

where $^{(j)}\tau_i$ is the $SU(2)$ generator in the spin-j representation.

Thus, $E_i(\mathcal{S})$ is a well-defined operator on $\mathcal{K}$. The fact that it is a *surface* integral of $E_i^a(\tau)$ which is well defined can be understood as follows. Geometrically, $E_i^a(\tau)$ is not a vector field, but rather a vector density. The natural associated geometric quantity is the two-form $E_i = \epsilon_{abc} E_i^a dx^b \wedge dx^c$. But a two-form can be naturally integrated over a surface, giving an object which is well behaved under diffeomorphisms. In fact, the operator we have defined corresponds precisely to the classical quantity

$$E_i(\mathcal{S}) = \int_\mathcal{S} E_i. \tag{6.69}$$

The same is true in the quantum theory. The functional derivative is a vector density, therefore a two-form and this is naturally integrated over a surface. Geometry and operator properties begin here to go nicely hand in hand.

The operators T_α and $E_i(\mathcal{S})$ defined on the Hilbert space $\mathcal{K}$ form a representation of the corresponding classical Poisson algebra. A major result of the mathematically rigorous approach to loop quantum gravity is the proof of a unicity theorem for this representation. The theorem is usually called the "LOST" theorem, from the initials of the people that have discovered it (more precisely, one of its versions; see the Bibliographical notes below.) The theorem is analogous to the Stone-vonNeuman theorem in nonrelativistic quantum mechanics, which shows the unicity of the Schrödinger representation. It relies heavily on the hypothesis of diffeomorphism invariance. The theorem shows, under certain general hypotheses, that the loop representation, which was built largely "by hand", is the only possible way of quantizing a diffeomorphism invariant theory. No such theorem hold in conventional quantum field theory. In particular, this shows that in the diffeomorphism invariant context the theory is rather tightly determined.

6.6 Operators on $\mathcal{K}_0$

To be well defined on $\mathcal{K}_0$, an operator must be invariant under internal gauge transformations. As far as the connection is concerned, this is very easy to obtain. We noticed above that any cylindrical function gives a well-defined operator. The cylindrical function needs simply to be gauge invariant to be well defined in $\mathcal{K}_0$. The operator T_α, for instance, defined in (6.58), is well defined on $\mathcal{K}_0$.

6.6.1 The operator $\mathbf{A}(\mathcal{S})$

The situation with E is slightly more complicated. The operator $E_i(\mathcal{S})$ clearly cannot be gauge invariant, as the index i transforms under internal gauges. On the other hand, we cannot obtain a gauge-invariant quantity by simply contracting this index as

$$E^2(\mathcal{S}) \equiv \sum_i E_i(\mathcal{S})E_i(\mathcal{S}), \tag{6.70}$$

because the transformation property of $E_i(\mathcal{S})$ is complicated by the integral over $\mathcal{S}$. Let us nevertheless compute its action on a spin network state S, since this is a crucial step for what follows. Let us assume that there is a single intersection P between the surface $\mathcal{S}$ and (the graph Γ) of the spin network S. Let j be the spin of the link at the intersection. Using (6.68), we see that the first operator $E_i(\mathcal{S})$ inserts a matrix $^{(j)}\tau_i$ at the intersection. So does the second, but $-^{(j)}\tau_i \, ^{(j)}\tau_i = j(j+1) \times \mathbf{1}$ is the Casimir operator of $SU(2)$. Therefore,

$$E^2(\mathcal{S})|\text{S}\rangle = \hbar^2 \, j(j+1) \, |\text{S}\rangle. \tag{6.71}$$

This beautiful result, however, is completely spoiled if Γ intersects $\mathcal{S}$ more than once, because in this case the τ_i matrices at different points get contracted, and we do not get a gauge-invariant state.

To circumvent this difficulty, let us define a gauge-invariant operator $\mathbf{A}(\mathcal{S})$ associated to the surface $\mathcal{S}$ as follows. For any N, partition the surface $\mathcal{S}$ into N small surfaces $\mathcal{S}_n$, that become smaller and smaller as $N \to \infty$, and such that for each N, $\bigcup_n \mathcal{S}_n = \mathcal{S}$. Then define

$$\mathbf{A}(\mathcal{S}) \equiv \lim_{N \to \infty} \sum_n \sqrt{E^2(\mathcal{S}_n)}. \tag{6.72}$$

Do not confuse the $\mathbf{A}$ chosen to denote this operator with the A that denotes the connection: there is no relation between these two quantities. The reason for choosing the letter $\mathbf{A}$ to denote the operator (6.72) will become clear shortly.

In the classical case, by Riemann's very definition of the integral, we have

$$\mathbf{A}(\mathcal{S}) = \int_{\mathcal{S}} \sqrt{n_a E_i^a n_b E_i^b} \; \mathrm{d}^2\sigma, \tag{6.73}$$

which is a well-defined gauge-invariant quantity. In the quantum case, the action of the operator (6.72) is easy to compute. Let us evaluate it on a spin network state, under the simplifying assumption that no spin network node lies on $\mathcal{S}$. For sufficiently high N, no $\mathcal{S}_n$ will contain more than one intersection with Γ, see Figure 6.4. Therefore, the sum over n reduces to a sum over the intersection points P between $\mathcal{S}$ and Γ, and is independent from N, for N sufficiently high. Using (6.71), we have then

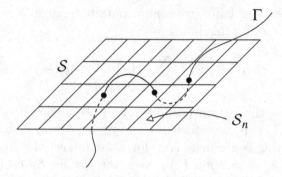

Fig. 6.4　A partition of $\mathcal{S}$.

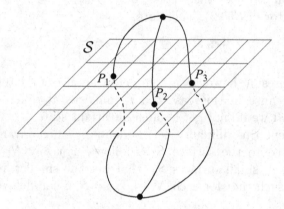

Fig. 6.5　A simple spin network S intersecting the surface $\mathcal{S}$.

immediately (see Figure 6.5)

$$\mathbf{A}(\mathcal{S})|S\rangle = \hbar \sum_{P \in (\mathcal{S} \cup \Gamma)} \sqrt{j_P(j_P + 1)} \, |S\rangle, \qquad (6.74)$$

where j_P is the color of the link that crosses $\mathcal{S}$ at P. This is a key result. First of all, the operator $\mathbf{A}(\mathcal{S})$ is well defined in $\mathcal{K}$. This is the operator corresponding to the classical quantity (6.73). Second, spin network states are eigenfunctions of this operator.

To summarize, we have obtained for each surface $\mathcal{S} \in M$ a well-defined $SU(2)$ gauge-invariant and self-adjoint operator $\mathbf{A}(\mathcal{S})$, which is diagonal on the spin networks that do not have a node on $\mathcal{S}$. The corresponding spectrum (with the restrictions mentioned) is labeled by multiplets $\vec{j} = (j_1, \ldots, j_n)$, $i = 1, \ldots, n$, and n arbitrary, of positive half-integers j_i. This is called the *main sequence* of the spectrum and is given by

$$\mathbf{A}_{\vec{j}} = \hbar \sum_i \sqrt{j_i(j_i + 1)}. \qquad (6.75)$$

We will compute the rest of the spectrum, which is also real and discrete, in Section 6.6.4.

Since the operator is diagonal on spin network states (with appropriate choices of intertwiners) and all its eigenvalues are real, it is also self-adjoint. In fact, this operator is well defined at the level of rigorous mathematical physics. For a completely rigorous detailed construction, see for instance [20] and [173]. The fact that this operator can be rigorously constructed at all, it is finite and its spectrum can even be computed, is a rather striking result, considering that its definition involves an operator product and a square root. This is the first remarkable pay-off of a well-defined diffeomorphism-invariant formalism for quantum field theory. As we will see in a moment, this result has major physical significance.

In principle, the operators T_α and $\mathbf{A}(\mathcal{S})$ are sufficient to define the quantum theory. In practice, it will be convenient later on to define other operators as well. Before doing so, however, let us discuss the physical meaning of the mathematical result achieved so far.

6.6.2 Quanta of area

In the previous section I have constructed and diagonalized the $SU(2)$ gauge-invariant and self-adjoint operator $\mathbf{A}(\mathcal{S})$. What is the physical interpretation of this operator? A direct comparison of (6.73) with (4.28) shows that $\mathbf{A}(\mathcal{S})$ is precisely the physical area of the surface $\mathcal{S}$!

Therefore, we obtain immediately an important physical result. The partial observable given by the area of a fixed two-dimensional surface is represented in the quantum theory by a self-adjoint operator with a discrete spectrum. But this yields immediately a physical prediction: any measurement of the area of any physical surface can only give an outcome which is in the spectrum of this operator. Since the spectrum is discrete, this means that the physical area is a quantized partial observable. A measurement of the area can only give a result contained in the spectrum (6.75)–(6.125) of $\mathbf{A}(\mathcal{S})$.

Restoring physical units for $8\pi G$ and c, the area operator is $8\pi Gc^{-3}$ times (6.72) and its eigenvalues are $8\pi Gc^{-3}$ times the ones in (6.75)–(6.125). The main sequence, for instance, gives

$$\mathbf{A}_{\vec{j}} = 8\pi\hbar Gc^{-3} \sum_i \sqrt{j_i(j_i + 1)} \,. \qquad (6.76)$$

Had we used the more general Barbero connection described in Section 4.2.3 instead of the real connection, then, from (4.46), we would have the operator E given by

$$\frac{c^3}{8\pi\gamma G}E_i^a(\tau)\Psi[A] = -i\hbar\frac{\delta}{\delta A_a^i(\tau)}\,\Psi[A] \qquad (6.77)$$

instead of (6.56). In this case the spectrum is modified by an overall constant factor:

$$\mathbf{A}_{\vec{j}} = 8\pi\gamma\hbar Gc^{-3}\sum_i \sqrt{j_i(j_i+1)}\,. \qquad (6.78)$$

Up to the single Immirzi parameter γ, this is a precise and quantitative prediction of LQG. It can, in principle, be verified or falsified. Alternatively, indirect consequences of this prediction could have observable effects. In fact, this quantization of the area is the basis of many results of the theory, such as the derivation of black-hole entropy.

The smallest (nonvanishing) eigenvalue in (6.76), taking the Immirzi parameter equal to 1, is

$$\mathbf{A}_0 = 4\sqrt{3}\pi\hbar Gc^{-3} \sim 10^{-66}\ \text{cm}^2. \qquad (6.79)$$

This is a sort of elementary quantum of area, of the order of the Planck area. It is the quantum of area carried by a link in the fundamental $j = 1/2$ representation. The fact that there is no area that can be measured below a minimum amount, indicates that there is a sort of minimal size of physical space, at the Planck scale.

An intrinsic discreteness of physical space at the Planck length has long been expected in quantum gravity. Notice that in the context of LQG this discreteness is not imposed or postulated. Rather, it is a direct consequence of a straightforward quantization of GR. Space geometry is quantized in the same manner in which the energy of an harmonic oscillator is quantized.

6.6.3 * n-hand operators and recoupling theory

The two-hand loop operator. The area operator can be defined also in a different manner, which is of interest because it employs a technique that we will use below. For each small surface $\mathcal{S}$, define $E^2(\mathcal{S})$ in an $SU(2)$ gauge-invariant manner as follows. Given a path γ with end points r and s, define the "two-handed loop operator"

$$T_\gamma^{ab} = E_i^a(r)R^{(1)}(U(A,\gamma))^{ij}E_j^b(s), \qquad (6.80)$$

where $R^{(1)}(U)^{ij}$ is the adjoint, $j = 1$ representation. Given two points r and s in a small surface $\mathcal{S}$, let γ_{rs} be a straight path (in the coordinate chosen) from r to s and

$$T^{ab}(r,s) = T_{\gamma_{rs}}^{ab}. \qquad (6.81)$$

Then define

$$E^2(\mathcal{S}) = \int_\mathcal{S} d^2\sigma \int_\mathcal{S} d^2\sigma' n_a(\sigma) n_b(\sigma') T^{ab}(\sigma, \sigma'). \qquad (6.82)$$

In the limit in which the surface is small, only the first term of the holonomy, which is the identity, survives, and therefore this definition converges to the one in (6.70) for small surfaces.

The advantage of using this kind of regularization in the quantum theory is that it simplifies the $SU(2)$ representation calculations. This is because the regularized operator is itself $SU(2)$ invariant.

The action of the quantum operator (6.81) on a spin network state is easy to compute. There is a contribution for each intersection of a spin network link with the surface $\mathcal{S}$ for each E. Each of the two intersections is called a grasp. For each of these contributions the spin network is modified by the creation of two nodes at the points r and s, one for each grasp, and the addition of the loop γ_{rs} to the spin network. We say that the "hands" of the operator "grasp" the spin network. Each node is trivalent, with two links being the ones of the grasped spin network, say in a representation j, and the other being γ_{rs}, in the representation $j = 1$. The intertwiner between these representations is the $SU(2)$ generator $^{(j)}(\tau_i)^\alpha_\beta$ in the representation j. This is not normalized. If we call the normalized intertwiner $i^\alpha_{i\beta}$, we have

$$^{(j)}(\tau_i)^\alpha_\beta = n_j \, i^\alpha_{i\beta}, \qquad (6.83)$$

where n_j can be computed easily by taking the norm of this equation. This gives

$$n_j^2 = \mathrm{tr}(^{(j)}\tau^i \, ^{(j)}\tau_i) = j(j+1)\,\mathrm{tr}(1) = j(j+1)\,(2j+1). \qquad (6.84)$$

Recoupling theory. In the limit in which the surface is small, the two grasps are on the same link and at the same point, and the line between them is infinitesimal. It is nevertheless useful to write the two grasps as separated, and the lines between them as finite lines, just remembering that the connection on these lines is trivial, namely that they are associated to identities. In this representation the result of the grasp on a link of spin j of the spin network can therefore be represented as follows

$$E^2(\mathcal{S}) \left| j \quad \sim \quad \hbar^2 j(j+1)\,(2j+1) \; ^j\!\left(\!\!\bigcirc\!\!\right)^{\!1}_j \right. . \qquad (6.85)$$

This picture can be directly interpreted in terms of recoupling theory, which is a simple graphical way of making calculations with $SU(2)$ representation theory. In this representation, lines represent contraction of representation indices and nodes represent normalized intertwiners

$$\left.\begin{matrix} \beta \\ \Big| j \\ \alpha \end{matrix}\right. \equiv \delta^\beta_\alpha, \qquad \overset{\gamma \quad\;\; \delta \quad\;\; \epsilon}{\underset{j \quad\; j' \quad\; j''}{\smile\!\Big|}} \equiv v^{\gamma\delta\epsilon}. \qquad (6.86)$$

Here, α, β and γ are indices in an orthonormal basis in the representation j, and δ and ϵ in the representations j' and j'' respectively. Index contraction is represented by joining open ends. Since the picture in (6.85) represents an overall intertwiner between the representation j and the representation j, it must be proportional to the identity in the representation j. Namely

$$j \ \bigD^{1} \ = c \ \Big| j \ .$$

$$\text{(6.87)}$$

The coefficient c can be computed by closing both sides, namely tracing the matrices. This gives

$$c = \frac{\text{(theta diagram with } j, j, 1)}{\text{(loop } j)} .$$

$$\text{(6.88)}$$

The theta-shaped diagram in the numerator has value unity, because it is the norm of an intertwiner, while the denominator is the trace of the identity, namely the dimension of the representation. Therefore

$$c = \frac{1}{2j + 1}.$$

$$\text{(6.89)}$$

Putting everything together, the action of E^2 gives

$$E^2(\mathcal{S}) \ \Big| j \quad \sim \hbar^2 \ j(j+1) \ \Big| j \ .$$

$$\text{(6.90)}$$

In the present case, this result was obtained earlier in a simpler way. But the idea can be used to define a general method of computing in LQG, which is very convenient. The general idea is that an operator such as (6.80) can be represented by the picture

$$\underset{r \ \ \alpha \ \ s}{\bullet\!\!-\!\!-\!\!-\!\!\bullet} ,$$

$$\text{(6.91)}$$

where the dots represent the operator E that can grasp a link. The result of a grasp is the formation of a node, and multiplication by the factor $\hbar\sqrt{j(j+1)\,(2j+1)}$. More precisely, if we include also the numerical part, the action of the grasp of a hand located at a point x over a link γ with spin j is

$$\underset{x}{-\!\!\!-\!\!\bullet} \ \Big| j \quad = \hbar \ n_j \Delta^a[\gamma, x] \ \underset{\gamma}{-\!\!\!-} \ \Big| j \ ,$$

$$\text{(6.92)}$$

where

$$\Delta^a[\gamma, x] \equiv \int ds \ \dot{\gamma}^a(s) \ \delta^3(\gamma(s), x).$$

$$\text{(6.93)}$$

Calculations have mostly been done with a slightly different notation, which derives from [174]. It is the Kauffman–Lins (KL) notation and is explained in Appendix A2.

Tables of formulas exist in this notation. Let us, therefore, now change to the KL notation in this section. In it, one uses the "color" $p = 2j$ of a link, which is twice its spin,[4] and is integer, and the trivalent nodes are not normalized to unity. The relation is given in (A.68). In the case of a vertex with spins $j, j, 1$, namely colors $p, p, 2$, as the one above, the normalization factor of the node, easily derived from the formulas of Appendix A2, is

$$\left(\begin{matrix} & j \quad j \\ & \vee \\ & {}_1 \end{matrix} \right)_{\text{spin network}} = \left(\sqrt{\frac{j}{(j+1)(2j+1)}} \begin{matrix} p=2j \; p=2j \\ \vee \\ {}_2 \end{matrix} \right)_{\text{KL}} \tag{6.94}$$

and, using (6.92), the action of the grasp operator in this notation is therefore

$$\begin{matrix} \bullet \\ {}_x \end{matrix}\Big|_{\gamma} p \;\; = \;\; p\hbar\, \Delta^a[\gamma,x] \;\; \Big|_{\gamma} p \;. \tag{6.95}$$

In the next section I give an example of a full calculation using this grasp operator, for computing the complete spectrum of the area operator.

Paths with many hands. The definition (6.80) can be generalized to paths with an arbitrary number of "hands." For instance, let

$$T^{abc}(x,r,s,t) = \frac{1}{3!}\epsilon_{ijk}R^{(1)}(U(A,\gamma_{xr}))^{il}\, E_l^a(r)$$
$$\times R^{(1)}(U(A,\gamma_{xs}))^{jm}\, E_m^b(s)R^{(1)}(U(A,\gamma_{xt}))^{kn}\, E_n^c(t). \tag{6.96}$$

Given a closed surface $\mathcal{S}$, define the three-hand generalization of the operator (6.90) as

$$E^3(\mathcal{S}) = \int_{\mathcal{S}} d^2\sigma \int_{\mathcal{S}} d^2\sigma' \int_{\mathcal{S}} d^2\sigma'' \, |n_a(\sigma)\, n_b(\sigma')\, n_c(\sigma'')\, T^{abc}(x,\sigma,\sigma',\sigma'')|, \tag{6.97}$$

where x is a point in the interior of $\mathcal{S}$ (whose exact position is irrelevant as we will always consider the limit of small $\mathcal{S}$). The absolute value in the definition is for later convenience. This can be represented by the picture

$$\tag{6.98}$$

As we will see in a moment, this operator also plays an important physical role.

---

[4]The expression "color" is routinely used with two distinct meanings. It indicates twice the spin, as here. Or it may designate any label of links or nodes (or, later, edges or faces of a spinfoam), as in: "the links of the spin network are colored with representations and the nodes with intertwiners."

6.6.4 * Degenerate sector

The simplification that we took above in order to compute the spectrum of the area was to assume that no node is on the surface. Here we drop this assumption, in order to find the full spectrum. If we drop this assumption, the regularization of the area operator considered above is not sufficient, because we obtain ill-defined expressions of the kind

$$\int_0^1 dx \, \delta(x) = ? \tag{6.99}$$

We need a better regularization of the operator. To this end, it is sufficient to smear the operator transverse to the surface. Introduce a smooth coordinate τ over a finite neighborhood of $\mathcal{S}$, in such a way that $\mathcal{S}$ is given by $\tau = 0$. Consider then the three-dimensional region around $\mathcal{S}$ defined by $-\delta/2 \leq \tau \leq \delta/2$. Partition this region into a number of blocks $\mathcal{D}$ of coordinate height δ and square horizontal section of coordinate side ϵ. For each fixed choice of ϵ and δ, we label the blocks by an index I. Later, we will send both δ and ϵ to zero. In order to have a one-parameter sequence, we now choose δ as a fixed function of ϵ. For technical reasons, the height of the block $\mathcal{D}$ must decrease more rapidly than ϵ in the limit; thus, we put $\delta = \epsilon^k$ with any k greater than 1 and smaller than 2.

Consider one of the blocks. The intersection of the block and a $\tau = constant$ surface is a square surface: let $A_I(\tau)$ be the area of such a surface. Let $A_{I\epsilon}$ be the average over τ of the areas of the surfaces in the block, namely

$$A_{I\epsilon} \equiv \frac{1}{\delta} \int_{-\delta/2}^{\delta/2} A_I(\tau) d\tau = \frac{1}{\delta} \int_{\mathcal{D}_I} d^3x \, \sqrt{E^{ai} E_i^b n_a n_b} \, . \tag{6.100}$$

Summing over the blocks yields the average of the areas of the $\tau = constant$ surfaces, and as ϵ (and therefore δ) approaches zero, the sum converges to the area of the surface $\mathcal{S}$. Therefore we have

$$\mathbf{A}(\mathcal{S}) = \lim_{\epsilon \to 0} \sum_I A_{I\epsilon} \equiv \lim_{\epsilon \to 0} A_\epsilon(\mathcal{S}). \tag{6.101}$$

The quantity $A_{I\epsilon}$ associated with each block can be expressed as follows. Write

$$A_{I\epsilon} = \sqrt{A_{I\epsilon}^2} \tag{6.102}$$

and notice that

$$A_{I\epsilon}^2 = \frac{1}{\delta^2} \int_{\mathcal{D} \otimes \mathcal{D}} d^3x \, d^3y \, n_a(x) n_b(y) T^{ab}(x, y) + O(\epsilon^5). \tag{6.103}$$

Equation (6.103) holds because of the following. We have

$$T^{ab}(x, y) = E^{ai}(x_I) E_i^b(x_I) + O(\epsilon) \tag{6.104}$$

for any three points x, y, and x_I, in $\mathcal{D}$. It follows that

$$\epsilon^4 \, n_a(x_I) n_b(x_I) E^{ai}(x_I) E_i^b(x_I) = \frac{1}{2\delta^2} \int_{\mathcal{D} \otimes \mathcal{D}} d^3x \, d^3y \, n_a(x) n_b(y) T^{ab}(x, y) + O(\epsilon^5). \tag{6.105}$$

Equation (6.103) follows from

$$A_I^2 = \left(\frac{1}{\delta} \int_{-\delta/2}^{\delta/2} A_I(\tau) d\tau \right)^2 = \left(\frac{1}{\delta} \int_{\mathcal{D}} d^3x \, \sqrt{E^{ai} E_i^b n_a n_b} \right)^2$$

$$= \epsilon^4 \, n_a(x_I) n_b(x_I) E^{ai}(x_I) E_i^b(x_I) + O(\epsilon^5). \tag{6.106}$$

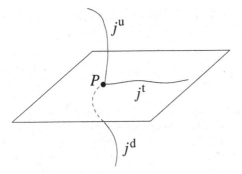

Fig. 6.6   The three classes of links that meet at a node on the surface.

Equations (6.101), (6.102) and (6.103) define the regularization of the area. The quantum operator $\mathbf{A}(\mathcal{S})$ is defined by (6.101) where

$$A_{I\epsilon}^2 \equiv \frac{1}{2\delta^2} \int_{\mathcal{D}\otimes\mathcal{D}} \mathrm{d}^3x\,\mathrm{d}^3y\, n_a(x)n_b(y)T^{ab}(x,y). \tag{6.107}$$

The action of $\mathbf{A}(\mathcal{S})$ on the quantum states is found from the action of the T^{ab} operators. The operator $T^{ab}(x,y)$ annihilates the state $|S\rangle$ unless its hands x and y fall on some links of the graph of S. If this happens, the action of the operator on the state gives the union of S and α, with two additional nodes at the points x and y. More precisely, if x and y fall over two edges of β with color p and q respectively, using the grasp operator (6.95) we have

$$T^{ab}(x,y) \quad \overset{p}{\underset{q}{{-}{-}{-}}} = \hbar^2\,p\,q\,\Delta^a[\beta,x]\,\Delta^b[\beta,y] \quad \overset{x}{\underset{y}{\overset{p}{\underset{q}{\rvert\,2}}}}. \tag{6.108}$$

Since the loop α runs back and forth between the intersection points x and y (the two grasps), it has spin one, or color 2.

Consider now the action of the operator $\mathbf{A}(\mathcal{S})$ on a generic spin network state $|S\rangle$. Due to the limiting procedure involved in its definition, the operator $\mathbf{A}(\mathcal{S})$ does not affect the graph of $|S\rangle$. Furthermore, since the action of T^{ab} inside a specific coordinate block $\mathcal{D}$ vanishes unless the graph of the state intersects $\mathcal{D}$, the action of $\mathbf{A}(\mathcal{S})$ ultimately consists of a countable sum of terms, one for each intersection P of the graph with the surface.

Consider an intersection P between the spin network and the surface. For the purpose of this discussion, we can consider a generic point on a link as a "bivalent node," and thus say, without loss of generality, that P is a node. In general, there will be n links emerging from P. Some of these will emerge upward(u), some downward(d) and some tangential(t) to the surface $\mathcal{S}$, see Figure 6.6. Since we are taking the limit in which the blocks shrink to zero, we may assume, without loss of generality, that the surface and the links are linear around P (see below for subtleties concerning higher derivatives). Due to the two integrals in (6.107), the positions of the two hands of the area operator are integrated over each block. As the action of T^{ab} is nonvanishing only when both hands fall on the spin network, we obtain n^2 terms, one for every couple of grasped links. Consider one of these terms, in which the grasped links have color p and q. Let us write the result of the action of T^{ab}, with a finite ϵ, on the links p and q of an n-valent intersection P (up to the prefactor) as

(6.109)

The irrelevant links are not shown. The links labeled p and q are generic, in the sense that their angles with the surface do not need to be specified at this point (the two links may also be identical). From the definition (6.101) and (6.107) of the area operator and the definition of the T^{ab} operator, each term in which the grasps run over two links of color p and q is of the form

$$T = \frac{1}{2\delta^2} \int_{\mathcal{D}\otimes\mathcal{D}} d^3x \, d^3y \, n_a(x) \, \Delta^a[\beta,x] \, n_b(y)\Delta^b[\beta,y] \, p \, q \quad \epsilon \, ,$$ (6.110)

giving

$$T = \frac{1}{2\delta^2} \int_{\mathcal{D}\otimes\mathcal{D}} \left(n_a(x) \int_\beta ds \, \dot\beta^a(s)\delta^3[\beta(s),x] \right.$$ (6.111)

$$\times \, n_b(y) \int_\beta dt \, \dot\beta^b(t)\delta^3[\beta(t),\vec y] \, p \, q \quad \epsilon \, \bigg) d^3x \, d^3y$$

$$= \frac{1}{2\delta^2} \int_\beta ds \, n_a(s) \, \dot\beta^a(s) \int_\beta dt \, n_b(t) \, \dot\beta^b(t) p \, q \quad \epsilon$$

$$= \frac{p \, q}{2\delta^2} \left(\int_\beta ds \, n_a(s) \, \dot\beta^a(s) \int_\beta dt \, n_b(t) \, \dot\beta^b(t) \right) \quad + O(\epsilon).$$

In the last step I have pulled the state out of the integral. This is possible because the ϵ-dependent states all have the same limit state as $\epsilon \to 0$. I write this limit simply as , without ϵ, that is

$$\quad \epsilon = \quad + O(\epsilon).$$ (6.112)

Hence, the substitution of the ϵ-dependent states with their limit in the integral is possible up to terms of order $O(\epsilon)$. Note that

$$\int_\beta dt \, n_b(t) \, \dot\beta^b(t) = \begin{cases} 0 & \text{if } \beta \text{ is tangent to } \mathcal{S} \\ \delta/2 & \text{otherwise.} \end{cases}$$ (6.113)

This result is independent of the angle the link makes with the surface because δ can always be chosen sufficiently small so that β crosses the top and bottom of the coordinate block $\mathcal{D}$. (This is the reason for requiring that δ goes to zero faster than ϵ.) Also, since we have chosen k smaller than 2, it follows that any link tangential to the

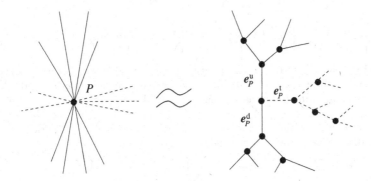

Fig. 6.7   Trivalent expansion of an n-valent node. The dashed lines indicate the lines tangent to the surface.

surface exits the box from the side, irrespective of its second (and higher) derivatives, for sufficiently small ϵ, and gives a vanishing contribution as ϵ goes to zero. Therefore, in the limit, the links tangent to the surface do not contribute to the action of the area whereas every nonvanishing term takes the form

$$\frac{\hbar^2\,p\,q}{8} \quad \vcenter{\hbox{$p\diagdown\,\,_2$}} \quad . \tag{6.114}$$

Generically, there will be several links above, below, and tangential to the surface $\mathcal{S}$. Expand the node P into a virtual trivalent spin network. We choose to perform the expansion in such a way that all links above the surface converge to a single "principal" virtual link e^{u}; all links below the surface converge to a single principal virtual link e^{d}; and all links tangential to the surface converge to a single principal virtual link e^{t}. The three principal links join in the principal trivalent node. This trivalent expansion is shown in Figure 6.7.

This choice simplifies the calculation of the action of the area, since the sum of the grasps of one hand on *all* real links above the surface is equivalent to a single grasp on e^{u} (and similarly for the links below the surface and e^{d}). This follows from the identity

$$p\quad \vcenter{\hbox{$^{2}\!\diagup^{p}\quad^{q}$}} \quad +q \quad \vcenter{\hbox{$^{p\;2\;q}$}} \quad = r \quad \vcenter{\hbox{$^{p}\quad^{q}$}}_{2} \quad , \tag{6.115}$$

which can be proven as follows. Using the recoupling theorem (A.65), the left-hand side of (6.115) can be written as

$$\sum_j \left(p \begin{Bmatrix} 2 & p & j \\ q & r & p \end{Bmatrix} - q\,\lambda_j^{2r} \begin{Bmatrix} r & p & j \\ q & 2 & q \end{Bmatrix} \right) \quad \vcenter{\hbox{$^{p}\diagdown\,\diagup^{q}$}}_{2\;\;\;\;r} \tag{6.116}$$

where j can take the values $r-2$, r and $r+2$. A straightforward calculation using

(A.61) gives

$$p \begin{Bmatrix} 2 & p & j \\ q & r & p \end{Bmatrix} - q \lambda_j^{2r} \begin{Bmatrix} r & p & j \\ q & 2 & q \end{Bmatrix} = r \, \delta_{jr}, \tag{6.117}$$

and (6.115) follows. A repeated application of the identity (6.115) allows us to slide all grasps from the real links down to the two virtual links e^u and e^d. Thus, each intersection contributes as a single principal trivalent node, regardless of its valence.

We are now in a position to calculate the action of the area on a generic intersection. From the discussion above, the only relevant terms are as follows

$$A_P^2 \;\; \vcenter{\hbox{(diagram)}} \;=\; \frac{\hbar^2}{8} \left(p^2 \; \vcenter{\hbox{(diagram)}} \;+\; q^2 \; \vcenter{\hbox{(diagram)}} \right.$$

$$\left. +\; 2pq^2 \; \vcenter{\hbox{(diagram)}} \right), \tag{6.118}$$

where the first term comes from grasps on the links above the surface, the second from grasps on two links below the surface and the third from the terms in which one hand grasps a link above and the other grasps a link below the surface. Each term in the sum is proportional to the original state (see (A.63), (A.64)). Therefore, we have

$$A_P^2 \;\; \vcenter{\hbox{(diagram)}} \;=\; -\frac{l_0^4}{8} \left(p^2 \, \lambda_u + q^2 \, \lambda_d + 2pq \, \lambda_t \right) \;\; \vcenter{\hbox{(diagram)}} \,. \tag{6.119}$$

The quantities λ_u, λ_d and λ_t are easily obtained from the recoupling theory. Using the formulas in Appendix A2, we obtain

$$\lambda_u = \frac{\theta(p,p,2)}{\Delta(p)} = -\frac{(p+2)}{2p}. \tag{6.120}$$

λ_d is obtained by replacing p with q in (6.120). λ_t has the value

$$\lambda_t = \frac{\mathrm{Tet} \begin{bmatrix} p & p & r \\ q & q & 2 \end{bmatrix}}{\theta(p,q,r)} = \frac{-2p(p+2) - 2q(q+2) + 2r(r+2))}{8pq}. \tag{6.121}$$

(Tet is defined in (A.60).)

Substituting in (6.118), we have

$$A_P^2 \;\; \vcenter{\hbox{(diagram)}} \;=\; \frac{\hbar^2}{16} \left(2p\,(p+2) + 2q\,(q+2) - r\,(r+2) \right) \;\; \vcenter{\hbox{(diagram)}} \,. \tag{6.122}$$

Since A_P^2 is diagonal, the square root can be easily taken:

$$
A_P \quad \begin{array}{c} p \\ \bullet\!\!-\!\!-\!\!-\!\!- \\ q \end{array}\!\!{}^{r} \;=\; \sqrt{A_P^2} \quad \begin{array}{c} p \\ \bullet\!\!-\!\!-\!\!-\!\!- \\ q \end{array}\!\!{}^{r}
$$

$$
= \sqrt{\frac{\hbar^2}{4}\left(2\frac{p}{2}\left(\frac{p}{2}+1\right) + 2\frac{q}{2}\left(\frac{q}{2}+1\right) - \frac{r}{2}\left(\frac{r}{2}+1\right)\right)} \quad \begin{array}{c} p \\ \bullet\!\!-\!\!-\!\!-\!\!- \\ q \end{array}\!\!{}^{r}. \tag{6.123}
$$

Adding over the intersections and getting back to the spin notation $p/2 = j^{\mathrm{u}}$, $q/2 = j^{\mathrm{d}}$ and $r/2 = j^{\mathrm{t}}$, the final result is

$$
\mathbf{A}(\mathcal{S})|S\rangle = \left(\frac{\hbar}{2}\sum_{P\in\{S\cap\mathcal{S}\}}\sqrt{2j_P^{\mathrm{u}}(j_P^{\mathrm{u}}+1)+2j_P^{\mathrm{d}}(j_P^{\mathrm{d}}+1)-j_P^{\mathrm{t}}(j_P^{\mathrm{t}}+1)}\right)|S\rangle. \tag{6.124}
$$

This expression provides the complete spectrum of the area. It reduces to the earlier result (6.75) for the case $j_P^{\mathrm{t}} = 0$ and $j_P^{\mathrm{d}} = j_P^{\mathrm{u}}$ (for every P).

The complete spectrum of $\mathbf{A}(\mathcal{S})$ is therefore labeled by n-tuplets of triplets of positive half-integers j_i, namely $\vec{j}_i = (j_i^{\mathrm{u}}, j_i^{\mathrm{d}}, j_i^{\mathrm{t}})$, $i = 1,\dots,n$, and n arbitrary. It is given, restoring natural units and the Immirzi parameter, as

$$
\mathbf{A}_{\vec{j}_i}(\mathcal{S}) = \frac{4\pi\hbar G\gamma}{c^3}\sum_i\sqrt{2j_i^{\mathrm{u}}(j_i^{\mathrm{u}}+1)+2j_i^{\mathrm{d}}(j_i^{\mathrm{d}}+1)-j_i^{\mathrm{t}}(j_i^{\mathrm{t}}+1)}. \tag{6.125}
$$

It contains the previous case (6.75) which corresponds to the choice $j_i^{\mathrm{u}} = j_i^{\mathrm{d}}$ and $j_i^{\mathrm{t}} = 0$. The eigenvalues which are contained in (6.125) but not in (6.75) are called the degenerate sector.

6.6.5 Quanta of volume

A second operator that plays a key role in the physical interpretation of the quantum states of the gravitational field is the operator $\mathbf{V}(\mathcal{R})$ corresponding to the volume of a region $\mathcal{R}$. As for the area operator constructed above, this quantity requires a bit of work to be defined in the quantum theory, because of the care to be taken in the definition of the operator products involved in $\det E$ and the square root. Consider a three-dimensional region $\mathcal{R}$. The volume of $\mathcal{R}$ is

$$
V(\mathcal{R}) = \int_{\mathcal{R}} d^3x\sqrt{\frac{1}{3!}\left|\epsilon_{abc}\epsilon_{ijk}E^{ai}E^{bj}E^{ck}\right|}. \tag{6.126}
$$

To construct a regularized form of this expression, consider the classical quantity (6.96). In the limit in which r, s and t converge to x we have:

$$
T^{abc}(x,s,t,r) \to 2\epsilon_{ijk}\,E^{ai}(x)E^{bj}(x)E^{ck}(x) = 2\,\epsilon^{abc}\,\det E(x). \tag{6.127}
$$

We can therefore use the 3-hand loop operator to regularize the volume.

Fix an arbitrary chart of the 3-manifold, and consider a small cubic region $\mathcal{R}_I$ of coordinate volume ϵ^3. Let x_I be an arbitrary but fixed point in $\mathcal{R}_I$. Since classical fields are smooth we have $E(s) = E(x_I) + O(\epsilon)$ for every $s \in \mathcal{R}_I$, and $H_\alpha(s,t)_A^{\;B} = 1_A^{\;B} + O(\epsilon)$ for any $s, t \in \mathcal{R}_I$ and straight segment α joining s and t. Consider the quantity

$$W_I = \frac{1}{16\epsilon^6 \, 3!} \; E^3(\partial\mathcal{R}_I), \tag{6.128}$$

where E^3 is defined in (6.97). Because of (6.127), we have, to lowest order in ϵ

$$W_I = \frac{1}{8\epsilon^6 \, 3!} \; \left|\det(E(x_I))\right| \left| \int_{\partial\mathcal{R}_I} d^2\sigma \int_{\partial\mathcal{R}_I} d^2\tau \int_{\partial\mathcal{R}_I} d^2\rho \left| n_a(\sigma) n_b(\tau) n_c(\rho) \epsilon^{abc} \right| \right.$$

$$= \left|\det E(x_I)\right|. \tag{6.129}$$

Thus, W_I is a nonlocal quantity that approximates the volume element for small ϵ. Using the Riemann theorem as in the case of the area, we can then write the volume $\mathbf{V}(\mathcal{R})$ of the region $\mathcal{R}$ as follows. For every ϵ, we partition $\mathcal{R}$ into cubes $\mathcal{R}_{I_\epsilon}$ of coordinate volume ϵ^3. Then

$$\mathbf{V}(\mathcal{R}) = \lim_{\epsilon \to 0} V_\epsilon(\mathcal{R}); \tag{6.130}$$

$$V_\epsilon(\mathcal{R}) = \sum_{I_\epsilon} \epsilon^3 W_{I_\epsilon}^{1/2}. \tag{6.131}$$

Volume operator. Returning now to the quantum theory, we have then immediately a definition of the volume operator as

$$\mathbf{V}(\mathcal{R}) = \lim_{\epsilon \to 0} V_\epsilon(\mathcal{R}); \tag{6.132}$$

$$V_\epsilon(\mathcal{R}) = \sum_{I_\epsilon} \epsilon^3 W_{I_\epsilon}^{1/2}; \tag{6.133}$$

$$W_{I_\epsilon} = \frac{1}{16\epsilon^6 \, 3!} \; E^3(\partial\mathcal{R}_I), \tag{6.134}$$

where these quantities now are operators. Notice the crucial cancellation of the ϵ^6 factor, when inserting (6.134) into (6.133).

The meaning of the limit in (6.132) needs to be specified. The specification of the topology in which the limit is taken is an integral part of the definition of the operator. As is usual for limits involved in the regularization of quantum field theoretical operators, the limit cannot be taken in the Hilbert space topology where, in general, it does not exist. The limit must be taken in a topology that "remembers" the topology in which the corresponding classical limit (6.130) is taken. This is easy to do in the present context. We say that a sequence of quantum states Ψ_n converges to the state Ψ if $\Psi_n[A]$ converges to $\Psi[A]$ for all smooth connections A. We use the corresponding operator topology: $O_n \to O$ if $O_n\Psi \to O\Psi$ for all Ψ in the domain.

An important consequence of the use of this topology is that a sequence of cylindrical functions converges to a cylindrical function defined on the limit graph. The graphs

Γ_n converge to Γ in the topology of the 3-manifold. This fact allows us to separate the study of a limit into two steps. First, we study the graph of the limit state. Second, we can study what happens to the coloring of states, in order to express the limit representation in terms of the spin network basis.

Let us now begin to compute the action of this operator on a spin network state. The three surface integrals on the surface of the cube and the line integrals along the loops combine, as in the case of the area, to give three intersection numbers, which select three intersection points between the spin network and the boundary of the cube. At these three points, which we denote as r, s and t, the small graph $\gamma_{\sigma\tau\rho}$ of the operator grasps the spin network.

Notice that the integration domain of the (three) surface integrals is a six-dimensional space – the space of the possible positions of three points on the surface of a cube. Let us denote this integration domain as D^6. The absolute value in (6.134) plays a crucial role here: contributions from different points of D^6 have to be taken as their absolute value, while contributions from the same point of D^6 have to be summed algebraically before taking the absolute value. The position of each hand of the operator is integrated over the surface, and therefore each hand grasps each of the three points r, s and t, producing 3^3 distinct terms. However, because of the absolute value, a term in which two hands grasp the same point, say r, vanishes. This happens because the result of the grasp is symmetric but the operator is antisymmetric in the two hands – as follows from the antisymmetry of the trace of three sigma matrices. Thus, only terms in which each hand grasps a distinct point give nonvanishing contributions. For each triplet of points of intersection r, s and t between spin network and cube surface, there are 3! ways in which the three hands can grasp the three points. These 3! terms have alternating signs because of the antisymmetry of the operator, but the absolute value prevents the sum from vanishing, and yields the same contribution for each of the 3! terms.

If there are only two intersection points between the boundary of the cube and the spin network, then there are always two hands grasping at the same point; contributions have to be summed before taking the absolute value, and thus they cancel. Thus, the sum in (6.133) reduces to a sum over the cubes I_ϵ^i whose boundaries have at least three distinct intersections with the spin network, and the surface integration reduces to a sum over the triple-grasps at *distinct* points. For ϵ small enough, the only cubes whose surfaces have at least three intersections with the spin network are cubes containing a node i of the spin network. Therefore, the sum over cubes reduces to a sum over the nodes $n \in \{\mathsf{S} \cap \mathcal{R}\}$ of the spin network, contained inside $\mathcal{R}$. Let us denote by I_ϵ^n the cube containing the node n. We then have

$$\mathbf{V}(\mathcal{R})|\mathsf{S}\rangle = \lim_{\epsilon \to 0} \sum_{n \in \{\mathsf{S} \cap \mathcal{V}\}} \epsilon^3 \sqrt{|W_{I_\epsilon^n}|} \, |\mathsf{S}\rangle$$

$$W_{I_\epsilon^n}|\mathsf{S}\rangle = \frac{1}{16 \, \epsilon^6 \, 3!} E^3(\partial \mathcal{R}_I)|\mathsf{S}\rangle. \tag{6.135}$$

The action of the operator $E^3(\partial\mathcal{R}_I)$ is the sum over the triplets (r, s, t) of distinct intersections between the spin network and the boundary of the cube. For each such triplet, let $\mathcal{T}(r, s, t)|S\rangle$ be the result of this action. Then

$$\mathbf{V}(\mathcal{R})|S\rangle = \lim_{\epsilon \to 0} \sum_{n \in \{S \cap \mathcal{V}\}} \epsilon^3 \sqrt{|W_{I_\epsilon^n}|} \ |S\rangle$$

$$W_{I_\epsilon^n}|S\rangle = \frac{1}{16 \ \epsilon^6 \ 3!} \sum_{rst} \mathcal{T}(r, s, t) \ |S\rangle. \tag{6.136}$$

Next, the key point now is that in the limit $\epsilon \to 0$, the operator does not change the graph of the spin network state, nor the coloring of the links. The only possible action of the operator is therefore on the intertwiners. Therefore,

$$\mathbf{V}(\mathcal{R}) \ |\Gamma, j_l, i_1 \dots i_N\rangle = (16\pi\hbar G)^{3/2} \sum_{n \in \{S \cap \mathcal{V}\}} \mathcal{V}_{i_n}^{\ i'_n} \ |\Gamma, j_l, i_1 \dots i'_n \dots i_N\rangle.$$

$$\tag{6.137}$$

The computation of the numerical matrices $\mathcal{V}_{i_n}^{\ i'_n}$ is an exercise in recoupling theory. For instance, for a trivalent node we have to compute W in

$$\tag{6.138}$$

and more complicated diagrams for higher-valence nodes. The complete calculation is presented in great detail in [175], where a list of eigenvalues is also given. One of the interesting outcomes of the detailed calculation is that the node must be at least quadrivalent in order to have a non-vanishing volume.

The operator can be shown to be a well-defined self-adjoint nonnegative operator, with discrete spectrum. For each given graph and labeling, we shall choose, from now on, a basis i_n of intertwiner that diagonalizes the matrices $\mathcal{V}_{i_n}$, and therefore the volume operator. We denote V_{i_n} the corresponding eigenvalues.

6.7 Quantum geometry

Physical interpretation of the spin network states. The essential property of the volume operator is that it has contribution only from the nodes of a spin network state $|S\rangle$. That is, the volume of a region $\mathcal{R}$ is a sum of terms, one for each node of S inside $\mathcal{R}$. Therefore, each node of a spin network represents a quantum of volume. That is, we can interpret a spin network with N nodes as an ensemble of N quanta of volume, or N "chunks" of space, located in the manifold "around" the node, each with a quantized volume V_{i_n}.

The elementary chunks of quantized volume are then separated from each other by surfaces. The area of these surfaces is governed by the area operator. The area operator $A(\mathcal{S})$ has contribution from each link of S that crosses $\mathcal{S}$. Therefore the following interpretation follows. Two chunks of space are contiguous if the corresponding nodes are connected by a link l. In this case, there is an elementary surface separating them, and the area of this surface is determined by the color j_l of the link l to be

$$\mathbf{A}_l = 8\pi c^{-3}\hbar G \ \sqrt{j_l(j_l+1)}. \tag{6.139}$$

Therefore, the intertwiners associated with the nodes are the quantum numbers of the volume, and the spins associated with the links are quantum numbers of the area. Volume is on the nodes and area is on the links separating them. The graph Γ of the spin network determines the adjacency relation between the chunks of space.

In other words, the graph Γ can be interpreted as the graph dual to a cellular decomposition of physical space, in which each cell is a quantum of volume.

Thus a spin network state $|S\rangle$ determines a discrete quantized 3d metric. This physical picture is beautiful and compelling. However, its full beauty reveals itself only in going to the space of the diffeomorphism-invariant states $\mathcal{K}_{\mathrm{diff}}$.

Physical interpretation of the s-knot states. Consider an s-knot state $|s\rangle$. For simplicity, consider the generic case in which its symmetry group is trivial, so that we can disregard the technicalities due to the diffeomorphisms that change orientation and ordering. Then we can view $|s\rangle$ as the projection under P_{diff} of a spin network state $|S\rangle$. In going from the spin network state $|S\rangle$ to the s-knot state $|s\rangle$ we preserve the entire information in $|S\rangle$ except for its localization on the 3d space manifold. This is precisely as the implementation of diffeomorphism invariance in the classical theory, where a physical geometry is an equivalence class of metrics under diffeomorphisms. In the quantum case, $|s\rangle$ retains the information about the volume and the adjacency of the chunks of volumes, and about the area of the surfaces that separate these volumes. But any information of the localization of the chunks of volume on the 3d manifold is lost under P_{diff}.

The physical interpretation of the resulting state $|s\rangle$ is therefore extremely compelling: it represents a discrete quantized geometry. This is formed by abstract chunks of space, which do not live on the 3d manifold:

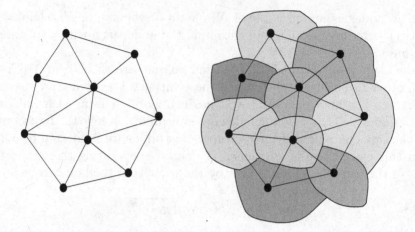

Fig. 6.8 The graph of an abstract spinfoam and the ensemble of "chunks of space," or quanta of volume, it represents. Chunks are adjacent when the corresponding nodes are linked. Each link cuts one elementary surface separating two chunks.

they are only localized with respect to one another. Their spatial relation is only determined by the adjacency defined by the links, see Figure 6.8. These are not quantum excitations *in* space: they are quantum excitations *of* space itself. Volume of the chunks and area of the surfaces are given by the coloring of the *s*-knot. The spins j_l are the quantum numbers of the area and the intertwiners i_n are the quantum numbers of the volume.

These are quantum states defined in a completely 3d diffeomorphism-invariant manner and with a simple physical interpretation. These are the quantum states of space.

Surfaces and regions on s-knots. Recall that in classical GR we distinguish between a metric g and a geometry $[g]$. A geometry is an equivalence class of metrics under diffeomorphism. For instance, in three dimensions, the euclidean metric $g_{ab}(x) = \delta_{ab}$ and a flat metric $g'_{ab}(x) \neq \delta_{ab}$ are different metrics, but define the same geometry $[g] = [g']$. The notion of geometry is diffeomorphism invariant, while the notion of metric is not. On a given manifold with coordinates x, we can define a surface by $\mathcal{S} = (\sigma^1, \sigma^2) \rightarrow x^a(\sigma^i)$. Then it makes sense to ask what is the area of $\mathcal{S}$ in a given metric $g_{ab}(x)$, but it makes no sense to ask what is the area of $\mathcal{S}$ in a given geometry, because the relative location of $\mathcal{S}$ and the geometry is not defined.

However, given a geometry, it is meaningful to define surfaces *on the geometry itself*. For instance, (in 2d) given the geometry of the surface of the Earth (an ellipsoid), the equator is a well-defined (1d) surface, and

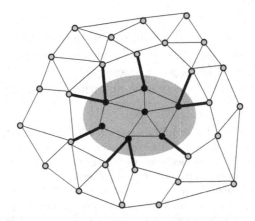

Fig. 6.9 Regions and surfaces defined on an s-knot: the set of the thick black nodes define a "region" of space; the set of the thick black links define the "surface" surrounding this region.

so is the parallel 1 km north of the equator: their location is determined with respect to the geometry itself. Concretely, a surface on a geometry can be defined in various manners. For instance, it can be defined by the couple $(\mathcal{S}, g)$ with $g \in [g]$. The couple $(\phi\mathcal{S}, \phi^*g)$ defines the same surface. Alternatively, the surface can be defined intrinsically (the equator is the longest geodesic.)

Now, in quantum gravity we find precisely the same situation. Above we have defined coordinate surfaces $\mathcal{S}$ and regions $\mathcal{R}$, and their areas and volumes. Such coordinate surfaces and regions are not defined at the diffeomorphism-invariant level. However, we can nevertheless define surfaces and regions *on the abstract quantum state* $|s\rangle$ itself, and associate areas and volumes with them. A region $\mathcal{R}$ is simply a collection of nodes. Its boundary is an ensemble of links and defines a surface; we can say that this surface "cuts" these links. A moment of reflection will convince the reader that this is precisely the same situation as in the classical theory.

For instance, consider an s-knot state with two four-valent nodes and four links with spins $1/2, 1/2, 1, 1$. On this quantum geometry we can identify a closed surface separating the two quanta of volume. This surface cuts the four links, and has area $\mathbf{A} = (8\sqrt{3} + 16\sqrt{2})\pi\hbar G c^{-3}$. A more complex situation is illustrated in Figure 6.9.

Eigenvalues and measurements. Suppose we had the technological capability to measure the area of a surface, or the volume of a region, with Planck-scale precision. An example of an area measurement, for instance, is the measurement of the cross section of an interaction. Shall we obtain one of the eigenvalues computed in this section? If the theory developed so far is physically correct, the answer is yes. In fact, area and volume are partial observables. Partial observables can be measured, and the theory predicts that the possible outcomes of a measurement are the numbers in the spectrum of the corresponding operator.

Therefore these spectra are precise quantitative physical predictions of LQG.

This prediction has raised a certain discussion. The objection has been made that in the classical theory area and volume of coordinate surfaces and regions are not diffeomorphism-invariant quantities, and therefore we cannot interpret them as true observables. The objection is not correct; it is generated by the obscurity of the constrained treatment of diffeomorphism-invariant systems. To clarify this point, consider the following simple example. Consider a particle moving on a circle, subject to a force. Let ϕ be the angular coordinate giving the position of the particle, and p_ϕ its conjugate momentum. As we know well, p_ϕ turns out to be quantized. Now, if we write the covariant formulation of this system, we have the Wheeler–DeWitt equation

$$H\psi(t,\phi) = \left(i\hbar\frac{\partial}{\partial t} - \hbar^2\frac{\partial^2}{\partial\phi^2} + V(\phi) \right) \psi(t,\phi) = 0, \qquad (6.140)$$

which, in the language of constrained systems theory, is the hamiltonian constraint equation. Notice that the momentum p_ϕ is *not* a gauge-invariant quantity: it does not commute with the operator H, that is $[p_\phi, H] \neq 0$. This happens precisely for the same reason for which area and volume are not gauge-invariant quantities in GR. But this does not affect the simple fact that we *can* measure p_ϕ and we *do* predict that it is quantized. The confusion originates from the distinction between complete observables and partial observables, which was explained in detail in Chapter 3: p_ϕ is not a complete observable, but it is nevertheless a partial observable. We cannot predict the physical value of p_ϕ from a physical state, namely from a solution of the Wheeler–DeWitt equation, because we do not know at which time this is to be measured. But we can nevertheless compute and predict its eigenvalues.

Alternatively, we can define the evolving constant of the motion $p_\phi(T)$, as in Section 5.1.5. This is a gauge-invariant quantity. The spectral properties of $p_\phi(T)$, however, are the same as those of p_ϕ. More importantly, they are not affected by the potential $V(\phi)$, namely they are not affected by the dynamics. Similarly, we could in principle use gauge-invariant definitions of areas of surfaces, that would be genuinely diffeomorphism invariant, but this complicated exercise is useless, because the spectral properties can be directly determined by the partial observable operators.

*   *Why Diff*?* It is now time to address the question of the choice of the precise functional space of coordinate transformations, or active gauge maps $\phi : M \to M$, and to justify the fact that I have chosen *Diff** instead of *Diff*. Notice that in the classical theory the precise functional space in which we choose the fields is dictated by mathematical convenience, not by the physics. In fact, we always make measurements smeared in spacetime, which cannot be directly sensitive to what happens at points. Indeed, in classical field theory we choose to change freely the class of functions when convenient. For instance, twice differentiable fields allow us to write the equations of motion more easily, but then we prefer to work with distributional fields in certain applications. Analytic fields are usually considered too rigid, because we do not like too much the idea that a field in a small finite neighborhood could uniquely determine the field everywhere. The smooth category (C^∞) is often easy and convenient, and it

has been generally taken as the natural point of departure in quantum gravity, but it is not God-given. If we use smooth fields, it is natural to consider smooth coordinate transformations and *Diff* as the gauge group. In fact, this was the traditional choice in quantum gravity. However, at the end of Section 6.4 I pointed out that if we choose *Diff* as the gauge group, then the knot classes are labeled by continuous parameters (moduli) and the space $\mathcal{K}_{\text{diff}}$ turns out to be nonseparable. At first, we may think that these moduli represent physical degrees of freedom. If they did, there would be observable quantities that are affected by them. However, none of the operators that we have constructed in this chapter is sensitive to these moduli. In particular, a moment of reflection shows that all the geometric operators are only sensitive to features of graphs (and surfaces and volumes) that are invariant under continuous deformations of the graphs (surfaces and regions). Therefore, it is possible that these moduli are an artifact of the mathematics: they have nothing to do with the physics. They just reflect the fact that we have not chosen the functional space of the maps ϕ appropriately. The ϕ in *Diff* are too "rigid," in the sense that they leave invariant the linear structure of the tangent space at a node, while this linear structure has no physical significance. The choice of $Diff^*$ as gauge group is a simple extension of the gauge group that gets rid of the redundant parameters. Accordingly, we have to work with a space of fields slightly larger that C^∞. Nothing changes in the classical theory, while the quantum theory is cured of the double problem of having a nonseparable Hilbert space and redundant, physically meaningless, moduli. Of course, choices other than $Diff^*$ are possible.

Noncommutativity of the geometry. I close this section with an observation. Consider a spin network state containing a four-valent node n. Let l_1, l_2, l_3, l_4 be the four links adjacent to the node n. Let i be the intertwiner on this node. Consider a surface $S_{(1,2)(3,4)}$ such that n is on $S_{(1,2)(3,4)}$, the links l_1 and l_2 are on one side of the surface while the links l_3 and l_4 are on the other side of it. We can choose a basis in the space of the intertwiners by splitting the node n into two trivalent nodes joined by a virtual link l (see Appendix A1). Let us do so by pairing the links as (l_1, l_2) and (l_3, l_4). That is, the two trivalent nodes are between (l_1, l_2, l) and (l, l_3, l_4). A basis in the space of intertwiners is then given by

$$v_j^{\alpha_1 \alpha_2 \alpha_3 \alpha_4} = v^{\alpha_1 \alpha_2 \alpha_j} v^{\alpha_j \alpha_3 \alpha_4}, \tag{6.141}$$

where the indices α_i are in the representations of the links and the index α_j is in the representation j. It is not hard to show that in order for the state to be an eigenstate of the area of S_a, the intertwiner must be one of these basis elements. In other words, the basis (6.141) diagonalizes the area of S_a. Now consider a surface $S_{(1,3)(2,4)}$ such that n is on it and the links l_1 and l_3 are on one side of the surface while the links l_2 and l_4 are on the other side of it. Clearly in this case it will be a different basis in the space of the intertwiners that diagonalizes the area. It will be the basis

$$w_k^{\alpha_1 \alpha_2 \alpha_3 \alpha_4} = v^{\alpha_1 \alpha_3 \alpha_k} v^{\alpha_k \alpha_2 \alpha_4}. \tag{6.142}$$

The two bases are related by a $6j$ symbol. In general, they are different. It follows that the operator $\mathbf{A}(S_{(1,2)(3,4)})$ and the operator $\mathbf{A}(S_{(1,3)(2,4)})$ do not commute (if they did, they would be diagonalized by the same basis). Therefore the 3d geometry is, in a sense, noncommutative: area operators of intersecting surfaces do not commute with each other.

6.7.1 The texture of space: weaves

What is the connection between the discrete and quantized geometry described above and the smooth structure of physical geometry that we perceive around us? The answer requires a few steps.

Weaves. Ordinary measurements of geometric quantities – that is, measurements of the gravitational field – are macroscopic: we observe the geometry of space at a scale l much larger than the Planck length l_P. At this large scale, planckian discreteness is smoothed out.

Consider the fabric of a T-shirt as an analogy. At a distance, it is a smooth curved two-dimensional geometric surface. At a closer look, it is composed of thousands of one-dimensional linked threads. The image of space given by LQG is similar. Consider a very large spin network formed by a very large number of nodes and links, each of Planck scale. Microscopically, it is a planckian-size lattice. But probed at a macroscopic scale, it appears as a three-dimensional continuous metric geometry. Physical space around us can therefore be described as a very fine weave. The hidden texture of reality is a weave of spins.

This intuitive picture can be made precise. Fix a classical macroscopic 3d gravitational field e, which determines a macroscopic 3d metric $g_{ab}(\vec{x}) = e_a^i(\vec{x})\, e_{ib}(\vec{x})$. It is possible to construct a spin network state $|S\rangle$ that approximates this metric, at a scale $l \gg l_P$. The precise relation between $|S\rangle$ and e is the following. Consider a region $\mathcal{R}$ (or a surface $\mathcal{S}$) with a size larger than l (in the metric g), and slowly varying at this scale. Require that $|S\rangle$ is an eigenstate of the volume operator $\mathbf{V}(\mathcal{R})$ (and of the area operator $\mathbf{A}(\mathcal{S})$), with eigenvalues equal to the volume of $\mathcal{R}$ (and of the area of $\mathcal{S}$) determined by e, up to small corrections in l_P/l. That is

$$\mathbf{V}(\mathcal{R})|S\rangle = \big(\mathbf{V}[e, \mathcal{R}] + O(l_P/l)\big)\, |S\rangle,$$
$$\mathbf{A}(\mathcal{S})|S\rangle = \big(\mathbf{A}[e, \mathcal{S}] + O(l_P/l)\big)\, |S\rangle, \tag{6.143}$$

where $\mathbf{V}[e, \mathcal{R}]$ (resp. $\mathbf{A}[e, \mathcal{S}]$) given in (2.73) (and (2.70)), is the volume (the area) of the region (the surface), determined by the gravitational field e.

A spin network state $|S\rangle$ that satisfies these equations for any large region and surface is called a "weave" state of the metric g. At large scale, the state $|S\rangle$ determines precisely the same volumes and areas as g.

This definition is given at the nondiff-invariant level; but it can be easily carried over to the diff-invariant level: the s-knot state $|s\rangle = P_{\text{diff}} |S\rangle$ is called the weave state of the 3-geometry $[g]$, the equivalence class of 3-metrics to which the metric g belongs.

Several weave states were constructed and studied in the early days of LQG, for various 3d metrics, including the ones of flat space, Schwarzschild and gravitational waves. They satisfied (6.143), or equations similar to these (at the time the area and volume operators were not known, and other operator functions of the gravitational field were used to play the same role). Most of these weave states were constructed before the discovery of the spin network basis, working with the more cumbersome loop basis. Equations (6.143) do not determine a weave state uniquely from a given 3-metric. There is a large freedom in constructing a weave state for a given metric, because only the averaged properties are constrained by (6.143). The weave states constructed should not be taken as realistic proposals for the microstates of a given macroscopic geometry. They are only a proof of existence of microstates that have specified macroscopic properties.

On the other hand, the weave states have played a very important role in the historical development of the LQG. I recall this role below, because it contains an important physical lesson on the physics of Planck-scale discreteness.

The failure of the $a \to 0$ limit and the emergence of Planck-scale discreteness. There is a gap of several years between the construction of the loop representation of quantum GR (*c.* 1988) and the calculation of the eigenvalues of area and volume (*c.* 1995), which revealed that the theory predicts a discrete structure of space. During these years, the fact that the loops have "Planck size" was not known, and at first not even suspected. The intuition was that a macroscopic geometry could be constructed by taking a *limit* of an infinitely dense lattice of loops – roughly as a conventional QFT can be defined by taking the limit of a lattice theory, as the lattice size a goes to zero. To construct a weave state approximating a classical metric, therefore, the aim was at first to satisfy equations like (6.143) by quantum states defined as limits, where the spatial density of the loops was taken to infinity. But something unexpected and very remarkable happened. With increasing density of loops, the accuracy of the approximation did not increase. Instead, the eigenvalue of the operator increased!

Let me be more precise. Suppose we start with a 3d manifold with coordinates $\vec{x}$. We want to define a weave on this manifold, that approximates the flat 3d metric $g^{(0)}{}_{ab}(\vec{x}) = \delta_{ab}$, that is, the field $e^{(0)i}{}_a(x) = \delta^i_a$.

We construct a spatially uniform weave state $|S_{a_0}\rangle$ formed by a tangle of loops of coordinate density $\rho = a_0^{-2}$. (The coordinate density ρ can be defined as the ratio between the total coordinate length L of the loops and the total coordinate volume V.) The loops are then at an average distance a_0 from each other. Therefore, one expects that the approximation (6.143) would break down at the scale $l \sim a_0$. The idea was therefore to improve the approximation by decreasing the "lattice spacing" a_0, namely by increasing the coordinate density of the loops. But decreasing a_0 to become $a < a_0$, the calculations yielded instead

$$A(\mathcal{S})\,|S_a\rangle \sim \frac{a_0^2}{a^2}\left(A[e^{(0)},\mathcal{S}] + O(l_P/l)\right)\,|S\rangle : \qquad (6.144)$$

instead of a decrease in the error, the area increases! In other words, by adding loops we do not obtain a better approximation. Rather, we approximate a different field. Since

$$\frac{a_0^2}{a^2}\,A[e^0,\mathcal{S}] = A[(a_0/a)\,e^{(0)},\mathcal{S}] = A[e^{(a)},\mathcal{S}], \qquad (6.145)$$

where $e^{(a)i}_{a}(x) = \frac{a_0^2}{a^2}\,\delta_{ai}$, the weave with increased loop density approximates the metric

$$g^{(a)}_{ab}(\vec{x}) = \frac{a_0^2}{a^2}\,\delta_{ab}. \qquad (6.146)$$

But notice that the *physical* density of the loops, ρ_a, does not change with decreasing a. The physical density ρ_a is the ratio between the total length of the loops and the total volume, determined by the metric $g^{(a)}$, namely by the metric that the state $|S_a\rangle$ itself determines via equation (6.143). This is

$$\rho_a = \frac{L_a}{V_a} = \frac{(a_0/a)L}{(a_0/a)^3V} = \frac{a^2}{a_0^2}\,\rho = \frac{a^2}{a_0^2}\,a^{-2} = a_0^{-2}. \qquad (6.147)$$

The physical density remains a_0^{-2}, irrespective of the density of the loops a chosen! But then, if a_0 is not determined by the density of the loops, it must be given by a dimensional constant of the theory itself, and since the only scale in the theory is the Planck scale, we have, necessarily, that up to numerical factors

$$a_0 \sim l_P. \qquad (6.148)$$

At first this result was disconcerting. The theory refused to approximate a smooth geometry at a physical scale lower than l_P. Then the reason became clear: there is no physical scale lower than l_P. Each loop carries a quantum of geometry of Planck size: more loops give more size, not a better approximation to a given geometry. This was the first unexpected hint that the loops themselves have an intrinsic geometric size, and that in the theory there is no spatial structure at physical scales smaller than the Planck scale.

Quantum and classical discreteness: superposition of weaves. The weave picture of space resembles the space of a lattice theory. But there is a strong difference between the two and the analogy should be taken with great care.

Planck-scale discreteness is *predicted* by loop quantum gravity, on the basis of a standard quantization procedure, in the same manner in which the quantization of the energy levels of an atom is predicted by nonrelativistic quantum mechanics, while the discretization of space in a lattice theory is *assumed*.

But the difference is far more substantial than this. In a lattice theory, the lattice is a fixed structure on which the theory is defined. A weave, on the other hand, is one of many quantum states that have a certain macroscopic property, and a very peculiar one, since it is a single element of the spin network basis. There is no reason for the physical state of space not to be in a generic state, and the *generic* quantum state that has this macroscopic property is not a weave state: it is a *quantum superposition* of weave states. Therefore it is reasonable to expect that, at small scale, space is a quantum superposition of weave states.

Therefore the picture of physical space suggested by LQG is not truly that of a small-scale lattice. Rather, it is a quantum probabilistic cloud of such lattices.

The Minkowski vacuum. To a certain approximation, macroscopic space around us is described by the Minkowski metric. We should therefore expect that in the quantum theory there is a state $|0_M\rangle$ (M for Minkowski) that reproduces the Minkowski metric at large scales.

At fixed time, Minkowski 3d space is described by a flat 3-metric g, namely by the gravitational field $e_a^i(x) = \delta_a^i$. Does this imply that we should expect $|0_M\rangle$ to be a weave state of e? It does not. In quantum Maxwell theory, the state that corresponds to the classical solution with vanishing electric and magnetic fields, $E = B = 0$, is the vacuum state $|0\rangle$. But $|0\rangle$ is not an eigenstate of the electric field E. Since E and B do not commute, E and B have no common eigenstates. Eigenstates of E are maximally spread in B. Instead, E and B have vanishing mean value and minimal spread on $|0\rangle$. The situation is precisely the same as for the vacuum state of an harmonic oscillator. The classical solution $x(t) = p(t) = 0$ corresponds to the vacuum state, a state which is neither an eigenstate of the position x nor an eigenstate of the momentum p. An eigenstate of x has Δp very large and would spread instantaneously.

Similarly, $|0_M\rangle$ cannot be an eigenstate of the gravitational field E. An eigenstate of E has maximum spread in the gravitational magnetic field and would spread instantaneously. Accordingly, $|0_M\rangle$ is not a weave state of the flat geometry, nor a superposition of weave states of the flat geometry.

$|0_M\rangle$ must be a state with vanishing mean value, and minimal spread, of the gravitational electric and magnetic fields. On the other hand, it should be concentrated around weave states, in the same sense in which the vacuum of the harmonic oscillator is concentrated around $x = 0$.

We do not yet know the form of the state $|0_M\rangle$ explicitly. In fact, the exploration of this functional is one of the main open problems in the theory. I will come back to this problem in Chapter 9.

We can obtain some hints about the state $|0_M\rangle$ from the classical field theory obtained linearizing GR around the Minkowski solution. In the classical theory, let us write

$$e_\mu^I(x) = \delta_\mu^I + h_\mu^I(x), \qquad (6.149)$$

and restrict our considerations to solutions of the Einstein equations where $h_\mu^I(x) \ll 1$. To first order in h, these solutions satisfy the linearized Einstein equations. The linearized Einstein equations are free wave equations on Minkowski space for a spin-2 field. The solutions are superpositions of plane waves. They can be gauge-fixed, fixing $h_\mu^0 = h_0^I = 0$ and restricting $h_a^i(x)$ to the sole transverse traceless components, that is $\partial_a h_a^i = \partial_i h_a^i = h_a^a = 0$. For each momentum $\vec{k}$ there are then two independent polarizations $\epsilon_\pm$. In this gauge, the linearized Einstein equations describe simply a collection of uncoupled harmonic oscillators of frequency $\omega(\vec{k}) = |\vec{k}| = \sqrt{k_a k^a}$, one for each polarization and for each Fourier mode

$$h_a^i(\vec{k}) = (2\pi)^{-3/2} \int d\vec{x} \, e^{i\vec{k}\vec{x}} \, h_a^i(\vec{x}). \qquad (6.150)$$

The quantum field theory of this free system is a fully conventional free QFT. In its vacuum state, $|0_{\text{lin}}\rangle$ (lin for linearized) all oscillators are in their ground state. The Hilbert state of the theory is the Fock space spanned by the states

$$|k_1, \epsilon_1, \ldots, k_n, \epsilon_n\rangle \qquad (6.151)$$

containing n quanta with momenta and polarizations (k_i, ϵ_i). These quanta are called gravitons. $h_a^i(x)$ is a conventional field operator on this Fock space (formed by creation and annihilation parts). For each (gauge-fixed) field configuration h, we can write the (generalized) eigenstates

$|h\rangle$. These allow us to write a generic state $|\psi\rangle$ of the Fock space in the Schrödinger representation

$$\Psi[h] = \langle h|\psi\rangle. \qquad (6.152)$$

In particular, the vacuum state is given in this representation by

$$\Psi_0[h] = \langle h|0_{\text{lin}}\rangle = Ne^{-\frac{1}{2\hbar}\int d\vec{k}\, |\vec{k}|\, h_i^a(-\vec{k})\, h_a^i(\vec{k})}. \qquad (6.153)$$

This is a gaussian functional concentrated around the field configuration $h = 0$. We can rewrite this state as a functional of the gravitational field as

$$\Psi_0[e] = \langle e - \delta|0_{\text{lin}}\rangle = Ne^{-\frac{1}{2\hbar}\int d\vec{k}\, |\vec{k}|\, (e_i^a(-\vec{k})-\delta_i^a)\, (e_a^i(\vec{k})-\delta_a^i)}, \qquad (6.154)$$

which is a functional concentrated around $g = \delta$.

It is then reasonable to suspect that $|0_{\text{M}}\rangle$ should satisfy

$$\Psi_{\text{M}}[S_e] \equiv \langle S_e|0_{\text{M}}\rangle \sim \Psi_0[e] = Ne^{-\frac{1}{2\hbar}\int d\vec{k}\, |\vec{k}|\, (e_i^a(-\vec{k})-\delta_i^a)\, (e_a^i(\vec{k})-\delta_a^i)} \qquad (6.155)$$

for all spin network states S_e that are weaves for the field e. This is a state concentrated around the flat weave S_0.

The "empty" state. The Minkowski vacuum state $|0_{\text{M}}\rangle$ should not be confused with the covariant vacuum state $|0\rangle$ and with the empty state $|\emptyset\rangle$ (see Section 5.4.2). The state

$$\Psi_\emptyset[A] = \langle A|\emptyset\rangle = 1 \qquad (6.156)$$

is an eigenstate of $\mathbf{A}(\mathcal{S})$ and $\mathbf{V}(\mathcal{R})$ with vanishing eigenvalue. Therefore, it describes a space with no volume and no area. Spin network states can be constructed acting on $|\emptyset\rangle$ with the holonomy operator. In this particular sense, $|\emptyset\rangle$ is analogous to the Fock vacuum. The state $|\emptyset\rangle$ is gauge invariant and diff invariant, hence this state is also in $\mathcal{K}_0$ and in $\mathcal{K}_{\text{diff}}$. In fact, it represents the quantum state of the gravitational field in which there is no physical space at all. As we shall see in the next chapter, $|\emptyset\rangle$ is a solution of the Wheeler–DeWitt equation, and therefore it is in $\mathcal{H}$ as well.

———

Bibliographical notes

The "loop representation of quantum general relativity" was introduced in [176, 177]. These papers present the first surprising results of the approach: solutions of the Wheeler–DeWitt equation, and general solutions of the diffeomorphism constraint. The loop transform, mapping between functionals of the connection and loop functionals, was illustrated in [170]. The approach was motivated by Ted Jacobson's and Lee Smolin's discovery [178] of loop solutions of the Wheeler–DeWitt equation written in Ashtekar variables. Rodolfo Gambini and his collaborators have independently developed a formal loop quantization for Yang–Mills theory [179]. The importance for the theory of the nodes – where loops intersect – was stressed by Jorge Pullin; nodes were studied in [180]. An account of this first stage of LQG can be found in the review [2].

The importance for the theory of the graphs was understood by Jerzy Lewandowski. The spin network basis was introduced in quantum gravity in [171], solving the longstanding difficulty of the overcompleteness of the loop basis. The inspiration came from Penrose's speculations on the combinatorial structure of space [181]. For motivations and the history of the idea of spin networks, see [182]. The mathematical systematization of this idea is due to John Baez [183]. The fact that the equivalence classes under *Diff* of graphs with intersections are not discrete and the associated problem of the nonseparability of the state space was pointed out in [171] and studied in [184], where the structure of the corresponding moduli spaces is analyzed. The solution of the problem in terms of *Diff** is discussed in [185]. A different approach to obtain a separable Hilbert space is in [186]. The uniqueness theorem for the loop representation in quantum gravity (the "LOST" theorem) has been proven by Jerzy Lewandowski, Andrzej Okolow, Hanno Sahlmann and Thomas Thiemann [187] and, in a slightly different version, by Christian Fleischhack [188].

The idea that LQG could predict a discrete Planck-scale geometry emerged in studying the weave states [189]. The first explicit claim that the eigenvalues of the area represent a physical prediction of the theory, observable in principle, is in [190]. The definition of the area and volume operators and the first calculations of their discrete eigenvalues are in [191]. The main sequence of the spectrum of the area was calculated in this work. The degenerate sector was computed in [173] and in [192], which I have followed here. A calculation mistake in the spectrum of the volume in the first version of [191] was soon found by Renate Loll [193] in developing the lattice version of the volume operator; Loll noticed that the node must be at least quadrivalent in order to have a nonvanishing volume. There exist a number of equivalent constructions of the volume operator in the literature; they all define the same operator, except for one possible variant. A systematic study of the variants in the definition of the volume operator has been completed by Jerzy Lewandowski in [194]. Systematic techniques to compute geometry eigenvalues were developed in [175], using the mathematics of [174]. For more details (and another application) see also [195]. A length operator (which is surprisingly far more complicated than area and volume to deal with) is studied by Thiemann in [196]. The noncommutativity of the areas of intersecting surfaces has been studied in detail in [197]. On angle operators, see [198]. An applealing and well written introduction to spin networks and their geometry is Seth Major's [199].

The mathematical-physics version of LQG started from the seminal work of Abhay Ashtekar, Chris Isham and Jerzy Lewandowski [200, 201]. See Ashtekar's 1992 Les Houches lectures [202] and John Baez [203]. This direction led to the paper [204], where the loop representation of [176, 177] was mathematically systematized; the relation between the different variants of the formalisms is still very confused in [204] and was elucidated by Roberto De Pietri [205] and Thomas Thiemann. For a detailed introduction and full references, see [9]. The relation between the full theory and the linearized theory has been explored in [206]. The notion of weave state was introduced in [189]; for a recent discussion and references, see [207].

7

Dynamics and matter

In the previous chapter, I constructed the Hilbert space and the basic operators of LQG. In this chapter I discuss the dynamical aspects of the theory. For this, we must write a well-defined version of the Wheeler–DeWitt equation (6.3) in order to construct the hamiltonian operator H.

The construction of the hamiltonian operator of LQG has taken a long time, proceeding through a number of steps. The first was to realize that the operator can be defined via a simple regularization and that any loop state Ψ_α solves $H\Psi_\alpha = 0$, provided that α is a loop without self-intersections. This observation opened the door to LQG. The result remains true in all subsequent definitions of H. With this first simple regularization, however, H diverges on intersections.

The second step was to realize that a finite operator H could be obtained provided that (i) its action is defined on diffeomorphism-invariant states and (ii) its density character is appropriately dealt with. By a sort of magic, the action of the operator with the correct density weight on a diff-invariant state *converges* trivially, in the limit in which the cut-off is removed. This result is the major pay-off of the background-independent approach to QFT. It is a manifestation of the relation between background independence and the absence of UV divergences.

The third step was the idea of writing H as a commutator, a technique that allows us to write the operator avoiding square roots and inversion of matrices. This technique solved at once the remaining roadblocks: on the one hand, it made it possible to use the real Barbero connection, thus avoiding the difficulties of implementing nontrivial reality conditions in the quantum theory; on the other hand, it made it possible to circumvent the difficulties due to the nonpolynomiality of H, such as the square root previously used to get a density-weight-one hamiltonian.

In this chapter, I do not follow the contorted historical path, but rather define the operator directly. I discuss only the operator corresponding to the first term of the hamiltonian (4.43) that defines lorentzian GR with a real connection. This first term alone defines euclidean quantum GR. The technique described here extends directly to the second term, for which I refer the reader to [20].

Following common GR parlance, I call "matter" anything which is not the gravitational field. At best as we know, the content of the Universe is the one described by GR and the standard model: fermions, Yang–Mills fields, gravitational field and, presumably, Higgs scalars. As explained in Chapter 1, LQG has no ambition of providing a naturally "unified" theory, or explaining the reasons of the content of the Universe.

In this chapter I assume that the four entities noted above make up the Universe, and I describe how the background-independent quantum theory of the gravitational field described thus far can be very naturally extended to a background-independent theory for all these fields.

The remarkable aspect of this extension is that the finiteness of the gravitational dynamics extends to matter. In the theory there is "no space" for UV divergences, neither for gravity nor for matter.

7.1 Hamiltonian operator

Regularization. The form of the hamiltonian H which is most convenient for the quantum theory is the one given in (4.16), namely

$$H = \int N \, \mathrm{tr}(F \wedge \{\mathbf{V}, A\}). \tag{7.1}$$

The reason is that we have already defined the quantum operator $\mathbf{V}$, and the operators F and A can be defined as limits of holonomy operators of small paths, while the classical Poisson bracket can readily be realized in the quantum theory as a quantum commutator.

Fix a point x and a tangent vector u at x; consider a path $\gamma_{x,u}$ of coordinate length ϵ that starts at x tangent to u. Then the holonomy can be expanded as

$$U(A, \gamma_{x,u}) = 1 + \epsilon \, u^a \, A_a(x) + O(\epsilon^2). \tag{7.2}$$

Similarly, fix a point x and two tangent vectors u and v at x, and consider a small triangular loop $\alpha_{x,uv}$ with one vertex at x, and two sides tangent to u and v at x, each of length ϵ. Then

$$U(A, \alpha_{x,uv}) = 1 + \frac{1}{2}\epsilon^2 \, u^a v^b \, F_{ab}(x) + O(\epsilon^3). \tag{7.3}$$

Using this, and writing $h_\gamma = U(A, \gamma)$ we can regularize the expression of the hamiltonian by writing it as

$$H = \lim_{\epsilon \to 0} \frac{1}{\epsilon^3} \int N \epsilon^{ijk} \mathrm{tr}\left(h_{\gamma_{x,u_k}^{-1}} \, h_{\alpha_{x,u_i u_j}} \{\mathbf{V}, h_{\gamma_{x,u_k}}\}\right) d^3 x. \tag{7.4}$$

Here (u_1, u_2, u_3) are any three tangent vectors at x whose triple product is equal to unity. Following the same strategy we used for the area and volume operators, let us partition the 3d coordinate space in small regions $\mathcal{R}_m$ of coordinate volume ϵ^3. We can then write the integral as a Riemann sum and write

$$H = \lim_{\epsilon \to 0} \frac{1}{\epsilon^3} \sum_m \epsilon^3 N_m \epsilon^{ijk} \mathrm{tr}\left(h_{\gamma_{xm,u_k}^{-1}} \, h_{\alpha_{xm,u_i u_j}} \{\mathbf{V}(\mathcal{R}_m), h_{\gamma_{xm,u_k}}\}\right), \tag{7.5}$$

where x_m is now an arbitrary point in $\mathcal{R}_m$ and $N_m = N(x_m)$. The fact that the limit is independent of the choice of this point is assured by the Riemann theorem. Notice that the ϵ^3 factors cancel, and can therefore be dropped.

Definition of the operator. Since $\mathbf{V}(\mathcal{R}_m)$ and h_γ are well-defined operators in $\mathcal{K}$, we can then consider the corresponding quantum operator

$$H = -\frac{i}{\hbar} \lim_{\epsilon \to 0} \sum_m N_m \, \epsilon^{ijk} \, \text{tr}\left(h_{\gamma_{xm,u_k}^{-1}} h_{\alpha_{xm,u_iu_j}} [\mathbf{V}(\mathcal{R}_m), h_{\gamma_{xm,u_k}}]\right). \quad (7.6)$$

To complete the definition of the operator we have yet to choose the point x_m, the three vectors (u_1, u_2, u_3), and the paths γ_{x,u_k} and α_{x,u_iu_j} in each region $\mathcal{R}_m$. This must be done in such a way that the resulting quantum operator is well defined, covariant under diffeomorphism, invariant under internal gauges, and nontrivial. These requirements are highly nontrivial. Remarkably, there is a choice that satisfies all of them.

The key observation to find it is the following. When acting on a spin network state, this operator acts only on the nodes of the spin network, because of the presence of the volume. (This is not changed by the presence of the term $h_{\gamma_{x,u}}$ in the commutator for the following reason. The volume operator vanishes on trivalent nodes. The operator $h_{\gamma_{x,u}}$ can at most increase the valence of a node by one. Therefore, there must be at least a trivalent node in the state for H not to vanish.)

Therefore, in the sum (7.6) only the regions $\mathcal{R}_m$ in which there is a node n give a nonvanishing contribution. Call $\mathcal{R}_n$ the region in which the node n of a spin network S is located. Then

$$H|S\rangle = \lim_{\epsilon \to 0} H_\epsilon|S\rangle, \quad (7.7)$$

where

$$H_\epsilon|S\rangle = -\frac{i}{\hbar} \sum_{n \in S} N_n \epsilon^{ijk} \, \text{tr}\left(h_{\gamma_{xn,u_k}^{-1}} h_{\alpha_{xn,u_iu_j}} [\mathbf{V}(\mathcal{R}_n), h_{\gamma_{xn,u_k}}]\right) |S\rangle. \quad (7.8)$$

The sum is now on the nodes. Now, the only possibility to have a nontrivial commutator is if the path γ_{xn,u_k} itself touches the node. We therefore demand this. This can be obtained by requiring that x_n is precisely the location of the node. Recall that the precise location of x_n is irrelevant in the classical theory, because of the Riemann theorem, but not so in the quantum theory. This fixes x_n.

Finally, there is a natural choice of the three vectors (u_1, u_2, u_3) and for the paths γ_{x,u_k} and α_{x,u_iu_j}: take (u_1, u_2, u_3) tangent to three links l, l', l'' emerging from the node n. (The condition that their triple product is

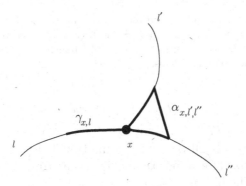

Fig. 7.1   The path $\gamma_{x,l}$ and the loop $\alpha_{x,l',l''}$ at a trivalent node in the point x.

unity can be satisfied by adjusting the length.) Take γ_{x,u_k} to be a path $\gamma_{x,l}$ of coordinate length ϵ along the link l. Take $\alpha_{x,u_i u_j}$ to be the triangle $\alpha_{x,l',l''}$ formed by two sides of coordinate length ϵ along the other two links l' and l'', and take the third side as a straight line (in the coordinates x) connecting the two end points. This straight line is called an "arc." The sum over i, j, k is a sum over all permutations of the three links, see Figure 7.1. If the node has valence higher than three, that is, if there are more than three links at the node n, we preserve covariance summing over all ordered triplets of distinct links. Thus, we pose

$$H_\epsilon |S\rangle = -\frac{i}{\hbar} \sum_{n \in S} N_n \sum_{l,l',l''} \epsilon_{ll'l''} \, \mathrm{tr} \left(h_{\gamma_{x_n,l}^{-1}} h_{\alpha_{x_n,l',l''}} [\mathbf{V}(\mathcal{R}_n), h_{\gamma_{x_n,l}}] \right) |S\rangle,$$

$$(7.9)$$

where $\epsilon_{ll'l''}$ is the parity of the permutation (determined by the sign of the triple product). This completes the definition of the hamiltonian constraint (up to one additional detail, that we add below). Two major questions are left open. First, whether the limit is finite. Second, whether it is well behaved under diffeomorphisms and gauge transformations. In particular, whether it is independent of the coordinates chosen. The two questions are intimately connected.

A side remark. There is a simple intuitive way of understanding why the hamiltonian acts only on nodes. Consider the naive (divergent) form (6.3) of the Wheeler–DeWitt operator and act with this on a spin network state, at a point where there are no nodes. Equation (6.59) shows that the result of this action (disregarding the divergence) is proportional to

$$F_{ab}^{ij}(\vec{\tau}) \frac{\delta}{\delta A_a^i(\vec{\tau})} \frac{\delta}{\delta A_b^j(\vec{\tau})} \, \Psi_S[A] \sim F_{ab}^{ij}(\vec{\tau}) \, \dot{\gamma}^a \dot{\gamma}^b, \qquad (7.10)$$

where $\dot{\gamma}^a$ is the tangent to the spin network links at the point $\vec{\tau}$ where the operator acts. But this expression vanishes because F_{ab} is antisymmetric in ab, while $\dot{\gamma}^a \dot{\gamma}^b$ is symmetric. On the other hand, acting on a node, the two functional derivatives give mixed terms of the kind $F_{ab}\dot{\gamma}_1^a\dot{\gamma}_2^b$, where $\dot{\gamma}_1^a$ and $\dot{\gamma}_2^a$ are the tangents of two different links emerging from the node. These terms may be nonvanishing. Hence, the operator has a nontrivial action only on nodes.

7.1.1 Finiteness

In general, the limit (7.7) does not exist. This is no surprise, as operator products are generally ill defined in quantum field theory: they can be defined in regularized form, but then a divergence develops when removing the regulator ϵ. The big surprise, however, is that the limit (7.7) *does* exist on a subclass of states: diffeomorphism-invariant states. As these are the physical states, this is precisely what we need and is sufficient to define the theory. Here is where the intimate interplay between diffeomorphism invariance and quantum field theoretical short-scale behavior begins to shine: we are here at the core of diffeomorphism-invariant QFT.

To compute H on diff-invariant states, recall these are in the dual space $\mathcal{S}'$. So far, we have only considered H on spin network states, or, by linearity, on $\mathcal{S}$. The action of H on $\mathcal{S}'$ is immediately defined by duality

$$(H\Phi)(\Psi) \equiv (\Phi)(H\Psi). \qquad (7.11)$$

(To be precise, I should call $H^\dagger$ the operator in the left-hand side, but for simplicity I do not.) Equivalently, for every spin network state

$$(H\Phi)(|S\rangle) = \Phi(H|S\rangle). \qquad (7.12)$$

The key point is that we want to consider the regularized operator on $\mathcal{S}'$ and take the limit there. Thus, instead of simply inserting (7.7) in the last equation and writing

$$(H\Phi)(|S\rangle) = \Phi(\lim_{\epsilon\to 0} H_\epsilon|S\rangle), \qquad (7.13)$$

I *define* the hamiltonian operator on $\mathcal{S}'$ by

$$(H\Phi)(|S\rangle) = \lim_{\epsilon\to 0} \Phi(H_\epsilon|S\rangle). \qquad (7.14)$$

Notice that the limit is now a limit of a sequence of numbers (not a limit of a sequence of Hilbert space vectors). I now show that the limit exists (namely is finite) if $\Phi \in \mathcal{K}_{\text{diff}}$, namely if Φ is a diffeomorphism-invariant state.

The key to see this is the following crucial observation. Given a spin network S, the operator in the parentheses modifies the state $|S\rangle$ in two ways: by changing its graph Γ as well as its coloring. The volume operator does not change the graph. The graph is modified by the two operators

$h_{\gamma_{x_n,l}}$ and $h_{\alpha_{x_n,l,l'}}$. The first superimposes a path of length ϵ to the link l of Γ. The second superimposes a triangle with two sides of length proportional to ϵ along the links l' and l'' of Γ, and a third side that is not on Γ, as in Figure 7.1. The fundamental observation is that for ϵ sufficiently small, changing ϵ in the operator changes the resulting state *but not its diffeomorphism equivalence class*. (The maximal ϵ_m for this to happen is the value of ϵ such that the added paths cross or link other nodes or links of S.) This is rather obvious: adding a smaller triangle is the same as adding a larger triangle and then reducing it with a diffeomorphism. Therefore, for $\epsilon < \epsilon_m$, the term in the parentheses remains in the same diffeomorphism equivalence class as ϵ is further reduced. But Φ is invariant under diffeomorphisms, and, therefore, the dependence on ϵ of the argument of the limit becomes *constant* for $\epsilon < \epsilon_m$!!

Therefore the value of the limit (7.14) is simply given by

$$(H\Phi)(|S\rangle) = \lim_{\epsilon \to 0} \Phi(H_\epsilon|S\rangle) = \Phi(H|S\rangle), \qquad (7.15)$$

where

$$H|S\rangle = -\frac{i}{\hbar} \sum_{n \in S} N_n \sum_{l,l',l''} \epsilon_{ll'l''} \mathrm{tr}\left(h_{\gamma_{x_n,l}^{-1}} h_{\alpha_{x_n,l',l''}}[\mathbf{V}(\mathcal{R}_n), h_{\gamma_{x_n,l}}] \right) |S\rangle,$$

$$(7.16)$$

and the size ϵ of the regularizing paths is simply taken to be small enough so that the added arc does not run over other nodes or link other links of S. The finiteness of the limit is then immediate.

Discussion: relation between regularization and background independence. This result is very important, and deserves a comment. The first key point is that the coordinate space $\vec{x}$ has no physical significance at all. The physical location of things is only location relative to one another, not the location with respect to the coordinates $\vec{x}$. The diffeomorphism-invariant level of the theory implements this essential general-relativistic requirement. The second point is that the excitations of the theory are quantized. This is reflected in the short-scale discreteness, or in the discrete combinatorial structure of the states. This is the result of the quantum mechanical properties of the gravitational field. When these two features are combined, there is literally no longer room for diverging short-distance limits. The limit $\epsilon \to 0$ is a limit of small coordinate distance: it becomes finite simply because making the regulator smaller cannot change anything below the Planck scale, as there is nothing below the Planck scale. Once the regulating small loop $\alpha_{x,l',l''}$ is smaller than the size needed to link or cross other parts of the spin network, any further decrease of its size is gauge, not physics. This is how diffeomorphism invariance cures in depth the ultraviolet pathologies of quantum field theory.

One last technicality. In the definition given, there is a residual (discrete) dependence on the coordinates x. In two different coordinate systems the arc may link the original links of the graph differently. For instance, a fourth link l''' may pass "over" or "under" the arc. If the node is n-valent, the possible alternatives are labeled by the homotopy classes of lines (without intersection) going from the north to the south pole on a sphere with $n - 2$ punctures. To have a fully diffeomorphism-invariant definition, we must therefore sum over these N_n alternatives. On the other hand, we can consistently exclude coordinate systems in which three links are coplanar and the arc intersects a link for all values of ϵ. We can do this because we are using extended diffeomorphisms. Thus, calling $\alpha^r_{x,l',l''}, r = 1, \ldots, N_n$ a representative of the rth homotopy class, we arrive at

$$(H\Phi)(|S\rangle) = \Phi(H|S\rangle) , \tag{7.17}$$

where

$$H|S\rangle = -\frac{i}{\hbar} \sum_{\substack{n \in S \\ l,l',l'',r}} N_n \epsilon_{ll'l''} \operatorname{tr} \left(h_{\gamma^{-1}_{x_n,l}} h_{\alpha^r_{x_n,l',l''}} [\mathbf{V}(\mathcal{R}_n), h_{\gamma_{x_n,l}}] \right) |S\rangle. \tag{7.18}$$

This is the final form of the operator.

7.1.2 Matrix elements

The resulting action of H on s-knot states is simple to derive and to illustrate. (i) The action gives a sum of terms, one for each node n of the state. (ii) For each node, H gives a further sum of terms, one term for each triplet of links arriving at the node, and, for each triplet, one term for every permutation of the three links l, l', l''. Each of these terms acts as follows on the s-knot state (see Figure 7.2). (iii) It creates two new nodes n' and n'' at a finite distance from n along the links l' and l''. The exact location of these nodes is of course irrelevant for the s-knot state. (iv) It creates a new link of spin-$1/2$ connecting n' and n'' (without linking any other node). This new link is called "arc." (v) It changes the coloring j' of the link connecting n and n' and the coloring j'' of the link connecting n and n''. These turn out to be the colors of the links l' and l'' increased or decreased by $1/2$. (vi) It changes the intertwiner at the node n; the new intertwiner is between the representations corresponding to the new colorings of the adjacent links.

Remark. Again, it is easy to understand the origin of this action of the hamiltonian operator, on the basis of the simple form (6.3) of the hamiltonian constraint. The two functional derivatives "grasp" a spin network, and, as explained before, the grasp vanishes except in the vicinity of a node. The curvature term F_{ab} is essentially an infinitesimal holonomy. Therefore, it creates a small loop next to the node. This loop

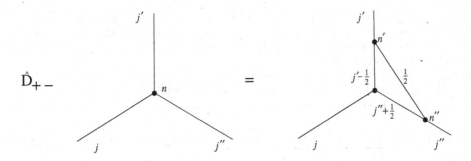

Fig. 7.2   Action of $D_{n,l',l'',r,\epsilon,\epsilon'}$.

must be in the plane of two grasped links, and can be identified with the triangle defined by the added arc.

Notice that the Clebsch–Gordan conditions always hold at the modified node. This follows immediately from the fact that the modified node is obtained from recoupling theory: the matrix element associated with nodes not satisfying the Clebsch–Gordan conditions turns out to vanish.

Call $D_{n,l',l'',r,\pm\pm}$ an operator that acts around the node n by acting as described in (iii), (iv), (v). This is illustrated in Figure 7.2. Then

$$H|S\rangle = \sum_{n\in S} N_n \sum_{l,l',l'',r} \sum_{\epsilon',\epsilon''=\pm} H_{n,l',l'',\epsilon',\epsilon''} \, D_{n,l',l'',r,\epsilon',\epsilon''} \, |S\rangle. \qquad (7.19)$$

The operator $H_{n,l',l'',\epsilon',\epsilon''}$ acts as a finite matrix on the space of the intertwiners at the node n. The explicit computation of its matrix elements is a straightforward problem in $SU(2)$ representation theory. It is discussed in detail in [208], where its matrix elements are explicitly given for simple nodes.

The operator H is defined on S' and, as we have seen, is finite when restricted to $\mathcal{K}_{\text{diff}}$. Notice however that the operator does not leave $\mathcal{K}_{\text{diff}}$ invariant. In general, the state $H|s\rangle$ is not a diffeomorphism-invariant state. This is because of its dependence on N. To see this, compute its action on a generic spin network state

$$\langle s|H|S\rangle = \sum_{n\in S} N_n \sum_{l,l',l'',r} \sum_{\epsilon',\epsilon''=\pm} \langle s|H_{n,l',l'',\epsilon',\epsilon''} \, D_{n,l',l'',r,\epsilon',\epsilon''} \, |S\rangle. \qquad (7.20)$$

On the right-hand side, the quantities N_n are the values of $N(\vec{x})$ at the points $\vec{x}_n$ where the nodes n of the spin network S are located. The rest of the expression is diff invariant, but these values obviously change if we perform a diffeomorphism on $|S\rangle$. This has of course to be expected, because the classical quantity H is itself not diffeomorphism invariant. Therefore, there is no reason for the corresponding quantum operator to be diff invariant, and preserve $\mathcal{K}_{\text{diff}}$. The theory this operator defines is nevertheless diffeomorphism invariant because the operator enters the theory via the Wheeler–DeWitt equation $H\Psi = 0$. This equation *is* well defined on $\mathcal{K}_{\text{diff}}$. Its solutions are the (possibly generalized) states in $\mathcal{K}_{\text{diff}}$ that are in the kernel of H. Since H is finite on the entire $\mathcal{K}_{\text{diff}}$ this is well defined. This is completely analogous to the classical theory, where H is not diff invariant but the equation $H = 0$ is perfectly sensible as an equation for diff-invariant equivalence classes of solutions. (An alternative strategy, yielding a version of the hamiltonian operator sending $\mathcal{K}_{\text{diff}}$ into itself, has been recently explored by Thiemann in [209].)

Recalling the definition of $|s\rangle$, we can write the matrix elements of H among spin network states

$$\langle S'|H|S\rangle = \sum_{n \in S} N_n \sum_{|\Psi\rangle = U_\phi|S'\rangle} \sum_{l,l',l''} \sum_{\epsilon',\epsilon''=\pm} \langle \Psi|H_{n,l',l'',\epsilon',\epsilon''} D_{n,l',l'',\epsilon',\epsilon''} |S\rangle.$$

(7.21)

The operator H is not symmetric. This is evident for instance from the fact that it adds arcs but does not remove them. Its adjoint $H^\dagger$ can be defined simply by the complex conjugate of the transpose of its matrix elements

$$\langle S'|H^\dagger|S\rangle = \overline{\langle S|H|S'\rangle},$$

(7.22)

and a symmetric operator is defined by

$$H_s = \frac{1}{2}(H + H^\dagger).$$

(7.23)

This is an operator that adds as well as removes arcs. It is reasonable to expect that this operator be better behaved for the classical limit. Therefore we take this operator as the basic operator defining the theory.

7.1.3 Variants

The striking fact about the hamiltonian operator is that it can be defined at all. But how unique is it? There are a number of possible variants of the operator that one may consider. These can be seen as quantization ambiguities, that is, they define different dynamics in the quantum theory, all of which, at least at first sight, have the same classical limit. So far, it is not clear if these are truly all viable, or whether there are physical or mathematical constraints that select among them.

Higher j. If we expand the matrix representing the holonomy of a connection in a representation j we obtain an expression analogous to that in (7.2). Therefore, we can regularize the terms F and A in the hamiltonian by using the holonomy in any arbitrary representation j. That is, we can write, up to an irrelevant numerical factor,

$$H = \lim_{\epsilon \to 0} \frac{1}{\epsilon^3} \int N \, \text{tr} \left(h^{(j)}_{\gamma_{x,w}^{-1}} h^{(j)}_{\alpha_{x,uv}} \{ \mathbf{V}, h^{(j)}_{\gamma_{x,w}} \} \right) d^3 x, \quad (7.24)$$

where $h^{(j)} = R^{(j)}(h)$. This has no effect on the classical limit. However, the corresponding quantum operator

$$H^{(j)} = -\frac{\mathrm{i}}{\hbar} \lim_{\epsilon \to 0} \sum_m N_m \, \mathrm{tr} \left(h^{(j)}_{\gamma_{xm},w}^{-1} \, h^{(j)}_{\alpha_{xm},uv} [\mathbf{V}(\mathcal{R}_m), h^{(j)}_{\gamma_{xm},w}] \right) \quad (7.25)$$

is different from the operator (7.6). This can be easily seen by noticing that the added arc does not have spin $1/2$, but rather spin j. In general, any arbitrary linear combination

$$\tilde{H} = \sum_j c_j \, H^{(j)} \quad (7.26)$$

defines an hamiltonian operator with a correct classical limit. There are indications that the coefficients c_j should be different from zero in a consistent theory, discussed for instance in [210]. In the same paper, matrix elements of the operator $H^{(j)}$ are computed, and some arguments on criteria to fix the coefficients c_j are discussed. As we shall see in Chapter 8, this quantization ambiguity plays a role in loop quantum cosmology.

H_s or H? Both the symmetric operator H_s and the nonsymmetric operator H define a quantum dynamics. While there are arguments for taking the symmetric one, there are also arguments for taking the nonsymmetric one [20].

Other regularizing loops. Loops different from $\alpha_{x,l',l''}$ and $\gamma_{x,l}$ could be chosen for the regularization. The freedom is strongly limited by diffeomorphism invariance and by the condition that the resulting operator is finite and nonvanishing. But other choices might be possible.

Ordering. A different ordering can be chosen between the volume and the holonomy operators.

Others? More generally, no uniqueness theorem exists so far.

This freedom in the definition of the hamiltonian operator is not a problem, it is an asset. No complete and completely consistent theory of quantum gravity with a well-understood low-energy limit exists so far. Having more than one is, for the moment, the very least of our worries. The remarkable result is the *existence* of a finite and interesting hamiltonian operator. The fact that we have a certain residual latitude in its definition might very well turn out to be helpful. The "correct" variant could be selected by some internal consistency requirement that has not yet been considered, or by requiring the correct classical limit. If these conditions turn out to be insufficient, we shall simply have nonequivalent quantum theories with the same classical limit. The physically correct one will have to be determined by experiments. I wish we were already there.

7.2 Matter: kinematics

The evolution of the mathematical description of the matter fields during the twentieth century has slowly converged with the evolution of the mathematical description of the gravitational field. In both cases, differential geometry notions such as fiber bundles, sections and automorphisms of the bundle play a role in the description of the classical fields. If we take, for instance, a coupled Einstein–Yang–Mills system, and describe the gravitational field by means of an $SO(3,1)$ connection, the structures of the two fields, the gauge and the gravitational field, are barely different from each other.

Indeed, as I have argued in Section 2.3.2, the distinction between matter and spacetime (gravity) is not profound: it is largely conventional. The gravitational field is not substantially different from the other matter fields. It is the full coupled gravity + matter theory which is profoundly different from a theory on a fixed background. When the dynamical gravitational field is not approximated by a fixed background, the full theory is generally covariant and the physical fields live only "on one another," as the animals and the whale of the metaphor.

Accordingly, the methods developed above for the gravitational field extend naturally to other fields. The theory of pure quantum gravity and the theory of quantum gravity and matter do not differ much from each other. The second has just some additional degrees of freedom. The similarity and compatibility of the classical mathematical structures makes the extension of the quantum theory to these additional degrees of freedom very natural.

7.2.1 Yang–Mills

The easiest extension of the theory described in the previous chapter is to Yang–Mills fields. Let G_{YM} be a compact Yang–Mills group, such as, in particular, the group $SU(3) \times SU(2) \times U(1)$ that defines the standard model. Let A_{YM} be a 3d Yang–Mills connection for this group, and A be the gravitational connection. The two 3d connections A (gravitational) and A_{YM} (Yang–Mills) can be considered together as a single connection $\mathcal{A} = (A, A_{\mathrm{YM}})$ for the group $G = SU(2) \times G_{\mathrm{YM}}$. The construction of $\mathcal{K}$, $\mathcal{K}_0$ and $\mathcal{K}_{\mathrm{diff}}$ of Chapter 6 extends immediately to this connection, without difficulties.

Holonomies of the Yang–Mills field can be defined as operators on $\mathcal{K}$, precisely as gravitational holonomies. Surface integrals of the Yang–Mills electric field can be defined precisely as for the E gravitational field.

The diffeomorphism-invariant quantum states of GR + Yang–Mills are then given by s-knot states labeled by abstract knotted graphs carrying irreducible representations of the group G on the links, and the corresponding intertwiners on the nodes. Since G is a direct product, its irreducibles are simply given by products of irreducibles of $SU(2)$ and G_{YM}. In other words, each link is labeled with a spin j_l and an irreducible representation of G_{YM}.

Notice how the quantum theory realizes the relational localization characteristic of GR: the position of the Yang–Mills field is well defined with respect to the quantum state of spacetime defined by the gravitational part of the spin network, *or, equivalently, vice versa.*

Notice also that in the absence of diffeomorphism invariance the above construction would not yield a sensible quantum state space of the Yang–Mills field, because it would yield a nonseparable Hilbert space.

7.2.2 Fermions

Let $\eta(x)$ be a Grassman-valued fermion field. It transforms under a representation k of the Yang–Mills group G_{YM}, and under the fundamental representation of $SU(2)$. It is more convenient for the quantum theory to take the densitized field $\xi \equiv \sqrt{|\det E|}\, \eta$ as the basic field variable. The Grassman-valued field ξ and its complex conjugate take value in a (finite-dimensional) superspace S. An integral over S is defined by the Berezin symbolic integral $d\xi$.

Define a cylindrical functional $\Psi[A, \psi]$ of the connection and the fermion field as follows. Given (i) a collection Γ of a finite number L of paths γ_l, (ii) a finite number N of points x_n, and (iii) a function f of L group elements and N Grassman variables, a cylindrical functional is defined by

$$\Psi_{\Gamma,f}[A, \psi] = f(U(A, \gamma_1), \ldots, U(A, \gamma_L), \xi(x_1), \ldots, \xi(x_N)). \qquad (7.27)$$

Since Grassman variables anticommute, cylindrical functionals can be at most linear in each (component of the) fermion fields in each point n.

A scalar product is defined on the space of these functions as follows. Given two cylindrical functionals defined by the same Γ, define

$$(\Psi_{\Gamma,f}, \Psi_{\Gamma,g}) \equiv \int_{G^L} dU_l \int_{S^N} d\xi_n \, \overline{g(U_1, \ldots, U_N, \xi_1, \ldots, \xi_N)}$$
$$\times f(U_1, \ldots, U_N, \xi_1, \ldots, \xi_N). \qquad (7.28)$$

The extension of this scalar product to any two cylindrical functions is then completely analogous to the purely gravitational case. This defines the extension of $\mathcal{K}$ to fermions.

Basis states are easily constructed by fixing the degree F_n of the monomials in ξ at each node n, which determine the fermion number in the region of the node.

In the absence of fermions, we constructed gauge-invariant functionals by contracting the indices of the holonomies among themselves with intertwiners. In the presence of fermions, we can also contract $SU(2)$ and Yang–Mills indices with the indices of the fermions. In other words, fermions live on the nodes of the graph Γ, and the gravitational and Yang–Mills lines of flux can end at a fermion. For this to happen, of course, generalized Clebsch–Gordan conditions must be satisfied. For instance, a single fermion cannot sit at the open end of a link in the trivial representations of $SU(2)$, because its $SU(2)$ index cannot be saturated. This physical picture is well known, for instance, in canonical lattice gauge theory. In fact, each Hilbert subspace $\mathcal{K}_\Gamma$ can be identified as the Hilbert space of a lattice Yang–Mills theory with fermions defined on a lattice Γ. We are back to the original intuition of Faraday: the lines of force can emerge from the charged particles.

7.2.3 Scalars

The present formulation of the standard model requires also a certain number of scalar fields, so far unobserved. Whether these fields – the Higgs fields – are in fact present in Nature and observable, or whether they represent a phenomenological description of some aspect of Nature we haven't yet fully understood is still unclear. Scalar fields can be incorporated in LQG, but in a less natural manner than Yang–Mills fields and fermions.

Let $\phi(x)$ be a suitable multiplet of scalar fields (that we can always take as real), transforming in a representation k of the gauge group G_{YM}, and therefore taking value in the corresponding vector space $\mathcal{H}_k$.

The complication is that $\mathcal{H}_k$ is noncompact – it has infinite volume under natural invariant measures. This makes the definition of the scalar product of the theory more difficult (because the Hilbert space associated to subgraphs is not a subspace of the Hilbert space associated to a graph). One way out of this difficulty is the following. Assume k is the adjoint representation of G_{YM}. We can then exponentiate the field $\phi(x)$ defining $U(x) = \exp\{\phi(\mathrm{x})\}$. The field $U(x)$ then takes values in G_{YM}, which is compact, and carries the Haar invariant measure.

Define a cylindrical functional $\Psi[A, \psi, \phi]$ of the connection, fermion, and scalar fields as follows. Given (i) a collection Γ of a finite number L of paths γ_l, (ii) a finite number N of points x_n, and (iii) a function f of L group elements, N fermion variables and N other group elements, a cylindrical functional is defined by

$$\Psi_{\Gamma,f}[A, \psi, \phi] = f(U(A, \gamma_1), \ldots, U(A, \gamma_L), \psi(x_1), \ldots, \psi(x_N), e^{\phi(x_1)}, \ldots,$$

$$e^{\phi(x_N)}). \quad (7.29)$$

A scalar product is defined on the space of these functions as follows. Given two cylindrical functionals defined by the same Γ, we define

$$(\Psi_{\Gamma,f}, \Psi_{\Gamma,g}) \equiv \int_{G^L} dU_l \int_{S^N} d\xi_n \int_{G^N} dU'_n$$
$$\times \ \overline{f(U_1, \ldots, U_L, \xi_1, \ldots, \xi_N, U'_1, \ldots, U'_N)}$$
$$\times \ g(U_1, \ldots, U_L, \xi_1, \ldots, \xi_N, U'_1, \ldots, U'_N). \quad (7.30)$$

and extend this to any graph as usual.

7.2.4 The quantum states of space and matter

A state $|s\rangle$ in $\mathcal{K}_{\text{diff}}$ can then be labeled by the following quantum numbers.

- An abstract knotted graph Γ with links l and nodes n.

- A spin j_l associated with each link l.

- An irreducible representation k_l of the Yang–Mills group G_{YM} associated with each link l.

- An integer F_n associated with each node.

- An irreducible representation S_n of the Yang–Mills group G_{YM} associated with each node n.

- An $SU(2)$ intertwiner i_n associated with each node n.

- A G_{YM} intertwiner w_n associated with each node n.

Thus we can write

$$|s\rangle = |\Gamma, j_l, k_l, F_n, S_n, i_n, w_n\rangle. \quad (7.31)$$

This state describes a quantum excitation of the system that has a simple interpretation as follows. There are N regions n, that have volume and where fermions and Higgs scalars can be located. These are separated by L surfaces l, that have area and are crossed by flux of the (electric) gauge field. The quantum numbers are related to observable quantities as in Table 7.1. This completes the definition of the kinematics of the coupled gravity+matter system.

7.3 Matter: dynamics and finiteness

The dynamics of the coupled gravity+matter system is simply defined by adding the terms defining the matter dynamics to the gravitational

Table 7.1 Quantum numbers of the spin network
states for gravity and matter.

Quantum number	Physical quantity
Γ	adjacency between the regions
i_n	volume of the node n
j_l	area of the surface l
F_n	number of fermions at node n
S_n	number of scalars at node n
w_n	field strength at node n
k_l	electric flux across the surface l

relativistic hamiltonian. The hamiltonian for the fields described is given
by

$$H = H_{\text{Einstein}} + H_{\text{Yang–Mills}} + H_{\text{Dirac}} + H_{\text{Higgs}}. \qquad (7.32)$$

H_{Einstein} is the gravity hamiltonian described in the previous chapters.
The other terms are

$$H_{\text{Yang–Mills}} = \frac{1}{2g_{\text{YM}}^2 e^3}\, \text{tr}[E_a E_b]\, \text{Tr}[\mathcal{E}^a \mathcal{E}^b + B^a B^b]$$

$$H_{\text{Dirac}} = \frac{1}{2e}\, E_i^a (\text{i}\pi\tau^i \mathcal{D}_a \xi + \mathcal{D}_a(\pi\tau^i \xi) + \frac{\text{i}}{2} K_a^i \pi\xi + c.c.)$$

$$H_{\text{Higgs}} = \frac{1}{2e}\left(p^2 + \text{tr}[E^a E^b]\, \text{Tr}[(\mathcal{D}_a\phi)(\mathcal{D}_b\phi)] + e^2 V(\phi^2) \right). \qquad (7.33)$$

Here: π is the momentum conjugate to the fermion field; p the momentum
conjugate to the scalar field; $\mathcal{E}$ is the momentum conjugate to the Yang–
Mills potential (the electric field); B the curvature of the Yang–Mills
potential (the magnetic field); $\mathcal{D}$ is the $SU(2) \times G_{\text{YM}}$ covariant derivative;
tr is the trace in the $SU(2)$ Lie algebra and Tr is the trace in the Lie
algebra of G_{YM}; $e \equiv \sqrt{|\det E|}$; V is the Higgs potential; $p^2 = \text{Tr}[pp]$
and $\phi^2 = \text{Tr}[\phi\phi]$; and g_{YM} is the Yang–Mills coupling constant. For a
derivation and a discussion of these expressions, see [211].

To define the quantum hamiltonian, the expressions in (7.33) must be
regulated and expressed in terms of the operators well defined on $\mathcal{K}$. This
can be done following the same strategy we used for the gravitational
quantum hamiltonian in Section 7.1. I do not present the detailed construction here, for which I refer the reader to Thiemann's work [211].

The essential result of this construction is that the total hamiltonian
(7.32) can be constructed as a well-defined operator on the Hilbert
space of gravity and matter $\mathcal{K}$. The operator acts on nodes, as does the
gravitational part, but its action is more complex than the pure gravity

operator, and codes the entire dynamics of the standard model and general relativity.

The fact that the total hamiltonian turns out to be finite is extremely remarkable. It is perhaps the major pay-off of the background-independent quantization strategy on which LQG is based.

I advise the reader to read the beautiful account by Thiemann in [211] and especially in [11] for an explanation of the internal reasons of this finiteness. Thiemann illustrates how the ultraviolet divergences of ordinary quantum field theory can be directly interpreted as a consequence of the approximation that disregards the quantized, discrete, nature of quantum geometry. For instance, Thiemann shows how the operator $\frac{1}{2\beta^2 e^3} \operatorname{tr}[E_a E_b] \operatorname{Tr}[\mathcal{E}^a \mathcal{E}^b]$, the kinetic term of the Yang–Mills hamiltonian, is well defined so long as we treat E as an operator, but becomes infinite as soon as we replace E with a smooth background field.

7.4 Loop quantum gravity

With the definition of the operator H for gravity and matter, finite on $\mathcal{K}_{\text{diff}}$, the formal definition of the quantum theory of gravity is completed. The theory is finite, provides a compelling intuitive description of the Planck-scale structure of space, has definite predictions such as the eigenvalues of area and volume, and reduces to classical GR in the naive $\hbar \to 0$ limit. The transition amplitudes of the theory are defined by

$$W(s, s') = \langle s|P|s' \rangle, \tag{7.34}$$

where s and s' are two s-knot states and P is the projector on the space of the solutions of the equation $H\Psi = 0$. The quantity $W(s, s')$ is interpreted as the probability amplitude of observing the discretized geometry with matter determined by the s-knot s if the geometry with matter determined by s' was observed.

If we consider a region of spacetime bounded by two disconnected surfaces, the diff-invariant boundary space is $\mathsf{K}_{\text{diff}} = \mathcal{K}^*_{\text{diff}} \otimes \mathcal{K}_{\text{diff}}$, and we can rewrite (7.34) in terms of the covariant vacuum state

$$\langle 0|s_{\text{out}}, s_{\text{in}} \rangle = \langle s_{\text{out}}|P|s_{\text{in}} \rangle, \tag{7.35}$$

where $|s_{\text{out}}, s_{\text{in}} \rangle = \langle s_{\text{out}}| \otimes |s_{\text{in}} \rangle \in \mathsf{K}_{\text{diff}}$.

More generally, we can consider a finite region of spacetime bounded by a 3d surface Σ. If s represents the outcome of the measurement of the gravitational field and matter fields on Σ, then

$$W(s) = \langle 0|s \rangle \tag{7.36}$$

gives the correlation probability amplitude of the measurement of the state s. This can also be viewed as the transition amplitude from the empty set to the full s, hence

$$W(s) = \langle \emptyset | P | s \rangle, \tag{7.37}$$

or

$$|0\rangle = P|\emptyset\rangle, \tag{7.38}$$

in K. This is loop quantum gravity.

Much remains to be done. Here are some issues that I have not addressed.

(i) *Lorentzian theory.* So far, I have dealt only with the euclidean theory. As already mentioned in Section 4.2.2, the lorentzian theory can be expressed in terms of the same kinematics as the euclidean theory, only adding a second term to the hamiltonian. I shall not discuss the quantization of this second term here. This is done in detail in [20]. Alternatively, the quantization has to be defined using the complex connection, but the full quantum state space with a complete operator algebra has not yet been constructed for the complex connection, as far as I know. Another alternative is to derive the amplitudes of the lorentzian theory from the amplitudes of the euclidean theory, as one can do in flat-space QFT. As mentioned, a naive reproduction of the flat-space technique is not viable in quantum gravity, but a suitable extension of this might work.

(ii) *Transition amplitudes.* The matrix elements of the projector P are not easy to compute.

(iii) *Scattering.* A general technique to connect the transition amplitudes $W(s, s')$ to particle observables such as graviton–graviton scattering must be developed.

(iv) *Classical limit.* Can we prove explicitly that classical GR can be recovered from LQG?

(v) *Form of the dynamics.* Is the proposed form of the hamiltonian constraint correct or does it have to be corrected?

(vi) *Physical consequences.* What does the theory say about the standard physical problems where quantum gravity is expected to be relevant, such as black-hole thermodynamics and early cosmology?

(vii) *Observable predictions.* Are there any?

Much is known on several of these issues. Some of them are discussed below. In particular, Chapter 8 deals with (vi) and (vii); Chapter 9 with (i), (ii), (iii) and (v).

Nevertheless, I emphasize the fact that, whatever its consequences and its physical correctness, the theory developed thus far provides a finite and consistent general covariant and background-independent quantum field theory, for the gravitational field and the matter fields. In it, the core physical insights of GR and QFT merge beautifully. Finding such a theory was our major aim.

7.4.1 * Variants

The LQG theory that I have described above is a standard version of the theory. There are a number of possible variants that have been considered in the literature.

Different regularization of H. I have described this possibility above, in Section 7.1.

q-deformed spin networks. An intriguing possibility is to replace the group $SU(2)$ with the quantum group $SU(2)_q$ in the quantum theory, and choose q to be given by $q^N = 1$, where N is a large number. It is possible to define q-deformed spin networks, labeled by representations of $SU(2)_q$ and build the rest of the theory as above. This is an interesting possibility for several reasons. Several of the spinfoam models studied in Chapter 9 are defined using quantum groups, and their states are q-deformed spin networks. The spinfoam models show that the use of q-deformed spin networks is naturally connected with a cosmological constant λ. The quantum group $SU(2)_q$ has a finite number of irreducible representations, which grows with N. This implies that the quantum of area has a maximum value, determined by N, and related to the (large) length determined by the cosmological constant. Finally, N works like a natural infrared cutoff, which is likely to cure any eventuality of infrared divergences. q-deformed LQG has been studied in the literature, but a systematic construction of LQG in terms of q-deformed spin networks is still missing.

Different ordering of the area operator. We can quantize the harmonic oscillator choosing an ordering of the hamiltonian such that the vacuum energy is zero instead of $\frac{1}{2}\hbar\omega$. Similarly, we can choose different orderings for the area operator and obtain a different spectrum. The ordering used in this book, and in most of the literature, is the natural one for the Casimir operator, but alternatives have been considered, and produce some intriguing effects. In particular, it is possible to order the operator to obtain an equally spaced spectrum. This would reintroduce the Bekenstein–Mukhanov effect studied below in Section 8.2.4, and, apparently, would automatically give a dominance of spin-1 quanta for a black hole, bringing the value of the Immirzi parameter to match the frequency of the black-hole ringing modes (see Section 8.2.3).

Different regularization of the area operator. The full spectrum of the area operator, given in (6.125), contains the main sequence (6.75). If we think that a diffeomorphism-invariant notion of a surface is truly the boundary of a region, and the region is an ensemble of quanta of volume, then we are led to the idea that physical surfaces are described by the mathematical surfaces that cut the links without touching the nodes. These surfaces have area given by the main sequence (6.75). Thus the degenerate sector might be physically spurious. To have the eigenvalues in the degenerate sector, we need a surface that cuts precisely through the node, and this is against the intuition that the location of the surfaces is only defined up to Planck scale. A different regularization of the area operator might get rid of the degenerate sector.

Unknotted spin networks. A very interesting possibility is to modify the definition of the spin network states of the theory, dropping the information on the knotting and linking of the graphs. That is, to define the graphs Γ that form the spin networks solely in terms of the adjacency relations between nodes, as is usually done in graph theory, and not, as is done above, as equivalence classes of embedded graphs under extended diffeomorphisms. The physical differences implied by the two definitions are not clear at present.

Different regularization of the volume. Two definitions of the volume have been given in the literature. The two turn out to be slightly different. Originally, the two were given in different mathematical languages, and it was thought that the difference had to do with the different formulations of the theory. Later, it became clear that both operators can be defined in either formulation. The volume operator defined here does not distinguish a node in which some tangents of the adjacent links are coplanar, from a node in which they are not coplanar. The other version of the volume operator, used for instance in [20], makes this distinction: its action on a node with coplanar links differs from the one given here. This second operator is covariant under *Diff*, but not under *Diff**.

Extended loop representation. Gambini and Pullin have developed a version of LQG in which loop states are not normalizable. Normalizable states are obtained smearing loop states. The main motivation is the fact that in flat-space QFT this is the case. I refer the reader to their book [7] for a discussion and details.

Lorentz spin networks. In the hamiltonian theory on which LQG is based, there is a partial gauge-fixing. One of the consequences of this gauge-fixing is that the connection with the covariant formalism used in the spinfoam models becomes technically more cumbersome. To avoid this difficulty, Sergei Alexandrov has studied the possibility of defining LQG without making this gauge-fixing, and keeping the full Lorentz group in the hamiltonian formalism [127, 212]. This might give a different regularization of the area operator as well [213].

––––

Bibliographical notes

The Wheeler–DeWitt equation appeared in [214]. The first version of the hamiltonian operator of LQG and its first solutions were constructed in [177]. Various other solutions were found, see for instance [215]. A review of early solutions of the hamiltonian operator is in [216]. The result that diffeomorphism-invariant states make the operator finite appeared in [217]. The general structure of the hamiltonian constraint and the operator D are illustrated in [218].

The idea of expressing the hamiltonian as a commutator, and therefore the first fully well-defined version of the hamiltonian operator was obtained by Thomas Thiemann in [133] and systematically developed by Thiemann in the remarkable "QSD" series of papers [201, 209]. Matrix elements of this operator were systematically studied in [208]. On the hamiltonian operator with positive cosmological constant, and the possibility of defining the theory in this case, see [220]. Thiemann and collaborators are developing an original and promessing approach to the definition of the quantum dynamics, called the "Master Program". The idea is to condensate a full set of constraints into a single one. For an introduction and references, see [221].

Fermions were introduced in LQG in [222] and in [223]. The key step for the present formulation of the fermion–LQG coupling was taken by Thomas Thiemann using half-density spinor fields [224]. A complete study of the matter hamiltonian in LQG is due to Thiemann. See his [20] and complete references therein. The fermionic contribution to the spectrum of the area operator was considered in [225]. The intriguing possibility that fermions are described by the linking of the spin networks has been recently explored in [226].

On the area operator with equispaced eigenvalues and its effect on the black-hole entropy, see [227]. A q-deformed version LQG was considered in [228]; see [220] and references therein. For q-deformed spin networks, see also [229], [230] and [231].

8
Applications

In this chapter I briefly mention some of the most successful applications of LQG to concrete physical problems. I have no ambition of completeness, and I will not present any detailed derivation. For these, I refer to original papers and review articles. I only illustrate the main ideas and the main results.

The two traditional applications of quantum gravity are early cosmology and black-hole physics. In both these fields, LQG has obtained interesting results. In addition, a certain number of tentative calculations concerning other domains where Planck-scale physical effects could perhaps be observable have also been performed.

8.1 Loop quantum cosmology

A remarkable application of LQG is to early cosmology. A direct treatment of semiclassical states in $\mathcal{K}_{\mathrm{diff}}$ representing cosmological solutions of the Einstein equations is not yet available. However, it is possible to impose homogeneity and isotropy on the basis states and operators of the theory and, in this way, restrict the theory to a finite-dimensional system describing a quantum version of the cosmological dynamics, that can be studied in detail.

The result is different from the traditional Wheeler–DeWitt minisuperspace quantization of the dynamics of Friedmann models. The key to the difference is the fact that the system inherits certain physical aspects of the full theory. In particular, the quantization of the geometry. These have a major effect on the dynamics of the early Universe. The main results are the following.

(i) *Absence of singularities.* Dynamics is well defined at the Big Bang, with no singular behavior. In particular, the inverse scale factor is bounded. In this sense the Universe has a minimal size.

(ii) *Semiclassical behavior.* Cosmological evolution approximates the standard Friedmann dynamics for large values of the scale factor $a(t)$, but differs from it at small values of $a(t)$.

(iii) *Quantization of the scale factor.* The scale factor – and the volume of the Universe – are quantized.

(iv) *Discrete cosmological evolution.* We can view the scale factor as a cosmological time parameter. Then we can say that cosmological time is quantized. Accordingly, the Wheeler–DeWitt equation is a difference equation, and not a differential equation in a.

(v) *Inflation.* Just after the Big Bang, the Universe underwent an inflationary phase $d^2a(t)/dt^2 > 0$. This is driven not by a scalar inflaton field, but by quantum properties of the gravitational field itself.

These are all remarkable results, but of different kinds. Results (i) and (ii) are what one would expect from a quantum theory of gravity giving a consistent description of the early Universe. Results (iii) and (iv) reflect the most characteristic aspect of LQG: the quantization of the geometry. Result (v) came as a big surprise. Let me briefly illustrate how these results are derived.

Consider a homogeneous and isotropic Universe. Its gravitational field is given by the well-known line element

$$ds^2 = -dt^2 + a^2(t)\left(\frac{dr^2}{1 - kr^2} + r^2(d\theta^2 + \sin\theta\ d\phi^2)\right), \qquad (8.1)$$

where $a(t)$ is the scale factor and k is equal to zero or ± 1 (see for instance [75]). The Einstein equations reduce to the Friedmann equation

$$\left(\frac{\dot{a}}{a}\right)^2 = \frac{8\pi G}{3}\rho - \frac{k}{a^2}, \qquad (8.2)$$

where ρ is the time-dependent matter energy density. Let us represent the matter content of the Universe in terms of a single field, that for simplicity can be taken as a scalar field ϕ. Homogeneity then demands that ϕ is a function of the sole time coordinate. The system is therefore described by $a(t)$ and $\phi(t)$. Assume for simplicity that $\phi(t)$ has a simple quadratic self-interaction (potential) term, namely its hamiltonian is

$$H_\phi = \frac{1}{2}(p_\phi^2 + \omega^2\phi^2), \qquad (8.3)$$

where p_ϕ is the momentum conjugate to ϕ. Therefore,

$$\phi(t) = A\sin(\omega t + \phi_0). \qquad (8.4)$$

Do not confuse the field ϕ with the inflaton: it has no inflationary potential. The energy density is related to the *conserved* matter energy H_ϕ by

$$\rho = a^{-3}H_\phi = a^{-3}\rho_0 = \frac{1}{2}a^{-3}\omega^2A^2. \qquad (8.5)$$

The constant ρ_0 is the density at $a = 1$. The Friedmann equation (8.2) can be derived from the hamiltonian

$$H = -\left(\frac{p_a^2}{8a} + 2ka\right) + 16\pi G H_\phi \qquad (8.6)$$

by simply computing $\dot{a} = \mathrm{d}H/\mathrm{d}p_a$ and using $H = 0$. In the simplest, spatially flat case $k = 0$, the Friedmann equation (8.2) reduces to

$$\left(\frac{\dot{a}}{a}\right)^2 = \frac{8\pi G}{3}\frac{\rho_0}{a^3}; \qquad (8.7)$$

by taking a derivative we obtain

$$\ddot{a} = -\frac{4}{3}\pi G\rho_0\frac{1}{a^2}, \qquad (8.8)$$

which is precisely equation (2.112) that we had obtained in the context of newtonian cosmology. The equation is solved by the well-known Friedmann evolution

$$a(t) = a_0(t - t_0)^{2/3}. \qquad (8.9)$$

Interpretation. These equations are written in a particular gauge-choice for the variable t, but the full theory is invariant under reparametrization in t. The relativistic configuration space is coordinatized by a and ϕ. The physical content of the theory is not in the dependence of these two quantities on t, but in their dependence on each other. The proper meaning of (8.4)–(8.9) concerns the relation between ϕ and a. For instance, we can interpret ϕ as a clock. That is, we can *define* its oscillations as isochronous. This defines a physical time variable. Then the scale factor grows in this time variable as described by (8.9). Alternatively, we can use the scale factor as a measure of time. In this cosmological time, all material physical processes slow down as in

$$\phi(a) = A\sin(\tilde{\omega}\,a^{3/2} + \tilde{\phi}_0). \qquad (8.10)$$

In other words, Friedmann evolution is the relative evolution of the rate of change of the material processes and the rate of change of the scale factor; it is the evolution of the ratio between the two rates of change. A solution of the Friedmann equation describes, therefore, the values that the scale factor can take for a given value ϕ of the matter variable, or, equivalently, the values $\phi(a)$ that the matter variable can take at a given value a of the scale factor (equation (8.10)).

Notice that there is no need to think in terms of "evolution in t" for these relations to make sense. As discussed in Section 3.4, time evolution, namely the idea of a physical "flow" of time with respect to which

a increases and ϕ oscillates, may simply derive from the physics of thermodynamical processes that happen in the presence of many (gravity and matter) variables. Therefore the question *"What happened before the Big Bang?"* might be as empty as the question *"What is there on the Earth's surface one meter north of the North pole?"*

Traditional quantum cosmology. In the traditional approach to quantum cosmology, one introduces a wave function $\psi(a, \phi)$. This is governed by the Wheeler–DeWitt equation obtained from (8.6). Up to factor ordering, this can be written as

$$\left(\frac{\hbar^2}{8a} \frac{\partial}{\partial a} \frac{\partial}{\partial a} + 2ka \right) \psi(a, \phi) = 16\pi G \, H_\phi^0 \, \psi(a, \phi), \qquad (8.11)$$

where H_ϕ^0 is the hamiltonian of an harmonic oscillator with angular frequency ω. This equation has semiclassical solutions which are wave packets that approximate the Friedmann evolution for large a. For small a, on the other hand, the singular behavior of the classical theory persists.

Loop quantum cosmology. What changes if we use LQG? The essential novelty is the quantization of the geometry. Recall that up to a constant

$$a \sim \sqrt[3]{V}, \qquad (8.12)$$

where V is the volume of the compact universe. But the volume has a discrete spectrum in the theory. Therefore, we should expect a to have a discrete spectrum. In fact, the detailed construction carried out in [232] shows that this is precisely the case. The observable a has a discrete spectrum, with eigenstates $|n\rangle$ labeled by an integer n and eigenvalues

$$a_n = a_1 \sqrt{n}, \qquad (8.13)$$

where the constant a_1 is

$$a_1 = \sqrt{\frac{4}{3} \gamma \pi \hbar G}. \qquad (8.14)$$

Therefore, in LQG the size of the universe is quantized. If a is quantized, we cannot represent states as functions $\psi(a, \phi)$ (for the same reason that we do not represent states of the harmonic oscillators as continuous functions of the energy). Rather, we can represent states in the form

$$\psi_n(\phi) = \langle n, \phi | \psi \rangle, \qquad (8.15)$$

where $|n, \phi\rangle$ is an eigenstate of a and ϕ. Accordingly, the partial derivatives with respect to a in the Wheeler–DeWitt equation (8.11) are replaced in loop quantum cosmology with finite-difference operators. In fact, the Wheeler–DeWitt equation is explicitly derived in [232]. It has the form

$$\alpha_n \psi_{n+4}(\phi) - 2\beta_n \psi_n(\phi) + \gamma_n \psi_{n+4}(\phi) = 16\pi G \, a^{-3} H_\phi \psi_n(\phi), \qquad (8.16)$$

where the constants $\alpha_n, \beta_n, \gamma_n$ are given in [232]. Notice the volume density factor a^{-3} on the right-hand side. It appears because the quantum constraint must be obtained in LQG from the *densitized* hamiltonian $a^{-3}H$.

The key point is now the meaning of the operator a^{-3} in the right-hand side of this equation. Recall that in the definition of the hamiltonian operator of the full theory it was essential to use a proper definition of the inverse volume element $1/\det E$ in order to define a well-behaved operator. This was obtained by expressing it via a Poisson bracket in the classical theory and via a commutator in the quantum theory. This procedure circumvents technical difficulties associated with the definition of the inverse of the volume element operator. The inverse scale factor a^{-3} in (8.16) is what remains of that term in the cosmological theory. But if the cosmological theory has to approximate the full theory, we have to better define this operator in the same way the inverse volume element was defined in the full theory. In fact, it is not hard to do so: writing

$$d = a^{-3} \qquad (8.17)$$

as a commutator of well-defined quantum operators. The resulting operator d is well defined. Its spectrum is however more complicated than the simple inverse of the spectrum of a^3:

$$d_n = \left(\frac{12}{j(j+1)(2j+1)} \sum_{k=-j,j} k\sqrt{V_n} \right)^6. \qquad (8.18)$$

In the definition of the operator d there is a quantization ambiguity. This is because d is defined using a holonomy, and this can be taken in any representation j. This is precisely the ambiguity in the definition of the hamiltonian operator that was discussed above in Section 7.1.3. Remarkably, its spectrum turns out to be bounded. In fact, for large n we have

$$d_n \sim a_n^{-3}; \qquad (8.19)$$

but for small n

$$d_n \sim a_n^{12}. \qquad (8.20)$$

There is a maximum value of d_n, whose value and location are determined by the free quantization parameter j. Using this operator in the Wheeler–DeWitt equation (8.16) yields a perfectly well-behaved evolution on and around $n = 0$. A numerical study of this equation shows easily that it gives the standard semiclassical behavior, and therefore standard Friedmann evolution for large n.

8.1.1 Inflation

The most surprising and intriguing aspect of LQC is the fact that it predicts an inflationary phase in the expansion of the early Universe. This can be seen by explicit numerical solutions of the Wheeler–DeWitt equation (8.16), or, more simply, as follows. For small n, the behavior of the operator d is governed by (8.20) instead of (8.19). The corresponding cosmological evolution can therefore be effectively approximated by a modification of the Friedmann equation in which H_ϕ is proportional to a^{12} instead of a^{-3}. This yields an accelerated initial expansion of the form

$$a(t) \sim (t_0 - t)^{-2/9}. \tag{8.21}$$

A numerical solution of the Wheeler–DeWitt equation confirms this result. The initial acceleration subsequently decreases smoothly and converges to a standard decelerating Friedmann solution. The duration of this inflationary expansion is governed by j.

What goes on physically can be understood as follows. The kinetic term of the matter hamiltonian contains effectively a coupling with gravity. When the gravitational field is strong, near the initial singularity, the matter field feels the quantum structure of the gravitational field, which affects its dynamics.

This scenario deserves to be explored in more detail and better understood.

8.2 Black-hole thermodynamics

The first hint that a black hole can have thermal properties came from classical GR. In 1972, Hawking proved a theorem stating that the Einstein equations imply that the area of the event horizon of a black hole cannot decrease. Shortly after, Bardeen, Carter and Hawking showed that in GR black holes obey a set of laws that strongly resembles the principles of thermodynamics; impressed by this analogy, Bekenstein suggested that we should associate an entropy

$$S_{\mathrm{BH}} = a \, \frac{k_{\mathrm{B}}}{\hbar G} \, A \tag{8.22}$$

to a Schwarzschild black hole of surface area A. (In this chapter, where the connection does not appear, the area is denoted A, not $\mathbf{A}$.) Here a is a constant of the order of unity, k_B the Boltzmann constant, and the speed of light is taken to be 1. The reason for the appearance of $\hbar$ in this formula is essentially to get dimensions right. Bekenstein's suggestion was that the second law of thermodynamics should be extended in the presence of black holes: the total entropy that does not decrease in time is the sum of the ordinary entropy with the black-hole entropy S_{BH}. Bekenstein presented several physical arguments supporting this idea, but the reaction of the physics community was very cold, mainly for the following reason. The area A of a Schwarzschild black hole is related to its energy M by

$$M = \sqrt{\frac{A}{16\pi G^2}}. \tag{8.23}$$

If (8.22) was correct, the standard thermodynamical relation $T^{-1} = dS/dE$ would imply the existence of a black-hole temperature

$$T = \frac{\hbar}{a32\pi k_B GM}, \tag{8.24}$$

and therefore a black hole would emit thermal radiation at this temperature: a consequence difficult to believe. However, shortly after Bekenstein's suggestion, Hawking derived precisely such a black-hole thermal emission, from a completely different perspective. Using conventional methods of quantum field theory in curved spacetime, Hawking studied a quantum field in a gravitational background in which a black hole forms (say a star collapses), and found that if the quantum field is initially in the vacuum state, after the star collapse we find it in a state that has properties of a thermal state. This can be interpreted by saying that the black hole emits thermal radiation. Hawking computed the emission temperature to be

$$T = \frac{\hbar}{8\pi k_B GM}, \tag{8.25}$$

which beautifully supports Bekenstein's speculation, and fixes the constant a at

$$a = \frac{1}{4}, \tag{8.26}$$

so that (8.22) becomes

$$S_{BH} = \frac{k_B A}{4\hbar G}. \tag{8.27}$$

Since then, the subscript BH in S_{BH} does not mean "black hole": it means "Bekenstein–Hawking." Hawking's theoretical discovery of black-hole emission has since been rederived in a number of different ways, and is today generally accepted as very credible.[1]

---

[1] Although perhaps some doubts remain about its interpretation. One can write a *pure* quantum state in which the energy distribution of the quanta is planckian. Is the

Hawking's beautiful result raises a number of questions. First, in Hawking's derivation the quantum properties of gravity are neglected. Are these going to affect the result? Second, we understand macroscopical entropy in statistical mechanical terms as an effect of the microscopical degrees of freedom. What are the microscopical degrees of freedom responsible for the entropy (8.22)? Can we derive (8.22) from first principles? Because of the appearance of $\hbar$ in (8.22), it is clear that the answer to these questions requires a quantum theory of gravity. The capability of answering these questions has since become a standard benchmark against which a quantum theory of gravity can be tested.

A detailed description of black-hole thermodynamics has been developed using LQG, and research is active in this direction. The major result is the derivation of (8.22) from first principles, for Schwarzschild and for other black holes, with a well-defined calculation where no infinities appear. As far as I know, LQG is the only detailed quantum theory of gravity where this result can be achieved.[2]

As I illustrate below, the result of LQG calculations gives (8.22) with

$$a \approx \frac{0.2375}{4\gamma}, \tag{8.28}$$

where γ is the Immirzi parameter. This agrees with Hawking's value (8.26) provided that the Immirzi parameter has the value

$$\gamma \approx 0.2375. \tag{8.29}$$

In fact, this is the way the value of γ is fixed in the theory nowadays. The calculation can be performed for different kinds of black holes, and the same value of γ is found, assuring consistency. An independent way of determining γ would make this result much stronger.

In what follows, I present the main ideas that underlie the derivation of this result.

8.2.1 The statistical ensemble

The degrees of freedom responsible for the entropy. Consider a black hole with no charge and no angular momentum. Its entropy (8.22) can originate

---

state of the quantum field after the collapse truly a thermal state, or a pure state that has the energy distribution of a thermal state? Namely, are the relative phases of the different energy components truly random, or are they fixed deterministically by the initial state? Do the components of the planckian distribution form a thermal or a quantum superposition? In the second case, the transition to a mixed state is just the normal result of the difficulty of measuring hidden correlations.

[2] So far, string theory can only deal with the highly unphysical extreme or nearly extreme black holes.

from horizon microstates, corresponding to a macrostate described by the Schwarzschild metric. Intuitively, we can think of this as an effect of fluctuations of the shape of the horizon.

One can raise an immediate objection to this idea: a black hole has "no hair," namely a black hole with no charge and no angular momentum is necessarily a spherically symmetric Schwarzschild black hole, leaving no free degrees of freedom to fluctuate.

This objection, however, is not correct. It is the consequence of a common confusion about the meaning of the term "black hole." The confusion derives from the fact that the expression "black hole" is used with two different meanings in the literature. In its first meaning a "black hole" is a region of spacetime hidden beyond an horizon, such as a collapsed star. In its second meaning "black hole" is used as a synonym of "*stationary* black hole." When one says that "a black hole is uniquely characterized by mass, angular momentum and charge," one refers to *stationary* black holes, not to arbitrary black holes. In particular, a black hole with no charge and no angular momentum is *not* necessarily a Schwarzschild black hole and is *not* necessarily spherical. Its rich dynamics is illustrated, for instance, by the beautiful images of the rapidly varying shapes of the horizon obtained in numerical calculations of, say, the merging of two holes. Generally a black hole has a large number of degrees of freedom and its event horizon can take arbitrary shapes. These degrees of freedom of the horizon can be the origin of the entropy.

To be sure, in the classical theory a realistic black hole with vanishing charge and vanishing angular momentum evolves very rapidly towards the Schwarzschild solution, by rapidly radiating away all excess energy. Its oscillations are strongly damped by the emission of gravitational radiation. But we cannot infer from this fact that the same is true in the quantum theory, or in a thermal context. In the quantum theory, the Heisenberg principle prevents the hole from converging exactly to a Schwarzschild metric, and fluctuations may remain. In fact, we will see that this is the case.

Recall that in the context of statistical mechanics, we must distinguish between the macroscopic state of a system and its microstates. Obviously the symmetry of the macrostate does not imply that the relevant microstates are symmetric. For instance, in the statistical mechanics of a sphere of gas, the individual motions of the gas molecules are certainly not confined by spherical symmetry. When the macrostate is spherically symmetric and stationary, the microstates are not necessarily spherically symmetric or stationary.

When we study the thermodynamical behavior of a Schwarzschild black hole, it is therefore important to remember that the Schwarzschild solution is just the macrostate. Microstates can be nonstationary and

non-spherically symmetric. Indeed, trying to explain black-hole thermodynamics from properties of stationary or spherically symmetric metrics
alone is a nonsense such as trying to derive the thermodynamics of an
ideal gas in a spherical box just from spherically symmetric motions of
the molecules.

Thermal fluctuations of the geometry. To make the case concrete, consider a realistic physical system containing a nonrotating and noncharged
black hole as well as other physical components such as dust, gas or radiation, which I denote collectively as "matter." We are interested in the statistical thermodynamics of such a system. Because of Einstein's equations,
at finite temperature the microscopic time-dependent inhomogeneities of
the matter distribution due to its thermal motion must generate time-
dependent microscopic thermal inhomogeneities in the gravitational field
as well. One usually safely disregards these ripples of the geometry. For
instance, we say that the geometry over the Earth's surface is given by the
Minkowski metric (or the Schwarzschild metric, due to the Earth's gravitational field), disregarding the inhomogeneous time-dependent gravitational field generated by each individual fast-moving air molecule. The
Minkowski geometry is therefore a "macroscopic" coarse-grained average of the microscopic gravitational field surrounding us. These thermal
fluctuations of the gravitational field are small and can be disregarded
for most purposes, but not when we are interested in the statistical–
thermodynamical properties of gravity: these fluctuations are precisely
the sources of the thermal behavior of the gravitational field, as is the
case for any other thermal behavior.

In a thermal context, the Schwarzschild metric represents therefore only
the coarse-grained description of a microscopically fluctuating geometry.
Microscopically, the gravitational field is nonstationary (because it interacts with nonstationary matter) and nonspherically symmetric (because
matter distribution is spherically symmetric on average only, and not
on individual microstates). Its microstate, therefore, is *not* given by the
Schwarzschild metric, but by some complicated time-dependent nonsymmetric metric.

Horizon fluctuations. Let us make the considerations above slightly more
precise. Consider first the classical description of a system at finite temperature in which there is matter, the gravitational field and a black hole.
Foliate spacetime into a family of spacelike surfaces Σ_t, labeled with a
time coordinate t. The intersection h_t between the spacelike surface Σ_t
and the event horizon (the boundary of the past of future null-infinity)
defines the instantaneous microscopic configuration of the event horizon
at coordinate time t. I loosely call h_t the surface of the hole, or the horizon. Thus, h_t is a closed 2d surface in Σ_t. As argued above, generally this

microscopic configuration of the event horizon is not spherically symmetric. Denote by g_t the intrinsic and extrinsic geometry of the horizon h_t. Let $\mathcal{M}$ be the space of all possible (intrinsic and extrinsic) geometries of a 2d surface. As t changes, the (microscopic) geometry of the horizon changes. Thus, g_t wanders in $\mathcal{M}$ as t changes.

Since the Einstein evolution drives the black hole towards the Schwarzschild solution, (we can choose the foliation in such a way that) g_t will converge towards a point g_A of $\mathcal{M}$ representing a sphere of a given radius A. However, as mentioned before, exact convergence may be forbidden by quantum theory, and quantum effects may keep g_t oscillating in a finite region around g_A.

Which microstates are responsible for S_{BH}? Let us assume that (8.22) represents a true thermodynamical entropy associated with the black hole. That is, let us assume that heat exchanges between the hole and the exterior are governed by S_{BH}. Where are the microscopical degrees of freedom responsible for this entropy located? The microstates that are relevant for the entropy are only the ones that can affect energy exchanges with the exterior. That is, only the ones that can be distinguished from the exterior. If I have a system containing a perfectly isolated box, the internal states of the box do not contribute to the entropy of the system, as far as the heat exchange of the system with the exterior is concerned. The state of matter and gravity inside a black hole has no effect on the exterior. Therefore the states of the interior of the black hole are irrelevant for S_{BH}.

To put it vividly, the black-hole interior may be in one out of an infinite number of states indistinguishable from the outside. For instance, the black-hole interior may, in principle, be given by an infinite Kruskal spacetime: on the other side of the hole there may be billions of galaxies that do not affect the side detectable by us. The potentially infinite number of internal states does not affect the interaction of the hole with its surroundings and is irrelevant here, because it cannot affect the energetic exchanges between the hole and its exterior which are the ones that determine the entropy. We are only interested in configurations of the hole that have distinct effects on the exterior of the hole.

Observed from outside, the hole is completely determined by the geometric properties of its surface. Therefore, the entropy (relevant for the thermodynamical description of the thermal interaction of the hole with its surroundings) is entirely determined by the geometry of the black-hole surface, namely by g_t.

The statistical ensemble. We have to determine the ensemble of the microstates g_t over which the hole may fluctuate. In conventional statistical thermodynamics, the statistical ensemble is the region of phase space over which the system could wander if it were isolated, namely if it did not exchange energy with its surroundings. Can we translate this condition to the case of a black hole? The answer is yes, because we know that in GR energy exchanges of the black hole are accompanied by a change in its area. Therefore, we must define the statistical ensemble as the ensemble of g_t with a given value A of the area.

To support the choice of this ensemble consider the following.[3] The ensemble must contain reversible paths only. In the classical theory, reversible paths conserve the area, because of the Hawking theorem. Quantum theory does not change this, because it allows area decrease only by emitting energy (Hawking radiance), namely violating the (counterfactual) assumption that defines the statistical ensemble: that the system does not exchange energy.

We can conclude that the entropy of a black hole is given by the number $N(A)$ of states of the geometry g_t of a 2d surface h_t of area A. The quantity $S(A) = k_B \ln N(A)$ is the entropy we should associate with the horizon in order to describe its thermal interactions with its surroundings.

Quantum theory. This number $N(A)$ is obviously infinite in the classical theory. But not in the quantum theory. The situation is similar to the case of the entropy of the electromagnetic field in a cavity, which is infinite classically and finite in quantum theory. To compute it, we have to count the number of (orthogonal) quantum states of the geometry of a two-dimensional surface, with total area A. The problem is now well defined, and can be translated into a direct computation.

Two objections. I have concluded that the entropy of a black hole is determined by the number of the possible states of a 2d surface with area A. The reader may wonder if something has got lost in the argument: does this imply that *any* surface has an associated entropy, just because it has an area? Where has the information about the fact that this is a black hole gone? And, where has the information about the Einstein equations gone? These objections have often been raised to the argument above. Here is the answer.

The first objection can be answered as follows. Given any arbitrary surface, we can of course ask the mathematical question of how many states exist that have a given area. But there is no reason, generally, to say that there is an entropy associated with the surface. In a general situation, energy, or, more generally, information, can flow across a surface.

---

[3]In this context, it is perhaps worthwhile recalling that difficulties to rigorously justifying a priori the choice of the ensemble plague conventional thermodynamics anyway.

The surface may emit heat without changing its geometry. Therefore, in general, the geometry of the surface and the number of its states have nothing to do with heat exchange or with entropy. But, in the special case of a black hole the horizon screens us from the interior and any heat exchange that we can have with the hole must be entirely determined by the geometry of the surface. It is only in this case that the counting is meaningful, because it is only in this case that the number of states of a geometry of a given area corresponds precisely to the number of states of a region which are distinguishable from the exterior. To put it more precisely, the future evolution of the surface of a black hole is completely determined by its geometry and by the exterior; this is not true for an arbitrary surface. It is because of these special properties of the horizon that the number of states of its geometry determine an entropy.

You can find out how much money you own by summing up the numbers written on your bank account. This does not imply that if you sum up the numbers written on an arbitrary piece of paper you get the amount of money you own. The calculation may be the same, but an arbitrary piece of paper is not a bank account, and only for a bank account does the result of the calculation have that meaning. Similarly, you can make the same calculation for any surface, but only for a black hole, because of its special properties, is the result of the calculation an entropy.

The second objection concerns the role of the Einstein equations that is, the role of the dynamics. This objection has been raised often, but I have never understood it. The role of the Einstein equations is precisely the usual role that the dynamical equations always play in statistical mechanics. Generally, the only role of the dynamics is that of defining the energy of the system, which is the quantity which is conserved if the system is isolated, and exchanged when heat is exchanged. The statistical ensemble is then determined by the value of the energy. In the case of a black hole, it is the Einstein equations that determine the fact that the *area* governs heat exchange with the exterior of the hole. If it wasn't for the specific dynamics of general relativity, the area would not increase for an energy inflow or decrease for energy loss. Thus, it is the Einstein equations that determine the statistical ensemble.

8.2.2 Derivation of the Bekenstein–Hawking entropy

Above we have found, on physical grounds, what the entropy of a black hole should be. It is given by

$$S_{BH} = k_B \ln N(A) \tag{8.30}$$

where $N(A)$ is the number of states that the geometry of a surface with area A can assume. It is now time to compute it.

Let the quantum state of the geometry of an equal-time spacelike 3d Σ_t be given by a state $|s\rangle$ determined by an s-knot s. The horizon is a 2d surface $\mathcal{S}$ immersed in Σ_t. Its geometry is determined by its intersections with the s-knot s.

Intersections can be of three types: (a) an edge crosses the surface; (b) a vertex lies on the surface; (c) a finite part of the s-knot lies on the surface. Here we are interested in the geometry as seen from the exterior of the surface, therefore the geometry we consider is, more properly, the limit of the geometry of a surface surrounding $\mathcal{S}$, as this approaches $\mathcal{S}$. This limit cannot detect intersections of the type (b) and (c), and we therefore disregard such intersections.

Let $i = 1, \ldots, n$ label the intersections of the s-knot with the horizon $\mathcal{S}$. Let $j_1, \ldots, j_n$ be the spins of the links intersecting the surface. The area of the horizon is

$$A = 8\pi\gamma\hbar G \sum_i \sqrt{j_i(j_i + 1)}. \tag{8.31}$$

The s-knot is cut into two parts by the horizon $\mathcal{S}$. Call s_{ext} the external part. The s-knot s_{ext} has n open ends, that end on the horizon. From the point of view of an external observer, a possible geometry of the surface is a possible way of "ending" the s-knot. A possible "end" of a link with spin j is simply a vector in the representation space $\mathcal{H}_j$. Therefore, a possible end of the external s-knot is a vector in $\otimes_i \mathcal{H}_{j_i}$. Thus, seen from the exterior, the degrees of freedom of the hole appear as a vector in this space. In the limit in which the area is large, any further constraint on these vectors becomes irrelevant. The possible states are obtained by considering all sets of j_i that give the area A and, for each set, the dimension of $\otimes_i \mathcal{H}_{j_i}$. Let us first assume that the number of possible states is dominated by the case $j_i = 1/2$. In this case, the area of a single link is

$$A_0 = 4\pi\gamma\hbar G \sqrt{3}. \tag{8.32}$$

This is the first value that was derived for the Immirzi parameter from back hole states counting. Later, Domagala and Lewandowski realized [233] that the assumption that the entropy is dominated by spin $1/2$ is wrong and found a higher value, that was evaluated by Meissner [234], giving (8.29).

Hence, there are

$$n = \frac{A}{A_0} = \frac{A}{4\pi\gamma\hbar G \sqrt{3}} \tag{8.33}$$

intersections, and the dimension of $\mathcal{H}_{1/2}$ is 2; so the number of states of the black hole is

$$N = 2^n = 2^{A/4\pi\gamma\hbar G \sqrt{3}}, \tag{8.34}$$

and the entropy is

$$S_{\text{BH}} = k_{\text{B}} \ln N = \frac{1}{\gamma} \frac{\ln 2}{4\pi\sqrt{3}} \frac{k_{\text{B}}}{\hbar G} A. \tag{8.35}$$

This is the Bekenstein–Hawking entropy (8.22). The numerical factor agrees with the Hawking value (8.26), and we get (8.27) if the Immirzi parameter is fixed at the value

$$\gamma = \frac{\ln 2}{\pi\sqrt{3}}. \tag{8.36}$$

A far more detailed account of this derivation is given in [238], where the Hilbert space of the states of the black-hole surface is carefully constructed by quantizing a theory that has an isolated horizon as a *boundary*.

Following [238], however, Thiemann [20] derives an equation that looks like (8.36) but in fact differs from it by a factor of 2. Why this discrepancy? The reason is that Thiemann observes that if the horizon is a boundary, then, in the absence of the "other side of the horizon" in the formalism, we have $j_d = 0$ in (6.125). With $j_u^i = j^i$ and $j_d^i = 0$ we have, by definition of j_t, $j_t^i = j^i$, and therefore (taking $\gamma = 1$) the area of a link with spin j entering the horizon contributes a quantum of area

$$A = 4\pi G\hbar\sqrt{j(j+1)}, \tag{8.37}$$

which is one-*half* the contribution to the area of a link of spin j that cuts a surface in the bulk. What is happening is that the area operator, in a sense, counts the area of a surface by summing contributions of links entering the two sides; in the absence of one side, this does not work. Therefore, the area of the boundary is one-half the area of a bulk surface infinitesimally close to it. This is not very convincing, on physical grounds, of course. The authors of [238] correct this discrepancy by effectively doubling the area of the horizon. This gives the same final result as in this book. Thiemann has observed that the need of this correction "by hand" is an inconvenience of a formalism that treats the horizon as a boundary. One way out is to define the horizon area as the limit of the area of a bulk surface approaching the horizon.

Notice that the black hole turns out to carry one bit of information per quantum of area A_0. This is precisely the "it from bit" picture that John Wheeler suggested should be at the basis of black-hole physics in [165, 166].

One remark before concluding. The reader may object to the derivation above (and the one in [238]) as follows: the states that we have counted are transformed into each other by a gauge transformation. Why, then, do I consider them distinct in the entropy counting? The answer to this objection is the following. When we break the system into components, gauge degrees of freedom may become physical degrees of freedom on the boundary. The reason is that if we let the gauge group act independently on the two components, it will act *twice* on the boundary. A holonomy of a connection across the boundary, for instance, will become ill defined. Therefore, there are degrees of freedom on the boundary that are not gauge; they tie the two sides to each other, so to say.

To illustrate this point, let us consider two sets A and B, and a group G that acts (freely) on A and on B. Then G acts on $A \times B$. What is the space $(A \times B)/G$? One might be tempted to say that it is (isomorphic to) $A/G \times B/G$, but a moment of reflection shows that this is not correct and the correct answer is

$$\frac{A \times B}{G} \sim \frac{A}{G} \times B. \tag{8.38}$$

(If G does not act freely over A, we have to divide B by the stability groups of the elements of A.) Now, imagine that A is the space of the states of the exterior of the black hole, B the space of the states of the black hole, and G the gauge group of the theory. Then we see that we must not divide B by the gauge group of the surface, but only by those internal gauges and diffeomorphisms that leave the rest of the spin network invariant.[4]

8.2.3 Ringing modes frequencies

Is there a way of understanding the peculiar numerical value (8.36)? The minimal quantum of area that plays a central role in black-hole thermodynamics is the one with spin $j = 1/2$, which, using (8.36), is

$$A_{1/2} = 8\pi\hbar G\gamma\sqrt{\tfrac{1}{2}\left(\tfrac{1}{2}+1\right)} = 4\ln 2\,\hbar G. \tag{8.39}$$

Because of (8.23), a change of one such quantum of area implies a change of energy

$$\Delta E = \Delta M = \frac{A_{1/2}}{32\pi GM} = \frac{\ln 2\,\hbar}{8\pi M}. \tag{8.40}$$

If we use the Bohr relation $\Delta E = \hbar\omega$ to interpret this as a quantum emitted by an oscillator with angular frequency ω, then the quantum gravity theory indicates that in the system there should be something oscillating with a proper frequency

$$\omega = \frac{\ln 2}{8\pi M}. \tag{8.41}$$

As far as I know, this frequency plays no role in the classical theory. However, suppose that for some reason the minimal quantum of area that plays a central role in black-hole thermodynamics was due not to $j = 1/2$ spins, as above, but to $j = 1$ spins. Then the calculation above would be slightly different. The minimal area is

$$A_1 = 8\pi\hbar G\gamma\sqrt{1(1+1)}, \tag{8.42}$$

---

[4]Actually, in [239] only the boundary degrees of freedom due to diff invariance were taken into account, while in [238] and here only the boundary degrees of freedom due to internal gauge invariance are taken into account. Perhaps by taking both into account (8.36) could change.

and since the spin-1 representation has dimension 3, the entropy is

$$S = \ln(3^{A/A_1}) = \frac{\ln 3}{8\pi\hbar G\gamma\sqrt{2}}A. \tag{8.43}$$

This agrees with the Bekenstein–Hawking entropy if

$$\gamma = \frac{\ln 3}{2\pi\sqrt{2}}, \tag{8.44}$$

which, in turn, fixes the minimal relevant quantum of area to be

$$A_1 = 4\ln 3\hbar G. \tag{8.45}$$

Using (8.23) and the Bohr relation we obtain the proper frequency

$$\omega = \frac{\ln 3}{8\pi M}. \tag{8.46}$$

Now, very remarkably, *there is* something oscillating precisely with this frequency in a *classical* Schwarzschild black hole! In fact, the frequency (8.46) is precisely the frequency of the most damped ringing mode of a Schwarzschild black hole! The calculation of this frequency from the Einstein equations is complicated. It was first computed numerically, then guessed to be (8.46) on the basis of the numerical value, and only recently derived analytically. Quite remarkably, the quantum theory of gravity appears to know rather directly about this frequency, hidden inside the nonlinearity of the Einstein equations.

This fact seems to support the idea that the ringing modes of the black hole are at the roots of its thermodynamics. On the other hand, it is not clear why we should not consider the spin $j = 1/2$. Several possibilities have been suggested, including dynamical selection rules and an equispaced area spectrum. Overall, this intriguing observation raises more questions than providing fully satisfactory answers.

8.2.4 The Bekenstein–Mukhanov effect

In 1995, Bekenstein and Mukhanov suggested that the thermal nature of Hawking's radiation may be affected by quantum properties of gravity quite dramatically. They observed that in some approaches to quantum gravity the area can take only quantized values. Since the area of the black-hole surface is connected to the black-hole energy, the latter is likely to be quantized as well. The energy of the black hole decreases when radiation is emitted. Therefore, emission happens when the black hole makes a quantum leap from one quantized value of the energy to a lower quantized value, very much as atoms do. A consequence of this picture is that radiation is emitted at quantized frequencies, corresponding to the differences between energy levels. Thus, quantum gravity implies a discrete emission spectrum for the black-hole radiation.

This result is not physically in contradiction with Hawking's prediction of an effectively continuous thermal spectrum. To understand this, consider the black-body radiation of a gas in a cavity, at high temperature. This radiation has a thermal planckian emission spectrum, essentially continuous. However, radiation is emitted by elementary quantum emission processes yielding a discrete spectrum. The solution of the apparent contradiction is that the spectral lines are so dense in the range of frequencies of interest that they give rise – effectively – to a continuous spectrum.

However, Bekenstein and Mukhanov suggest that the case of a black hole may be quite different from the case of the radiation of a cavity. They consider a simple ansatz for the spectrum of the area: that the area is quantized in multiple integers of an elementary area A_0. Namely, that the area can take the values

$$A_n = nA_0, \tag{8.47}$$

where n is a positive integer, and A_0 is an elementary area of the order of the Planck area

$$A_0 = \alpha\hbar G, \tag{8.48}$$

where α is a number of the order of unity. The ansatz (8.47) agrees with the idea of a quantum picture of a geometry made up of elementary "quanta of area." Since the black-hole mass (energy) is related to the area by (8.23), it follows from this relation and the ansatz (8.47) that the energy spectrum of the black hole is given by

$$M_n = \sqrt{\frac{n\alpha\hbar}{16\pi G}}. \tag{8.49}$$

Consider an emission process in which the emitted energy is much smaller than the mass M of the black hole. From (8.49), the spacing between the energy levels is

$$\Delta M = \frac{\alpha\hbar}{32\pi GM}. \tag{8.50}$$

From the quantum mechanical relation $E = \hbar\omega$ we conclude that energy is emitted in frequencies that are integer multiples of the fundamental emission frequency

$$\bar{\omega} = \frac{\alpha}{32\pi GM}. \tag{8.51}$$

This is the fundamental emission frequency of Bekenstein and Mukhanov. Let us now assume that the emission amplitude is correctly given by

Hawking's thermal spectrum. Then the full emission spectrum is given by spectral lines at frequencies that are multiples of $\bar{\omega}$, whose envelope is Hawking's thermal spectrum. Now, this spectrum is drastically different than the Hawking spectrum. Indeed, the maximum of the planckian emission spectrum of Hawking's thermal radiation is around

$$\omega_{\mathrm{H}} \sim \frac{2.82 k_{\mathrm{B}} T_{\mathrm{H}}}{\hbar} = \frac{2.82}{8\pi GM} = \frac{2.82 \times 4}{\alpha} \bar{\omega} \approx \bar{\omega}. \tag{8.52}$$

The fundamental emission frequency $\bar{\omega}$ is of the same order of magnitude as the maximum of the Planck distribution of the emitted radiation! It follows that there are only a few spectral lines in the regions where emission is appreciable. The Bekenstein–Mukhanov spectrum and the Hawking spectrum have the same envelope, but while the Hawking spectrum is continuous, the Bekenstein–Mukhanov spectrum is formed by just a few lines in the interval of frequencies where emission is appreciable. This is the Bekenstein–Mukhanov effect.

Is this Bekenstein–Mukhanov effect truly realized in LQG? At first sight, one is tempted to say yes, since the spectrum of the area in LQG, given in (6.78), is quite similar to the ansatz (8.47). If we disregard the $+1$ under the square root in (6.78), we obtain the ansatz (8.47), and thus the Bekenstein–Mukhanov effect. But, the $+1$ is there and the difference turns out to be crucial.

Let us study the consequences of the presence of the $+1$. Consider a surface Σ; in the present case, the event horizon of the black hole. The area of Σ can take only a set of quantized values. These quantized values are labeled by unordered n-tuples of positive half-integers $\vec{j} = (j_1, \ldots, j_n)$ of arbitrary length n.

We estimate the number of area eigenvalues between the value $A \gg \hbar G$ and the value $A + \mathrm{d}A$ of the area, where we take $\mathrm{d}A$ much smaller than A but still much larger than $\hbar G$. Since the $+1$ in (6.78) affects in a considerable way only the terms with low spin j_i, we can neglect it for a rough estimate. It is more convenient to use integers rather than half-integers. Let us therefore define $p_i = 2j_i$. We must estimate the number of unordered strings of integers $\vec{p} = (p_1, \ldots, p_n)$ such that

$$\sum_{i=1,n} p_i = \frac{A}{8\pi\gamma\hbar G} \gg 1. \tag{8.53}$$

This is a well-known problem in number theory. It is called the partition problem. It is the problem of computing the number N of ways in which an integer I can be written as a sum of other integers. The solution for large I is a classic result by Hardy and Ramanujan [240]. According to the Hardy–Ramanujan formula, N grows as the exponent of the square root of I. More precisely, we have for large I that

$$N(I) \sim \frac{1}{4\sqrt{3I}} e^{\pi\sqrt{\frac{2}{3}I}}. \tag{8.54}$$

Applying this result in our case we have that the number of eigenvalues between A and $A + dA$ is

$$\rho(A) \approx e^{\sqrt{\frac{\pi A}{12\gamma\hbar G}}}. \tag{8.55}$$

Using (8.23) we have that the density of the states is

$$\rho(M) \approx e^{\sqrt{\frac{4G}{3\gamma\hbar}}\pi M}. \tag{8.56}$$

Now, without the $+1$ term there is a high degeneracy due to the fact that all states with the same value of $\sum_n p_n$ are degenerate. The presence of the $+1$ term kills this degeneracy and eigenvalues can overlap only accidentally: generically, all eigenvalues will be distinct. Therefore, the average spacing between eigenvalues will be the inverse of the density of states, and will decrease *exponentially* with the inverse of the square of the area. This result is to be contrasted with the fact that this spacing is constant and of the order of the Planck area in the case of the ansatz (8.47). It follows that for a macroscopic black hole the spacing between energy levels is infinitesimal, and the spectral lines are virtually dense in frequency. We effectively recover in this way Hawking's thermal spectrum.[5]

The conclusion is that the Bekenstein–Mukhanov effect disappears if we replace the naive ansatz (8.47) with the area spectrum (6.78) computed from LQG.

8.3 Observable effects

Possible low-energy effects of LQG have been studied using a semiclassical approximation. Gambini and Pullin have introduced the idea to study the propagation of matter fields over a weave state, taking expectation values of smeared geometrical operators. For suitable weave states, this may lead to the possibility of having quantum gravitational effects on the dispersion relations. In particular, they have studied light propagation and pointed out the possibility of an intriguing birefringence effect. Alfaro, Morales-Tecotl and Urrita have developed this technique. In particular, for a

---

[5]Mukhanov subsequently suggested that discretization could still occur, as a consequence of dynamics. For instance, transitions in which a single Planck unit of area is lost could be strongly favored by the dynamics.

fermion of mass m they derived dispersion relations between energy E and momentum p of the general form

$$E^2 = p^2 + m^2 + f(p, l_P),\qquad(8.57)$$

where the last, Lorentz-violating, term may be helicity dependent. The possibility that LQG could yield observable effects indeed has raised much interest. Suggestions that these effects may be connected to observable, or even already observed, effects have been put forward. These regard cosmic-ray energy thresholds, gamma-ray bursts, pulsar velocities, and others.

In fact, the old idea that quantum gravitational effects are certainly unobservable at present has been strongly questioned in recent years. Some have even expressed the hope that we could be "at the dawn of quantum gravity phenomenology" [241]. It is too soon to understand if these hopes will be realized, but the possibility is fascinating, and the development of LQG calculations that could relate to these possible observations is an important direction of development.

Lorentz invariance in LQG. Lorentz-violating effects might *not* be present in LQG. There are two reasons for expecting Lorentz violations in LQG. One is that the short-scale structure of a macroscopically Lorentz-invariant weave might break Lorentz invariance. However, it is not clear whether all weave states break Lorentz invariance or not. A single spin network state cannot be Lorentz invariant, but this does not imply that a state which is a quantum superposition of spin network states cannot either.

The second reason, which is often mentioned, is the observation that a minimal length (or a minimal area) *necessarily* breaks Lorentz invariance. The reason would be the following: if an observer measures the minimal length l_P, then a boosted observer will observe the Lorentz-contracted length $l' = \gamma^{-1} l_P$, which is shorter than l_P, and therefore l_P cannot be a minimal length. Here $\gamma = 1/\sqrt{1 - v^2/c^2}$ is the Lorentz–Fitzgerald contraction factor. This observation is wrong, because it ignores quantum mechanics.

Length, area and volume are not classical quantities. They are quantum observables. If an observer measures the length l_P of some system, this means that the system is in an eigenstate of the length operator. A boosted observer who measures the length of the same system is measuring a different observable L', which generally does not commute with L. If the system is in an eigenstate of L, generally it will not be in an eigenstate of L'. Therefore, there will be a distribution of probabilities of observing different eigenvalues of L'. The eigenvalues of L' will be the same as the

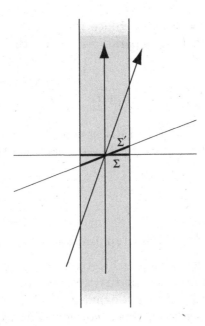

Fig. 8.1 The grey region represents the world-history of an object. The two arrows represent the worldlines of two observers. The bold segments Σ and Σ' are the intersections between the object world-history and the observers' equal-time surfaces. Two observers in relative motion measuring the lengths of the same object measure the gravitational field on these two *distinct* surfaces. The gravitational field on Σ does not commute with the gravitational field on Σ'. Hence the two lengths do not commute.

eigenvalues of L: it is the expectation value of L' that will be Lorentz contracted.

The situation is the same as for the L_z component of angular momentum. Consider a quantum system with total spin $= 1$. Say an observer measures L_z and obtains the eigenvalue $L_z = \hbar$. Does this mean that a second observer, rotated by an angle α, will observe the eigenvalue $L'_z = \cos\alpha \ \hbar$? Of course not. The second observer will still measure $L'_z = 0, \pm\hbar$, with a probability distribution such that the mean value is $L'_z = \cos\alpha \ \hbar$. States and mean values transform continuously with a rotation, but eigenvalues stay the same. In the same fashion, in a Lorentz boost states and values transform continuously, while the eigenvalues stay the same.

To understand why a length L and boosted length L' do not commute, consider Figure 8.1. It represents the world-history of an object, and the two lengths measured by two observers in relative motion. Notice that the two observers measure the gravitational field on two distinct segments Σ

and Σ' with a time separation. Since no quantum field operator commutes with itself at timelike separations, clearly the two functions of the gravitational field $e(x)$

$$L = \int_\Sigma \sqrt{|e|} \qquad \text{and} \qquad L' = \int_{\Sigma'} \sqrt{|e|} \qquad (8.58)$$

do not commute. For a detailed discussion of this point and the actual construction of boosted geometrical operators, see [242].

Bibliographical notes

Loop quantum cosmology has been mostly developed by Martin Bojowald. The absence of the initial singularity was derived in [243] and the suggestion about the possibility of a quantum-driven inflation presented in [244]. For a review, see [232]. Loop cosmology is rapidly developing; a recent introduction with up-to-date (2007) results and references is [235].

Another powerful recent application of loop quantum gravity is to resolve the $r = 0$ singularity at the center of a black hole. The idea has been proposed by Leonardo Modesto [236] and has been independently considered and developed by Ashtekar and Bojowald [237].

The Hawking theorem on the growth of the black-hole area is in [245]. Bardeen, Carter and Hawking, presented their "four laws of black-hole mechanics" in [246]. Bekenstein entropy was presented in [247], and Hawking black-hole radiance in [248]. For a review of the field, see [28].

The first suggestions on the possibility of using the counting of the quanta of area of LQG to describe black-hole thermodynamics were proposed by Kirill Krasnov [249]. For discussions on black-hole entropy in LQG, I have followed here [239]. The derivation with the horizon as a boundary is in [238]. The idea that black-hole entropy originates from the fluctuations of the shape of the horizon was suggested by York [250]. The relevance of the horizon surface degrees of freedom for the entropy has since been emphasized from different perspectives, see for instance [251]. On the notion of an isolated horizon, see [252]. The relevance of the Chern–Simon boundary theory for the description of the horizon surface degrees of freedom was noticed in [253], the importance of the gauge degrees of freedom on the boundary has been emphasized in [254]. A recent review including the new value of the Immirzi parameter, is [?].

The relation between the ringing-mode frequencies and the spectrum has been studied by S Hod in [255] and the fact that spin-1 area excitations in LQG are related to the ringing-mode frequency has been pointed out by Olaf Dreyer in [256]. On the reason why spin-1 could contribute to the horizon area more than spin-1/2, see [257] for a possible dynamical selection rule and [227] for the role of the ordering of the area operator.

The Bekenstein–Mukhanov effect was presented in [258] (with $\alpha = 4\ln 2$). For a review of earlier suggestions in this direction, see [259]. The argument presented here on the absence of this effect in loop gravity appeared in [260]. The same result was derived in [173].

The possibility of a semiclassical description of the propagation over a weave, that leads to estimates of quantum gravity effects, was introduced in [261] for photons, and developed in [262] for photons and neutrinos. For a review, see [232]. On the fact that violations of Lorentz invariance are not necessarily implied by LQG, see [242] and [263]. On the difficulties implied by the breaking of Lorentz invariance in quantum gravity, see also [264]. On the possibility that Planck-scale observation might be within observation reach, see [241, 265, 266].

9

Quantum spacetime: spinfoams

Classical mechanics admits two different kinds of formulations: hamiltonian and lagrangian (I never understood why). Feynman realized that so does quantum mechanics: it can be formulated canonically, with Hilbert spaces and operators, or covariantly as a sum-over-paths. The two formulations have different virtues, and calculations that are simple in one can be hard in the other. Generically, the lagrangian formalism is simpler, more transparent and intuitive and keeps symmetries and covariance manifest. The hamiltonian formalism is more general, more powerful, and, in the case of quantum theory, far more rigorous. The ideal situation, of course, is to be able to master a theory in both formalisms.

So far I have discussed the hamiltonian formulation of LQG. In this chapter I discuss the possibility of a lagrangian sum-over-paths formulation of the same theory. This formulation has been variously called sum-over-surfaces, state-sum and goes today under the name of "spinfoam formalism." The spinfoam formalism can be viewed as a mathematically well-defined and possibly divergence-free version of Stephen Hawking's formulation of quantum gravity as a sum-over-geometries.

The spinfoam formalism is less developed than the hamiltonian version of loop theory. Also, while the general structure of the spinfoam models matches nicely with the hamiltonian loop theory, the precise relation between the two formalisms (each of which exists in several versions) has been rigorously established only in 3d. But research is currently moving fast in this direction.

The aim of the spinfoam formalism is to provide an explicit tool to compute transition amplitudes in quantum gravity. These are expressed as a sum-over-paths. The "paths" summed over are "spinfoams." A spinfoam can be thought of as the worldsurface swept out by a spin network. Spinfoams are background-independent combinatorial objects, and do not need a spacetime to live in. A spinfoam itself represents a spacetime, in the same sense in which a spin network represents a space.

The most remarkable aspect of the spinfoam approach is the fact that a surprising number of independent research directions converge towards the same formalism. I illustrate some of these converging research paths below. This convergence seems to indicate that the spinfoam formalism is a sort of natural general language for sum-over-paths formulations of general-relativistic quantum field theories.

There are several good review articles on this subject – see the bibliographical notes at the end of the chapter. I do not repeat here what is done in detail in other reviews and I recommend the studious reader to refer to these reviews to complement the introduction provided here, since this is a subject which can be approached from a variety of points of view. Here I focus on the overall significance of these models for our quest for a complete theory of quantum gravity.

9.1 From loops to spinfoams

Sum-over-paths. Consider a nonrelativistic one-dimensional quantum system. Let x be its dynamical variable. The propagator $W(x, t, x', t')$ of the system is defined in (5.10). As emphasized by Feynman, $W(x, t, x', t')$ contains the full dynamical information about the quantum system. It is simply related to the exponential of the Hamilton function $S(x, t, x', t')$, which, as illustrated in Chapter 3, codes the classical dynamics of the system. As illustrated in Chapter 5, in the relativistic formalism $W(x, t, x', t')$ can be obtained as the matrix element of the projection operator P between states $|x, t\rangle$ that are eigenstates of the operators corresponding to the partial observables x and t:

$$W(x, t; x', t') = \langle x, t | P | x', t' \rangle_{\mathcal{K}}, \tag{9.1}$$

where $\mathcal{K} = L_2[R^2, dxdt]$ is the kinematical Hilbert space on which the operators corresponding to the partial observables x and t are defined.

A key intuition of Richard Feynman was that the propagator can be expressed as a path integral

$$W(x, t, x', t') \sim \int_{\substack{x(t)=x \\ x(t')=x'}} D[x(t)] \, e^{iS[x]}, \tag{9.2}$$

in which the sum is over the paths $x(t)$ that start at (x', t') and end at (x, t), and $S[x] = \int_{t'}^{t} \mathcal{L}(x(t), \dot{x}(t))dt$ is the action of this path. There are several techniques to define and manipulate this integral. Feynman and many after him have suggested that (9.2) can be taken as the basic definition of the quantum formalism. This can therefore be based on a sum of complex amplitudes $e^{iS[x]}$ over the paths $x(t)$.

The idea of utilizing a sum-over-paths formalism in quantum gravity is old (see Appendix B) and has been studied extensively. The idea is to try to define a path integral over 4d metrics

$$\int D[g_{\mu\nu}(x)] \, e^{iS_{\mathrm{GR}}[g]}. \tag{9.3}$$

In particular, we can consider 3d metric g' and a final 3d metric g and study the transition amplitude between these defined by

$$W[g, g'] = \int_{\substack{g_{|t=1}=g \\ g_{|t=0}=g'}} D[g_{\mu\nu}(x)] \, e^{iS_{\mathrm{GR}}[g]}, \tag{9.4}$$

where S_{GR} is the action of the strip between $t = 0$ and $t = 1$. The specific value chosen for the coordinate t is irrelevant, if the functional integral is defined in a diffeomorphism-invariant manner. The techniques used in quantum mechanics and quantum field theory to give meaning to the functional integral, however, fail in gravity, and the effectiveness of the path integral approach has long remained confined to crude approximations. The reason is that nonperturbative definitions of the measure $\mathrm{D}[g_{\mu\nu}(x)]$ leading to a complete theory are not known, and perturbative definitions around a background metric lead to nonrenormalizable divergences.

The situation has changed with loop gravity, because of the discovery of the discreteness of physical space. To understand why, a few general considerations are in order.

From the hamiltonian theory to the sum-over-paths. Generally, we do not have a very good control of the *direct* definition of the Feynman sum-over-paths. For many issues, such as the choice of the measure, the canonical theory often provides the best route toward the correct definition of the integral. Feynman of course derived the functional integral from the canonical theory in the first place. He did so by writing the evolution operator $e^{-iH_0 t}$ as a product of small time-step evolution operators, inserting resolutions of the identity $1 = \int \mathrm{d}x \, |x\rangle\langle x|$, and taking the limit for the time interval $\mathrm{d}t = (t - t')/N \to 0$. The functional integral is then defined as

$$\int_{\substack{x(t)=x \\ x(t')=x'}} \mathrm{D}[x(t)] \, e^{iS[x]} \equiv \lim_{N \to \infty} \int \mathrm{d}x_1 \ldots \mathrm{d}x_{N-1}$$

$$\langle x| \, e^{-iH_0 \frac{(t-t')}{N}} |x_{N-1}\rangle\langle x_{N-1}| \, e^{-iH_0 \frac{(t-t')}{N}} |x_{N-2}\rangle$$

$$\ldots \langle x_2| e^{-iH_0 \frac{(t-t')}{N}} |x_1\rangle\langle x_1| \, e^{-iH_0 \frac{(t-t')}{N}} |x'\rangle.$$

$$(9.5)$$

The canonical theory can therefore be used to construct a sum-over-paths. Can this be done in quantum gravity? The problem is to obtain in quantum gravity the analog of the Feynman–Kac formula, which provides a precise definition of the path integral. So far, this problem is unsolved, and a sum-over-paths formalism has not yet been rigorously derived from the canonical theory. However, we can still proceed in this direction, using the following strategy.

First, we may write formal equations that indicate the *general structure* that a sum-over-paths in quantum gravity should have. That is, we can derive the basics of the gravitational sum-over-paths formalism from the canonical theory. Indeed, as we shall see in a moment, a few loose formal manipulations show immediately that a generally covariant sum-over-paths formalism for quantum gravity must have quite peculiar properties, because of the discreteness of space. This is done below. Second, we can study specific theoretical models having this general structure. This is done in the subsequent sections.

Transition amplitudes are between spin networks. A key observation is that the quantity x in the argument of the propagator is not the classical variable, but rather a label of an eigenstate of this variable. The difference is irrelevant so long as x is an observable with a continuous spectrum such as the position. But it becomes relevant if the spectrum of x is nontrivial. For instance, consider a harmonic oscillator subjected to an external force, or to a small nonlinear perturbation. Instead of asking for the amplitude $W(x, t; x', t')$ of measuring x given x', ask for the probability amplitude $W(E, t, E', t')$ of measuring the (unperturbed) energy E. This is given by

$$W(E, t, E', t') = \langle E | e^{-iH_0(t-t')} | E' \rangle, \tag{9.6}$$

where $|E\rangle$ is the eigenstate of the unperturbed energy, with eigenvalue E. (H_0 is the nonrelativistic hamiltonian, not the unperturbed one.) But the eigenvalues E are quantized: $E = E_n$, and $|n\rangle = |E_n\rangle$. Therefore, $W(E, t, E', t')$ is defined only for the values of $E = E_n$ in the spectrum. The argument of the propagator must be discrete energy levels, not classical energies. Thus, (9.6) only makes sense in the special form

$$W(n, t, n', t') \equiv W(E_n, t, E_{n'}, t') = \langle n | e^{-iH_0(t-t')} | n' \rangle. \tag{9.7}$$

Consider now the integral (9.4). This must define transition amplitudes between eigenstates of the 3-geometry. We have seen in Chapter 6 that the eigenvalues of the 3-geometry are not 3d continuous metrics. Rather, they have a discrete structure and are labeled by spin networks. *Therefore, the propagator in quantum gravity must be a function of spin networks.* For instance, we should study, instead of (9.4), a quantity $W(s, s')$, giving the probability amplitude of measuring the quantized 3-geometry described by the spin network s if the quantized 3-geometry described by the spin network s' has been measured.

Histories of spin networks. Consider the propagator $W(s, s')$. As discussed at the end of Section 7.4, we can express $W(s, s')$ in the form

$$W(s, s') = \langle s | P | s' \rangle_{\mathcal{K}}. \tag{9.8}$$

I now write some formal expressions that may indicate the way to express $W(s, s')$ as a sum-over-paths. Loosely speaking, the operator P is the projector on the kernel of the hamiltonian operator H. If we tentatively assume that H has a nonnegative spectrum, we can formally write this projector as

$$P = \lim_{t\to\infty} e^{-Ht} \tag{9.9}$$

because if $|n\rangle$ is a basis that diagonalizes H and E_n are the corresponding eigenvalues, then

$$P = \lim_{t\to\infty} \sum_n |n\rangle e^{-E_n t}\langle n| = \sum_n \delta_{0,E_n}|n\rangle\langle n|. \tag{9.10}$$

Since H is a function of a spatial coordinate $\vec{x}$, we can formally write

$$P = \lim_{t\to\infty} \prod_x e^{-H(x)t} = \lim_{t\to\infty} e^{-\int d^3x H(x)t}. \tag{9.11}$$

Hence,

$$W(s,s') = \lim_{t\to\infty} \langle s|e^{-\int d^3x\, H(x)t}|s'\rangle_K. \tag{9.12}$$

If we can define the 4d propagation generated by H in a diff-invariant manner, the limit is irrelevant, and we can write, again quite loosely,

$$W(s,s') = \langle s|e^{-\int_0^1 dt \int d^3x\, H(x)}|s'\rangle_K. \tag{9.13}$$

We can now expand this expression in the same manner as in the right-hand side of (9.5). Inserting resolutions of the identity $1 = \sum_s |s\rangle\langle s|$, we obtain an expression of the form

$$W(s,s') = \lim_{N\to\infty} \sum_{s_1\ldots s_N} \langle s|e^{-\int d^3x\, H(x)\, dt}|s_N\rangle_K \,\langle s_N|e^{-\int d^3x\, H(x)\, dt}|s_{N-1}\rangle_K$$

$$\ldots \langle s_1|e^{-\int d^3x\, H(x)\, dt}|s'\rangle_K. \tag{9.14}$$

For small dt we can expand the exponentials. At fixed N, the first term in the expansion ($e^x = 1 + 0(x)$) produces a term equivalent to histories with lower N. Therefore, we can view the sum as a sum-over-sequences of spin networks, where the sequences can have arbitrary lengths. The result is that the transition amplitude is not expressed as an *integral* over 4d fields, but rather as a discrete *sum-over-histories* σ of spin networks:

$$W(s,s') = \sum_\sigma A(\sigma). \tag{9.15}$$

A history is a discrete sequence $\sigma = (s, s_N, \ldots, s_1, s')$ of spin networks. The amplitude associated with a single history is a product of terms

$$A(\sigma) = \prod_v A_v(\sigma), \tag{9.16}$$

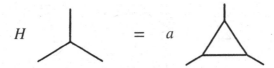

Fig. 9.1   Scheme of the action of H on a node of a spin network.

where v labels the steps of the history, and $A_v(\sigma)$ is determined by the matrix elements

$$\langle s_{n+1} | e^{-\int d^3 x H(x) dt} | s_n \rangle_{\mathcal{K}}. \tag{9.17}$$

For small dt we can keep only the linear term in H in these matrix elements, as Feynman did in (9.5). Now, the action of the hamiltonian operator H is given in (7.21). It is a sum over individual terms acting on the nodes of the spin network. Therefore, it has nonvanishing matrix elements only between spin networks s_{n+1} and s_n that differ at one node by the action of H. A typical term in the action (for instance) of H on a trivalent node is illustrated in Figure 7.2, or, more schematically, in Figure 9.1.

Summarizing, we can write $W(s, s')$ as a sum-over-paths of spin networks; the paths are generated by individual steps such as the one illustrated in Figure 7.2; the amplitude of each step is determined by the corresponding matrix element of H: the amplitude of the history is the product of the individual amplitudes of the steps.

Spinfoams. A history $\sigma = (s, s_N, \ldots, s_1, s')$ of spin networks is called a spinfoam. A more precise definition is given below. A spinfoam admits a natural representation as follows. Imagine a 4d space (representing coordinate spacetime) in which the graph of a spin network s is embedded. Now, imagine that this graph moves upward along a "time" coordinate of the 4d space, sweeping a worldsheet, changing at each step under the action of H. Call "faces" the worldsurfaces of the links of the graph, and denote them f. Call "edges" the worldlines of the nodes of the graph, and denote them as e. Figure 9.2 illustrates the worldsheet of a theta-shaped spin network.

Since the hamiltonian acts on nodes, the individual steps in the history of a spin network can be represented as the branching of the edges which, locally, changes the number of nodes. We call "vertices" the points where edges branch, and denote them as v.

For instance, an edge can branch to form three edges as in Figure 9.3, representing the action of the hamiltonian constraint illustrated in Figure 9.1.

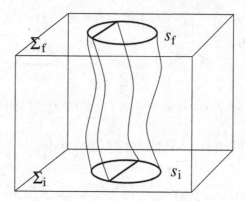

Fig. 9.2 The worldsheet of a spin network s_i on an initial surface Σ_i evolving without intersections into the spin networks s_f on the final surface Σ_f, forming a spinfoam.

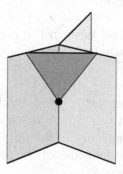

Fig. 9.3 A vertex of a spinfoam.

The resulting worldsheet is illustrated in Figure 9.4, which represents a spinfoam with a single vertex. A spinfoam with two vertices is represented in Figure 9.5. What we obtain in this manner is a collection of faces f, meeting at edges e which, in turn, meet at vertices v. The combinatorial object Γ defined by the set of these elements and their adjacency relations is called a "two-complex."

A spin network is not defined solely by its graph, but also by the coloring of its links (representations) and nodes (intertwiners). Accordingly, the two-complex Γ determined by a sequence of spin networks is colored with irreducible representations j_f associated with faces, and intertwiners i_e associated with edges. A spinfoam $\sigma = (\Gamma, j_f, i_e)$ is a two-complex Γ, with colored faces and edges. That is, it is a two-complex with a representation j_f associated with each face f and an intertwiner i_e associated with each edge e.

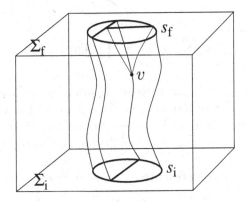

Fig. 9.4   A spinfoam with one vertex.

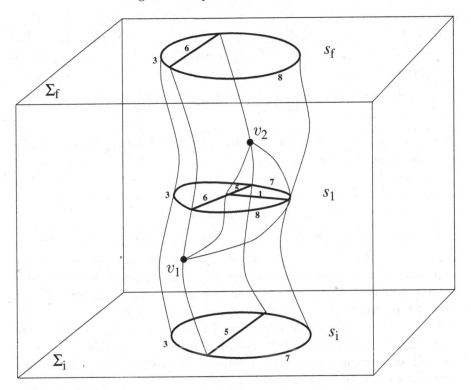

Fig. 9.5   A spinfoam with two vertices.

9.2 Spinfoam formalism

We are now ready to give a general definition of a spinfoam theory. The discussion in the previous section indicates that a sum-over-paths formulation of quantum gravity can be cast in the form of a sum-over-spinfoams of amplitudes given by products of individual vertex amplitudes.

Consider a sum defined as follows

$$Z = \sum_\Gamma w(\Gamma) \sum_{j_f, i_e} \prod_v A_v(j_f, i_e). \tag{9.18}$$

The sum is over a set of two-complexes Γ, and a set of representations and intertwiners j and i. The function $A_v(j_f, i_e)$, called the vertex amplitude, is an amplitude associated with each vertex v. It is a function of the colors adjacent to that vertex. $w(\Gamma)$ is a weight factor that depends only on the two-complex.

It is often convenient to rewrite expression (9.18) in the extended form

$$Z = \sum_\Gamma w(\Gamma) \sum_{j_f, i_e} \prod_f A_f(j_f) \prod_e A_e(j_f, i_e) \prod_v A_v(j_f, i_e), \tag{9.19}$$

with amplitudes A_f and A_e associated to faces and edges as well, although these can, in principle, be included in a redefinition of A_v. For most models so far considered, $A_f(j_f)$ is simply the dimension $\dim(j_f)$ of the representation j_f. Hence we have (using the notation $\sigma = (\Gamma, j_f, i_e)$),

$$Z = \sum_\sigma w(\Gamma(\sigma)) \prod_f \dim(j_f) \prod_e A_e(j_f, i_e) \prod_v A_v(j_f, i_e). \tag{9.20}$$

This is the general expression that we take as the definition of the spinfoam formalism.

A choice of

(i) a set of two-complexes Γ, and associated weight $w(\Gamma)$,

(ii) a set of representations and intertwiners j and i,

(iii) a vertex amplitude $A_v(j_f, i_e)$ and an edge amplitude $A_e(j_f, i_e)$,

defines a "spinfoam model." We shall study several of these models and their relation with gravity in the next section. Generally speaking, the choice (iii) of the vertex amplitude corresponds to the choice of a specific form of the hamiltonian operator in the canonical theory.

What is remarkable about expression (9.20) is that *many* very different approaches and techniques have converged precisely to this formula, as we shall see later on. Perhaps an expression of this sort can be taken as a general definition of a background-independent covariant QFT formalism.

In Table 9.1 I have summarized the terminology used to denote the elements of spin networks, spinfoams, and triangulations (which play a role later on.)

9.2.1 Boundaries

The boundary of a spinfoam σ is a spin network s. This follows easily from the very way we have constructed spinfoams. If σ is bounded by

Table 9.1 Terminology

	0d	1d	2d	3d	4d
spin network:	*node*	*link*			
spinfoam:	*vertex*	*edge*	*face*		
triangulation:	*point*	*segment*	*triangle*	*tetrahedron*	*four-simplex*

the spin network s, we write this as $\partial\sigma = s$. The relation between (9.20) and transition amplitudes is obtained by summing over spinfoams with a given boundary

$$W(s) = \sum_{\partial\sigma=s} w(\Gamma(\sigma)) \prod_f \dim(j_f) \prod_e A_e(j_f, i_e) \prod_v A_v(j_f, i_e). \quad (9.21)$$

In particular, if the spin network s is connected, it can be interpreted as the state of the gravitational field on the connected boundary of a space-time region. For instance, the boundary of a finite region of spacetime.

If the boundary spin network is composed of two connected components s and s', we write

$$W(s, s') = \sum_{\partial\sigma=s\cup s'} w(\Gamma(\sigma)) \prod_f \dim(j_f) \prod_e A_e(j_f, i_e) \prod_v A_v(j_f, i_e), \quad (9.22)$$

and we interpret the spinfoam model sum as a sum-over-paths definition of the transition amplitude between two quantum states of the gravitational field, in analogy with (9.2).

So far, the situation is *not* that we can compute $W(s, s')$ in the two formalisms and prove the two to be equal. In the spinfoam framework, there is uncertainty in the definition of the model but, as we shall see, transition amplitudes can be computed (order by order). In the hamiltonian formalism, on the other hand, even disregarding the uncertainty in the definition of the hamiltonian, we are not yet able to compute transition amplitudes.

9.3 Models

I shall now illustrate a few key examples of spinfoam models. Each of these is a realization of equation (9.20). Namely, each of these is obtained from (9.20) by choosing a set of two-complexes, a set of representations and intertwiners, and vertex and edge amplitudes. These models are related to different theories: general relativity without and with cosmological

Table 9.2   Spinfoam models (λ: cosmological constant; Tr: triangulation)

Model	Class theory	Two-complexes	Representation	Vertex
Ponzano–Regge	3d GR	fixed, dual 3d Tr	$SU(2)$	$\{6j\}$
Turaev–Viro	3d GR $+\lambda$	fixed, dual 3d Tr	$SU(2)_q$	$\{6j\}_q$
Ooguri (TOCY)	4d BF	fixed, dual 4d Tr	$SO(4)$	$\{15j\}$
Crane–Yetter	4d BF $+\lambda$	fixed, dual 4d Tr	$SO(4)_q$	$\{15j\}_q$
Barrett–Crane A	cut-off 4d GR	fixed, dual 4d Tr	simple $SO(4)$	$\{15j\}$
Barrett–Crane B	cut-off 4d GR	fixed, dual 4d Tr	simple $SO(4)$	$\{10j\}_{\mathrm{BC}}$
GFT A	4d GR	Feynman graphs	simple $SO(4)$	$\{15j\}$
GFT B	4d GR	Feynman graphs	simple $SO(4)$	$\{10j\}_{\mathrm{BC}}$

constant in 3d and 4d, and in BF theory (see Section 9.3.2). They are listed in Table 9.2.

These models form a natural sequence that has historically led to the present formulation of spinfoam quantum gravity. This is represented by the last of them, constructed using an auxiliary field theory defined on a group manifold. This represents a complete tentative sum-over-paths formalism for quantum gravity in 4d. The sequence of these models does not have just historical interest: rather, it allows me to introduce, step by step, the ingredients that enter the complete model. The models have increasing complexity. Each of them has introduced and illustrates an important peculiar aspect of the complete theory. Here is a condensed preview of the way each model has contributed to the construction of the formalism (terms and concepts will be clarified in the course of the chapter).

(i) The *Ponzano–Regge* theory is a quantization of gravity in 3d. It illustrates how a sum-over-paths in quantum gravity naturally takes the form of a spinfoam model, and why the gravitational vertex amplitude can be expressed in terms of simple invariant objects from group representation theory. For a long time, it was assumed that these simple features were characteristic of 3d and/or reflected the fact that the theory has no local degrees of freedom (is topological). The other models show that this assumption was wrong.

(ii) The *Ooguri, or TOCY* (*Turaev–Ooguri–Crane–Yetter*) model extends the formalism to 4d.

(iii) The *Barrett–Crane* models extend the formalism to a theory with local degrees of freedom. The key for doing this is the realization that GR can be obtained from a topological theory by adding certain constraints, and these constraints can be implemented in the model as a restriction on the set of representations summed over.

(iv) As soon as the model is no longer topological, the sum over two-complexes becomes nontrivial. A way to implement it is provided by the *Group Field Theory* (*GFT*).

9.3.1 3d quantum gravity

3d GR. Consider riemannian general relativity in three dimensions. This can be defined by minor modifications of the definition of 4d GR given at the beginning of Chapter 2. In 3d, the gravitational field e is a one-form

$$e^i(x) = e^i_a(x)\mathrm{d}x^a \tag{9.23}$$

with values in R^3. The spin connection ω is a one-form with values in the $so(3)$ Lie algebra

$$\omega^i(x) = \omega^i_a(x)\mathrm{d}x^a, \tag{9.24}$$

and we denote its curvature two-form as R^i. The action that defines the theory is

$$S[e, \omega] = \int e_i \wedge R[\omega]^i. \tag{9.25}$$

Varying ω gives the Cartan structure equation $De = 0$. Varying e gives the equation of motion $R = 0$, which implies that spacetime is flat.[1] That is, locally in spacetime there is a single solution of the equations of motion up to gauge.

The theory is nevertheless nontrivial if the space manifold has nontrivial global topology. For instance, there are distinct nonisometric flat tori. A flat torus is characterized, say, by its volume and its two radii, which are global variables. The dynamics of 3d general relativity is reduced to the dynamics of this kind of global variables. A theory of this sort, that has

---

[1] The Einstein–Hilbert action $S[g] = \int \mathrm{d}^3x \sqrt{g} R$, where g_{ab} is the 3d metric and R its Ricci scalar, gives the same equations of motion as (9.25) but differs from (9.25) by a sign when the determinant of e^i_a is negative. Hence, quantizations of the two actions might lead to inequivalent theories. Which of the two is to be called 3d GR is a matter of taste, as they have the same classical solutions.

Table 9.3 Relation between a triangulation Δ and its dual Δ^*, in 3d (left) and 4d (right). In italic, the two-complex. In parentheses: adjacent elements

Δ_3	Δ_3^*		Δ_4	Δ_4^*	
tetrahedron	*vertex*	(4 edges, 6 faces)	4-simplex	*vertex*	(5 edges, 10 faces)
triangle	*edge*	(3 faces)	tetrahedron	*edge*	(4 faces)
segment	*face*		triangle	*face*	
point	3d region		segment	3d region	
			point	4d region	

no local degrees of freedom, but only global ones, is called a topological theory.[2]

Discretization. I now give a concrete definition of the functional integral (9.3) for 3d GR by discretizing the theory. To this end, fix a triangulation Δ of the spacetime manifold. It is more convenient to work with the dual Δ^* of the triangulation, and, in particular, with the two-skeleton of Δ^*. These are defined as follows; see Table 9.3.

To obtain Δ^*, we place a vertex v inside each tetrahedron of Δ; if two tetrahedra bound the same triangle e, we connect the two corresponding vertices by an edge e, dual to the triangle e: for each segment f of the triangulation we have a face f of Δ^*, bounded by the edges corresponding to the triangles of Δ that are bounded by the segment f; finally, to each point of Δ we have a 3d region of Δ^*, bounded by the faces dual to the segments that are bounded by the point. In 4d, Δ^* is obtained by placing a vertex in each four-simplex, and so on. The collection of the sole vertices, edges and faces of Δ^* (with their boundary relations) is called the two-skeleton of Δ^*, and is precisely a two-complex.

Let g_e be the holonomy of ω along each edge e of Δ^*. (The connection ω is in the algebra, and the algebra of $SO(3)$ is the same as the algebra of $SU(2)$. In defining the holonomy, we have to decide whether we interpret ω as an $SO(3)$ or an $SU(2)$ connection. Let us choose the $SU(2)$ interpretation. That is, we define $g_e = \mathcal{P} \exp\{\int_e \omega^i \tau_i\} \in SU(2)$, where τ_i are Pauli matrices.)

Let l_f^i be the line integral of e^i along the segment f of Δ. We choose these as basic variables for the discretization. The variables of the

---

[2]The expression "topological field theory" is used with different meanings in the literature. In [267], for instance, it is used to designate any diffeomorphism-invariant theory, with a finite or infinite number of degrees of freedom. This is done to emphasize the similarities among all these theories, and their difference from QFT on a background. Here there is no risk of underemphasizing this difference, which is stressed throughout this book, and I prefer to follow the common usage.

discretized theory will therefore be an $SU(2)$ group element g_e associated with each edge e of Δ^* and a variable l_f^i in R^3, associated with each segment of Δ, or, equivalently, to each face f of Δ^*. Accordingly, we can discretize the action as

$$S[l_f, g_e] = \sum_f l_f^i \text{tr}[g_f \tau_i], \tag{9.26}$$

where

$$g_f = g_{e_1^f} \cdots g_{e_n^f} \tag{9.27}$$

is the product of the group elements associated with the edges $e_1^f \dots e_n^f$ that bound the face f. If we vary this action with respect to l_f^i we obtain the equation of motion $g_f = 1$, namely the lattice connection is flat. Using this, if we vary this action with respect to g_e we obtain the equation of motion $l_{f_1}^i + l_{f_2}^i + l_{f_3}^i = 0$ for the three sides f_1, f_2, f_3 of each triangle. This is the discretized version of the Cartan structure equation $De = 0$.

Path integral. Using this discretization, we can define the path integral as

$$Z = \int dl_f^i \, dg_e \, e^{iS[l_f, g_e]}, \tag{9.28}$$

where the measure on $SU(2)$ is the invariant Haar measure. The integral over l_f gives immediately (up to an overall normalization factor that we absorb in the definition of the measure),

$$Z = \int dg_e \, \prod_f \delta(g_{e_1^f} \cdots g_{e_n^f}). \tag{9.29}$$

We can now expand the delta function over the group manifold using the expansion

$$\delta(g) = \sum_j \dim(j) \, \text{tr} R^j(g), \tag{9.30}$$

where the sum is over all unitary irreducible representations of $SU(2)$. Inserting this in (9.29), and exchanging the sum and the product, we have

$$Z = \sum_{j_1 \dots j_N} \prod_f \dim(j_f) \int dg_e \prod_f \text{tr} R^{j_f}(g_{e_1^f} \cdots g_{e_n^f}). \tag{9.31}$$

It is not difficult to perform the integrations over the group. There is one integral per edge. Since every edge bounds precisely three faces, each integral is of the form

$$\int dU \, R^{j_1}(U)^{\alpha}_{\alpha'} \, R^{j_2}(U)^{\beta}_{\beta'} \, R^{j_3}(U)^{\gamma}_{\gamma'} = v^{\alpha\beta\gamma} \, v_{\alpha'\beta'\gamma'}, \tag{9.32}$$

where $v^{\alpha\beta\gamma}$ is the (unique) normalized intertwiner between the representations of spin j_1, j_2, j_3. The reader should not confuse the symbol v that denotes vertices with the tensor $v^{\alpha_1\cdots\alpha_n}$ used to denote intertwiners.

Each of the two invariant tensors on the right-hand side is associated with one of the two vertices that bound the edge (whose group element is integrated over). Its indices get contracted with those coming from the other edges at this vertex. A moment of reflection shows that at each vertex we have four of these tensors, that contract, giving a function of the six spins associated with the six faces that bound the vertex

$$\{6j\} \equiv \begin{pmatrix} j_1 & j_2 & j_6 \\ j_4 & j_3 & j_5 \end{pmatrix} \equiv \sum_{\alpha_1\ldots\alpha_6} v^{\alpha_3\alpha_6\alpha_2} \, v^{\alpha_2\alpha_1\alpha_5} \, v^{\alpha_6\alpha_4\alpha_1} \, v^{\alpha_4\alpha_3\alpha_5}. \quad (9.33)$$

The pattern of the contraction of the indices reproduces the structure of a tetrahedron: if we (i) represent each 3-tensor $v^{\alpha_1\alpha_2\alpha_3}$ using the representation (6.86), namely we write it as a trivalent vertex, (ii) represent index contraction by joining legs (open ends), and (iii) indicate the representation to which the index belongs, the $6j$ symbol is represented as

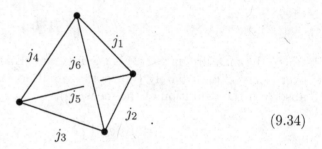

$$\hspace{11cm} (9.34)$$

This function, denoted $\{6j\}$, is a well-known function in the representation theory of $SU(2)$. It is a natural object that one can construct given six irreducible representations. It is called the Wigner $6j$ symbol, see Appendix A1.

Bringing all this together, we obtain the following form for the partition function of 3d GR

$$Z_{\text{PR}} = \sum_{j_1\ldots j_N} \prod_f \dim(j_f) \, \prod_v \{6j\}_v. \quad (9.35)$$

This is the Ponzano–Regge spinfoam model. Using the notation (6.86), we can write it in the form

$$Z_{\text{PR}} = \sum_j \prod_f \dim(j_f) \, \prod_v \quad\quad\quad\quad\quad\quad (9.36)$$

This expression has the general form (9.21), with the following choices:

- The set of two-complexes summed over is formed by a single two-complex. This is chosen to be the two-skeleton of the dual of a 3d triangulation.
- The representations summed over are the unitary irreducibles of $SU(2)$. The intertwiners are trivial.
- The vertex amplitude is $A_v = \{6j\}$.

Remarkably, these simple choices define the Ponzano–Regge quantization of 3d GR.

The fact that the vertex amplitude is simply the Wigner $6j$ symbol is perhaps surprising. The Wigner $6j$ symbol is a simple algebraic construct of $SU(2)$ representation theory. The action of general relativity is a complicated expression coding the complexity of the gravitational interaction. Even more remarkable is the fact that, as we shall see, this is not a strange coincidence of the particularly simple form of GR in 3d: rather, the same connection between simple algebraic group theoretical quantities and the gravitational action holds in four dimensions as well. This connection is one of the "miracles" that nurtures the interest in the spinfoam approach.

The original derivation: the Ponzano–Regge ansatz. The derivation above is not the original one of Regge and Ponzano. It is instructive to mention also the general lines of the original derivation. Regge introduced a discretization of classical general relativity, called Regge calculus, defined over a fixed triangulation. Consider the triangulation Δ of the spacetime manifold, and denote as f its segments (the choice of the letter f will be clear below). A gravitational field associates length l_f to each segment f. In turn, these lengths l_f can be taken as a discrete set of variables that replace the continuous metric. The action of a given gravitational field can be approximated by an action functional $S_{\text{Regge}}(l_f)$ of these lengths. As the continuous action is an integral over spacetime, the Regge action is a sum over the n-simplices v of the triangulation of the action of a single n-simplex

$$S = \sum_v S_v. \tag{9.37}$$

We can then write a discretized version of (9.3) in the form

$$Z = \int dl_1 \dots dl_N \ e^{iS_{\text{Regge}}(l_f)}. \tag{9.38}$$

Hence we can write

$$Z = \int dl_1 \dots dl_N \ \prod_v e^{iS_v(l_f)} \tag{9.39}$$

where $S_v(l_f)$ is the Regge action of an individual n-simplex.

Ponzano and Regge studied the integral (9.39) in the case of general relativity in 3d, under one additional assumption: that the length of each link can take only the discrete values

$$l_n = j_n, \qquad j_n = \frac{1}{2}, 1, \frac{3}{2}, 2, \ldots \qquad (9.40)$$

in units in which the Planck length $l_P = 1$. This assumption is called the Ponzano–Regge ansatz. Ponzano and Regge did not provide any justification for it. They introduced it just as discretization of the multiple integral over lengths. Notice that this is a *second* discretization, in addition to the triangulation of spacetime. Its physical meaning was clarified much later, and I will come back to it later on. Under the Ponzano–Regge ansatz (9.40), equation (9.39) becomes

$$Z_{\text{PR}} = \sum_{j_1 \ldots j_N} \prod_v e^{iS_v(j_n)}. \qquad (9.41)$$

In 3d, the Regge action of a 3-simplex (a tetrahedron) v can be written as a sum over the segments f in v, as

$$S_v = \sum_f l_f \theta_f(l_f), \qquad (9.42)$$

where θ_f is the dihedral angle of the segment f, that is, the angle between the outward normals of the triangles incident to the segment. One can show that this action is an approximation to the integral of the Ricci scalar curvature. Under the Ponzano–Regge ansatz, therefore, S_v is a function $S_v(j_n)$ of *six* spins $j_1, \ldots, j_6$. (A tetrahedron has six edges.)

The "miracle": GR dynamics in a symbol. The surprising discovery of Ponzano and Regge was that the Wigner $6j$ symbol approximates the action of general relativity. More precisely, they were able to show that in the limit of large js, we have the asymptotic formula

$$\{6j\} \sim \left(e^{iS_v(j_n)} + e^{-iS_v(j_n)} \right) + \frac{\pi}{4}. \qquad (9.43)$$

The term $\pi/4$ does not affect classical dynamics. The two exponential terms in (9.43) are analogous to the two terms that we found in Section 5.2.3; see in particular the discussion at the end of that section. The classical theory does not distinguish between forward and backward propagation in coordinate time, and the path integral sums over the two. The two terms in (9.43) correspond to these two propagations. Inserting (9.43) in (9.41) and fixing the normalization factors by imposing triangulation independence (see below), we get (9.35), which is the expression that Ponzano and Regge proposed as a discretized path integral for 3d quantum gravity.

Physical meaning of the Ponzano–Regge ansatz. As noted at the beginning of this chapter, the path integral defines transition amplitudes between eigenstates of field operators, not between classical fields. The Ponzano–Regge path integral (9.35) defines transition amplitudes between triangulated 2d surfaces where the links of the 2d triangulation have a length l_f. These lengths, under the Ponzano–Regge ansatz, are quantized. Therefore, the Ponzano–Regge ansatz is equivalent to the physical assumption that length is quantized in 3d quantum gravity. Now, this length quantization is not an ansatz, but a *result* in loop quantum gravity. Indeed, it is not hard to see that in 3d the result that the area is quantized, described in Chapter 6, translates into the quantization of *length*. Therefore, the key additional input provided by the Ponzano–Regge ansatz is physically justified by loop gravity.

In the previous derivation, the discretization of the length derives from the following manipulation that we did over the integral. First, we integrated over the continuous variable l_f^i, obtaining a delta function (equation (9.29)). Then, second, we expressed this delta function as a sum in equation (9.30). To understand the sense of this back and forward transformation, consider the following example. Let x be a variable in the interval $(-\pi, \pi)$. We have

$$\int \mathrm{d}p\, \mathrm{e}^{\mathrm{i}px} = 2\pi\delta(x), \qquad (9.44)$$

and also

$$\sum_n \mathrm{e}^{\mathrm{i}p_n x} = 2\pi\delta(x), \qquad (9.45)$$

where $p_n = n$. Here the distribution $\delta(x)$ is over functions on the compact $(-\pi, \pi)$ interval. Therefore, as long as we deal with functions on this compact interval, we can replace an integral with a sum. This is precisely what we did between (9.29) and (9.31). The compact space is the group manifold over which the holonomy takes values.

The mathematical fact that the "Fourier components" over a compact interval are discrete is of course strictly connected to the physical fact that quantities conjugate to variables that take value in a compact space are quantized. In fact, this is the origin of the quantization of the area in 4d and the quantization of length in 3d.

Indeed, we can view p_n in (9.45) as the quantized value of the continuous variable p in (9.44), conjugate to the compact variable x. Similarly, the representation j_e can be considered as quantization of the continuous variable l_e. More precisely, we can identify l_f^i with the generator of the spin-j_f representation and identify the length $|l_f|$ of the segment f with the square root of the quadratic Casimir operator of the representation j_f.

Sum-over-surfaces. The sum over representations in (9.35) has a nice interpretation as a sum-over-surfaces. Consider a triangulation Δ and a specific assignment of representations j_f to the faces of Δ^*. Consider an edge e that bounds the three faces f, f', f''. The intertwiner on e is nonvanishing (and therefore the amplitude of the spinfoam is nonzero) only if the three representations j_f, $j_{f'}$ $j_{f''}$ satisfy the Clebsch–Gordan relations (A.10)–(A.11). Assuming this is the case, associate $2j_f$ elementary parallel surfaces with each face f. Join these elementary surfaces across each edge. The constraint (A.10)–(A.11) is precisely the condition under which the surfaces can be joined, and there is only one way of joining them across each edge. (There are a, b and c surfaces crossing over from j_f to $j_{f'}$, from $j_{f'}$ $j_{f''}$, and from j_f to $j_{f''}$, respectively, and

$$2j_f = a + c, \quad 2j_{f'} = a + b, \quad 2j_{f''} = b + c.) \qquad (9.46)$$

In this way we obtain surfaces without boundaries that wrap around the two-skeleton of Δ^*. Each such surface carries a spin $= 1/2$.

The sum over representations in (9.35) can therefore be viewed as a sum over all the ways of wrapping these spin-1/2 surfaces around the two-skeleton of Δ^*. The coloring j_f of the face f is the total spin on the face, that is, half the number of spin-1/2 surfaces passing on f.

Divergences: bubbles. The Ponzano–Regge model suffers from infrared divergences. These have a peculiar structure, which is reproduced in all spinfoam models.

In the Feynman diagrams of ordinary QFT, divergences are associated with loops, that is, closed curves within the Feynman diagram. In the absence of loops, divergences do not appear, because momentum conservation at the vertices constrains the value of the momenta on the internal propagators. In a spinfoam model, the role of the internal momenta (integrated over) is played by the representations (summed over). These are constrained at edges by the Clebsch–Gordan conditions at the edge, which play the role of momentum conservation. Accordingly, divergences are not associated with loops, as in ordinary Feynman diagrams, but rather with "bubbles." A bubble is a collection of faces f in Δ^* that form a closed two-surface. If we increase each of the representations associated with the faces forming the bubble by the same amount j, then the relations (A.10)–(A.11) remain satisfied, because if $j_f, j_{f'}, j_{f''}$ satisfy (A.10)–(A.11), so do $j_f + j, j_{f'} + j, j_{f''}$.

The sum-over-surfaces picture described in the previous section gives us a clear understanding of the structure of the resulting divergences: we can always add an arbitrary number of spin-1/2 surfaces wrapped around the bubble.

The minimal bubble configuration is the "elementary bubble." This is formed by four triangular faces connected to each other as in a tetrahedron. This elementary bubble appears if four vertices are connected to one another in Δ^*, and, equivalently, if four tetrahedra are connected to one another in Δ. This configuration can be obtained, for instance, by decomposing a single tetrahedron into four tetrahedra, by picking an interior point and connecting it to the four vertices. Within the representation (9.29), this configuration gives the contribution

$$A_{\text{bubble}} = \int dg_1 \dots dg_6 \; \delta(g_1 g_2 g_6^{-1})\delta(g_3 g_4 g_6)\delta(g_4 g_1 g_5^{-1})\delta(g_2 g_3 g_5) \quad (9.47)$$

to Z (see (9.34)). Integration is immediate, giving the divergent expression

$$A_{\text{bubble}} = \delta(0). \quad (9.48)$$

The same result can be obtained in the sum over representations. For instance, assume the spins of all faces connected to the elementary bubble vanish. Then

$$A_{\text{bubble}} = \sum_j (\dim(j))^4 \; (\{6j\}(j,j,j,0,0,0))^4. \quad (9.49)$$

From the definition (9.33),

$$\{6j\}(j,j,j,0,0,0) = v^{\alpha_1 \alpha_2} \, v^{\alpha_1 \alpha_3} \, v^{\alpha_2 \alpha_3}. \quad (9.50)$$

The normalized intertwiner $v^{\alpha_1 \alpha_2}$ is $v^{\alpha_1 \alpha_2} = \delta^{\alpha_1 \alpha_2}/(\dim(j))^{1/2}$, giving

$$\{6j\}(j,j,j,0,0,0) = \dim(j)^{-3/2} \dim(j) = \dim(j)^{-1/2}. \quad (9.51)$$

Inserting this in (9.49) yields

$$A_{\text{bubble}} = \sum_j (\dim(j))^2, \quad (9.52)$$

which is equal to (9.48), as is clear from (9.30).

In the Regge triangulation picture, this divergence corresponds to the case in which there is a point of the Regge lattice connected to the rest of the lattice by four segments. We can then make the lengths of these four segments arbitrarily large. Geometrically, this describes a long and narrow "spike" emerging from the 3d (discretized) manifold.

Notice that this is an infrared divergence, since it regards large j_fs, namely large lengths. It is not related to ultraviolet divergences, absent in the theory. Ponzano and Regge have developed a renormalization procedure to divide away this divergence.

Turaev–Viro. An appealing way to get rid of the divergence of the model is to replace the representation theory of the group $SU(2)$ with the representation theory of the quantum group $SU(2)_q$, with q chosen to be a root of unity. Both the dimension and the Wigner $6j$ symbol are well defined for this quantum group. The irreducible representations of this quantum group are finite in number, and therefore the sum is finite. This sum is called the Turaev–Viro invariant. Furthermore, it can be argued that the deformation of the group from $SU(2)$ to $SU(2)_q$ corresponds simply to the addition of a cosmological term to the classical action. Namely, to the action

$$S[e, \omega] = \int \epsilon_{ijk} \, e^i \wedge (R[\omega]^{jk} - \frac{\lambda}{3} e^j \wedge e^k). \qquad (9.53)$$

A good discussion on this issue can be found in [231].

Triangulation independence. A remarkable result by Ponzano and Regge is triangulation independence: the quantity Z defined by (9.35) depends on the global topology of the 3-manifold chosen but not on Δ, namely not on the way the manifold is triangulated. In particular, whether we choose a minimal triangulation or a very fine triangulation, the partition function does not change.

The proof of triangulation independence is tricky in the Ponzano–Regge case, where a renormalization procedure is needed. On the other hand, it is a clean theorem in the Turaev–Viro case, where everything is finite. The Turaev–Viro sum is defined in terms of a triangulation of a compact 3-manifold, but it is a well-defined 3-manifold invariant.

Triangulation independence is the consequence of the fact that 3d GR is a topological theory. Since the theory has no local degrees of freedom, we are not really losing degrees of freedom in the discretization. Usually, discretization loses the short-scale degrees of freedom, but there are no short-scale degrees of freedom in this theory. Hence, the triangulated version of the theory has the same number of degrees of freedom as the full field theory, and refining the triangulation does not change this number.

The triangulation independence of the expression (9.35) is an important mathematical property. It has inspired much mathematical work. But we do not expect triangulation independence to hold for 4d GR, which is not a topological theory since it has local degrees of freedom. Therefore, from the point of view of the problem of quantum gravity, triangulation independence is a less interesting aspect of the Ponzano–Regge theory. It will not survive the generalization that we will study later on.

9.3.2 BF theory

Let us extend the above construction to four dimensions. As a first step we do not consider GR, but a much simpler 4d theory, called BF theory, which is topological and is a simple extension to 4d of the topological 3d GR. Consider BF theory for the group $SO(4)$. This is defined by two fields: a two-form B^{IJ} with values in the Lie algebra of $SO(4)$, and an $SO(4)$ connection ω^{IJ}. The action is a direct generalization of (9.25)

$$S[B, \omega] = \int B_{IJ} \wedge F^{IJ}[\omega], \qquad (9.54)$$

where F is the curvature two-form of ω. I use here the notation F instead of R, as this is standard in this context (and is the origin of the name "BF" of theory).

We can discretize this theory and define a path integral following the very same steps we took for 3d GR. We obtain precisely equation (9.29) again and, from this, equation (9.31) with the sole difference that the sum is over irreducible representations of $SO(4)$, and that the two-complex is the two-skeleton of the dual of a four-dimensional triangulation.

Again it is not difficult to perform the integrations over the group. But now every edge bounds four faces, not three. We have then, instead of (9.32), the integral

$$\int dU R^{j_1}(U)^{\alpha}_{\alpha'} \, R^{j_2}(U)^{\beta}_{\beta'} \, R^{j_3}(U)^{\gamma}_{\gamma'} \, R^{j_4}(U)^{\delta}_{\delta'} = \sum_i v_i^{\alpha\beta\gamma\delta} \, v^i_{\alpha'\beta'\gamma'\delta'}. \quad (9.55)$$

Here the index i labels the orthonormal basis $v_i^{\alpha\beta\gamma\delta}$ in the space of the intertwiners between the representations of spin j_1, j_2, j_3, j_4. We have therefore a sum over intertwiners for each edge, in addition to the sums over representations for each face. At each vertex, we have now ten representations (because the vertex bounds ten faces) and five intertwiners. These define the function

$$\{15j\} \equiv \mathcal{A}(j_1, \dots, j_{10}, \ i_1, \dots, i_5)$$
$$\equiv \sum_{\alpha_1 \dots \alpha_{10}} v_{i_1}^{\alpha_1\alpha_6\alpha_9\alpha_5} \, v_{i_2}^{\alpha_2\alpha_7\alpha_{10}\alpha_1} \, v_{i_3}^{\alpha_3\alpha_7\alpha_8\alpha_2} \, v_{i_4}^{\alpha_4\alpha_9\alpha_7\alpha_3} \, v_{i_5}^{\alpha_5\alpha_{10}\alpha_8\alpha_4}, \quad (9.56)$$

where the indices a_n are in the representation j_n. The pattern of the contraction of the indices reproduces the structure of (the one-skeleton of) a four-simplex:

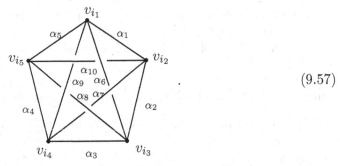

$$(9.57)$$

The function (9.56) is denoted $\{15j\}$. The name comes from the fact that if the group is $SU(2)$ the intertwiners can be labeled with the representation of the internal virtual link, hence this function depends on 15 spins.

Combining everything, we obtain the following form for the partition function of 4d BF theory

$$Z_{\text{TOCY}} = \sum_{j_f, i_e} \prod_f \dim(j_f) \, \prod_v \{15j\}_v, \qquad (9.58)$$

which we can write as

$$Z_{\text{TOCY}} = \sum_{j_f, i_e} \prod_f \dim(j_f) \, \prod_v \quad . \qquad (9.59)$$

This expression is called the TOCY model (from Turaev, Ooguri, Crane and Yetter), or the Ooguri model. It has the general form (9.21), with the following choices:

- The set of two-complexes summed over is formed by a single two-complex. This is chosen to be the two-skeleton of the dual of a 4d triangulation.
- The representations summed over are the unitary irreducibles of $SO(4)$.
- The vertex amplitude is $A_v = \{15j\}$.

These choices define the quantization of 4d BF theory.

Divergences and Crane–Yetter model. As for the Ponzano–Regge model, the sum (9.58) suffers from infrared divergences. The typical divergence is again associated with a "bubble." The elementary bubble is now formed by five vertices of Δ^* connected to each other. In Δ, this corresponds to five four-simplices connected to each other. Namely, the configuration obtained by subdividing a single four-simplex into five four-simplices, adding a single point inside and connecting it to the vertices. We can compute the degree of this divergence as we did for the Ponzano–Regge model, starting from the expression (9.29). In the present case, the pattern of the integration variables and the delta functions are given by (9.57). This gives

$$A_{\text{bubble}} = \int dg_1 \dots dg_{10} \, \delta(g_1 g_2 g_6^{-1}) \delta(g_2 g_3 g_7^{-1}) \delta(g_3 g_4 g_8^{-1}) \delta(g_4 g_5 g_9^{-1})$$
$$\times \, \delta(g_5 g_1 g_{10}^{-1}) \delta(g_1 g_7 g_8) \delta(g_2 g_8 g_9) \delta(g_3 g_9 g_{10}) \delta(g_4 g_{10} g_7) \delta(g_5 g_6 g_8).$$
$$(9.60)$$

Integration is immediate, giving the divergent expression

$$A_{\text{bubble}} = \delta^4(0). \tag{9.61}$$

The divergences can be cured by passing to the quantum group $SO(4)_q$. The definition of the quantum $15j$ symbol requires care but can be given. The resulting model is finite and triangulation independent. It is called the Crane–Yetter model. Its classical limit can be shown to be related to BF theory plus a cosmological term.

9.3.3 The spinfoam/GFT duality

There is a surprising duality between the Ponzano–Regge and TOCY models on the one hand, and certain peculiar QFTs defined over a group (Group Field Theory, or GFT) on the other. This duality will play an important role in what follows. I illustrate it here, in the 4d case.

Consider a real field $\phi(g_1, g_2, g_3, g_4)$ over the cartesian product of four copies of $G = SO(4)$. Require that ϕ is symmetric and $SO(4)$ invariant, in the sense

$$\phi(g_1, g_2, g_3, g_4) = \phi(g_1 g, g_2 g, g_3 g, g_4 g), \qquad (\forall\, g \in SO(4)). \tag{9.62}$$

Consider the QFT defined by the action

$$S[\phi] = \frac{1}{2} \int \prod_{i=1}^{4} \mathrm{d}g_i \; \phi^2(g_1, g_2, g_3, g_4)$$

$$+ \frac{\lambda}{5!} \int \prod_{i=1}^{10} \mathrm{d}g_i \; \phi(g_1, g_2, g_3, g_4)\phi(g_4, g_5, g_6, g_7)\phi(g_7, g_3, g_8, g_9)$$

$$\times \phi(g_9, g_6, g_2, g_{10})\phi(g_{10}, g_8, g_5, g_1). \tag{9.63}$$

The potential (fifth-order) term has the structure of a 4-simplex: if we represent each of the five fields in the product as a node with 4 legs – one for each g_i – and connect pairs of legs corresponding to the same argument, we obtain (the one-skeleton of) a 4-simplex, see Figure 9.6.

The remarkable fact about this field theory is the following. The Feynman expansion of the partition function of the GFT

$$Z = \int \mathcal{D}\phi \; e^{-S[\phi]} \tag{9.64}$$

turns out to be given by a sum over Feynman graphs

$$Z = \sum_{\Gamma} \frac{\lambda^{v[\Gamma]}}{\text{sym}[\Gamma]} Z[\Gamma], \tag{9.65}$$

where the amplitude of a Feynman graph is

$$Z[\Gamma] = \sum_{j_f, i_e} \prod_f \dim(j_f) \prod_v \{15j\}_v. \tag{9.66}$$

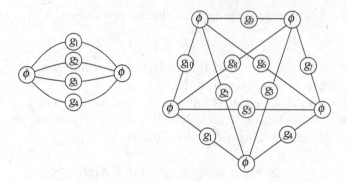

Fig. 9.6   The structure of the kinetic and potential terms in the action.

Here Γ is a Feynman graph, $v[\Gamma]$ the number of its vertices and sym$[\Gamma]$ its symmetry factor. The Feynman graphs Γ of the theory have a natural additional structure as two-complexes. The Feynman integrals over momenta are discrete sums (because the space on which the QFT is defined is discrete) over $SO(4)$ representations j_f and over intertwiners i_e associated with faces and edges of the two-complex. Furthermore, for each given two-complex Γ, the Feynman sum over momenta is *precisely* the TOCY model defined on that two-complex. Indeed, the right-hand side of (9.66) is equal to the right-hand side of (9.58)! That is,

$$Z[\Gamma] = Z_{\text{TOCY}}. \tag{9.67}$$

The proof of these results is a straightforward application of perturbative expansion methods in QFT, and the use of the Peter–Weyl theorem that allows us to mode-expand functions on a group in terms of a basis given by the unitary irreducible representations of the group. This is done in some detail below.

Mode expansion. First, expand the field $\phi(g_1, g_2, g_3, g_4)$ into modes and rewrite the action in terms of these modes (in "momentum space"). Consider a square integrable function $\phi(g)$ over $SO(4)$. The Peter–Weyl theorem tells us that we can expand this function in the matrix elements $R_{\alpha\beta}^{(j)}(g)$ of the unitary irreducible representations j

$$\phi(g) = \sum_j \phi_{\alpha\beta}^j \, R_{\alpha\beta}^{(j)}(g). \tag{9.68}$$

The indices α, β label basis vectors in the corresponding representation space. Accordingly, the field can be expanded into modes as

$$\phi(g_1, \dots, g_4) = \sum_{j_1 \dots j_4} \phi_{\alpha_1\beta_1 \dots \alpha_4\beta_4}^{j_1 \dots j_4} \, R_{\alpha_1\beta_1}^{(j_1)}(g_1) \dots R_{\alpha_4\beta_4}^{(j_4)}(g_4). \tag{9.69}$$

Using the invariance (9.62) under the left group action we can write

$$\phi(g_1 \dots g_4) = \int_{SO(4)} dg \, \phi(gg_1, \dots, gg_4). \tag{9.70}$$

Substituting here the mode expansion (9.69) and using the expression (9.55) for the integral of the product of four group elements, we can write

$$\phi(g_1,\ldots,g_4) = \sum_{j_1\cdots j_4} \phi^{j_1\ldots j_4}_{\alpha_1\ldots\alpha_4 i} \, R^{(j_1)}_{\alpha_1\beta_1}(g_1)\ldots R^{(j_4)}_{\alpha_4\beta_4}(g_4) \, v^i_{\beta_1\ldots\beta_4}, \tag{9.71}$$

where we have defined

$$\phi^{j_1\ldots j_4}_{\alpha_1\ldots\alpha_4 i} = \phi^{j_1\ldots j_4}_{\alpha_1\beta_1\ldots\alpha_4\beta_4} \, v^i_{\beta_1\ldots\beta_4}. \tag{9.72}$$

We use the quantities $\phi^{j_1\ldots j_4}_{\alpha_1\ldots\alpha_4 i}$ as the Fourier components of the field. Written in terms of these, the kinetic term of the action reads

$$\frac{1}{2}\int\prod_{i=1}^{4} \mathrm{d}g_i \, \phi^2(g_1,\ldots,g_4) = \frac{1}{2}\sum_{j_1\cdots j_4}\sum_{i} \phi^{j_1\ldots j_4\, i} \phi^{j_1\ldots j_4\, i}. \tag{9.73}$$

The interaction term becomes

$$\frac{\lambda}{5!}\int\prod_{i=1}^{10} \mathrm{d}g_i \phi(g_1,g_2,g_3,g_4)\phi(g_4,g_5,g_6,g_7)\phi(g_7,g_3,g_8,g_9)$$

$$\times\, \phi(g_9,g_6,g_2,g_{10})\phi(g_{10},g_8,g_5,g_1)$$

$$= \frac{\lambda}{5!}\sum_{j_1\cdots j_{10}}\sum_{i_1\cdots i_5} \phi^{j_1 j_2 j_3 j_4\, i_1}\phi^{j_4 j_5 j_6 j_7\, i_2}\phi^{j_7 j_3 j_8 j_9\, i_3}\phi^{j_9 j_6 j_2 j_{10}\, i_4}\phi^{j_{10} j_8 j_5 j_1\, i_5}$$

$$\times\, \mathcal{A}(j_1,\ldots,j_{10},i_1,\ldots,i_5)\,, \tag{9.74}$$

where $\mathcal{A}$ is given in (9.56).

Feynman graphs. The partition function is given by the integral over modes

$$Z = \int \left[\mathrm{D}\phi^{j_1\ldots j_4\, i}\right] e^{-S[\phi_\mathcal{A}]}. \tag{9.75}$$

We expand Z in powers of λ. The gaussian integrals are easily computed, giving the propagator

$$P^{j_1\ldots j_4\, i,\; j_1'\ldots j_4'\, i'} \equiv \langle \phi^{j_1\ldots j_4\, i}, \phi^{j_1'\ldots j_4'\, i'}\rangle = \frac{1}{4!}\sum_{\sigma} \delta^{j_1 j_{\sigma(1)}'}\ldots\delta^{j_4 j_{\sigma(4)}'} \delta^i_{i'}, \tag{9.76}$$

where σ are the permutations of $\{1,2,3,4\}$. There is a single vertex, of order five, which is:

$$\langle\phi^{j_1 j_2 j_3 j_4\, i_1}\ldots\phi^{j_{10} j_8 j_5 j_1\, i_5}\rangle = \lambda\,\mathcal{A}(j_1,\ldots,j_{10},i_1,\ldots,i_5). \tag{9.77}$$

The set of Feynman rules one gets is as follows. First, we obtain the usual overall factor $\lambda^{v[\Gamma]}/\mathrm{sym}[\Gamma]$ (see for instance [268], page 93). Second, we represent each of the terms in the right-hand side of the definition (9.76) of the propagator by four parallel strands, as in Figure 9.7, carrying the indices at their ends. We can represent the propagator itself by the symmetrization of the four strands. In addition, edges e are labeled by a representation j_e.

The Feynman graphs we get are all possible "4-strand" five-valent graphs, where a "4-strand graph" is a graph whose edges are collections of four strands, and whose vertices are those shown in Figure 9.8. Each strand of the propagator can be connected to a single strand in each of the five "openings" of the vertex. Orientations in the vertex and in the propagators should match (this can always be achieved by changing a representation to its conjugate). Each strand of the 4-strand graph goes through several vertices and several propagators, and then closes, forming a cycle. A particular

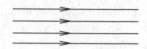

Fig. 9.7   The propagator can be represented by a collection of four strands, each carrying a representation.

Fig. 9.8   The structure of the vertex generated by the Feynman expansion.

strand can go through a particular edge of the 4-strand graph more than once. Cycles get labeled by the simple representations of the indices. For each graph, the abstract set formed by the vertices, the edges, and the cycles forms a two-complex, in which the faces are the cycles. The labeling of the cycles by simple representations of $SO(4)$ determines a coloring of the faces by spins. Thus, we obtain a colored two-simplex, namely a spinfoam.

Edges e are labeled by an intertwiner with index i_e. Vertices v contribute a factor λ times $\mathcal{A}$, which depends on the ten simple representations labeling the cycles that go through the vertex, and on the five intertwiners, basis elements in $K_{\vec{i}_e}$, labeling the edges that meet at v. The weight of two-complex Γ is then given by (9.66).[3]

---

[3]The following remarks may be useful for a reader who wants to compare the formulas given with others in the literature. For each given permutation, namely for each two-complex, and for each edge e, I have chosen a fixed orthonormal basis in the space of the intertwiners associated with the edge e. Alternatively, one can choose a basis associated with a decomposition of the four faces adjacent to e in two couples. If we do so, the propagator (9.76) contains also a matrix of change of basis. It reads

$$P^{j_1 \cdots j_4\, i,\ j_1' \cdots j_4'\, i'} \equiv \langle \phi^{j_1 \cdots j_4\, \Lambda}, \phi^{j_1' \cdots j_4'\, i'} \rangle = \frac{1}{4!} \sum_\sigma \delta^{j_1 j_{\sigma(1)}'} \cdots \delta^{j_4 j_{\sigma(4)}'}\ M_\sigma^{j_1 \cdots j_4\, i}{}_{i'},$$

(9.78)

where the matrix M_σ is given by a $6j$ symbol. Each edge contracts two vertices, say v and v', and contributes a matrix M_σ. This is the matrix of the change of basis from the intertwiner basis used at v and that used at v'. Since here I have fixed a basis of intertwiners for every e once and for all for each fixed two-complex, the matrix M_σ is automatically included in the vertex amplitude and the propagator is the identity.

Transition amplitudes. Next, consider $SO(4)$-invariant transition amplitudes in the GFT. That is, let $f[\phi]$ be an $SO(4)$-invariant polynomial
functional of the field, and consider the amplitude

$$W(f) = \int \mathcal{D}\phi \, f[\phi] \, e^{-S[\phi]} \tag{9.79}$$

and its expansion in Feynman graphs

$$W(f) = \sum_\Gamma \frac{\lambda^{v[\Gamma]}}{\mathrm{sym}[\Gamma]} \, Z_f[\Gamma]. \tag{9.80}$$

It is simplest to construct all $SO(4)$-invariant polynomial functionals of
the field in momentum space, namely as functions of the Fourier modes
$\phi_{\alpha_1 \ldots \alpha_4, i}^{j_1 \ldots j_4}$ defined in (9.69)–(9.71). To obtain an $SO(4)$ scalar, we must contract the indices α_n. We start with n field variables $\phi_{\alpha_1 \ldots \alpha_4, i}^{j_1 \ldots j_4}$, and contract
the indices pairwise in all possible manners. The resulting functional is
determined by a four-valent graph Γ giving the pattern of the indices contraction, colored with representations j_l on the links and the intertwiners
i_n on the nodes. The set of data $s = (\Gamma, j_l, i_n)$ forms precisely a spin
network. In other words, the $SO(4)$-invariant observables of the GFT are
labeled by spin networks!

Writing $n_1, \ldots, n_4$ to indicate four links adjacent to the node n, we
have

$$f_s[\phi] = \prod_n \phi_{\alpha_{n_1} \ldots \alpha_{n_4}, i_n}^{j_{n_1} \ldots j_{n_4}} \prod_l \delta_{l_1 l_2}, \tag{9.81}$$

where $n_i = l_1$ (or $n_i = l_2$) if the ith link of the node n is the outgoing (or
ingoing) link l.

For instance the spin network $s = (\Gamma, j_1 \ldots j_4, i_1 i_2)$ on a graph with two nodes
connected by four links determines the function of the field

$$f_s[\phi] = \sum_{\alpha_1 \ldots \alpha_4} \phi_{\alpha_1 \ldots \alpha_4, i_1}^{j_1 \ldots j_4} \phi_{\alpha_1 \ldots \alpha_4, i_2}^{j_1 \ldots j_4}. \tag{9.82}$$

I leave to the reader the simple exercise to show that expressions of this kind correspond
to coordinate space expressions such as

$$f_s[\phi] = \int \prod_l dg_l \prod_n \phi(g_{n_1}, \ldots, g_{n_4}) f_s(g_{n_i}), \tag{9.83}$$

where the spin network function is

$$f_s(g_{n_i}) = \prod_n v_{i_n}^{\alpha_{n_1} \ldots \alpha_{n_4}} \prod_l R^{j_l}(g_l)_{\alpha_{l_1} \alpha_{l_2}}. \tag{9.84}$$

All transition amplitudes of the GFT can therefore be expressed in terms of the spin network amplitudes

$$W(s) = \int \mathcal{D}\phi \ f_s[\phi] \ e^{-S[\phi]}. \tag{9.85}$$

Consider the Feynman expansion of these. As usual in feynmanology, the expectation value of a polynomial of order n in the fields has n external legs. In the Feynman expansion of the GFT we have, in addition, to consider the faces. I leave to the reader the simple and instructive exercise to show that these turn out to be bounded precisely by the links of the spin network. In other words, the Feynman expansion of $W(s)$ is given by

$$W(s) = \sum_{\partial\Gamma=s} \frac{\lambda^{v[\Gamma]}}{\mathrm{sym}[\Gamma]} \ Z_s[\Gamma], \tag{9.86}$$

where the sum is over all two-complexes bounded by s and the amplitude of the Feynman graph is

$$Z_s[\Gamma] = \sum_{j_f, i_e} \prod_f \dim(j_f) \prod_v \{15j\}_v. \tag{9.87}$$

The coloring on the external nodes and links is determined by s and not summed over.

Expressing this the other way around: the spinfoam sum at a fixed spin network boundary s is determined by the GFT expectation value (9.85)!

As far as the TOCY model is concerned, the duality I have just illustrated is not particularly useful. BF theory has a large invariance group that implies that the theory is topological. This implies that the corresponding spinfoam model is triangulation invariant, up to a divergent factor. Therefore, the GFT amplitudes are given by divergent sums of equal terms. On the other hand, the spinfoam/GFT duality will play a crucial role in the context of the BC models.

9.3.4 BC models

It is time to begin the return towards 4d GR. There is a strict relation between $SO(4)$ BF theory and euclidean GR. If we replace B^{IJ} in (9.54) by

$$B^{IJ} = \epsilon^{IJ}{}_{KL} \ e^K \wedge e^L, \tag{9.88}$$

we get precisely the GR action. We can therefore identify the B field of BF theory with the gravitational field $e \wedge e$. The constraint (9.88) on B, sometimes called the Plebanski constraint, transforms BF theory into GR. Can we implement the constraint (9.88) directly in the quantum theory?

An immediate consequence of (9.88) is

$$\epsilon_{IJKL}\, B^{IJ}B^{KL} = 0. \tag{9.89}$$

In 3d, the continuous variable l_f^i can be identified with the generators of the representation j_f. In 4d, it is the variable B_f^{IJ} that we can identify with a generator of the $SO(4)$ representation. If we do so, (9.89) becomes simply a restriction on the representations summed over.

Recall that the Lie algebra of $SO(4)$ is $\sim su(2) \oplus su(2)$. The irreducible representations of $SO(4)$ are therefore labeled by couples of representations of $SU(2)$, namely by two spins $j = (j_+, j_-)$. If B^{IJ} is the generator of $SO(4)$, the generators of the two $SU(2)$ groups are

$$B_\pm^i = P_{\pm IJ}^i B^{IJ}, \tag{9.90}$$

where the projectors $P_{\pm IJ}^i$ are the euclidean analogs of the projectors defined in (2.17). That is

$$P_{\pm jk}^i = \frac{1}{2}\epsilon_{jk}^i, \qquad P_{\pm 0j}^i = -P_{\pm j0}^i = \pm\frac{1}{2}\delta_j^i. \tag{9.91}$$

$SO(4)$ has two Casimirs: the scalar Casimir

$$C = B_{IJ}B^{IJ} = |B|^2, \tag{9.92}$$

and the pseudo-scalar Casimir

$$\tilde{C} = \epsilon_{IJKL}\, B^{IJ}B^{KL}. \tag{9.93}$$

This last one is precisely the quantity which is constrained to zero by (9.89). The $SO(4)$ representations in which the pseudo-scalar Casimir (9.89) vanishes are called "simple" or "balanced." The value of $\tilde{C}$ in the representation (j_+, j_-) is easy to compute because

$$\epsilon_{IJKL}\, B^{IJ}B^{KL} = B_+^i B_{+i} - B_-^i B_{-i} = j_+(j_+ + 1) - j_-(j_- + 1). \tag{9.94}$$

From (9.89) and (9.94) we have $j_+ = j_-$. The representations that satisfy this constraint, namely those of the kind $(j_+, j_-) = (j, j)$ are the simple representations; they are labeled by a single spin j. Some mathematical facts about representation theory of $SO(4)$ and about simple representations are included in Appendix A3.

This suggests that quantum GR can be obtained by restricting the sum over representations in (9.58) to the simple representations. This procedure defines a class of models denoted the BC models (from John Barrett and Louis Crane, who introduced the use of simple representations in the spinfoam formalism).

Quantum spacetime: spinfoams

Relation with loop gravity. In the discretization of BF theory, we have discretized the B field, which is a two-form, by assigning a variable B_f to each triangle f of the triangulation. B_f can be taken to be the surface integral of B on f. We can discretize the gravitational field e, which is a one-form, by assigning a variable e_s to each segment s of the triangulation. e_s can be taken to be the line integral of e along the segment s. Equation (9.88) then relates the variable B_f on a triangle with the variables e_s on two sides of the triangle. The scalar Casimir (9.92) can be expressed in terms of the gravitational field using (9.88), obtaining

$$C = |e \wedge e|^2 \tag{9.95}$$

which is the square of the area of the triangle. Hence, the Casimir of the representation j_f gives the area of the triangle f. The spin j_f can therefore be interpreted as the quantum number of the area of the triangle f.

Consider the case in which the triangulated manifold has a boundary, and the triangle f belongs to the boundary. In the two-complex picture, f is a face dual to the f triangle, and j_f is associated with this face. The face f cuts the boundary along a link, which is one of the links of the boundary spin network. The color of the link is j_f. This link intersects once and only once the triangle f. Hence, we conclude that the representation associated with the link that intersects the triangle f is the quantum number that determines the area of the triangle f. But this is precisely the result that we have obtained in hamiltonian theory in Chapter 6!

The boundary states of a BC spinfoam model is a spin network whose links carry quanta of area labeled by a spin j. This is *precisely* the structure of the states of hamiltonian loop quantum gravity. Spinfoam models and hamiltonian LQG "talk" very nicely to each other.

The other constraints. Equation (9.89) does not imply (9.88). On the other hand, the system formed by (9.89) and the two equations

$$\epsilon_{IJKL} \, B_{\mu\nu}^{IJ} B_{\nu\rho}^{KL} = 0, \tag{9.96a}$$

$$\frac{4! \epsilon_{IJKL} \, B_{\mu\nu}^{IJ} B_{\rho\sigma}^{KL}}{\epsilon^{\mu\nu\rho\sigma} \epsilon_{IJKL} \, B_{\mu\nu}^{IJ} B_{\rho\sigma}^{KL}} = \epsilon_{\mu\nu\rho\sigma} \tag{9.96b}$$

(no sum over repeated spacetime indices in the first equation) does. It follows that if we consider BF theory plus the additional constraint (9.96) the resulting theory has precisely all the solutions of GR.

Two remarks. First, GR has far more solutions than BF theory, which is topological. The fact that adding a constraint on B increases the number of solutions should not surprise us. In BF theory, B plays the role of Lagrange multiplier, enforcing the vanishing of the curvature. By constraining B in the action we reduce the number of independent components of the Lagrange multiplier. Hence we get less constraints on the curvature. Hence more solutions. If you jail some guards, more thieves go free.

Second, the system (9.96) has other classes of solutions, beside (9.88). In particular,

$$B^{IJ} = -\epsilon^{IJ}{}_{KL}\, e^K \wedge e^L, \qquad \text{and} \qquad B^{IJ} = \pm e^I \wedge e^J. \qquad (9.97)$$

The first simply redefines orientation. The others give a topological term in the action, that has no effect in the classical equations of motion. I refer the reader to the literature for a full discussion of this point.

The constraint (9.96a) can be implemented in the quantum theory by identifying the couple of variables $B^{IJ}_{\mu\nu} B^{KL}_{\nu\rho}$ that share an index with generators of the representations associated with faces that share an edge. In turn, for the constraint (9.96b) we identify the two $B^{KL}_{\rho\sigma} B^{KL}_{\nu\rho}$ variables without common indices with generators of the representations associated to opposite faces of a tetrahedron.

There are different ways of discretizing these constraints that have been considered in the literature, yielding different BC models. There are also different possible choices of the face and edge amplitudes A_f and A_e that have been considered in the literature. The situation is still unclear as to which variants correspond to discretized GR. I shall not enter into the details of motivations of the different choices here (see [19] for a detailed discussion). Rather, I simply illustrate some of the models, referring to the literature for their motivation.

BCA. The simplest of the BC models, denoted BCA model (Barrett–Crane model, version A), is simply obtained by choosing $A_e = A_f = 1$. It is given by

$$Z_{\text{BCA}} = \sum_{\text{simple } j_f} \sum_{i_e} \prod_v \qquad (9.98)$$

BCB. A second model, denoted BCB (Barrett–Crane model, version B), assumes that the intertwiners are constrained by (9.96) to the form

$$i^{(aa')(bb')(cc')(dd')}_{\text{BC}} = \sum_j (2j+1)\, v^{abf} v^{fcd}\, v^{a'b'f'} v^{f'c'd'}, \qquad (9.99)$$

where indices in an $SO(4)$ representation are given by couples of indices in an $SU(2)$ representation, and the indices f and f' are in the representation j. This is called the BC intertwiner. See [269] for details on the relation between this intertwiner and (9.96). The BC intertwiner has the property of being formed by a simple virtual link in any decomposition. It is the unique state with this property. Using the representation (6.86), it is given by

$$i_{bc} = \sum_j (2j+1) \; j \Big\rangle\!\!\Big\langle \qquad j \Big\rangle\!\!\Big\langle \quad . \tag{9.100}$$

Choosing $A_e = 1$ we obtain the sum

$$Z_{\mathrm{BCB}} = \sum_{\mathrm{simple}\; j_f} \prod_f \dim(j_f) \prod_v \qquad . \tag{9.101}$$

The vertex amplitude

$$A(j_1,\ldots,j_{10}) = \tag{9.102}$$

$$= \sum_{i_1\ldots i_5} \tag{}$$

depends on ten spins, and is called a $10j$ symbol. Thus we write (9.101) as

$$Z_{\mathrm{BCB}} = \sum_{\mathrm{simple}\; j_f} \prod_f \dim(j_f) \prod_v \{10j\}. \tag{9.103}$$

Therefore, the BCB model has the general form (9.21), with the following choices:

- The set of two-complexes summed over is formed by a single two-complex. This is chosen to be the two-skeleton of the dual of a 4d triangulation.
- The representations summed over are the *simple* unitary irreducible representations of $SO(4)$. All intertwiners are fixed to be i_{BC}.
- The vertex amplitude is $A_v = \{10j\}$.

These choices define a tentative quantization of 4d riemannian GR on a fixed two-complex, namely with a cut-off at the high-frequency modes. Clearly, a fixed two-complex can accommodate only a finite number of degrees of freedom, and cannot capture all the degrees of freedom of the theory, which are infinite.

BCC. A variant of the model B is of particular interest since, as we shall see, it is perturbatively finite. This is obtained by adding an edge amplitude to the BCB model:

$$Z_{\text{BCC}} = \sum_{\text{simple } j_f} \prod_f \dim(j_f) \prod_e A_e(j_{e_1}, \dots, j_{e_4}) \prod_v \{10j\}, \qquad (9.104)$$

where A_e is a function of the four representations $j_{e_1}, \dots, j_{e_4}$ associated to the four faces bounded by the edge e and is defined as follows. Let $\mathcal{H}_{j_1,\dots,j_4}$ be the tensor product of the four representations $j_1, \dots, j_4$ and $\mathcal{H}^0_{j_1,\dots,j_4}$ its invariant subspace. Then A_e is the ratio of the dimensions of these spaces

$$A_e(j_1, \dots, j_4) = \frac{\dim \mathcal{H}^0_{j_1,\dots,j_4}}{\dim \mathcal{H}_{j_1,\dots,j_4}}. \qquad (9.105)$$

As we shall see below, this amplitude emerges naturally in the group field theory (GFT) context.

At present it is not clear which of these variants, or others, is the most physically interesting one. Interpreting the choice as a choice of sum-over-paths measure and imposing diff invariance, Bojowald and Perez have obtained indications in favor of certain models [270]. In [271], the different statistical properties of these models have been analyzed numerically; however, it is not yet clear which are the "correct" statistical properties to be expected.

Geometrical interpretation of the Plebanski constraints. The constraints on the representations and intertwiners that define the BC models can be given a geometrical interpretation. In fact, they were first obtained from an independent set of considerations, based on this geometrical interpretation. Consider a tetrahedron embedded in R^4. Denote vectors in

R^4 as $v = (v^I), I = 1, \ldots, 4$. Label the four vertices of the tetrahedron as $v_i, i = 1, \ldots, 4$. A vector

$$v_{(ij)} = v_j - v_i \qquad (9.106)$$

describes the edge (ij) of the tetrahedron. The triangle (ijk) can be described by the "bivector"

$$v_{(ijk)}^{IJ} = v_{(ij)}^I \, v_{(jk)}^J - v_{(ij)}^J \, v_{(jk)}^I, \qquad (9.107)$$

often written as

$$v_{(ijk)} = v_{(ij)} \wedge v_{(jk)}. \qquad (9.108)$$

For instance, the triangle determined by the three vertices (123) is described by the bivector

$$v_{(123)}^{IJ} = v_1^I v_2^J - v_2^I v_1^J + v_2^I v_3^J - v_3^I v_2^J + v_3^I v_1^J - v_1^I v_3^J = \sum_i \epsilon^{ijk} v_j^I v_k^J \qquad (9.109)$$

obtained by inserting (9.106) in (9.107). The area of the triangle (ijk) is the norm of the bivector

$$A_{(ijk)}^{IJ} = v_{(ijk)}^{IJ} v_{(ijk) IJ}. \qquad (9.110)$$

From the definition,

$$\epsilon_{IJKL} \, v_{(ijk)}^{IJ} v_{(ijk)}^{KL} = 0, \qquad (9.111)$$

$$\epsilon_{IJKL} \, v_{(ijk)}^{IJ} v_{(ijl)}^{KL} = 0. \qquad (9.112)$$

Furthermore, consider a four-simplex embedded in R^4, with vertices v_i where $i = 1, \ldots, 5$. This defines the four tetrahedra $(ijkl)$, and the ten triangles (ijk). Then for two triangles sharing a vertex i

$$\epsilon_{IJKL} \, v_{(ijk)}^{IJ} v_{(ilm)}^{KL} = \sum_i \epsilon^{ijklm} \epsilon_{IJKL} \, v_j^I v_k^J v_l^K v_m^L, \qquad (9.113)$$

which is independent of i. Hence we can write

$$\frac{\epsilon_{IJKL} \, v_{(ijk)}^{IJ} v_{(ilm)}^{KL}}{\sum_i \epsilon^{ijklm} \epsilon_{IJKL} v_j^I v_k^J v_l^K v_m^L} = \epsilon_{jklm}. \qquad (9.114)$$

Notice the remarkable similarity of the bivector equations (9.111), (9.112) and (9.114) with the Plebanski constraints (9.89)–(9.96). The bivector equations can be interpreted as a discretization of the Plebanski constraints. Expressed the other way around, the Plebanski constraints can be interpreted as the requirement that the B field of the BF theory is an infinitesimal area element of elementary triangles in spacetime.

Quantum tetrahedron. The historical path to the BC models has been the one described above. I remember long hours of discussion with Louis Crane searching unsuccessfully for a way to implement the Plebanski constraint (9.89) within his TOCY model. But in the seminal work [269] which brilliantly solves the problem in terms of simple representations, no reference is made to the Plebanski constraint and the relation between BF and GR. The paper is based on a description of the intrinsic metric degrees of freedom of a single tetrahedron embedded in R^4, and a "quantization" of these degrees of freedom, as I now describe.

Consider a single tetrahedron embedded in R^4. View it as a physical system, whose dynamical variables are given by its geometry. We can expect that the quantum properties of this system are described by a quantum state space $\mathcal{H}$ and dynamical variables be represented by operators in this state space. The geometry of the tetrahedron can be described in terms of the bivectors $v_{(ijk)}^{IJ}$ defined in the previous section. Thus, bivectors will be represented by operators on $\mathcal{H}$. This construction defines a sort of "quantum tetrahedron." Since $SO(4)$ acts naturally on the bivectors, we expect that $\mathcal{H}$ carries a representation of $SO(4)$. Since classical bivectors transform in the adjoint representation, we expect the quantum operators to do the same. It follows that bivector operators $v_{(ijk)}^{IJ}$ are the infinitesimal generators of a representation j of $SO(4)$ in $\mathcal{H}$. The quadratic expression (9.110) giving the area of the triangle is precisely one of the two Casimirs of j. Hence, $\mathcal{H}$ will carry a representation j_{ijk} such that its Casimir is the area of the triangle (ijk). The other Casimir of $SO(4)$ is given by (9.111) constrained to vanish. Hence, $\mathcal{H}$ will contain only representations in which this Casimir vanishes. These are the simple representations. The sum (9.101) can then be constructed as a sum-over-states in the Hilbert space where the bivector operators live.

This method of arriving at the BC models has the shortcoming of hiding its relation with GR as well as with conventional quantization procedures. On the other hand, it has the virtue of opening an entirely new interesting perspective on quantum spacetime. The convergence of different ways of thinking of a model, which shed light on each other, is always valuable.

In fact, the idea of the quantum tetrahedron as an elementary system can perhaps be taken seriously on physical grounds. Ordinary QFT on a background has two natural physical interpretations. The two correspond to two different choices of families of observables. For instance, in free electromagnetism we can measure the electric field or the magnetic field and interpret the theory as a theory for a continuous field. Alternatively, we can measure energies and momenta, and interpret the theory as describing particles moving in spacetime, the photons. There is no contradiction, of course, between the two descriptions, for the same reason that there is no contradiction between the continuity of the elongation of a harmonic oscillator and the discreteness of its energy.

Similarly, we can construct the QFT starting from the quantization of the classical field theory and deriving the existence of the particles. Or we can construct the QFT starting from the particles: define the quantum theory of a single particle, then the "many-particle" quantum theory of an arbitrary number of particles, and so on. The two roads yield the same theory, as is well known. In an interacting QFT, nontrivial dynamics can be expressed by simple interaction vertices between the particle states.

In GR, loop quantization shows that space has a granular structure at short scale. Space can be thought of as made up of individual quanta of space (which can be connected to each other). These quanta of space are, like the particles, eigenstates of certain measurable quantities. It is then not unreasonable to think that we can reinterpret quantum GR as a "many-particle" theory built up from a quantum theory of a single quantum of space. The mathematics of the "quantum tetrahedron" can perhaps be seen as a first step in this direction. If we take this point of view, then dynamics can be expressed by simple interaction vertices between these "particle" states. For instance, an elementary vertex such as the one in Figure 9.3 can be interpreted as one quantum of space (connected to three other quanta, not represented) decaying into three quanta of space (connected to each other and to the three other quanta not represented), and so on. Spacetime is then an history of interactions of a variable number of quanta of space.

9.3.5 Group field theory

The remaining step to arrive at a model with some chance of describing quantum GR is to implement a sum over two-complexes, so that the infinite number of degrees of freedom of the theory could be captured.

Notice that in order to capture all degrees of freedom, we do not have the option of *refining* the triangulation (or the two-complex) as one does in lattice QCD. The reason is that there is in fact nothing to refine, no parameter such as the lattice spacing of lattice QCD, which can be set to zero.

In fact, the cut-off introduced by the Barrett–Crane models is not an ultraviolet cut-off. The theory does not have an ultraviolet sector because there are no degrees of freedom beyond Planck scale. Rather, it is a sort of infrared cut-off, in the sense that a fixed triangulation cannot capture configurations that can be written on a larger triangulation (a triangulation with more n-simplices). The sum includes arbitrary large geometries because it includes arbitrary high j_f (not in the quantum group case). But on a fixed Δ a large geometry can be represented only by large j_f, and not by small j_f over a larger triangulation. Clearly, this restriction reduces dramatically the class of continuous fields that the spinfoam can approximate.

We could think of defining the model in the limit of large Δ. This is certainly an interesting direction to explore. On the other hand, in summing over colorings of a large Δ we have to include configurations in which the representations are trivial except on a subset Δ' of Δ. The amplitude of this configuration can be viewed as associated with Δ' rather than Δ. Hence, we naturally fall back to a sum-over-triangulations.

For fixed boundary conditions yielding a classical geometry of volume $N l_P^4$, it may be reasonable to assume that triangulations with a number of four-simplices much larger than N would not contribute much. Hence the expansion in the size of the triangulation might be of physical interest.

How do we sum over two-complexes? The problem is to select a class of two-complexes to sum over, and to fix the relative weight. Now, the duality that I have illustrated above in Section 9.3.3 provides precisely a prescription for summing over two-complexes. It is therefore natural to take a dual formulation of the BC models as a natural ansatz for the complete sum over two-complexes. But is there a dual formulation of the BC models, or is duality a feature of the much simpler topological BF model?

Remarkably, a dual formulation of the BC models exists. BC models are obtained from the TOCY models by restricting representations to the simple ones. This restriction implements the constraints that transform BF theory into GR. In the dual picture, the sum over representations is obtained as an expansion of the field over the group in modes. A generic field can be expanded in a sum over all unitary irreducible representations. How can we pick a field whose expansion contains only simple representations? The answer turns out to be easy.

Pick a fixed $SO(3)$ subgroup H of $SO(4)$. Then the following holds. A field $\phi(g)$ on $SO(4)$ is invariant under the action of H, namely satisfies

$$\phi(g) = \phi(gh), \qquad \forall h \in H, \qquad (9.115)$$

if and only if its mode expansion contains only *simple* irreducible representations. This is an elementary result, described in Appendix A3.

Consider the field theory defined in Section 9.3.3. It is useful to slightly simplify this notation. First, write the action in the shorthand notation

$$S[\phi] = \frac{1}{2} \int \phi^2 + \frac{\lambda}{5!} \int \phi^5. \qquad (9.116)$$

Second, instead of demanding the field to satisfy (9.62), we can take an arbitrary field ϕ, not necessarily satisfying (9.63), and use the projection operator P_G defined by

$$P_G \phi(g_1, g_2, g_3, g_4) = \int_{SO(4)} dg \ \phi(g_1 g, g_2 g, g_3 g, g_4 g). \qquad (9.117)$$

We can also define the projector $_G P$ acting on the left

$$_G P \phi(g_1, g_2, g_3, g_4) = \int_{SO(4)} dg \ \phi(g g_1, g g_2, g g_3, g g_4). \qquad (9.118)$$

Define now the projector P_H on the simple representations

$$P_H \phi(g_1, g_2, g_3, g_4) = \int_{H^4} dh_1 \ldots dh_4 \ \phi(g_1 h_1, g_2 h_2, g_3 h_3, g_4 h_4). \qquad (9.119)$$

GFT/TOCY. The action

$$S[\phi] = \frac{1}{2}\int (P_G\phi)^2 + \frac{\lambda}{5!}\int (P_G\phi)^5 \tag{9.120}$$

for a generic field is equivalent to the action (9.63) and yields the TOCY model, as discussed in Section 9.3.3. Now, by simply inserting the projector P_H into this action we obtain the following surprising results.

GFT/A. Consider the action

$$S_A[\phi] = \frac{1}{2}\int ({}_GPP_H\phi)^2 + \frac{\lambda}{5!}\int ({}_GPP_H\phi)^5. \tag{9.121}$$

The Feynman expansion of the partition function of this theory gives

$$Z_A = \int \mathcal{D}\phi\, e^{-S[\phi]} = \sum_\Gamma \frac{\lambda^{v[\Gamma]}}{\mathrm{sym}[\Gamma]}\, Z_A[\Gamma]. \tag{9.122}$$

The amplitude of a Feynman graph turns out to be precisely the partition function (9.98) of the BCA model, where the model is over the two-complex determined by the Feynman graph! That is,

$$Z_A[\Gamma] = \sum_{\text{simple } j_f} \prod_v \{15j\} = Z_{\text{BCA}}. \tag{9.123}$$

GFT/B. The action

$$S_B[\phi] = \frac{1}{2}\int (P_G P_H\phi)^2 + \frac{\lambda}{5!}\int (P_G P_H\phi)^5, \tag{9.124}$$

gives the partition function (9.101) of the BCB model

$$Z_B[\Gamma] = \sum_{\text{simple } j_f} \prod_f \dim(j_f) \prod_v \{10j\} = Z_{\text{BCB}}. \tag{9.125}$$

GFT/C. The action

$$S_C[\phi] = \frac{1}{2}\int (P_G\phi)^2 + \frac{\lambda}{5!}\int (P_H P_G\phi)^5 \tag{9.126}$$

yields the partition function (9.104) of the BCC model

$$Z_C[\Gamma] = \sum_{\text{simple } j_f} \prod_f \dim(j_f) \prod_e A_e(j_{e_1}, \dots, j_{e_4}) \prod_v \{10j\}$$

$$= Z_{\text{BCC}}. \tag{9.127}$$

The derivation of these relations is a rather straightforward application of the mode expansion described in Section 9.3.3. I leave it as a good exercise for the reader. It can be found in the original papers quoted at the end of the chapter.

Now, in the TOCY case, the sum over Feynman graphs is trivial, divergent, and without physical motivation. It is trivial in the sense that all terms are equal, due to triangulation invariance. It is divergent because we sum an infinite number of equal terms. It has no physical motivation, because all the degrees of freedom of the classical theory are already captured by a finite triangulation.

In the case of the BC models, on the other hand, a choice of a fixed two-complex reduces the number of degrees of freedom of the theory. Therefore, a sum over two-complexes is necessary, if we hope to define quantum GR, and the Feynman expansion provides precisely such a sum. Since the BC models are not triangulation invariant, the sum is not trivial: each two-complex contributes in a different manner to the sum. The sum over two-complexes defined by the GFT defines in these cases a new spinfoam model, where the number of degrees of freedom are not cut off. We denote these as GFT spinfoam models. In particular, we denote the sum defined in (9.122) as the group field theory version A or GFT/A, and the corresponding ones in cases B and C as the GFT/B and GFT/C.

What about finiteness? Remarkably, the GFT/C appears to be finite at all orders in λ. The proof has been completed, up to certain degenerate two-complexes, which, however, are likely not to spoil the result. Therefore the GFT/C model, or some variant of it, can be taken as a tentative ansatz for a covariant definition of transition amplitudes in euclidean quantum gravity.

In particular, we can consider expectation values of spin networks as in (9.85)

$$W(s) = \int \mathcal{D}\phi \; f_s[\phi] \; e^{-\frac{1}{2} \int (P_G \phi)^2 + \frac{\lambda}{5!} \int (P_G P_H \phi)^5}. \qquad (9.128)$$

These quantities are well defined and are likely to be convergent at every order in λ. We can tentatively interpret them as transition amplitudes of euclidean quantum gravity.

9.3.6 Lorentzian models

The fact that expression (9.128) provides a finite tentative definition of quantum gravitational transition amplitudes is certainly exciting, but the GFT models described above are all euclidean. There are two possible directions to recover the physical lorentzian theory from here.

SO(3) and SO(2,1) lorentzian GFT. One direction for defining lorentzian amplitudes is to study the lorentzian analogs of these models. These can be obtained by simply replacing $SO(4)$ with the Lorentz group $SO(3,1)$. Let $\phi(g_1, \ldots, g_4)$ be a field on $[SO(3,1)]^4$. Define the projectors P_G and

P_H as in (9.117) and (9.119). To define the projector H there are two natural choices for the subgroup $H \subset SO(3,1)$. The first is to take it to be a fixed $H = SO(3)$ subgroup of $SO(3,1)$. This is the subgroup that keeps a chosen timelike vector invariant. The second is to take it as an $H = SO(2,1)$ subgroup of $SO(3,1)$. This is the subgroup that keeps a chosen spacelike vector invariant. Consider the action

$$S_H[\phi] = \frac{1}{2} \int (P_G \phi)^2 + \frac{\lambda}{5!} \int (P_G P_H \phi)^5 \qquad (9.129)$$

for the two cases. The Feynman expansion of the partition function defines two lorentzian spinfoam sums, denoted the $SO(3)$ and the $SO(2,1)$ lorentzian GFT, respectively.

The set of the representations that appear in these spinfoam sums is determined by the mode expansion of the field. This implies that the representations summed over are the unitary irreducible representations. These are infinite-dimensional for $SO(3,1)$, which is noncompact, and labeled also by a continuous variable. Therefore, in the lorentzian spinfoam models, sums over internal indices are replaced by integrals, and the spinfoam sum contains an integral over a continuous set of representations. Still, the technology developed above extends to this case quite well. There is an extensive literature on this, to which I refer the interested reader; see bibliographical notes at the end of the chapter. Most of the features of the euclidean theory persist in the lorentzian case. Most remarkably, finiteness results have been extended to the lorentzian $SO(3)$ GFT.

Unitary representations of $SO(3,1)$ fall naturally into two classes: spacelike and timelike ones, distinguished by the sign of the Casimir representing the square of the area. The action $S_{SO(3)}$ above gives rise to a sum over just the timelike representations; while the action $S_{SO(2,1)}$ gives rise to a sum over both kinds of representations. The kind of the representation associated with a triangle f gives a spacelike or timelike character to the triangles of the triangulation. This is physically appealing, but several aspects of this issue are unclear at this time. For instance, the sign of the square of the area appears to be opposite to what one would expect on the basis of the hamiltonian theory. This is an intriguing open problem that I signal to the reader.

Analytic continuation. The second possible direction for the construction of the physical theory is to define the lorentzian transition amplitudes from the euclidean ones by analytic continuation, as is done in conventional QFT. Standard theorems relating euclidean transition amplitudes to a lorentzian QFT are grounded on Poincaré invariance and do not extend to gravity. However, this does not imply that the project of defining physical transition amplitudes from the euclidean theory by some form of analytic continuation must necessarily fail. In particular, analytic continuation in the time coordinate is likely to be completely nonappropriate in a background-independent context; but analytic continuation in a physical time might be viable. I discuss this possibility below in Section 9.4.1.

9.4 Physics from spinfoams

A spinfoam model can be used to compute an amplitude $W(s)$ associated to any boundary spin network s. How can we relate these amplitudes to physical measurements?

Relation with hamiltonian theory. The spinfoam formalism has formal similarities with lattice gauge theory. The interpretation of the two formalisms, however, is quite different. In the case of lattice theories, the discretized action depends on a parameter, the lattice spacing a. The physical theory is recovered as a is taken to zero. In this limit, the discretization introduced by the lattice is removed. In particular, we can approximate continuous boundary fields in terms of sequences of lattice discretization. In gravity, on the other hand, there is no lattice spacing parameter a in the discretized action. Therefore, there is no sense in the $a \to 0$ limit. The discrete structure of the spinfoams must reflect actual features of the physical theory. In particular, boundary states are given by spin networks, not by continuous field configurations approximated by sequences of discretized fields.

The hamiltonian theory of Chapter 6 provides a physical interpretation for the boundary spin network s. Spin networks are eigenstates of area and volume operators, and can therefore be interpreted as describing the result of measurements of the geometry of a 3d surface. Such measurements do not correspond to complete observables, since they do not commute with the hamiltonian operator. However, they correspond to partial observables, a notion explained in Chapter 3. Therefore, they are still described by operators in the kinematical quantum state space, as argued in Chapter 5. In particular, the discrete spectrum of these operators *can* be interpreted as a physical prediction on the possible outcome of a physical measurement of these observables. As recalled at the beginning of this chapter, transition amplitudes do not, in general, depend on classical configurations: they depend on quantum eigenstates. This is why we expect quantum gravity transition amplitudes to be functions of spin networks, and not of continuous three-geometries. The fact that the spinfoam formalism yields precisely such functions of spinfoams is, therefore, satisfying. There is thus a strong and encouraging consistency between the physical picture of nonperturbative quantum gravity provided by spinfoam theory and by hamiltonian LQG.

It would be very good if we were able to translate directly between the two formalisms. A sketch of a formal derivation of a spinfoam sum from the loop formalism was given at the beginning of this chapter. Expressing this the other way around, it would be very interesting to reconstruct in detail the hamiltonian Hilbert space, as well as kinematical and dynamical operators of the loop theory, starting from the covariant spinfoam definition of the theory. At present, neither of these two paths is under complete control. The first would amount to a derivation of a Feynman–Kac formula (see for instance [272]) valid in the diffeomorphism-invariant context.

The second would amount at an extension to the diffeomorphism-invariant context of the Wightman and Osterwalder–Schrader reconstruction theorems [273]. There are two ways in which we can derive the Hilbert state space from a spinfoam model. One is to identify $\mathcal{K}$ with a linear closure of the set of the boundary data. This is the philosophy I have used above. The other is to directly reconstruct $\mathcal{H}$ from the amplitudes $W(s)$ using the Gelfand–Neimark–Siegal construction. This approach has been developed in [274] (see also [275]).

At an even more naive level, there are several gaps between the hamiltonian loop theory and the spinfoam models that have been discussed so far. These concern: the role of the selfdual connection; the role of the $SO(3,1) \to SO(3)$ gauge-fixing used in the hamiltonian framework; the fact that in the spinfoam models so far considered there are only four-valent nodes; the fact that in the GFT/B and C models there is no free boundary intertwiner variable associated with the node; the eventual role of the quantum group deformation in the hamiltonian theory [228]; and others. All these aspects of the relation between the two formalisms need to be clarified, before being able to cleanly translate between the two.

On the one hand, there is much latitude in the definition of the hamiltonian operator, as explained in Chapter 7. In general, covariant methods deal more easily with symmetries, and interaction vertices have a simple form in the covariant picture. For instance, compare the complexity of the full QED hamiltonian with the simplicity of the QED single vertex, which compactly summarizes all hamiltonian interaction terms. The spinfoam formalism could suggest the correct form of the hamiltonian operator.

On the other hand, hamiltonian methods are more precise and rigorous than covariant sums-over-paths. As we have seen, the hamiltonian picture provides a clean and well-motivated physical interpretation for the boundary spin network states, as well as a general justification of the discreteness of spacetime. Therefore, the two formalisms shed light on each other and their relation needs to be studied in detail.

(*Note added in the paperback edition:* The precise equivalence of the spinfoam and hamiltonian LQG formalisms has been proven rigorously by Karim Noui and Alejandro Perez. See Bibliographical notes below.)

9.4.1 Particles' scattering and Minkowski vacuum

Finally, I sketch here a way to compute the Minkowski vacuum state from the spinfoam formalism, following the general theory described in Section 5.4. This is the first step to define particle states. Consider a three-sphere formed by two "polar" *in* and *out* regions and one "equatorial" *side* region. Let the matter + gravity field on the three-sphere be split as $\varphi = (\varphi_{\text{out}}, \varphi_{\text{in}}, \varphi_{\text{side}})$. Fix the equatorial field φ_{side} to take the special value φ_{RT} defined as follows. Consider a cylindrical surface Σ_{RT} of radius R and height T in R^4, as defined above. Let Σ_{in} (and Σ_{out}) be a (3d) disk located within the lower (and upper) basis of Σ_{RT}, and let Σ_{side} be the part of Σ_{RT} outside these disks, so that $\Sigma_{RT} = \Sigma_{\text{in}} \cup \Sigma_{\text{out}} \cup \Sigma_{\text{side}}$. Let g_{RT} be the metric of Σ_{side} and let $\varphi_{RT} = (g_{RT}, 0)$ be the boundary field on Σ_{side} determined by the metric being g_{RT} and all other fields being zero. Given arbitrary values φ_{out} and φ_{in} of all the fields, including the metric, in the two disks, consider $W[(\varphi_{\text{out}}, \varphi_{\text{in}}, \varphi_{RT})]$. In writing the boundary field as composed of three parts $\varphi = (\varphi_{\text{out}}, \varphi_{\text{in}}, \varphi_{\text{side}})$ we are in fact splitting $\mathcal{K}$ as $\mathcal{K} = H_{\text{out}} \otimes H_{\text{in}}^* \otimes H_{\text{side}}$. Fixing $\varphi_{\text{side}} = \varphi_{RT}$ means contracting the covariant vacuum state $|0_\Sigma\rangle$ in $\mathcal{K}$ with the bra state $\langle\varphi_{RT}|$ in H_{side}. For large enough R and T, we expect the resulting state in $H_{\text{out}} \otimes H_{\text{in}}^*$ to reduce to the Minkowski vacuum. That is (again, bra/ket mismatch is apparent only)

$$\lim_{R,T\to\infty} \langle\varphi_{RT}|0_\Sigma\rangle = |0_M\rangle \otimes \langle 0_M|. \tag{9.130}$$

For a generic *in* configuration, and up to normalization,

$$\Psi_M[\varphi] = \lim_{R,T\to\infty} W[(\varphi, \varphi_{\text{in}}, \varphi_{RT})]. \tag{9.131}$$

(Below, I shall use a simpler geometry for the boundary.) These formulas allow us to extract the Minkowski vacuum state from a euclidean spinfoam formalism. n-particle scattering states can then be obtained by generalizations of the flat-space formalism, and, if this is well defined, by analytic continuation in the single variable T.

Consider a spin network that we denote as $s' = s \# s_T$, composed of two parts s and s_T connected to each other, where s is arbitrary and s_T is a weave state (6.143) of the three-metric g_T defined as follows. Take a 3-sphere of radius T in R^4. Remove a spherical 3-ball of unit radius. g_T is the three-metric of the three-dimensional surface (with boundary) formed by the sphere with removed ball. The quantity

$$\psi_M[s] = \lim_{T\to\infty} \int D\Phi \; f_{s\#s_T}[\Phi] \; e^{-\frac{1}{2}\int (P_G\phi)^2 + \frac{\lambda}{5!}\int (P_G P_H \phi)^5} \tag{9.132}$$

represents an ansatz for the Minkowski vacuum state in a ball of unit radius.

(*Note added in the paperback edition:* A technique for computing n-point functions from background independent quantum gravity has been proposed and developed. This has allowed the derivation the graviton propagator, which is, so to say, the derivation of "Newton law" from the theory without space and time. See Bibliographical notes below.)

Bibliographical notes

John Baez gives a nice and readable introduction to BF theory and spinfoams in [17], which contains also an invaluable, carefully annotated bibliography. A good general review is Daniele Oriti's [18]. In [19], Alejandro Perez describes the group field theory in detail and gives an extensive overview on different models. On the derivation from hamiltonian loop theory, see [11].

The idea of describing generally covariant QFT in terms of a "sum-over-surfaces" was initially discussed in [276–278]; the formal derivation from LQG in [279–281]. The relation spinfoams/triangulated spacetime was clarified by Fotini Markopoulou [282].

The Ponzano–Regge model was introduced in [283]. On the precise relation between $6j$ symbols and Einstein action, see [284]. Regge calculus is introduced in [285]; the Turaev–Viro model in [286]. The duality between spinfoam models and QFT on groups was pointed out by Boulatov in [287], as a duality between a QFT on $[SU(2)]^4$ and the 3d Ponzano–Regge model. Boulatov's aim was to extend from 2d to 3d the duality between the "matrix models" and 2d quantum gravity [21], or "zero dimensional string theory" [288]. (For a while, it was hoped that matrix models would provide a background-independent definition of string theory. More recently, they have been extensively developed and used in a range of applications.) The result was extended by Ooguri to 4d in [289], yielding the TOCY model described in this chapter; see also [290]. (BF theories were discussed in [291].) The precise construction of the quantum deformed version of this model and the proof of its triangulation independence were given in [292]. The relation between Ponzano–Regge model, LQG and length quantization was pointed out in [293].

For the construction of the BC models, I have followed Roberto DePietri and Laurent Freidel [294]; see also [295, 296]. The idea of the quantum tetrahedron was discussed by Andrea Barbieri in [297] and by John Baez and John Barrett in [298], and used to construct the spinfoam model for GR in [269] and [299], where the cute term "spinfoam" was introduced. The models BCA and BCB were defined in [300]. The model BCC was defined in [301] and [302]. The statistical behavior of different models is tentatively explored in [271]. Arguments for preferring some of these models on the basis of diff invariance are considered in [270]. The fact that the duality between spinfoam models and QFT on groups extends to BC models was noticed by DePietri and is presented in [300]. The remarkable fact that it can be extended to arbitrary spinfoam models was noticed by Michael Reisenberger and presented in [303]; see also [304]. The GFT finiteness proof appeared in [305]. For recent (2007) discussions on the group field theory approach, and updated references, see [306] and [307]. An intriguing recent result is the observation by Laurent Freidel and Etera Livine that a quantum theory of gravity plus matter in 3d is equivalent to a matter theory over a noncommutative spacetime [308].

The convergence of the full series in 3d is discussed in [309]. Lorentzian Ponzano–Regge models are discussed in [310]; lorentzian BC models in [311]. Quantum group versions of lorentzian models have been studied in [231], a paper I recommend for a detailed introduction to the subject and its extensive references. As for the euclidean case, the quantum deformation of the group controls the divergences and can be related to a cosmological term in the classical action. The lorentzian GFT model $S_{SO(3)}$ is defined in [312]. The lorentzian GFT finiteness proof for this model appeared in [313]. The model $S_{SO(2,1)}$ is defined in [314]. On the problem of the timelike/spacelike character of the representations, see [315] and references therein. Functional integral methods to define spinfoam models are discussed in [316].

A different approach to the lorentzian sum-over-histories is developed in [317]. An interesting variant of the spinfoam formalism, in which propagation forward and backward in time can be distinguished, has been introduced in [318].

On the reconstruction of the Hilbert space from the amplitudes of a spinfoam model, see [274] and [275]. On the relation between spinfoam and canonical LQG, see also [319, 320]. A complete proof of the equivalence in 3d has been obtained by Karim Noui and Alejandro Perez in [321]. On the role of the Immirzi parameter in BF theory, see [322]. The ansatz for the derivation of the Minkowski vacuum from spinfoam amplitudes appeared in [145]. A hamiltonian approach to the construction of coherent states representing classical solutions of the Einstein equations is being developed by Thiemann and Winkler [323]. A causal version of the spinfoam formalism has been studied by Fotini Markopoulou and Lee Smolin [324]. The extension of the spinfoam models to matter coupling is still embryonic, see [325, 326]. On the Peter–Weyl theorem and harmonic analysis on groups in general, see for instance [327, 328].

A background independent difinition of n-point functions has been given in [329]. The graviton propagator has been derived in [330]. See also [331].

10
Conclusion

In this book I have tried to present a compact and unified perspective on quantum gravity, its technical aspects and its conceptual problems, the way I understand them. There are a great number of other aspects of quantum gravity that I would have wished to cover, but my energies are limited. Just to mention a few of the major topics I have left out: $2 + 1$ gravity, the Kodama state and related results, quantum gravity phenomenology, supergravity in LQG, coherent states... Here I briefly summarize the physical picture that emerges from LQG and the solution it proposes to the characteristic conceptual issues of quantum gravity. I conclude with a short summary of the main open problems and the main results of the theory.

10.1 The physical picture of loop gravity

The effort to develop a quantum theory of gravity forces us to revise some conventional physical ideas. This was expected, given the conceptual novelty of the two ingredients, GR and QM, and the tension between the two. The physical picture presented in this book has emerged from several decades of research, and is a tentative solution of the puzzle. Here is a brief summary of its conceptual consequences.

10.1.1 GR and QM

The first conclusion of loop gravity is that GR and QM do not contradict each other. A quantum theory that has GR as its classical limit appears to exist. In order to merge, both QM and classical GR have to be suitably formulated and interpreted. More precisely, they both modify some aspects of classical prerelativistic physics, and therefore some aspects of each other.

GR changes the way we understand dynamics. It changes the structure of mechanical systems, classical and quantum. Classical and quantum mechanics admit formulations consistent with this conceptual novelty. These formulations have been studied in Chapters 3 and 5 respectively.

366

Both classical and quantum mechanics are well defined, consistent and predictive in this generalized form, although the relativistic formulation of QM has aspects that need to be investigated further.

The main novelty is that dynamics treats all physical variables (partial observables) on equal footing and predicts their correlations. It does not single out a special variable called "time," to describe evolution with respect to it. Dynamics is not about time evolution, it is about relations between partial observables.

QM modifies the picture of the world of classical GR as well. In quantum theory a physical system does not follow a trajectory. A classical trajectory of the gravitational field is a spacetime. Therefore, QM implies that continuous spacetime is ultimately unphysical, in the same sense in which the notion of the trajectory of a Schrödinger particle is meaningless. GR remains a meaningful theory even giving up the notion of classical spacetime, as Maxwell theory remains a physically meaningful theory in the quantum regime, even if the notion of classical Maxwell field is lost.

To make sense of GR in the absence of spacetime, we must read Einstein's major discovery in a light which is different from the conventional ones. Einstein's major discovery is that spacetime and gravitational field are the same object. A common reading of this discovery is that there is no gravitational field, there is just a dynamical spacetime. In the view of quantum theory, it is more illuminating and more useful to say that there is no spacetime, there is just the gravitational field. From this point of view, the gravitational field is very much a field like any other field. Einstein's discovery is that the fictitious background spacetime introduced by Newton does not exist. Physical fields and their relations are the only components of reality.

10.1.2 Observables and predictions

What does a physical theory predict, in the absence of time evolution and spacetime? Any physical measurement that we perform is ultimately a measurement of some local property of a quantum field. These measurements are represented by an operator in a suitable kinematical quantum space. The theory then gives two kinds of predictions.

- First, the spectral properties of the operator predict the quantization properties of the corresponding physical quantity. They determine the list of the values that the quantity may take.

- Second, quantum dynamics predicts correlation probabilities between observations. That is, it associates a correlation probability amplitude to ensembles of measurement outcomes.

This conceptual structure is sufficient to formulate a meaningful and predictive theory of the physical world, even in the absence of a background spacetime, and in particular in the absence of a background time.

There are no specific "quantum gravity observables." Any measurement involving the gravitational field is also a quantum gravitational measurement. Any measurement with which we test classical GR, and whose outcomes we predict using classical GR, is, in principle, a "quantum gravity measurement" as well. The distinction is one of experimental accuracy.

A measurement of a quantity that depends on the gravitational field is represented by an operator in the quantum theory of gravity. In particular, geometric measurements, such as measurements of volumes and areas, are of this kind. The theory illustrated in Chapter 6 constructs well-defined operators corresponding to these measurements. Their spectra are known, and provide quantitative quantum gravitational predictions.

Some experiments can be viewed as "scattering" experiments happening in a finite region surrounded by detectors. The traditional description of this setting is based on two distinct kinds of measurements:

(i) clocks and meters, which measure the relative position of the detectors;

(ii) particle detectors or other instruments, which measure field properties.

In prerelativistic physics, the measurements in class (i) refer to the location on background spacetime, while the measurements of class (ii) refer to the dynamical variables of the field theory. The distinction between (i) and (ii) disappears in gravity. This is because distances and time intervals are nothing other than properties of the gravitational field, and, therefore, the measurements of class (i) fall into class (ii): both refer to the value of the field on the boundary of the experimental region. Given an ensemble of detectors having measured a certain ensemble of field properties, and their relative distances, the theory should yield an associated correlation probability amplitude, that allows us to compute the relative frequency of a given outcome with respect to a different outcome of the same measurement.

10.1.3 Space, time and unitarity

Space. The disappearance of conventional physical space is a characteristic feature of the LQG picture. There are quantum excitations of the gravitational field that have given probability amplitudes of transforming into each other. These "quanta of gravity" do not live immersed in a spacetime. They *are* space. The idea of space as the inert "container" of the physical world has disappeared.

Instead, the physical space that surrounds us is an aggregate of individual quanta of the gravitational field, represented by the nodes of a spin network. More precisely, it is a quantum superposition of such aggregates.

As observed in Chapter 2, the disappearance of the space-container is not very revolutionary after all: it amounts to a return to the view of space as a relation between things, which was the dominant traditional way of understanding space in the Western culture before Newton.

Perhaps I could add that in the pre-copernican world the cosmic organization was quite hierarchical and structured. Hence, objects were located only with respect to one another but this was sufficient to grant every object a rather precise position in the grand scheme of things. This position marked the "status" of each object: vile objects down here, noble objects above in the heavens. With the copernican revolution, this overall grand structure was lost. Objects no longer knew "where" they were. Newton offered reality a global frame. It is a frame that, for Newton, was grounded in God: space was the "sensorium" of God, the World as perceived by God. With or without such an explicit reference to God, space has been held for three centuries as a preferred entity with respect to which all other entities are located. Perhaps with the twentieth century and with GR we are learning that we do not need this frame to hold reality. Reality holds itself. Objects interact with other objects, and this is reality. Reality is the network of these interactions. We do not need an external entity to hold the net.

Time. The disappearance of conventional physical time is the second characteristic feature of nonperturbative quantum gravity. This is perhaps a more radical step than the disappearance of space. This book is as much about time as about quantum gravity. A central idea defended in this book is that in order to formulate the quantum theory of gravity we must abandon the idea that the flow of time is an ultimate aspect of reality. We must not describe the physical world in terms of time evolution of states and observables. Instead, we must describe it in terms of correlations between observables.

This shift of point of view is already forced upon us by classical GR, but in classical GR each solution of the Einstein equations still provides a notion of continuous spacetime. It is only in the quantum theory of gravity, where classical solutions disappear, that we truly confront the absence of time at the fundamental level. Basic physics without time is viable. The formalism and its interpretation remain consistent. In fact, as soon as we give up the idea that the "time" partial observable is special, mechanics takes a far more compact and elegant form, as shown in Chapter 3.

Unitarity. In conventional QM and QFT, unitarity is a consequence of the time translation symmetry of the dynamics. In GR there isn't, in general, an analogous notion of time translation symmetry. Therefore, there is no sense in which conventional unitarity is necessary in the theory. One often hears that without unitarity a theory is inconsistent. This is a misunderstanding that follows from the erroneous assumption that all physical theories are symmetric under time translations.

Some people find the absence of time difficult to accept. I believe this is just a sort of nostalgia for the old newtonian notion of the absolute "Time" along which everything flows. But this notion has already been shown to be inappropriate for understanding the real world by special relativity. Holding on to the idea of the necessity of unitary time evolution, or to Poincaré invariance, is an anchorage to a notion that is inappropriate to describe general-relativistic quantum physics.

10.1.4 Quantum gravity and other open problems

All sorts of open problems in theoretical physics (and outside it) have been related to quantum gravity. For many of these I see no connection with quantum gravity. In particular:

- *Interpretation of quantum mechanics.* I see no reason why a quantum theory of gravity should not be sought within a standard interpretation of quantum mechanics (whatever one prefers). Several arguments have been proposed to connect these two problems. A common one is that in the Copenhagen interpretation the observer must be *external*, but it is not possible to be external from the gravitational field. I think that this argument is wrong: if it was correct it would apply to the Maxwell field as well. We can consistently use the Copenhagen interpretation to describe the interaction between a macroscopic classical apparatus and a quantum gravitational phenomenon happening, say, in a small region of (macroscopic) spacetime. The fact that the notion of spacetime breaks down at short scale within this region does not prevent us from having the region interacting with an external Copenhagen observer.[1]

- *Quantum mechanical collapse.* Roger Penrose has proposed a subtle argument to relate GR and the QM collapse issue. The argument is based on the fact that in the Schrödinger equation there is a time variable, but the flow of physical time is affected by the gravitational field. I think that this argument is correct but it only shows that the Schrödinger picture with an external time variable is not viable in quantum gravity.

---

[1] However, see Section 5.6.4.

- *Unification of all interactions.* To quantize the electromagnetic field, we did not have to unify it with other fields. And to find the quantum theory of the strong interactions, we do not have to unify them with other interactions. The only vague hint that the problem of quantum gravity and the problem of the unification of all interactions might be related is the fact that the scale at which the running coupling constants of the standard model meet is not very far from the Planck scale. But it is not very close either.

- *Particle masses, cosmological constant, standard model's families ... consciousness. ...* There are many aspects of the Universe which we do not understand; There is no reason why all of these have to be related to the problem of quantizing gravity, or the problem of understanding background-independent QFT. We are far from the "end of physics," and there is much we do not yet understand.

On the other hand, there are two important open problems to which quantum gravity *is* strongly connected:

- *Ultraviolet divergences.* The disappearance of the ultraviolet divergences is one of the major successes of loop gravity. This is achieved in a physically clear and compelling way, via the short-scale quantization of space.

- *Spacetime singularities.* There is no general result so far. But loop quantum cosmology, mentioned in Chapter 8, shows that the classical initial singularity can be controlled by the theory.

10.2 What has been achieved and what is missing?

The formalism presented in Chapters 6 and 7 provides a well-defined background-independent quantum theory of gravity and matter. The theory exists in euclidean and lorentzian versions. In more detail:

- *Background independence.* The main ambition of LQG was to combine GR and QM into a theory capable of merging the insights on Nature gathered by the two theories. The problem was to understand what is a general-relativistic QFT, or a QFT constructed without using a fictitious background spacetime. LQG achieves this goal. Whether or not it is physically correct, it proves that a QFT can be general-relativistic and background independent. It provides a nontrivial example of a background-independent QFT.

- *A physical picture.* LQG offers a novel tentative unitary picture of the world, that incorporates GR and QM. In the book I have tried to spell out in detail this picture, its assumptions and implications. The picture of the background-independent structure of the quantum gravitational field and matter is simple and compelling. Spin network states describe Planck-scale quantum excitations which themselves define localization and spatial relations, as the solutions of the Einstein equations do. Physical space is a quantum superposition of spin networks. Spin networks are not primary concrete "objects" like particles in classical mechanics: rather, they describe the way the gravitational field interacts, like the energy quanta of an oscillator do. The elements of the theory with a direct physical interpretation are elements of the algebra of the partial observables, of which spin networks characterize the spectrum.

- *Quantitative physical predictions.* The spectra of area and volume described in Chapter 6 provide a large body of precise *quantitative* physical predictions. These are unambiguous up to a single overall multiplicative factor, the Immirzi parameter, or, equivalently, the bare value of the Newton constant. Today's technology is not capable of directly testing these spectra. Indirect testing is not necessarily ruled out. What is interesting about these predictions, on the other hand, is the fact that they exist at all. A theory is not a scientific theory unless it can provide a large body of precise quantitative predictions capable, at least in principle, of being verified or falsified. As far as I know, no other current tentative quantum theory of gravity provides a similar large set of predicted numbers.

- *Ultraviolet divergences.* LQG appears to be free from ultraviolet divergences, even when coupled with the standard model.

- *Black-hole thermodynamics.* Although some aspects of the picture are still unclear (in particular, the determination of the Immirzi parameter), LQG provides a compelling explanation of black-hole entropy, as described in Section 8.2.

- *Big-Bang singularity.* The classical initial cosmological singularity is controlled in the application of LQG to cosmology described in Section 8.1.

The main aspects of LQG that are still missing or not sufficiently developed are the following:

- *Scattering amplitudes.* Having a well-defined physical theory is different from knowing how to extract physics from it. We can be capable of writing the full Schrödinger equation for the iron atom and be confident that this equation could predict the iron spectrum. But computing this spectrum is a different matter. In a sense, we are in a similar situation in LQG. We have a well-defined theory, but so far we do not have a great capacity of systematic calculations of observable amplitudes from the basic formalism of the theory. What is missing is a systematic formalism for doing so, in some appropriate form of perturbation expansion.

The difficulty in developing this formalism is of course due to the fact that a perturbative expansion around a classical solution of the gravitational field does not work. The reason why this happens is clear: nonperturbative effects dominate at the Planck scale, yielding the discrete quantized structure of space. We have to find an alternative way for performing perturbative calculations. One direction of research on this issue utilizes the covariant spinfoam formalism described in Chapter 9. But this formalism is not yet at the point of providing a systematic technique for computing transition amplitudes. (*Note added in the paperback edition:* Research is developing rapidly in this direction. General covariant n-point functions have been defined and computed. See the Bibliographical notes at the end of the previous chapter.)

- *Semiclassical limit.* The description of a macroscopic configuration of the electromagnetic field within quantum electrodynamics is not trivial, but it can be achieved, using for instance coherent-state techniques. Describing a macroscopic solution of the Einstein equations with LQG is a similar problem. Can we find a state in $\mathcal{K}_{\mathrm{diff}}$ that approximates a given macroscopic solution? Research programs in this direction are being pursued in particular by the groups of Thomas Thiemann and Abhay Ashtekar. I refer the reader to their works for a description of the state of the art in this rapidly developing direction of research.

LQG is in a peculiar specular position with respect to many traditional approaches to quantum gravity. The most common difficulty is to arrive at a description of Planck-scale physics: many formalisms tend to diverge or break down in some other way at the background-independent Planck-scale level. LQG, on the contrary, provides a formalism that gives a simple and compact description of the background-independent Planck-scale physics; but the recovery of low-energy physics appears more difficult. (*Note added in the paperback edition:* The recent computation of the n-point functions mentioned above provide a way of testing the large distance limit of the theory. In particular, the correct large-distance behavior of the propagator obtained in [330] can be interpreted as the recovery of the Newton law from the background independent theory. See the Bibliographical notes at the end of the previous chapter.)

(*Note added in the paperback edition:* An important recent result in this direction is the proof of the precise equivalence of the two formalishs in 3d [321].)

- *The Minkowski vacuum.* The most important state that we need is the coherent state $|0_M\rangle$ corresponding to Minkowski space. This is essential to connect the theory to the usual formalism of QFT and to define particle-scattering amplitudes. A direction for computing this state was suggested at the end of Chapter 9, but it is too early to see if this will work.

- *The form of the hamiltonian.* As discussed in Section 7.1.3, the precise form of the quantum hamiltonian is not yet settled. There are a number of quantization ambiguities and a number of possible variants that have been proposed. The difficulty of selecting the correct form of the hamiltonian is not only due to the lack of direct empirical guidance on the Planck scale, but also to the little control that we have in extracting physical predictions from the theory, as discussed above.

- *Relation between the spinfoam and the hamiltonian formalism.* Finally, the relation between the lagrangian approach of Chapter 9 and the hamiltonian approach of Chapters 6 and 7 is not yet sufficiently clear.

There are many problems that we have to solve before we can say we have a credible and complete quantum theory of spacetime. I hope that among the readers that have followed this book until this point there are those that will be able to complete the journey.

I close borrowing Galileo's marvelous prose:

Ora, perché è tempo di por fine ai nostri discorsi, mi resta a pregarvi, che se nel riandar più posatamente le cose da me arrecate incontraste delle difficoltà o dubbi non ben resoluti, scusiate il mio difetto, si per la novità del pensiero, si per la debolezza del mio ingegno, si per la grandezza del suggetto, e si finalmente perché io non pretendo né ho preteso da altri quell'assenso ch'io medesimo non presto a questa fantasia.[2]

---

[2] "Now, since it is time to end our discussion, it remains for me to pray of you that if, in reconsidering more carefully what I have presented, you find difficulties or doubts that were not well resolved, you excuse my deficiency: either because of the novelty of the ideas, the weakness of my understanding, the magnitude of the subject or, finally, because I neither ask nor have I ever asked from others that they attach to this imagination that certainty which I myself do not have." [332]

Part III
Appendices

Appendix A
Groups and recoupling theory

A1 $SU(2)$: spinors, intertwiners, n-j symbols

$SU(2)$ is the group of the unitary 2×2 complex matrices with determinant 1. We write these matrices as $U^A{}_B$, where the indices A and B take the values $A, B = 0, 1$. The fundamental representation of the group is defined by the natural action of these matrices on C^2. The representation space is therefore the space of complex vectors with two components. These are called spinors and denoted

$$\psi^A = \begin{pmatrix} \psi^0 \\ \psi^1 \end{pmatrix}. \tag{A.1}$$

Consider the space formed by completely symmetric spinors with n indices $\psi^{A_1 \cdots A_n}$. This space transforms into itself under the action of $SU(2)$ on all the indices. Therefore, it defines a representation of $SU(2)$

$$\psi^{A_1 \cdots A_n} \rightarrow U^{A_1}{}_{A'_1} \cdots U^{A_n}{}_{A'_n} \psi^{A'_1 \cdots A'_n}. \tag{A.2}$$

This representation is irreducible, has dimension $2j + 1$ and is called the spin-j representation of $SU(2)$, where $j = \frac{1}{2}n$. All unitary irreducible representations have this form.

The antisymmetric tensor ϵ^{AB} (defined with $\epsilon^{01} = 1$) is invariant under the action of $SU(2)$

$$U^A{}_C U^B{}_D \epsilon^{CD} = \epsilon^{AB}. \tag{A.3}$$

Contracting this equation with ϵ_{AB} (defined with $\epsilon_{01} = 1$) we obtain the condition that the determinant of U is 1

$$\det U = \frac{1}{2} \epsilon_{AC} \epsilon^{BD} U^A{}_B U^C{}_D = 1 \tag{A.4}$$

since

$$\epsilon_{AB} \epsilon^{AB} = 2. \tag{A.5}$$

The inverse of an $SU(2)$ matrix can be written simply as

$$(U^{-1})^A{}_B = -\epsilon_{BD} U^D{}_C \epsilon^{CA}. \tag{A.6}$$

Most of $SU(2)$ representation theory follows directly from the invariance of ϵ_{AB}. For instance, consider the tensor product of the fundamental representation $j = 1/2$ with itself. This defines a reducible representation on the space of the two-index spinors ψ^{AB}

$$(\psi \otimes \phi)^{AB} = \psi^A \phi^B. \tag{A.7}$$

We can decompose any two-index spinor ψ^{AB} into its symmetric and its antisymmetric part

$$\psi^{AB} = \psi_0 \epsilon^{AB} + \psi_1^{AB}, \tag{A.8}$$

where

$$\psi_0 = \frac{1}{2} \epsilon_{AB} \psi^{AB} \tag{A.9}$$

and ψ_1^{AB} is symmetric. Because of the invariance of ϵ_{AB}, this decomposition is $SU(2)$ invariant. The one-dimensional invariant subspace formed by the scalars ψ_0 defines the trivial representation $j = 0$. The three-dimensional invariant subspace formed by the symmetric spinors ψ_1^{AB} defines the adjoint representation $j = 1$. Hence the tensor product of two spin-1/2 representations is the sum of a spin-0 and a spin-1 representation: $1/2 \otimes 1/2 = 0 \oplus 1$.

In general, if we tensor a representation of spin j_1 with a representation of j_2 we obtain the space of spinors with $2j_1 + 2j_2$ indices, symmetric in the first $2j_1$ and in the last $2j_2$ indices. By symmetrizing all the indices, we obtain an invariant subspace transforming in the representation j_1+j_2. Alternatively, we can contract k indices of the first group with k indices of the second, using k times the tensor ϵ_{AB}, and then symmetrize the remaining $2(j_1 + j_2 - k)$ indices. This defines an invariant subspace of dimension $2(j_1 + j_2 - k)$. The maximum value of k is clearly the smallest between $2j_1$ and $2j_2$. Hence, the tensor product of the representations j_1 and j_2 gives the sum of the representations $|j_1 - j_2|, |j_1 - j_2| + 2, \ldots, (j_1 + j_2)$.

Thus, each irreducible j_3 appears in the product of two representations at most once and if and only if

$$j_1 + j_2 + j_3 = N \tag{A.10}$$

is integer and

$$|j_1 - j_2| \leq j_3 \leq (j_1 + j_2). \tag{A.11}$$

These two conditions are called the Clebsh–Gordon conditions. They are equivalent to the requirement that there exist three nonnegative integers

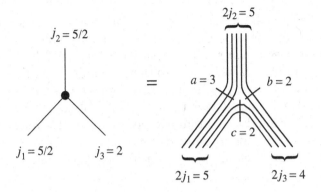

Fig. A.1 Clebsh–Gordon condition.

a, b and c such that

$$2j_1 = a + c, \quad 2j_2 = a + b, \quad 2j_3 = b + c. \tag{A.12}$$

If we have three representations j_1, j_2, j_3, the tensor product of the three contains the trivial representation if and only if one is in the product of the other two, namely, only if the Clebsch–Gordon conditions are satisfied. The invariant subspace in the product of the three is formed by invariant tensors with $2(j_1 + j_2 + j_3)$ indices, symmetric in the first $2j_1$, in the second $2j_2$ and in the last $2j_3$ indices. There is only one such tensor up to scaling, because it must be formed by combinations of the sole invariant tensor ϵ^{AB}. It is given by simply taking a tensors ϵ^{AB}, b tensors ϵ^{BC}, and c tensors ϵ^{CA}, that is

$$v^{A_1,\ldots,A_{2j_1},\; B_1,\ldots,B_{2j_2},\; C_1,\ldots,C_{2j_3}}$$
$$= \left(\epsilon^{A_1 B_1} \ldots \epsilon^{A_a B_a}\right) \left(\epsilon^{B_{a+1} C_1} \ldots \epsilon^{B_{a+b} C_b}\right) \left(\epsilon^{C_{b+1} A_{a+1}} \ldots \epsilon^{C_{b+c} A_{a+c}}\right). \tag{A.13}$$

This is an intertwiner between the representations j_1, j_2, j_3. We can choose a preferred intertwiner by demanding that the intertwiner is normalized, namely multiplying $v^{A_1,\ldots,A_{2j_1},\; B_1,\ldots,B_{2j_2},\; C_1,\ldots,C_{2j_3}}$ by a normalization factor K (which I give below). The normalized intertwiner is called the Wigner $3j$ symbol.

There is a simple graphical interpretation to the tensor algebra of the $SU(2)$ irreducibles, suggested by the existence of the three integers a, b, c, see Figure A.1. A representation of spin j is the symmetrized product of $2j$ fundamentals. When three representations come together, all fundamentals must be contracted among themselves. There will be a fundamentals contracted between j_1 and j_2, and so on. Let us represent each irreducible of spin j as a line formed by $2j$ strands. An invariant tensor is a trivalent node where three such lines meet and all strands are connected across the node: a strands flow from j_1 to j_2, and so on. The meaning of the Clebsh–Gordon conditions is then readily apparent: (A.10) simply demands that the total number of strands is even, so they can pair; (A.11) demands that j_3 is neither larger than $j_1 + j_2$, because then some strands of j_3 would remain unmatched, nor smaller than $|j_1 - j_2|$, because then the largest among j_1 and j_2 would remain unmatched. Indeed, this relation between the lines and the strands reproduces precisely the relation between spin networks and loops. Below, this graphical representation is developed in detail.

Orthonormal basis. The space of the symmetric spinors with n indices has (complex) dimension $2j + 1$. It is often convenient to choose a basis formed by $2j+1$ orthonormal vectors $e_\alpha^{\alpha_1 \cdots \alpha_n}$ in this space. For instance, if $j = 1$, the basis $e_i^{AB} = \frac{1}{\sqrt{2}}\sigma_i^{AB}$ defined using the Pauli matrices transforms under $SU(2)$ in the fundamental representation of $SO(3)$. The matrices σ_i^{AB} are obtained from the Pauli matrices

$$\sigma_{iB}^A = \left\{ \begin{pmatrix} 0 & 1 \\ 1 & 0 \end{pmatrix}, \begin{pmatrix} 0 & -i \\ i & 0 \end{pmatrix}, \begin{pmatrix} 1 & 0 \\ 0 & -1 \end{pmatrix} \right\} \tag{A.14}$$

by raising an index with ϵ^{CB}

$$\sigma_i^{AB} = \sigma_{iC}^A \epsilon^{CB} = \left\{ \begin{pmatrix} -1 & 0 \\ 0 & 1 \end{pmatrix}, \begin{pmatrix} i & 0 \\ 0 & i \end{pmatrix}, \begin{pmatrix} 0 & 1 \\ 1 & 0 \end{pmatrix} \right\}. \tag{A.15}$$

In general, if the spin j is integer then the real section of the representation defines a real irreducible representation.

Wigner $3j$ symbols. In an arbitrary orthonormal basis, we write the normalized invariant tensors (A.13) as

$$K\, v^{\alpha_1 \alpha_2 \alpha_3} = \begin{pmatrix} j_1 & j_2 & j_3 \\ \alpha_1 & \alpha_2 & \alpha_3 \end{pmatrix}. \tag{A.16}$$

If we chose the basis that diagonalizes the third component of the angular momentum ($\alpha \equiv m$), which we do below, then these are proportional to the Wigner $3j$ symbols. The normalization K is fixed by

$$\overline{K\, v^{\alpha_1 \alpha_2 \alpha_3}}\, K\, v_{\alpha_1 \alpha_2 \alpha_3} = 1. \tag{A.17}$$

For instance, we have easily

$$\begin{pmatrix} 1/2 & 1/2 & 1 \\ A & B & i \end{pmatrix} = \frac{1}{\sqrt{6}}\, \sigma^{iAB} \tag{A.18}$$

and

$$\begin{pmatrix} 1 & 1 & 1 \\ i & j & k \end{pmatrix} = \frac{1}{\sqrt{6}}\, \epsilon^{ijk}. \tag{A.19}$$

Wigner $6j$ symbols. Contracting four $3j$ symbols with the invariant tensor $(-1)^{j-\alpha}$ defines a $6j$ symbol.

$$\begin{pmatrix} j_1 & j_4 & j_6 \\ j_3 & j_2 & j_5 \end{pmatrix} = \sum_{\alpha_1 \ldots \alpha_6} (-1)^{\sum_A (j_1 - \alpha_1)} \begin{pmatrix} j_3 & j_6 & j_2 \\ \alpha_3 & \alpha_6 & \alpha_2 \end{pmatrix} \begin{pmatrix} j_2 & j_1 & j_5 \\ \alpha_2 & \alpha_1 & \alpha_5 \end{pmatrix}$$
$$\times \begin{pmatrix} j_6 & j_4 & j_1 \\ \alpha_6 & \alpha_4 & \alpha_1 \end{pmatrix} \begin{pmatrix} j_4 & j_3 & j_5 \\ \alpha_4 & \alpha_3 & \alpha_5 \end{pmatrix}. \tag{A.20}$$

Since the indices are all contracted this quantity does not depend on the basis chosen in the representation space. The pattern of contraction is dictated by the geometry of a tetrahedron. There is one $3j$ symbol for each vertex of the tetrahedron and one representation for each edge

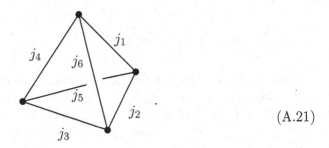

$$\tag{A.21}$$

Consider a tetrahedron with six spins $j_1, \ldots, j_6$, associated with its edges, as above. Denote $\mathcal{H}_j$ the representation space of the $SU(2)$ irreducible representation of spin j. Given a reducible representation $\mathcal{H}$ of $SU(2)$, denote by $[\mathcal{H}]_j$ its component of spin j. The Wigner $6j$ symbol is the dimension of the intersection between the subspaces

$$[[\mathcal{H}_{j_1} \otimes \mathcal{H}_{j_2}]_{j_6} \otimes \mathcal{H}_{j_3}]_{j_4} \quad \text{and} \quad [\mathcal{H}_{j_1} [\otimes \mathcal{H}_{j_2} \otimes \mathcal{H}_{j_3}]_{j_5}]_{j_4} \tag{A.22}$$

of the space $\mathcal{H}_{j_1} \otimes \mathcal{H}_{j_2} \otimes \mathcal{H}_{j_3}$.

Intertwiners. All intertwiners can be built starting from the three-valent ones. For instance a four-valent intertwiner between representations j_1, j_2, j_3 and j_4, can be defined (up to the normalization) by contracting two three-valent intertwiners:

$$v_i^{\alpha_1 \alpha_2 \alpha_3 \alpha_4} = i^{\alpha_1 \alpha_2 \alpha} \, i_\alpha^{\alpha_3 \alpha_4}, \tag{A.23}$$

where α is an index in a representation i. The space of the intertwiners is then spanned by the tensors v_i as i ranges over all representations that satisfy the two relevant Clebsh–Gordon conditions, namely such that the three-valent intertwiners exist. The representation i is said to be associated with a "virtual link" joining the two three-valent nodes into which the four-valent node has been decomposed.

A different basis on this same intertwiner space is obtained by coupling the first and the third leg instead of the first and the second. That is

$$w_i^{\alpha_1\alpha_2\alpha_3\alpha_4} = i^{\alpha_1\alpha_3\alpha} \, i_\alpha{}^{\alpha_2\alpha_4}. \tag{A.24}$$

The change of basis between the v_i and the w_i is given by the Wigner $6j$ symbols, as we will show below (equation (A.65))

$$v_i = \sum_j (2j+1) \begin{pmatrix} j_1 & j_2 & j \\ j_3 & j_4 & i \end{pmatrix} w_j. \tag{A.25}$$

Graphically,

$$j_2 \underset{j_1}{\overset{i}{\diagdown}} \overset{j_3}{\diagup}{}_{j_4} = \sum_j (2j+1) \begin{pmatrix} j_1 & j_2 & j \\ j_3 & j_4 & i \end{pmatrix} \begin{smallmatrix} j_2 \\ \\ j_1 \end{smallmatrix} \diagdown j \diagup \begin{smallmatrix} j_3 \\ \\ j_4 \end{smallmatrix}. \tag{A.26}$$

Five-valent intertwiners can be constructed contracting a three-valent and a four-valent intertwiner, and can thus be labeled with two irreducibles, and so on. In general, we can decompose an n-valent node into $n-2$ three-valent nodes, connected by $n-2$ virtual links, and construct intertwiners accordingly:

$$v_{i_1\dots i_{n-3}}^{\alpha_1\dots\alpha_n} = i^{\alpha_1\alpha_2\beta_1} \, i_{\beta_1}{}^{\alpha_3\beta_2} \, i_{\beta_2}{}^{\alpha_4\beta_3} \dots i_{\beta_{n-3}}{}^{\alpha_{n-1}\alpha_n}, \tag{A.27}$$

where the indices β_n belong to the representation i_n.

Pauli matrices identities. Define $\tau_i = -\frac{i}{2}\,\sigma_i$, where σ_i are the Pauli matrices (A.14). We have the following identities:

$$\operatorname{tr}[\tau_i\tau_j] = -\frac{1}{2}\delta_{ij}, \tag{A.28}$$

$$\operatorname{tr}[\tau_i\tau_j\tau_k] = -\frac{1}{4}\epsilon_{ijk}, \tag{A.29}$$

$$\delta^{ij}\tau_i{}^A{}_B\tau_j{}^C{}_D = -\frac{1}{4}\left(\delta^A{}_D\delta^B{}_C - \epsilon^{AC}\epsilon_{BD}\right), \tag{A.30}$$

$$\delta^{ij}\operatorname{tr}[A\tau_i]\operatorname{tr}[B\tau_j] = -\frac{1}{4}\left\{\operatorname{tr}[AB] - \operatorname{tr}[AB^{-1}]\right\}, \tag{A.31}$$

$$A^{-1A}{}_B = \epsilon^{AC}\epsilon_{BD}A^D{}_C, \tag{A.32}$$

$$\delta^A{}_B\delta^D{}_C = \delta^A{}_C\delta^D{}_B + \epsilon^{AD}\epsilon_{BC}, \tag{A.33}$$

$$\operatorname{tr}[A]\operatorname{tr}[B] = \operatorname{tr}[AB] + \operatorname{tr}[AB^{-1}], \tag{A.34}$$

where A and B are $SL(2,C)$ matrices.

A2 Recoupling theory

A2.1 Penrose binor calculus

In his doctoral thesis, Roger Penrose introduced the idea of writing tensor expressions in which there are sums of indices in a graphical way, a beautiful idea that is at the root of spin networks.[1] Consider in particular the calculus of spinors. Penrose represents the basic element of spinor calculus as

$$\psi^A = \boxed{\psi}^{\,A} \tag{A.35}$$

$$\psi_A = \boxed{\psi}_{\,A} \tag{A.36}$$

$$\delta_C^{\,A} = \big|_C^A \tag{A.37}$$

$$\epsilon_{AC} = \bigcap_{A\ C} \tag{A.38}$$

$$\epsilon^{AC} = \bigcup^{A\ C} \tag{A.39}$$

and, generally, any tensor object as

$$X_{AB}^{C} = \boxed{X}^{\,C}_{\,A\,B}. \tag{A.40}$$

The idea is then to represent index contraction by simply joining the open ends of the lines and dropping the index. This convention provides the possibility of writing the product of any two tensors in a graphical way. For example:

$$\epsilon_{AB}\psi^A\psi^B = \boxed{\psi}\,\boxed{\psi}. \tag{A.41}$$

However, notice that the meaning of a diagram is not invariant if we smoothly deform the lines. For instance

$$\epsilon_{AB}\eta^A\eta^B = \boxed{\eta}^{\,A}\;\boxed{\eta}^{\,B} \tag{A.42}$$

$$= -\epsilon_{AD}\epsilon_{BC}\epsilon^{CD}\eta^A\eta^B = -\boxed{\eta}^{\,A}\;\boxed{\eta}^{\,B\ C\ D} \tag{A.43}$$

$$= -\epsilon_{CD}\delta_A^D\delta_B^C\eta^A\eta^B = -\boxed{\eta}^{\,A}\;\boxed{\eta}^{\,B}. \tag{A.44}$$

Penrose introduced a modification of this graphical spinor calculus, which he denoted as binor calculus, that makes it invariant under deformations

---

[1] I have recently learned that Penrose called this notation "loop notation"!

of the lines. The binor calculus is obtained by adding two conventions to the calculus above. In translating a diagram into tensor notation, we must also ensure the following.

(i) We assign a minus sign to each minimum; and

(ii) assign a minus sign to each crossing.

(iii) Maxima and minima are taken with respect to a fixed direction in the plane. (This direction is conventionally taken to be the vertical direction on the written page.)

(iv) A vertical segment represents a Kronecker delta.

The advantage of these additional rules is that they make the calculus topologically invariant, namely one can arbitrarily smoothly deform a graphical expression without changing its meaning.

Expressed the other way around, any curve can now be decomposed into a product of δs and ϵs and any two curves that are ambient isotopic, i.e., that can be transformed one into the other by a sequence of Reidemeister moves, represent the tensorial expression as products of epsilons and deltas.

A closed loop with this convention has value -2, because

$$ \bigcirc = -\epsilon_{AB} \; \epsilon^{AB} = -2, \tag{A.45} $$

and we have the basic binor identity, which reads:

$$ \times + |\,| + \overset{\cup}{\cap} = (-1)\,\delta^C_B\,\delta^D_A + \delta^C_A\,\delta^D_B + (-1)\,\epsilon_{AB}\,\epsilon^{CD} = 0. \tag{A.46} $$

Remarkably, the two graphical identities

$$ \bigcirc = -2, \tag{A.47} $$

$$ \times = -|\,| - \overset{\cup}{\cap} \tag{A.48} $$

are sufficient to generate a very rich graphical calculus.

Kauffman brackets. Equations (A.47)–(A.48) can be seen as a particular case of a richer structure. In the context of knot theory, Lou Kauffman has defined a function of tangles, namely planar graphical representations of knots, which is now denoted the Kauffman brackets. A planar tangle is a set of lines on a plane that overcross or undercross at intersections. It represents the 2d projection of a 3d node. The Kauffman brackets of a tangle K are indicated as $\langle K \rangle$ and are completely determined by the two relations

$$ \langle \times \rangle = A \langle \overset{\cup}{\cap} \rangle + A^{-1} \langle |\,| \rangle \tag{A.49} $$

and

$$ \langle \bigcirc \cup K \rangle = d \langle K \rangle, \tag{A.50} $$

where $d = -A^2 - A^{-2}$ and K is any diagram that does not intersect the added loop. By applying equation (A.49) to all crossings, the Kauffman brackets of the tangle can be reduced to a linear combination of Kauffman brackets of nonintersecting tangles. By repeated application of (A.50) we can then associate a number to the tangle. Penrose binor calculus and (A.47)–(A.48) are recovered for $A = -1$. In this case, undercrossing and overcrossing are not distinct.

A2.2 KL recoupling theory

Following Kauffman and Lins' book [174], one can pose the following definitions.

The antisymmetrizer. Write n parallel lines as a single line labeled with n

$$\Big| n \equiv \Big|\cdots\Big| . \tag{A.51}$$

(The precise relation between the graphical calculus defined by this and the following equations, and the graphical calculus used in Chapter 6 defined by equation (6.86), is discussed below in Section A2.3.) Define the antisymmetrizer as

$$\rotatebox{90}{$\dashv$}\!{}^{\,n} = \frac{1}{n!} \sum_p (-1)^{|p|} \, P_n^{(p)}, \tag{A.52}$$

where $P_n^{(p)}, p = 1, \ldots, n!$ represents all the possible ways of connecting n incoming lines with n outgoing lines, obtained as $n!$ permutations, and $|p|$ is the sign of the permutation.

The 3-vertex. A special sum of tangles is indicated by a 3-vertex. Each line of the vertex is labeled with a positive integer n, m or p

$$\tag{A.53}$$

and it is assumed that $n = a + b$, $m = a + c$ and $p = b + c$ where a, b, c are positive integers. This last condition is called the *admissibility condition* for the 3-vertex (m, n, p). The 3-vertex is then defined as:

$$\tag{A.54}$$

Compare this definition with the discussion of the Clebsh–Gordon coefficients and Wigner $3j$ symbols given above in A1. It is clear that the 3-vertex in Penrose binor notation represents precisely the nonnormalized intertwiner (A.13). In turn, the Wigner $3j$ symbol can be obtained by normalizing this intertwiner. (Take care! The KL 3-vertex represents the nonnormalized intertwiner (A.13) while the spin network vertex defined in (6.86) corresponds to the Wigner $3j$ symbol. Therefore the two vertices differ by a normalization. (See A2.3 below.))

Chromatic evaluation. If we join several trivalent vertices by their edges, we obtain a trivalent spin network. Thus, in the present context a trivalent network is defined as a trivalent graph with links labeled by an admissible coloring. Notice that in this context networks are not embedded in a three-dimensional space. A link of color n represents n parallel lines and an antisymmetrizer. Thus, a trivalent spin network determines a closed tangle. The Penrose evaluation (or the Kauffman bracket with $A = -1$) of this tangle is called the chromatic evaluation, or network evaluation.

Contractions of intertwiners and Wigner $3j$ symbols can, therefore, be computed as chromatic evaluations of colored diagrams, using only the relations (A.47)–(A.48).

As an example, consider the spin network formed by two trivalent vertices joined to each other. This is called the θ network. Consider the case with edges of color $2, 1, 1$. Applying the definitions given, we have

$$= \frac{1}{2}(-2)^2 - \frac{1}{2}(-2) = 3. \tag{A.55}$$

Therefore,

$$= 3. \tag{A.56}$$

The general formula of the chromatic evaluation of a generic θ network is given below in (A.59).

Formulas from KL recoupling theory. Direct computation using the definitions above give the following formulas. (See also the appendix of [195].)
(1) The dimension

$$\Delta_n = \underset{n}{\underbrace{}} = (-1)^n (n + 1). \tag{A.57}$$

Notice that if we write $n = 2j$ then $(n + 1) = (2j + 1)$ is the dimension of the $SU(2)$ spin-j representation.

(2) The exchange of lines in a 3-vertex

$$
\begin{array}{c}
\includegraphics \\
\end{array}
= \lambda_c^{ab} \; \begin{array}{c} \includegraphics \end{array}, \tag{A.58}
$$

where $\lambda_c^{ab} = (-1)^{(a+b-c)/2} (-1)^{(a'+b'-c')/2}$, and $x' = x(x+2)$.

(3) The θ evaluation

$$
\theta(a, b, c) = \begin{array}{c} \includegraphics \end{array}
$$

$$
= \frac{(-1)^{m+n+p}(m+n+p+1)! \; m! \; n! \; p!}{a! \; b! \; c!}, \tag{A.59}
$$

where $m = (a + b - c)/2$, $n = (b + c - a)/2$, $p = (c + a - b)/2$.

(4) The tetrahedral net

$$
\text{Tet} \begin{bmatrix} A & B & E \\ C & D & F \end{bmatrix} = \begin{array}{c} \includegraphics \end{array} \tag{A.60}
$$

$$
= \frac{\mathcal{I}}{\mathcal{E}} \sum_{m \leq S \leq M} \frac{(-1)^S (S+1)!}{\prod_i (S - a_i)! \; \prod_j (b_j - S)!}, \tag{A.61}
$$

where

$$
a_1 = \frac{A + D + E}{2}, \qquad b_1 = \frac{B + D + E + F}{2},
$$

$$
a_2 = \frac{B + C + E}{2}, \qquad b_2 = \frac{A + C + E + F}{2},
$$

$$
a_3 = \frac{A + B + F}{2}, \qquad b_3 = \frac{A + B + C + D}{2},
$$

$$
a_4 = \frac{C + D + F}{2},
$$

$$
m = \max\{a_i\}, \qquad M = \min\{b_j\},
$$

$$
\mathcal{E} = A!B!C!D!E!F!, \qquad \mathcal{I} = \prod_{ij}(b_j - a_i)!.
$$

(5) The reduction formulas

$$\text{(diagram)} \qquad (A.62)$$

$$\text{(diagram)} \qquad (A.63)$$

Strictly speaking, the identity factors on the right-hand side (the nonintersecting a-tangles) of both of these equations should include the antisymmetrizer. When embedded in spin networks this gets absorbed into the nearest vertex.

Also

$$\text{(diagram)} = \frac{\text{Tet}\begin{bmatrix} p & p & r \\ q & q & 2 \end{bmatrix}}{\theta(p,q,r)} \ \text{(diagram)}. \qquad (A.64)$$

(6) The recoupling theorem:

$$\text{(diagram)} = \sum_i \left\{ \begin{matrix} a & b & i \\ c & d & j \end{matrix} \right\} \ \text{(diagram)} \qquad (A.65)$$

$$\left\{ \begin{matrix} a & b & i \\ c & d & j \end{matrix} \right\} = \frac{\Delta_i \ \text{Tet}\begin{bmatrix} a & b & i \\ c & d & j \end{bmatrix}}{\theta(a,d,i)\theta(b,c,i)}. \qquad (A.66)$$

These formulas are sufficient for most computations performed in loop quantum gravity.

A2.3 Normalizations

Finally, it is time to relate the Kauffman–Lins recoupling theory diagrams given in this appendix with the spin networks recoupling diagrams that we have used in Chapter 6, and which are defined by equation (6.86). There are two main differences. The first is trivial: lines are labeled by the spin j in spin network recoupling diagrams, while they are labeled by the color $n = 2j$ in the Kauffman–Lins diagrams. Thus,

$$\left(\left| \begin{matrix} \\ j \\ \\ \end{matrix} \right. \right)_{\text{spin network}} = \left(\left| \begin{matrix} \\ n = 2j \\ \\ \end{matrix} \right. \right)_{\text{Kauffman–Lins}} . \tag{A.67}$$

The second and more important difference is that the trivalent nodes of the spin network diagrams represent *normalized* intertwiners. Therefore, they are proportional to the recoupling theory trivalent nodes, and the proportionality factor is easily obtained from (A.59). Thus, up to possible phase factors

$$\left(\begin{matrix} j & j' \\ & \\ & j'' \end{matrix} \right)_{\text{spin network}} = \left(\frac{1}{\sqrt{\theta(a,b,c)}} \begin{matrix} a=2j & b=2j' \\ & \\ & c=2j'' \end{matrix} \right)_{\text{Kauffman–Lins}} .$$

$$\tag{A.68}$$

These two equations provide the complete translation rules. It follows, for instance, that the Wigner $6j$-symbol is given by

$$\begin{Bmatrix} j_1 & j_2 & j_5 \\ j_3 & j_4 & j_6 \end{Bmatrix} = \left(\begin{matrix} j_2 & j_6 & j_3 \\ & & j_5 \\ j_1 & & j_4 \end{matrix} \right)_{\text{spin network}}$$

$$= \frac{\left(\begin{matrix} 2j_2 & 2j_6 & 2j_3 \\ & & 2j_5 \\ 2j_1 & & 2j_4 \end{matrix} \right)_{\text{Kauffman–Lins}}}{\sqrt{\begin{matrix} 2j_1 & 2j_3 & 2j_1 & 2j_2 \\ 2j_2 & 2j_4 & 2j_4 & 2j_3 \\ 2j_6 & 2j_6 & 2j_5 & 2j_5 \end{matrix}}}$$

$$= \frac{\text{Tet} \begin{bmatrix} 2j_1 & 2j_2 & 2j_5 \\ 2j_3 & 2j_4 & 2j_6 \end{bmatrix}}{\sqrt{\theta(2j_1, 2j_2, 2j_6)\theta(2j_3, 2j_4, 2j_6)\theta(2j_1, 2j_4, 2j_5)\theta(2j_2, 2j_3, 2j_5)}} .$$

$$\tag{A.69}$$

And the recoupling theorem (A.26) follows from (A.65).

The reason I have described two distinct normalization conventions for the diagrams is that they are both utilized in the loop quantum gravity literature, and both turn out to be useful. The diagrammatic notation developed in this appendix is the one used by Kauffman and Lins in

[174]. This notation has often been used for computing matrix elements of loop operators.

On the other hand, one can find many results on the Wigner $3nj$ symbols and a well-developed graphical calculus for the representation theory of $SU(2)$ in the general literature. These results are routinely used for instance in atomic and nuclear physics. A standard reference is, for instance, Brink and Satcheler [333]. The Brink–Satcheler diagrams are written with the convention I have used in Chapter 6 for the spin networks.[2]

A3 $SO(n)$ and simple representations

I collect here some facts on the representation theory of $SO(n)$. Label finite-dimensional irreducible representations of $SO(n)$ by their highest weight Λ. Here, Λ is a vector of length $n = [d/2]$ ([·] is the integral part): $\Lambda = (N_1, \cdots, N_n)$, where N_i are integers and $N_1 \geq \ldots \geq N_n$. If we are interested in representations of $Spin(n)$ we let the N_i be half-integers. The representation labeled by the highest weight $\Lambda = (N, 0, \ldots, 0)$ are called *simple* or *spherical*. Let $X_{ij}, 1 \leq i, j \leq n$ be a basis of the Lie algebra of $SO(n)$. The simple representations are those for which the "simplicity" relations

$$X_{[ij}X_{ij]} \cdot V_N = 0 \tag{A.70}$$

are satisfied. The representation space V_N of a simple representation can be realized as a space of spherical harmonics, that is, harmonic homogeneous polynomials on R^n. Any L_2 function on the sphere can be uniquely decomposed in terms of these spherical harmonics

$$L_2(S^{n-1}) = \oplus_{N=0}^{\infty} V_N. \tag{A.71}$$

In the case of $SO(4)$, since $Spin(4) = SU(2) \times SU(2)$, there is an alternate description of the representation as products of two representations j' and j'' of $SU(2)$. The relation with the highest weight presentation is given by

$$N_1 = j' + j'', \qquad N_2 = j' - j''. \tag{A.72}$$

The simple representations are, therefore, the representation in which $j' = j'' \equiv j$. Thus, we can label simple representations with a half-integer spin j. Notice that the integer "color" $N = 2j$ is also the (nonvanishing component of the) highest weight of the representation.

---

[2]With a minor difference: in the angular momentum literature, Wigner's $6j$ symbols are generally indicated with curl brackets while, following Kauffman, I have denoted $6j$ symbols by round brackets, reserving the curl brackets, in (A.65), to denote the recoupling matrix of a four-valent node, which differs from the $6j$ symbol by the normalization factor Δ_i.

An elementary illustration of simple representations can be given as follows. The vectors v^α of the representation $\Lambda = (j', j'')$ can be written as spinors $\psi^{A_1 \cdots A_{j'} \, \dot{B}_1 \cdots \dot{B}_{j''}}$ with j' "undotted" symmetrized indices $A_i = 1, 2$, transforming under one of the $SU(2)$, and j'' "dotted" antisymmetrized indices $\dot{B}_i = 1, 2$ transforming under the other. Consider the particular vector w^α which (in the spinor notation, and in a given basis) has components $\psi^{A_1 \cdots A_{j'} \, \dot{B}_1 \cdots \dot{B}_{j''}} = \epsilon^{(A_1 \dot{B}_1} \cdots \epsilon^{A_j) \dot{B}_j}$, where $\epsilon^{A\dot{B}}$ is the unit antisymmetric tensor, and the symmetrization is over the A_i indices only. The subgroup of $SO(4)$ that leaves this vector invariant is an $SO(3)$ subgroup of $SO(4)$ (which depends on the basis chosen). Clearly, since $\epsilon^{A\dot{B}}$ is the only object invariant under this $SU(2)$, a normalized $SO(3)$-invariant vector (and only one) exists in these simple representations only.

Equivalently, the simple representations of $SO(4)$ are those defined by the completely symmetric traceless 4d tensors of rank N. The invariant vector w is then the traceless part of the tensor with (in the chosen basis) all components vanishing except $w^{444\cdots}$. The $SO(3)$ subgroup is given by the rotations around the fourth coordinate axis. The relation between the vector and spinor representation is obtained contracting the spinor indices with the (four) Pauli matrices: $v^{\mu_1 \cdots \mu_j} = \psi^{A_1 \cdots A_{j'} \, \dot{B}_1 \cdots \dot{B}_{j''}} \sigma^{\mu_1}_{A_1 \dot{B}_1} \cdots \sigma^{\mu_j}_{A_j \dot{B}_j}$.

Let V_Λ be a representation of $SO(n)$; we say that $\omega \in V_\Lambda$ is a spherical vector if it is invariant under the action of $SO(n-1)$. Such a vector exists if and only if the representation is simple. In that case this vector is unique up to normalization.

Let ω be a vector of V_Λ, and consider an orthonormal basis v_i of V_Λ. We can construct the following functions on G:

$$\Theta_i(g) = \langle \omega | R^{-1}(g) | v_i \rangle. \tag{A.73}$$

These functions span a subspace of $L_2(G)$. The group acts on this subspace by the right regular representation, and the corresponding representation is equivalent to the representation V_Λ. If ω is spherical, then these functions are in fact L_2 functions on the quotient space $SO(n)/SO(n-1) = S^{n-1}$, and V_Λ is therefore a spherical representation. On the other hand, if the representation is spherical then we can construct a spherical vector: $\Theta_\omega(g) = \sum_i \Theta_i(g) \Theta_i(1)$. When ω is spherical, the spherical function Θ_ω is a function on the double-quotient space $SO(n-1)\backslash SO(n)/SO(n-1) = U(1)$. It is now a standard exercise to show that there is a unique harmonic polynomial on R^n invariant by $SO(n-1)$ of a given degree, hence there is a unique spherical function.

The space of the intertwiners of three representations of $SO(4)$ is at most one-dimensional. The dimension $n_{\Lambda_1, \Lambda_2, \Lambda_3}$ of the space of the intertwiners between three representations $\Lambda_1, \Lambda_2, \Lambda_3$ is given by the integral

$$n_{\Lambda_1,\Lambda_2,\Lambda_3} = \int dg \ \chi_{\Lambda_1}(g)\chi_{\Lambda_2}(g)\chi_{\Lambda_3}(g), \qquad (A.74)$$

where χ_Λ are the characters of the representation Λ. Since any representation of $SO(4)$ is the product of two representations of $SU(2)$, the intertwining number of three $SO(4)$ representations is the product of $SU(2)$ intertwining numbers. These numbers can take the values 0 or 1 for $SU(2)$.

Representations of $SO(n)$ are real. This means that it is always possible to choose a basis of V_Λ such that the representation matrices are real. For the half-integer spin representations of $Spin(n)$, it is still true that Λ is equivalent to its complex conjugate or dual, but the isomorphism is nontrivial.

Bibliographical notes

On Penrose calculus, see [181] and [334]. The basic reference on recoupling theory that I have used here is [174], where the formulas given here are derived in the general case with arbitrary A; see also the appendix of [195] for more details. On the relation between recoupling theory and spin networks, see [335, 336]. Chromatic evaluation is used to compute $SU(2)$ Clebsh–Gordon coefficients in, for instance, [337]. A widely used graphical method for angular momentum theory is the one of Levinson [338], developed by Yutsin, Levinson and Vanagas [339] and the slightly modified version of Brink and Satcheler [333].

Appendix B
History

In this appendix I sketch the main lines of development of the research in quantum gravity, from the first explorations in the early 1930s (the thirties) to nowadays.

I have no ambition of presenting complete references to all the important works on the subject; some of the references are to original works, others to reviews where references can be found. Errors and omissions are unfortunately unavoidable and I apologize for these. I have made my best efforts to be balanced, but in a field that has not yet succeeded in finding consensus, my perspective is obviously subjective. Trying to write history in the middle of the developments is hard. Time will pass, dust will settle, and it will slowly become clear if we are right, if some of us are right, or – a possibility never to disregard – if we all are wrong.

I am very much indebted to the many friends that have contributed to this historical perspective. I am particularly grateful to John Stachel, Augusto Sagnotti, Gary Horowitz, Ludwig Faddeev, Alejandro Corichi, Jorge Pullin, Lee Smolin, Joy Christian, Bryce DeWitt, Cecile DeWitt, Giovanni Amelino-Camelia, Daniel Grumiller, Nikolaos Mavromatos, Stanley Deser, Ted Newman and Gennady Gorelik.

B1 Three main directions

The quest for quantum gravity can be separated into three main lines of research. The relative weight of these lines has changed, there have been important intersections and connections between the three, and there has been research that does not fit into any of the three lines. Nevertheless, the three lines have maintained a distinct individuality across 70 years of research. They are often denoted "covariant," "canonical," and "sum-over-histories," even if these names can be misleading and are often used interchangeably. They cannot be characterized by a precise definition, but within each line there is a certain methodological unity, and a certain consistency in the logic of the development of the research.

- *The covariant line of research* is the attempt to build the theory as a quantum field theory of the fluctuations of the metric over a flat

Minkowski space, or some other background metric space. The program was started by Rosenfeld, Fierz and Pauli in the thirties. The Feynman rules of GR were laboriously found by DeWitt, Feynman and Faddeev in the sixties. t'Hooft and Veltman, Deser and Van Nieuwenhuizen, and others, found increasing evidence of nonrenormalizability at the beginning of the seventies. Then, a search for an extension of GR giving a renormalizable or finite perturbation expansion started. Through high-derivative theory and supergravity, the search converged successfully to string theory in the late eighties.

- *The canonical line of research* is the attempt to construct a quantum theory in which the Hilbert space carries a representation of the operators corresponding to the full metric, or some functions of the metric, without any background metric to be fixed. The program was initiated by Bergmann and Dirac in the fifties. Unraveling the canonical structure of GR turned out to be laborious. Dirac, Bergmann and his group, and Peres completed the task in the fifties. Their cumbersome formalism was drastically simplified by the introduction of new variables: first by Arnowit Deser and Misner in the sixties and then by Ashtekar in the eighties. The formal equations of the quantum theory were written down by Wheeler and DeWitt in the middle sixties, but turned out to be too ill defined. A well-defined version of the same equations was successfully found only in the late eighties, with the formulation of LQG.

- *The sum-over-histories line of research* is the attempt to use some version of Feynman's functional integral quantization to define the theory. The idea was introduced by Misner in the fifties, following a suggestion by Wheeler, and developed by Hawking in the form of euclidean quantum gravity in the seventies. Most of the discrete (lattice-like, posets, . . .) approaches and the spinfoam formalism, introduced more recently, belong to this line as well.

- *Others.* There are of course other ideas that have been explored.

  - Noncommutative geometry has been proposed as a key mathematical tool for describing Planck-scale geometry, and has recently obtained very surprising results, particularly with the work of Connes and collaborators.

  - Twistor theory has been more fruitful on the mathematical side than on the strictly physical side, but it is still developing.

  - Finkelstein, Sorkin, and others, pursue courageous and intriguing independent paths.

- Penrose's idea of a gravity-induced quantum state reduction has recently found new life with the perspective of a possible experimental test.

- ...

So far, however, none of these alternatives has been developed into a detailed quantum theory of gravity.

B2 Five periods

Historically, the evolution of the research in quantum gravity can roughly be divided into five periods, summarized in Table B.1.

- *The Prehistory: 1930–1957.* The basic ideas of all three lines of research appear as early as the thirties. By the end of the fifties, the three research programs are clearly formulated.

- *The Classical Age: 1958–1969.* The sixties see the strong development of two of the three programs, the covariant and the canonical. At the end of the decade, the two programs have both achieved the basic construction of their theory: the Feynman rules for the gravitational field on one side and the Wheeler–DeWitt equation on the other. To get to these beautiful results, an impressive amount of technical labour and ingenuity proves necessary. The sixties close – as they did in many regards – with the promise of a shining new world.

- *The Middle Ages: 1970–1983.* The seventies disappoint the hopes of the sixties. It becomes increasingly clear that the Wheeler–DeWitt equation is too ill defined for genuine field theoretical calculations. And evidence for the nonrenormalizability of GR piles up. Both lines of attack have found their stumbling block.

In 1974, Stephen Hawking derives black-hole radiation. Trying to deal with the Wheeler–DeWitt equation, he develops a version of the sum-over-histories as a sum over "euclidean" (riemannian) geometries. There is excitement with the idea of the wave function of the Universe and the approach opens the way for thinking of and computing topology change. But for field theoretical quantities the euclidean functional integral will prove as weak a calculation tool as the Wheeler–DeWitt equation.

On the covariant side, the main reaction to nonrenormalizability of GR is to modify the theory. Strong hopes, then disappointments, motivate extensive investigations of supergravity and

Table B.1   The search for a quantum theory of the gravitational field.

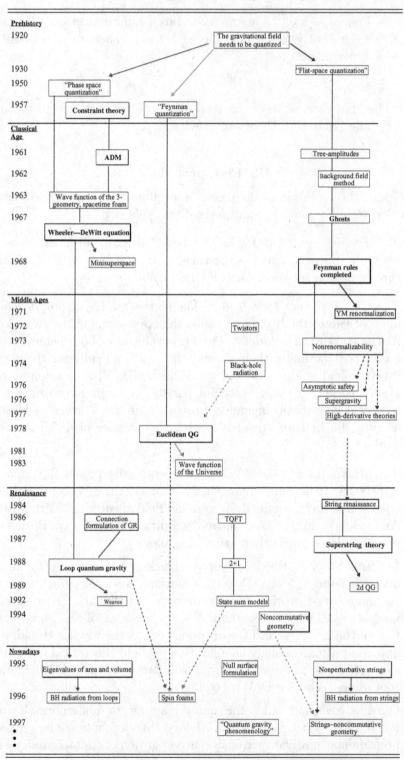

Prehistory
1920

The gravitational field needs to be quantized

1930
1950   "Phase space quantization"   "Flat-space quantization"
1957   **Constraint theory**   "Feynman quantization"

Classical Age
1961   **ADM**
1962   Tree-amplitudes
1963   Wave function of the 3-geometry, spacetime foam   Background field method
1967   **Wheeler—DeWitt equation**   Ghosts
1968   Minisuperspace   **Feynman rules completed**

Middle Ages
1971   YM renormalization
1972   Twistors
1973   Nonrenormalizability
1974   Black-hole radiation
1976   Asymptotic safety
1976   Supergravity
1977   High-derivative theories
1978   **Euclidean QG**
1981
1983   Wave function of the Universe

Renaissance
1984   String renaissance
1986   Connection formulation of GR   TQFT
1987   **Superstring theory**
1988   **Loop quantum gravity**   2+1
1989   2d QG
1992   Weaves   State sum models
1994   Noncommutative geometry

Nowadays
1995   Eigenvalues of area and volume   Null surface formulation   Nonperturbative strings
1996   BH radiation from loops   Spin foams   BH radiation from strings
1997   "Quantum gravity phenomenology"   Strings–noncommutative geometry
.
.
.

higher-derivative actions for GR. The landscape of quantum gravity is gloomy.

- *The Renaissance: 1984–1994.* Light comes back in the middle of the eighties. In the covariant camp, the various attempts to modify GR to get rid of the infinities merge into string theory. Perturbative string theory finally delivers on the long search for a computable perturbative theory for quantum gravitational scattering amplitudes. To be sure, there are prices to pay, such as the wrong dimensionality of spacetime, and the introduction of supersymmetric particles, which, year after year, are expected to be discovered but, so far, are not. But the result of a finite perturbation expansion, long sought after, is too good to be discarded just because the world insists in looking different from our theories.

  Light returns to shine on the canonical side as well. Twenty years after the Wheeler–DeWitt equation, LQG finally provides a version of the theory sufficiently well defined for performing explicit computations. Here, as well, we are far from a complete and realistic theory, and scattering amplitudes, for the moment, can't be computed at all, but the excitement for having a rigorously defined, nonperturbative, generally covariant and background-independent quantum field theory, in which physical expectation values can be computed, is strong.

- *Nowadays: 1995c.* Both string theory and LQG grow strongly for a decade, until, in the middle of the nineties, they begin to deliver physical results. The Bekenstein–Hawking black-hole entropy formula is derived within both approaches, virtually simultaneously. LQG leads to the computation of the first Planck-scale quantitative physical predictions: the spectra of the eigenvalues of area and volume.

  The sum-over-histories tradition, in the meantime, is not dead. In spite of the difficulties of the euclidean integral, it remains as a reference idea, and guides the development of several lines of research, from the discrete lattice-like approaches, to the "state sum" formulation of topological theories. Eventually, the last motivate the spinfoam formulation, a translation of LQG into a Feynman sum-over-histories form.

  Other ideas develop in the meanwhile, most notably noncommutative geometry, which finds intriguing points of contact with string theory towards the end of the decade.

  The century closes with two well-developed contenders for a quantum theory of gravity: string theory and LQG, as well as a set of intriguing novel new ideas that go from noncommutative geometry to the null surfaces formulation of GR, to the attempt to merge strings and loops. And even on a very optimistic note: the birth of a new line of research, the self-styled "quantum gravity phenomenology" which investigates the possibility that Planck-scale type measurements might be within reach. And thus that perhaps we could finally know which of the theoretical hypotheses, if any, make sense.

I now describe the various periods and their main steps in more detail.

B2.1 The Prehistory: 1930–1957

General relativity is discovered in 1915; quantum mechanics in 1926. A few years later, around 1930, Born, Jordan and Dirac are already capable of formalizing the quantum properties of the electromagnetic field. How long did it take to realize that the gravitational field should, presumably, behave quantum mechanically as well? Almost no time: already in 1916 Einstein points out that quantum effects must lead to modifications in the theory of general relativity [340]. In 1927, Oskar Klein suggests that quantum gravity should ultimately modify the concepts of space and time [341]. In the early thirties, Rosenfeld [342] writes the first technical papers on quantum gravity, applying Pauli's method for the quantization of fields with gauge groups to the linearized Einstein field equations. The relation with a linear spin-2 quantum field is soon unraveled in the works of Fierz and Pauli [343] and the spin-2 quantum of the gravitational field is already a familiar notion in the thirties. Its name, "graviton," is already in use in 1934, when it appears in a paper by Blokhintsev and Gal'perin [344] (published in the ideological magazine *Under the Banner of Marxism*). Bohr considers the idea of identifying the neutrino and the graviton. In 1938, Heisenberg [345] points out that the fact that the gravitational coupling constant is dimensional is likely to cause problems with the quantum theory of the gravitational field.

The history of these early explorations of the quantum properties of spacetime has recently been reconstructed by John Stachel [346]. In particular, John describes in his paper the extensive, but largely neglected, work conducted in the middle thirties by a Russian physicist, Matvei Petrovich Bronstein. Persistent rumors claim that Bronstein was a nephew of Leon Trotsky, and that he hid this relation that became dangerous, but Gennady Gorelik (of the Center for Philosophy and History of Science at Boston University and Institute for the History of Science and Technology of the Russian Academy of Sciences) assures me that this rumor is false. Bronstein re-derives the Rosenfeld–Pauli quantization of the linear theory, but realizes that the unique features of gravitation require a special treatment, when the full nonlinear theory is taken into account.

He realizes that field quantization techniques must be generalized in such a way as to be applicable in the absence of a background geometry. In particular, he realizes that the limitation posed by general relativity on the mass density radically distinguishes the theory from quantum electrodynamics and would ultimately lead to the need to "reject riemannian geometry" and perhaps also to "reject our ordinary concepts of space and time" [347]. The reason Bronstein has remained unknown for so long has partly to do with the fact that he was executed by the Soviet State Security Agency (the NKVD) at the age of 32. I am told that in Russia some still remember Bronstein as "smarter than Landau" (but Gorelik doubts this opinion could be shared by a serious physicist). For a discussion of Bronstein's early work in quantum gravity, see [348].

References and many details on these pioneering times are in the fascinating paper by John Stachel mentioned above. Here, I pick up the historical evolution after World War II. In particular, I start from 1949, a key year for the history of quantum gravity.

1949

– Peter Bergmann starts his program of phase space quantization of nonlinear field theories [349]. He soon realizes that physical quantum observables must correspond to coordinate-independent quantities only [350]. The search for these gauge-independent observables is started in the group that forms around Bergmann, at Brooklyn Polytechnic and then in Syracuse. For instance, Ted Newman develops a perturbation approach for finding gauge-invariant observables order by order [351]. The group studies the problems raised by systems with constraints and reaches a remarkable clarity, unfortunately often forgotten later on, on the problem of what are the observables in general relativity. The canonical approach to quantum gravity is born.

– Bryce DeWitt completes his thesis. He applies Schwinger's covariant quantization to the gravitational field.

– Dirac presents his method for treating constrained hamiltonian systems [113].

1952

– Following the pioneering works of Rosenfeld, Fierz and Pauli, Gupta [352] develops systematically the "flat-space quantization" of the gravitational field. The idea is simply to introduce a fictitious "flat space," that is, Minkowski metric $\eta_{\mu\nu}$, and quantize the small fluctuations of the metric around Minkowski, $h_{\mu\nu} = g_{\mu\nu} - \eta_{\mu\nu}$. The covariant approach is fully born. The first difficulty appears immediately when searching for the propagator: because of gauge invariance, the quadratic term of the lagrangian is singular, as for the electromagnetic field. Gupta's treatment uses an indefinite norm state space, as for the electromagnetic field.

1957

– Charles Misner introduces the "Feynman quantization of general relativity" [353]. He quotes John Wheeler for suggesting the expression

$$\int \exp\{(i/\hbar)(\text{Einstein action})\} \, d(\text{field histories}), \qquad \text{(B.1)}$$

and studies how to have a well-defined version of this idea. Misner's paper [353] is very remarkable in many respects. It explains with complete clarity notions such as why the quantum hamiltonian must be zero, why the individual spacetime points are not defined in the quantum theory, and the need of dealing with gauge invariance in the integral. Even more remarkably, the paper opens with a discussion of the possible directions for quantizing gravity, and lists the three lines of directions – covariant, canonical, and sum-over-histories – describing them almost precisely with the same words we would today![1]

At the end of the fifties, all the basic ideas and the research programs are clear. It is only a matter of implementing them, and seeing if they work. The implementation, however, turns out to be a rather herculean task, that requires the ingenuity of people of the caliber of Feynman and DeWitt on the covariant side, and of Dirac and DeWitt on the canonical side.

B2.2 The Classical Age: 1958–1969

1958

– The Bergmann group [129], and Dirac [113, 114], work out the general hamiltonian theory of a constrained system. For a historical reconstruction of this achievement, see [354]. At the beginning, Dirac and the Bergmann group work independently. The present double classification into primary and secondary constraints, and into first- and second-class constraints, still reflects this original separation.

1959

– By 1959, Dirac has completely unraveled the canonical structure of GR [130].

1961

– Arnowitt, Deser and Misner complete what we now call the ADM formulation of GR, namely its hamiltonian version in appropriate variables,

---

[1] To be sure, Misner lists a fourth approach as well, based on the Schwinger equations for the variation of the propagator, but notices that "this method has not been applied independently to general relativity," a situation that might have changed only very recently [145, 146].

which greatly simplify the hamiltonian formulation and make its geometric reading transparent [131].

In relation to the quantization, Arnowitt, Deser and Misner present an influential argument for the finiteness of the self-energy of a point-particle in classical GR and use it to argue that nonperturbative quantum gravity should be finite.

– Tullio Regge defines the Regge calculus [285].

1962

– Feynman attacks the task of computing transition amplitudes in quantum gravity. He shows that tree-amplitudes lead to the physics one expects from the classical theory [355].

– DeWitt starts developing his background field methods for the computation of perturbative transition amplitudes [356].

– Bergmann and Komar clarify what one should expect from a Hilbert space formulation of GR [357].

– Following the ADM methods, Peres writes the Hamilton–Jacobi formulation of GR [358]

$$G^2\left(q_{ab}q_{cd} - \frac{1}{2}q_{ac}q_{bd}\right)\frac{\delta S(q)}{\delta q_{ac}}\frac{\delta S(q)}{\delta q_{bd}} + \det q\, R[q] = 0, \qquad (\text{B.2})$$

which is our fundamental equation (4.9) written in metric variables, and will soon lead to the Wheeler–DeWitt equation; here q_{ab} is the ADM 3-metric.

1963

– John Wheeler realizes that the quantum fluctuations of the gravitational field must be short-scale fluctuations of the geometry and introduces the physical idea of spacetime foam [359]. Wheeler's *Les Houches* lecture notes are remarkable in many respects, and are the source of many of the ideas still current in the field. To mention two others: "Problem 56" suggests that gravity in $2+1$ dimensions may not be so trivial after all, and indicates it may be an interesting model to explore. "Problem 57" suggests studying quantum gravity by means of a Feynman integral over a spacetime lattice.

Julian Schwinger introduces the tetrad spin-connection formulation in quantum gravity [80]. On the strict relation between this formalism and Yang–Mills theories, he writes:

> Weyl, the originator of the electromagnetic gauge invariance principle, also recognized that the gravitational field can be characterized by a kind of gauge transformation [79]. This is the possibility of altering freely at each point the orientation of a local Lorentz coordinate

frame, while suitably transforming certain gravitational potentials. In a subsequent development, Yang and Mills introduced an arbitrarily oriented 3d isotopic space at each spacetime point. The occasional remark that the gravitational field can be viewed as a Yang–Mills field is thus rather anachronistic.

1964

– Penrose introduces the idea of spin networks, and of a discrete structure of space controlled by $SU(2)$ representation theory. The construction exists only in the form of a handwritten manuscript. It gets published only in 1971 [181]. The idea will surprisingly re-emerge 25 years later, when spin networks will be found to label the states of LQG.

– Beginning to study loop corrections to GR amplitudes, Feynman observes that unitarity is lost for naive diagrammatic rules. DeWitt [360] develops the combinatorial means to correct the quantization (requiring independence of diagrams from the longitudinal parts of propagators). These correction terms can be put in the form of loops of fictitious fermionic particles, the Faddeev–Popov ghosts [361]. The key role of DeWitt in this context was emphasized by Veltman in 1974 [362]:

> ...Essentially due to this, and some deficiencies in his combinatorial methods, Feynman was not able to go beyond one closed loop. DeWitt in his 1964 Letter and in his subsequent monumental work derived most of the things that we know of now. That is, he considered the question of a choice of gauge and the associated ghost particle. Indeed, he writes the ghost contribution in the form of a local Lagrangian containing a complex scalar field obeying Fermi statistics. Somewhat illogically this ghost is now called the Faddeev–Popov ghost.

The designation "Faddeev–Popov ghost" is far from illogical: in comparison with the complicated combinatorics of DeWitt, the Faddeev–Popov approach has the merit of a far greater technical simplicity and a transparent geometric interpretation, which justifies its popularity. It is only in the work of Faddeev that the key role played by the gauge orbits as the true dynamical variables is elucidated [363].

1967

– Bryce DeWitt publishes the "Einstein–Schrödinger equation" [214]

$$\left((\hbar G)^2 (q_{ab}q_{cd} - \frac{1}{2}q_{ac}q_{bd}) \frac{\delta}{\delta q_{ac}} \frac{\delta}{\delta q_{bd}} - \det q \, R[q] \right) \Psi(q) = 0, \qquad \text{(B.3)}$$

which is the main quantum gravity equation (6.1), in metric variables. Bryce will long denote this equation as the "Einstein–Schrödinger equation," attributing it to Wheeler – while John Wheeler called it the DeWitt equation – until, finally, in 1988, at an Osgood Hill conference, DeWitt gives up and calls it what everybody else had been calling it since the beginning: the "Wheeler–DeWitt equation."

The story of the birth of the Wheeler–DeWitt equation is worth telling. In 1965, during an air trip, John had to stop for a short time at the Raleigh–Durham airport in North Carolina. Bryce lived nearby. John phoned Bryce and proposed to meet at the airport during the wait between two flights. Bryce showed up with the Hamilton–Jacobi equation for GR, published by Peres in 1962 and mumbled the idea of doing precisely what Schrödinger did for the hydrogen atom: replace the square of the derivative with a second derivative. Surprising Bryce, John was enthusiastic (John is often enthusiastic, of course), and declared immediately that *the* equation of quantum gravity had been found. The paper with the equation, the first of Bryce's celebrated 1967 quantum gravity trilogy [214, 364], was submitted in the spring of 1966, but its publication was delayed until 1967. Among the reasons for the delay, apparently, were difficulties with publication charges.

– John Wheeler discusses the idea of the wave function $\Psi(q)$ on the space of the "3-geometry" q, and the notion of superspace, the space of the 3-geometries, in [38].

– Roger Penrose starts twistor theory [365].

– The project of DeWitt and Feynman is concluded. A complete and consistent set of Feynman rules for GR are written down [361, 364].

1968
– Ponzano and Regge define a quantization of 3d euclidean GR [283]. The model will lead to major developments.

1969
– Developing an idea in Bryce's paper on canonical quantum gravity, Charles Misner starts quantum cosmology: the game of truncating the Wheeler–DeWitt equation to a finite number of degrees of freedom [366]. The idea is beautiful, but it will develop into a long-lasting industry from which, after a while, little new will be understood.

The decade closes with the main lines of the covariant and the canonical theory clearly defined. It will soon become clear that neither theory works.

B2.3 The Middle Ages: 1970–1983

1970
– The decade of the seventies opens with a word of caution. Reviving a point made by Pauli, a paper by Zumino [367], suggests that the quantization of GR may be problematic and might make sense only by viewing

GR as the low-energy limit of a more general theory. More than thirty years later opinions still diverge on whether this is true.

1971

– Using the technology developed by DeWitt and Feynman for gravity, t'Hooft and Veltman decide to study the renormalizability of GR. Almost as a warm-up exercise, they consider the renormalization of Yang–Mills theory, and find that the theory is renormalizable – a result that won them the Nobel prize [368]. In a sense, one can say that the first physical result of the research in quantum gravity is the proof that Yang–Mills theory is renormalizable.

– David Finkelstein writes his inspiring "spacetime code" series of papers [369] (which, among other ideas, discuss quantum groups).

1973

– Following the program initiated with Veltman in 1971, t'Hooft finds evidence of nonrenormalizable divergences in GR with matter fields. Shortly after, t'Hooft and Veltman, as well as Deser and Van Nieuwenhuizen, confirm the evidence [370].

1974

– Hawking announces the derivation of black-hole radiation [248]. A (macroscopic) Schwarzschild black hole of mass M emits thermal radiation at the temperature (8.25). The result comes as a surprise, anticipated only by the observation by Bekenstein, a year earlier, that entropy is naturally associated with black holes, and thus they could be thought of, in some obscure sense, as "hot" [247], and by the Bardeen–Carter–Hawking analysis of the analogy between laws of thermodynamics and dynamical behavior of black holes [246]. Hawking's result is not directly connected to quantum gravity – it is a skillful application of quantum field theory in curved spacetime – but has a very strong impact on the field. It fosters an intense activity in quantum field theory in curved spacetime, it opens a new field of research in "black-hole thermodynamics," and it opens the quantum-gravitational problems of understanding the statistical origin of the black-hole (the Bekenstein–Hawking) entropy (8.27).

An influential, clarifying and at the same time intriguing paper is written two years later by Bill Unruh. The paper points out the existence of a general relation between accelerated observers, quantum theory, gravity and thermodynamics [371]. Something deep about Nature should be hidden in this tangle of problems, but we do not yet know what.

1975

– It becomes generally accepted that GR coupled to matter is not renormalizable. The research program started with Rosenfeld, Fierz and Pauli is dead.

1976

– A first attempt to save the covariant program is made by Steven Weinberg, who explores the idea of asymptotic safety [39], developing earlier ideas from Giorgio Parisi [372], Kenneth Wilson and others, suggesting that nonrenormalizable theories could nevertheless be meaningful.

– To resuscitate the covariant theory, even if in modified form, the path has already been indicated: find a high-energy modification of GR. Preserving general covariance, there is not much one can do to modify GR. An idea that attracts much enthusiasm is supergravity [373]: it seems that by simply coupling a spin-3/2 particle to GR, namely with the action (in first-order form)

$$S[g, \Gamma, \psi] = \int d^4x \ \sqrt{-g} \ \left(\frac{1}{2G} R - \frac{i}{2} \ \epsilon^{\mu\nu\rho\sigma} \ \psi_\mu \gamma_5 \gamma_\nu D_\rho \psi_\sigma \right), \qquad \text{(B.4)}$$

one can get a theory finite even at two loops.

– Supersymmetric string theory is born [349].

1977

– Another, independent, idea is to keep the same kinematics and change the action. The obvious thing to do is to add terms proportional to the divergences. Stelle proves that an action with terms quadratic in the curvature

$$S = \int d^4x \ \sqrt{-g} \ \left(\alpha R + \beta R^2 + \gamma R^{\mu\nu} R_{\mu\nu} \right), \qquad \text{(B.5)}$$

is renormalizable for appropriate values of the coupling constants [375]. Unfortunately, precisely for these values of the constants the theory is bad. It has negative energy modes that make it unstable around the Minkowski vacuum and not unitary in the quantum regime. The problem becomes to find a theory renormalizable and unitary at the same time, or to circumvent nonunitarity.

1978

– The Hawking radiation is soon re-derived in a number of ways, strongly reinforcing its credibility. Several of these derivations point to thermal techniques [376], thus motivating Hawking [40] to revive the Wheeler–Misner "Feynman quantization of general relativity" [353] in the form of a "euclidean" integral over *riemannian* 4-geometries g

$$Z = \int Dg \ e^{-\int \sqrt{g} R}. \qquad \text{(B.6)}$$

Time-ordering and the concept of positive frequency are incorporated into the "analytic continuation" to the euclidean sector. The hope is double: to deal with topology change, and that the euclidean functional integral will prove to be a better calculation tool than the Wheeler–DeWitt equation.

1980

– Within the canonical approach, the discussion focuses on understanding the disappearance of the time coordinate from the Wheeler–DeWitt theory. The problem has actually nothing to do with *quantum* gravity, since the time coordinate disappears in the *classical* Hamilton–Jacobi form of GR as well; and, in any case, physical observables are coordinate independent, and thus, in particular, independent from the time coordinate, in whatever correct formulation of GR. But in the quantum context there is no single spacetime, as there is no trajectory for a quantum particle, and the very concepts of space and time become fuzzy. This fact raises much confusion and a vast interesting discussion (whose many contributions I can not possibly summarize here) on the possibility of doing meaningful fundamental physics in the absence of a fundamental notion of time. For early references on the subject see, for instance, [42, 44].

1981

– Polyakov [377] shows that the cancellation of the conformal anomaly in the quantization of the string action

$$S = \frac{1}{4\pi\alpha'} \int \mathrm{d}^2\sigma \sqrt{g} \, g^{\mu\nu} \partial_\mu X^a \partial_\nu X^b \eta_{ab} \qquad (\text{B.7})$$

leads to the critical dimension. A new problem is created: how to recover our 4-dimensional world from a string theory which is defined in the critical dimension.

1983

– The hope is still high for supergravity, now existing in various versions, as well as for higher-derivative theories, whose rescue from nonunitarity is explored using a number of ingenious ideas (large-N expansions, large-d expansions, Lee–Wick mechanisms, ...). At the tenth GRG conference in Padova in 1983, two physicists of indisputable seriousness, Gary Horowitz and Andy Strominger, summarize their contributed paper [378] with the words

> In sum, higher-derivative gravity theories are a viable option for resolving the problem of quantum gravity ...

At the same conference, supergravity is vigorously advertised as the final solution of the quantum gravity puzzle. But very soon it becomes clear that supergravity is nonrenormalizable at higher loops and that higher-derivatives theories do not lead to viable perturbative expansions. Excitement, hope and hype fade away.

In its version in 11 dimensions, supergravity will find new importance in the late 1990s, in connection with string theory. High-derivative corrections will also re-appear, in the low-energy limit of string theory.

– Hartle and Hawking [157] introduce the notion of the "wave function of the Universe" and the "no-boundary" boundary condition for the Hawking integral, opening up a new intuition on quantum gravity and quantum cosmology. But the euclidean integral does not provide a way of computing genuine field theoretical quantities in quantum gravity any better than the Wheeler–DeWitt equation, and the atmosphere at the middle of the eighties is again rather gloomy. On the other hand, Jim Hartle [26] develops the idea of a sum-over-histories formulation of GR into a fully fledged extension of quantum mechanics to the generally covariant setting. The idea will later be developed and formalized by Chris Isham [379].

– Sorkin introduces his poset approach to quantum gravity [380].

B2.4 The Renaissance: 1984–1994

1984

– Green and Schwarz realize that strings might describe "our Universe" [381]. Excitement starts to build up around string theory, in connection with the unexpected anomaly cancellation and the discovery of the heterotic string [382].

– The relation between the ten-dimensional superstrings theory and four-dimensional low-energy physics is studied in terms of compactification on Calaby–Yau manifolds [383] and orbifolds. The dynamics of the choice of the vacuum remains unclear, but the compactification leads to 4d chiral models resembling low-energy physics.

– Belavin, Polyakov and Zamolodchikov publish their analysis of conformal field theory [384].

1986

– Goroff and Sagnotti [29] finally compute the two-loop divergences of pure GR, definitely nailing the corpse of pure GR perturbative quantum field theory into its coffin: the divergent term in the effective action is

$$\Delta S = \frac{209}{737\,280\pi^4} \, \frac{1}{\epsilon} \int \mathrm{d}^4 x \, \sqrt{-g} \, R^{\mu\nu}{}_{\rho\sigma} R^{\rho\sigma}{}_{\epsilon\theta} R^{\epsilon\theta}{}_{\mu\nu}. \tag{B.8}$$

– Penrose suggests that the wave function collapse in quantum mechanics might be of quantum-gravitational origin [385]. The idea is radical and implies a re-thinking of the basis of mechanics. Remarkably, the idea may be testable: work is today in progress to study the feasibility of an experimental test.

– String field theory represents a genuine attempt to address the main problem of string theory: finding a fundamental, background-independent definition of the theory [386]. The string field path, however, turns out to be hard.

– The connection formulation of GR is developed by Abhay Ashtekar [132], on the basis of some results by Amitaba Sen [82]. At the time, this is denoted the "new variables" formulation. It is a development in classical general relativity, but it has long-ranging consequences on quantum gravity, as the basis of LQG.

1987

– Fredenhagen and Haag explore the general constraint that general covariance puts on quantum field theory [387].

– Green, Schwarz and Witten publish their book on superstring theory. In the gauge in which the metric has no superpartner, the superstring action is

$$S = \frac{1}{4\pi\alpha'} \int \mathrm{d}^2\sigma \sqrt{g} \left(g^{\mu\nu} \partial_\mu X^a \partial_\nu X^b - i\psi^a \gamma^\mu \partial_\mu \psi^b \right) \eta_{ab}. \qquad (B.9)$$

Interest in the theory grows very rapidly. To be sure, string theory still obtains a very small place at the 1991 Marcel Grossmann meeting [388]. But the research in supergravity and higher-derivative theories has merged into strings, and string theory is increasingly viewed as a strong competing candidate for the quantum theory of the gravitational field. As a side product, many particle physicists begin to study general relativity, or at least some bits of it. Strings provide a consistent perturbative theory. The covariant program is fully re-born. The problem becomes understanding why the world described by the theory appears so different from ours.

1988

– Ted Jacobson and Lee Smolin find loop-like solutions to the Wheeler–DeWitt equation formulated in the connection formulation [178], opening the way to LQG.

– The "loop representation of quantum general relativity" is introduced, in [176, 177]. It is based on the new connection formulation of GR [132], on the Jacobson–Smolin solutions [178], and on Chris Isham's ideas on the need of nongaussian, or nonFock representations in quantum gravity [43]. Loop quantization had been previously and independently developed by Rodolfo Gambini and his collaborators for Yang–Mills theories [179]. In the gravitational context, the loop representation leads immediately to two surprising results: an infinite family of exact solutions of the Wheeler–DeWitt equation is found, and knot theory controls the physical quantum states of the gravitational field. Classical knot theory, with its extensions, becomes a branch of mathematics relevant to describe the diff-invariant states of quantum spacetime [215]. The theory transforms the old Wheeler–DeWitt theory into a formalism that can be concretely used to compute physical quantities in quantum gravity. The canonical program is fully re-born. Nowadays, the theory is called "loop quantum gravity."

– Ed Witten introduces the notion of topological quantum field theory (TQFT) [389]. In a celebrated paper [390], he uses a TQFT to give a field theoretical representation of the Jones polynomial, a knot theory invariant. The expression used by Witten has an interpretation in LQG: it can be seen as the "loop transform" of a quantum state given by the exponential of the Chern–Simon functional [215].

Formalized by Atiyah [391], the idea of TQFT will have beautiful developments, and will strongly influence later developments in quantum gravity. General topological theories in any dimensions, and in particular BF theory, are introduced by Gary Horowitz shortly afterwards [392].

– Witten finds an ingenious way of quantizing GR in $2+1$ spacetime dimensions [393] (thus solving "Problem 56" of the 1963 Wheeler's *Les Houches* lectures), opening up a big industry of analysis of the theory (for a review, see [394]). The quantization method is partially a sum over histories and partially canonical. Covariant perturbative quantization seemed to fail for this theory. The theory had been studied a few years earlier by Deser, Jackiw, t'Hooft, Achucarro, Townsend, and others [395].

1989

– Amati, Ciafaloni and Veneziano find evidence that string theory implies that distances smaller than the Planck scale cannot be probed [396].

– In the string world, there is excitement for some nonperturbative models of strings "in 0 dimension," equivalent to 2d quantum gravity [397]. The excitement dies fast, as is often the case, but the models will re-emerge in the nineties [398], and will also inspire the spinfoam formulation of quantum gravity [278].

1992

– Turaev and Viro [286] define a state sum that, on the one hand, is a rigorously defined TQFT and, on the other hand, can be seen as a regulated and well-defined version of the Ponzano–Regge [283] quantization of $2+1$ gravity. Turaev, and Ooguri [289] soon find a 4d extension, which will have a remarkable impact on later developments.

– The notion of *weave* is introduced in LQG [189]. It is evidence of a discrete structure of spacetime emerging from LQG. The first example of a weave which is considered is a 3d mesh of intertwined rings. Not surprisingly, the intuition was already in Wheeler! (See Figure B.1, taken from Misner, Thorne and Wheeler [399]).

1993

– Gerard 't Hooft introduces the idea of holography, developed by Lenny Susskind [375]. According to the "holographic principle," the information on the physical state in the interior of a region can be represented on the region's boundary and is limited by the area of this boundary. This

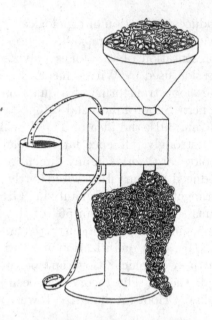

Fig. B.1   The weave, in Wheeler's vision. From Ref. [374].

principle can be also interpreted as referring to the information on the system *accessible from the outside* of the region, in which case it makes much more sense to me.

1994

– Noncommutative geometry, often indicated as a tool for describing certain aspects of Planck-scale geometry, finds a strict connection to GR in the work of Alain Connes. Remarkably, the Connes–Chamseddine "spectral action," just the trace of a simple function of a suitably defined Dirac-like operator D

$$S = \mathrm{tr}[f(\mathrm{D}^2/(\hbar G))], \tag{B.10}$$

where f is the characteristic function of the $[0, 1]$ interval, turns out to include the standard model action, as well as the Einstein–Hilbert action [401].

B2.5 Nowadays: 1995–

1995

– Nonperturbative aspects of string theory begin to appear: branes [402], dualities [403], the matrix model formulation of M theory [404], ... (for a review, see, for instance, [405]). The interest in strings booms. At the plenary conference of a meeting of the American Mathematical Society in Baltimore, Ed Witten claims that

> The mathematics of the next millennium will be dominated by string theory

causing a few eyebrows to raise.

The various dualities appear to relate the different versions of the theory, pointing to the existence of a unique fundamental theory. The actual construction of the fundamental background-independent theory, however, is still missing, and string theory exists so far only in the form of a number of (related) expansions over assigned backgrounds.

– Two results in loop gravity appear: (i) the overcompleteness of the loop basis is resolved by the discovery of the spin network basis [171]; (ii) eigenvalues of area and volume are computed [191]. The latter result is rapidly extended and derived in a number of alternative ways.

The rigorous mathematical framework for LQG starts to be developed [200, 201].

– Ted Newman and his collaborators introduce the Null Surface Formulation of GR [406].

1996

– The Bekenstein–Hawking black-hole entropy (8.22) is computed within LQG as well as within string theory, almost at the same time.

The loop result is obtained by computing the number of (spin network) states which endow a 2-sphere with a given area [239, 249], as well as by loop quantizing the classical theory of the field outside the hole and studying the boundary states [238]. These gravitational surface states [254] can be identified with the states of a Chern–Simon theory on a surface with punctures [253]. The computation is valid for various realistic black holes. The 1/4 factor in (8.27) is obtained by fixing the Immirzi parameter.

In string theory, the computation exploits a strong coupling/weak coupling duality, which, in certain supersymmetric configurations, preserves the number of states: the physical black hole is in a strong coupling situation, but the number of its microstates can be computed in a weak-field configuration that has the same charges at infinity. One obtains precisely the 1/4 factor of (8.27), as well as other aspects of the Hawking radiation phenomenology [407]. However, the calculation method is indirect, and works only for extremal or near-extremal black holes.

– A rigorously defined, finite and anomaly-free hamiltonian constraint operator is constructed by Thomas Thiemann in LQG [133]. Some doubts are raised on whether the classical limit of this theory is in fact GR (the issue is still open), but the construction defines a consistent generally covariant quantum field theory in 4d.

– Intriguing state sum models obtained modifying a TQFT are proposed by Barrett and Crane, Reisenberger, Iwasaki and others as a tentative model for quantum GR. All these models appear as sums of "spinfoams": branched surfaces carrying spins.

– The loop representation is "exponentiated", *à la* Feynman, giving rise, again, to a spinfoam model, corresponding to canonical LQG. These developments revive the sum-over-histories approach.

1997

– Intriguing connections between noncommutative geometry and string theory appear [408].

– There is a lively discussion on the difficulties of the lattice approaches in finding a second-order phase transition [409].

1998

– Juan Maldacena shows [410] that the large-N limit of certain conformal field theories includes a sector describing supergravity on the product of anti-deSitter spacetimes and spheres. He conjectures that the compactifications of M/string theory on an anti-deSitter spacetime is dual to a conformal field theory on the spacetime boundary. This leads to a new proposal for defining M theory itself in terms of the boundary theory: an effort to reach background independence (for M theory) using background-dependent methods (for the boundary theory).

A consequence of this "Maldacena conjecture" is an explosion of interest for 't Hooft, and Susskind holographic principle (see year 1993).

– Two papers in the influential journal *Nature* [265] raise the hope that seeing spacetime-foam effects, and testing quantum gravity theories might not be as forbidding as usually assumed. The idea is that there are a number of different instances (the neutral kaon system, gamma-ray burst phenomenology, interferometers, . . .) in which presently operating measurement or observation devices, or instruments that are going to be soon constructed, involve sensitivity scales comparable to – or not too far from – the Planck scale [266]. If this direction fails, testing quantum gravity might require the investigation of very early cosmology [411].

1999 . . .

I stop here, because too-recent history is not yet history.

B3 The divide

The lines of research that I have summarized in Appendix B2 have found many points of contact in the course of their development and have often intersected. For instance, there is a formal way of deriving a sum-over-histories formulation from a canonical theory and vice versa; the perturbative expansion can also be obtained by expanding the sum-over-histories; string theory today faces the problem of finding its nonperturbative formulation, and thus the typical problems of a canonical theory; and LQG has mutated into the spinfoam models, a sum-over-histories formulation, using techniques that can be traced to a development of string theory of the early nineties. However, in spite of this continuous cross-fertilization, the three main lines of development have kept their essential separateness.

The three directions of investigation were already clearly identified by Charles Misner in 1959 [353]. In the concluding remark of the *Conférence Internationale sur les Théories Relativistes de la Gravitation*, in 1963, Peter Bergmann notes [412]

> In view of the great difficulties of this program, I consider it a very positive thing that so many different approaches are being brought to bear on the problem. To be sure, the approaches, we hope, will converge to one goal.

This was 40 years ago ...

The divide is particularly strong between the covariant line of research, more connected to the particle-physics tradition, and the canonical/sum-over-histories one, more connected to the relativity tradition. This divide has remained through over 70 years of research in quantum gravity. Here is a typical comparison, arbitrarily chosen among many. On the particle-physics side, at the *First Marcel Grossmann Meeting*, Peter van Neuwenhuizen writes [413]

> ... gravitons are treated on exactly the same basis as other particles such as photons and electrons. In particular, particles (including gravitons) are always in flat Minkowski space and move *as if* they followed their geodesics in curved spacetime because of the dynamics of multiple graviton exchange ... : Pure relativists often become somewhat uneasy at this point because of the following two aspects entirely peculiar to gravitation: (1) ... One must decide before quantization which points are spacelike separated, but it is only after quantization that the fully quantized metric field can tell us this spacetime structure ... (2) ... In a classical curved background one needs positive and negative frequency solutions, but in non-stationary spacetimes it is not clear whether one can define such solutions. *The strategy of particle physicists has been to ignore these problems for the time being, in the hope that they will ultimately be resolved in the final theory. Consequently we will not discuss them any further.*

On the relativity side, Peter Bergmann comments [414]

> The world-point by itself possesses no physical reality. It acquires reality only to the extent that it becomes the bearer of specific properties of the physical fields imposed on the spacetime manifold.

The conceptual divide is huge. Partially, it reflects the different understanding of the world held by the particle-physics community, on the one hand, and the relativity community, on the other. The two communities have made repeated and sincere efforts to talk to each other and understand each other. But the divide remains. Both sides have the feeling that the other side is incapable of appreciating something basic and essential: the structure of QFT as it has been understood over half a century of investigation, on the particle-physics side; the novel physical understanding of space and time that has appeared with GR, on the relativity side. Both sides expect that the point of view of the other will turn out, at the end of the day, to be not very relevant. One side because the experience with QFT is on a fixed metric spacetime, and thus is irrelevant in a genuinely background-independent context. The other because GR is only a low-energy limit of a much more complex theory, and thus cannot be taken too seriously as an indication about the deep structure of Nature. Hopefully, the recent successes of both lines will force the two sides, finally, to face the problems that the other side considers prioritary: background independence on the one hand, control of a perturbation expansion on the other.

After 70 years of research, there is no consensus, no established theory, and no theory that has yet received any direct or indirect experimental support. In the course of 70 years, many ideas have been explored, fashions have come and gone, the discovery of the Holy Grail has been several times announced, with much later scorn. *Ars longa, vita brevis.*

However, in spite of its age, the research in quantum gravity does not seem to have been meandering meaninglessly, when seen in its entirety. On the contrary, one sees a logic that has guided the development of the research, from the early formulation of the problem and the research directions in the fifties to nowadays. The implementation of the programs has been extremely laborious, but has been achieved. Difficulties have appeared, and solutions have been proposed which, after much difficulty, have led to the realization, at least partial, of the initial hopes. It was suggested in the early seventies that GR could perhaps be seen as the low-energy theory of a theory without uncontrollable divergences; today, 30 years later, such a theory – string theory – is known. In 1957, Charles Misner indicated that in the canonical framework one should be able to compute eigenvalues; and in 1995, 37 years later, eigenvalues were computed – within loop quantum gravity. The road is not yet at the end, much remains to be understood, some of the current developments might lead nowhere. But looking at the entire development of the subject, it is difficult to deny that there has been progress.

Appendix C
On method and truth

I collect in this appendix some simple reflections on scientific methodology, and on the content of scientific theories, relevant for quantum gravity. In particular, I try to make more explicit the methodological assumptions at the root of some of the research described in this book, and to give it some justification.

I am no professional philosopher, and what follows has no ambition in that sense. I am convinced, however, of the utility of a dialog between physics and philosophy. This dialog has played a major role during the other periods in which science has faced fundamental problems. I think that most physicists underestimate the effect of their own epistemological prejudices on their research. And many philosophers underestimate the effect – positive or negative – they have on fundamental research. On the one hand, a more acute philosophical awareness would greatly help the physicists engaged in fundamental research. As I have argued in Chapter 1, during the second half of the twentieth century fundamentals were clear in theoretical physics and the problems were technical, but today foundational problems are back on the table, as they were at the time of Newton, Faraday, Heisenberg and Einstein. These physicists couldn't certainly have done what they have done if they weren't nurtured by (good or bad) philosophy. On the other hand, I wish contemporary philosophers concerned with science would be more interested in the ardent lava of the fundamental problems science is facing today. It is here, I believe, that stimulating and vital issues lie.

C1 The cumulative aspects of scientific knowledge

Part of the reflection about the science of the last decades has emphasized the "noncumulative" aspect in the development of scientific knowledge: the evolution of scientific theories is marked by large or small breaking points, in which, to put it crudely, empirical facts are reorganized within new theories, which are, to some extent, "incommensurable" with respect to their antecedent. These ideas – correctly understood or misunderstood – have had a strong influence on the physicists.

The approach to quantum gravity described in this book assumes a different reading of the evolution of scientific knowledge. Indeed, I have based the discussion on quantum gravity on the expectation that the central

415

physical tenets of QM and GR represent our best guide for accessing even the extreme and unexplored territories of the quantum-gravitational regime. In my opinion, the emphasis on the incommensurability between theories has clarified an important aspect of science, but risks to obscure something of the internal logic according to which, historically, physics has extended knowledge. There is a subtle, but very definite cumulative aspect in the progress of physics, which goes far beyond the growth in the validity and precision of the empirical content of the theories. In moving from a theory to the theory that supersedes it, we do not save only the verified empirical content of the old theory, but more. This "more" is a central concern for good physics. It is the source, I think, of the spectacular and undeniable predictive power of theoretical physics. I think that by playing it down one risks misleading theoretical research into a less effective methodology.

Let me illustrate this point with a historical case. There was a problem between Maxwell equations and Galileo transformations. There were two obvious ways out. To consider Maxwell theory as a theory of limited validity: a phenomenological theory of some yet-to-be-discovered aether's dynamics. Or to consider that galilean equivalence of inertial systems had limited validity, thus accepting the idea that inertial systems are not equivalent in electromagnetic phenomena. Both directions are logical and were pursued at the end of the nineteenth century. Both are sound applications of the idea that a scientific revolution changes in depth what old theories teach us about the world. Which of the two ways did Einstein, successfully, take?

Neither. For Einstein, Maxwell theory was a source of awe. He rhapsodizes about his admiration for the theory. For him, Maxwell had opened a new window on the world. Given the astonishing success of Maxwell theory, empirical (electromagnetic waves), technological (radio) as well as conceptual (understanding the nature of light), Einstein's admiration is comprehensible. But Einstein had a tremendous respect for Galileo's insight as well. Young Einstein was amazed by a book with Huygens' derivation of collision theory virtually out of galilean invariance alone. Einstein understood that Galileo's great intuition – that the notion of velocity is only relative – *could not be wrong*. I am convinced that in this faith of Einstein in the core of the great galilean discovery there is very much to learn, for the philosophers of science, as well as for the contemporary theoretical physicists. So, Einstein *believed the two theories, Maxwell and Galileo, and assumed that their tenets would hold far beyond the regime in which they had been tested.* He assumed that Galileo had grasped something about the physical world which was, simply, *correct*. And so had Maxwell. Of course, details had to be adjusted. The core of Galileo's insight was that all inertial systems are equivalent and that velocity is relative, not the details of the galilean transformations. Einstein knew the Lorentz transformations (found by Poincaré), and was able to see that they do not contradict Galileo's insight. If there was contradiction in putting the two together, the problem was ours: we were surreptitiously sneaking some incorrect assumption into our deductions. He found the incorrect assumption, which, of course, was that simultaneity could be well defined. It was Einstein's faith in the *essential physical correctness* of the old theories that guided him to his spectacular discovery.

There are very many similar examples in the history of physics, that could equally well illustrate this point. Einstein found GR "out of pure thought," having Newton theory on the one hand and special relativity – the understanding that any interaction is mediated by a field – on the other; Dirac found quantum field theory from Maxwell equations and quantum mechanics; Newton combined Galileo's insight that acceleration governs dynamics with Kepler's insight that the source of the force that governs the motion of the planets is the Sun . . . The list could be long. In all these cases, confidence in the insight that came with some theory, or "taking a theory seriously," led to major advances that greatly extended the original theory itself. Far be it from me to suggest that there is anything simple, or automatic, in figuring out where the true insights are and in finding the way of making them work together. But what I am saying is that figuring out where the true insights are and finding the way of making them work together is the work of fundamental physics. This work is grounded on *confidence* in the old theories, not on a random search for new ones.

One of the central concerns of the modern philosophy of science is to face the apparent paradox that scientific theories change, but are nevertheless credible. Modern philosophy of science is to some extent an aftershock reaction to the fall of newtonian mechanics. A tormented recognition that an extremely successful scientific theory can nevertheless be untrue. But I think that a notion of truth which is challenged by the event of a successful physical theory being superseded by a more successful one is a narrow-minded notion of truth.

A physical theory is a conceptual structure that we develop and use in order to organize, read and understand the world, and make predictions about it. A successful physical theory is a theory that does so effectively and consistently. In the light of our experience, there is no reason not to expect that a more effective conceptual structure might always exist. An effective theory may always show its limits and be replaced by a better one. However, a novel conceptualization cannot but rely on what the previous one has already understood. Thought is in constant evolution and in constant reorganization. It is not a static entity. Science is itself the process of the evolution of thinking.

When we move to a new city, we are at first confused about its geography. Then we find some reference points, and we make a first rough mental map of the city in terms of these points. Perhaps there is part of the city on the hills and part on the plain. As time goes on, the map improves. There are moments in which we suddenly realize that we had it wrong. Perhaps there were two areas with hills, and we were previously confusing the two. Or we had mistaken a large square, called Earth square, for the downtown, while downtown was farther away, around a square called Sun square. So we update the mental map. Sometime later, we have learned names and features of neighborhoods and streets, and the hills, as references, fade away. The neighborhood structure of knowledge is more effective than the hill/plain one ... The structure changes, but the knowledge increases. And Earth square, now we know it, is not downtown, and we know it forever.

There are discoveries that are forever. That the Earth is not the center of the Universe, that simultaneity is relative, that absolute velocity is meaningless. That we do not get rain by dancing. These are steps humanity takes, and does not take back. Some of these discoveries amount simply to clearing from our thinking wrong, encrusted, or provisional credences. But also discovering classical mechanics, or discovering electromagnetism, or quantum mechanics, are discoveries forever. Not because the details of these theories cannot change, but because we have discovered that a large portion of the world admits to being understood in certain terms, and this is a *fact* that we will have to keep facing forever.

One of the main theses of this book is that general relativity is the expression of one of these insights, which will stay with us "forever." The insight is that the physical world does not have a stage, that localization and motion are relational only, that this background independence is required for any fundamental description of our world.

How can a theory be effective even outside the domain for which it was found? How could Maxwell predict radio waves, Dirac predict antimatter and GR predict black holes? How can theoretical thinking be so magically powerful?

It has been suggested that these successes are due to chance, and seem grand only because of the historically deformed perspective. A sort of darwinian natural selection for theories has been suggested: there are hundreds of theories proposed, most of them die, the ones that survive are the ones remembered. There is alway somebody who wins the lottery, but this is not a sign that humans can magically predict the outcome of the lottery. My opinion is that such an interpretation of the development of science is unjust, and, worse, misleading. It may explain something, but there is more in science. There are tens of thousands of persons playing the lottery, there were only two relativistic theories of gravity, in 1916, when Einstein predicted that light would be deflected by the Sun precisely by an angle of 1.75 seconds of arc. Familiarity with the history of physics, I feel confident to claim, rules out the lottery picture.

I think that the answer is simpler. Say somebody predicts that the Sun will rise tomorrow, and the Sun rises. This successful prediction is not a matter of chance: there aren't hundreds of people making random predictions about all sorts of strange objects appearing at the horizon. The prediction that tomorrow the Sun will rise, is sound. However, it cannot be taken for granted. A neutron star could rush in, close to the speed of light, and sweep the Sun away. Who, or what, grants the right of induction? Why should I be confident that the Sun will rise, just because it has been rising so many times in the past? I do not know the answer to *this* question. But what I know is that the predictive power of a theory beyond its own domain is *precisely of the same sort*. Simply, we learn something about Nature and what we learn is effective in guiding us to predict Nature's behavior. Thus, the spectacular predictive power of theoretical physics is nothing less and nothing more than common induction: it follows from the successful assumption that there are regularities in Nature, at all levels. The spectacular success of science in making predictions about territories not yet explored is as comprehensible (or as incomprehensible) as my ability to predict that the Sun will rise tomorrow. Simply, Nature around us happens to be full of regularities *that we recognize*, whether or not we understand why regularities exist at all. These regularities give us strong confidence – although not certainty – that the Sun will rise tomorrow, as well as that the basic facts about the world found with QM and GR will be confirmed, not violated, in the quantum-gravitational regimes that we have not yet empirically probed.

This view is not dominant in theoretical physics nowadays. Other attitudes dominate. The "pessimistic" scientist has little faith in the possibilities of theoretical physics, because he worries that all possibilities are open, and anything might happen between here and the Planck scale. The "wild" scientist observes that great scientists had the courage to break with "old and respected assumptions" and to explore some novel "strange" hypotheses. From this observation, the "wild" scientist concludes that to do great science one has to explore all sorts of strange hypotheses, and to *violate respected ideas*. The wilder the hypothesis, the better. I think wildness in physics is sterile. The greatest revolutionaries in science were extremely, almost obsessively, conservative. So was certainly the greatest revolutionary, Copernicus, and so was Planck. Copernicus was pushed to the great jump from his pedantic labor on the minute technicalities of the ptolemaic system (fixing the equant). Kepler was forced to abandon the circles by his extremely technical work on the details of Mars' orbit. He was using ellipses as approximations to the epicycle-deferent system, when he began to realize that the approximation was fitting the data better than the supposedly exact curve. And Einstein and Dirac were also extremely conservative. Their vertigo-inducing leaps ahead were not pulled out of the blue sky. They did not come from the thrill of violating respected ideas, or trying a new pretty idea. They were forced out of respect towards previous physical insights. Today, instead, we have plenty of seminars on "a new pretty idea," regularly soon forgotten and superseded by a new fad. In physics, novelty has always emerged from new data or from a humble, devoted interrogation of the old theories. From turning these theories over and over in one's mind, immersing oneself in them, making them clash, merge, talk, until, through them, the missing gear can be seen.

Finally, the "pragmatic" scientist ignores conceptual questions and physical insights, and only cares about developing a theory. This is an attitude successful in the sixties in arriving at the standard model. But in the sixties empirical data were flowing in daily, to keep research on track. Today theoreticians have no new data. The "pragmatic" theoretician does not care. He does not trust the insight of the old theories. He focuses only on the development of the novel theory, and cannot care less if the world predicted by the theory resembles less and less the world we see. He is even excited that the theory looks so different from the world, thinking that this is evidence of how much ahead he has advanced in knowledge, which is a complete nonsense. Theoretical physics becomes a mental game closed in on itself and the connection with reality is lost.

In my opinion, precious research energies are today wasted in these attitudes. A philosophy of science that downplays the component of factual knowledge in physical theories might have part of the responsibility.

C2 On realism

A scientific theory is a conceptual structure that we use to read, organize and understand the world, at some level of our knowledge. It is one step along a process, since knowledge increases. In my view, scientific thinking is not much different from common-sense thinking. In fact, it is only a better instance of the same activity: thinking about the world, and updating our mental schemes. Science is the organized enterprise of continuously exploring the possible ways of thinking about the world, and constantly selecting the ones that work best.

If this is correct, there cannot be any qualitative difference between the theoretical notions introduced in science and the terms in our everyday language. A fundamental intuition of classical empiricism is that nothing grants us that the concepts that we use to organize our perceptions refer to "real" entities. Some modern philosophy of science has emphasized the application of this intuition to the concepts introduced by science. Thus, we are warned to doubt the "reality" of the theoretical objects (electrons, fields, black holes, . . .). I find these warnings incomprehensible. Not because they are ill founded, but because they are not applied consistently. The fathers of empiricism consistently applied this intuition to *any* physical object. Who grants me the reality of a chair? Why should a chair be more than a theoretical concept organizing certain regularities in my perceptions? I will not venture here in disputing nor in agreeing with this doctrine. What I find incomprehensible is the position of those who grant the solid status of reality to a chair, but not to an electron. The arguments against the reality of the electron apply to the chair as well. The arguments in favor of the reality of the chair apply to the electron as well. A chair, as well as an electron, is a concept that we use to read, organize and understand the world. They are equally real. They are equally volatile and uncertain.

Perhaps this curious schizophrenic attitude of being antirealist with electrons and iron-realist with chairs is the result of a tortuous historical evolution, initiated by the rebellion against "metaphysics," and, with it, the granting of confidence to science alone. From this point of view, metaphysical questioning on the reality of chairs is sterile – true knowledge is in science. Thus, it is to scientific knowledge that we apply empiricist rigor. But understanding science in empiricists' terms required making sense of the raw empirical data on which science is based. With time, the idea of raw empirical data showed more and more its limits. The common-sense view of the world was reconsidered as a player in our picture of knowledge. This common-sense view should give us a language and a foundation from which to start – the old antimetaphysical prejudice still preventing us, however, from applying empiricist rigor to this common-sense view of the world. But if one is not interested in questioning the reality of chairs, for the very same reason why should one be interested in questioning the "reality of the electrons?"

Again, I think this point is important for science itself. The factual content of a theory is our best tool. The faith in this factual content does not prevent us from being ready to question the theory itself, if sufficiently compelled to do so by novel empirical evidence or by putting the theory in relation to other things *we know* about the world or *we learn* about it. Scientific antirealism, in my opinion, is not only a shortsighted application of a deep classical empiricist insight; it is also a negative influence over the development of science. H. Stein (private communication) has recently beautifully illustrated a case in which a great scientist, Poincaré, was blocked from getting to a major discovery (special relativity) by a philosophy that restrained him from "taking seriously" his own findings.

Science teaches us that our naive view of the world is imprecise, inappropriate, biased. It constructs better views of the world. (Better for some use, worse for others, of course, which is why it is silly to think of our girlfriend as a collection of electrons.) Electrons, if anything, are "more real" than chairs, not "less real," in the sense that they underpin a way of conceptualizing the world which is in many respects more powerful. On the other hand, the process of scientific discovery, and the experience of the twentieth century in particular, has made us painfully aware of the provisional character of *any* form of knowledge. Our mental and mathematical pictures of the world are only mental and mathematical pictures. Between our images of reality and our experience of reality there is always an hiatus. This is true for abstract scientific theories as well as for the image we have of our dining room (not to even mention the image we have of our girlfriend). Nevertheless, the pictures are effective and we can't do any better.

C3 On truth

So, is there anything we can say with confidence about the "real physical world?" A large part of the recent reflection on science has taught us that raw data do not exist, that any information about the world is already deeply filtered and interpreted by the theory, and that theories are all likely to be superseded. It has been useful and refreshing to learn this. Far more radically, the European reflection, and part of the American as well, has emphasized the fact that truth is always internal to the theory, that we can never leave language, that we can never exit the circle of discourse within which we are speaking. As a scientist, I appreciate and share these ideas.

But the fact that the only notion of truth is internal to our discourse does not imply that we should lose confidence in it. If truth is internal to our discourse, then *this internal truth* is what we mean by truth. Indeed there may be no valid notion of truth outside our own discourse, but it is precisely "from within" this discourse, not from without it, that we can and do assert the truth of the reality of the world and the truth of what we have learned about it. More significantly still: it is structural to our language to be a language *about* the world, and to our thinking to be a thinking *of* the world.[1]

Therefore, there is no sense in denying the truth of what we have learned about the world, precisely because there is no notion of truth except the

---

[1]The rational investigation of the world started with the pre-Socratic λόγος (logos), which is the principle (that we seek) governing the cosmos, as well as human reasoning and speaking about the cosmos: it is at the same time the truth, and our reasoning about it.

one within our own discourse. If there is no place we can go which is outside our language, in which place are they standing, those who question the truth we find? It can only be a pleasant short dreamy place where we are happy to stay for a short while, smiling as if we were wise, and then come back to reality. The world is real, solid, and understandable, precisely because the language, our only home, states so. The best we can say about the physical world, and about what is physically real out there, is what good physics says about it.[2]

At the same time, there is no reason that our perceiving, understanding, and conceptualizing the world should not be in continuous evolution. Science is the form of this evolution. At every stage, the best we can say about the reality of the world is precisely what we are saying. The fact we will understand it better later on does not make our present understanding less valuable, or less credible. When we walk in the mountains, we do not dismiss our map just because there may exist a better map which we don't have. Searching for a fixed point on which to rest our restlessness, is, in my opinion, naive, useless and counterproductive for the development of knowledge. It is only by believing our insights and, at the same time, questioning our mental habits, that we can go ahead. I believe that this process of cautious faith and self-confident doubt is the core of scientific thinking. Science is the human adventure that consists in exploring possible ways of thinking of the world. Being ready to subvert, if required, anything we have been thinking so far.

I think this is among the best of human adventures. Research in quantum gravity, in its effort to conceptualize quantum spacetime, and thus modify in depth the notions of space and time, is a step in this adventure.

---

[2] I certainly do not wish to suggest that the physical description of the world exhausts it. It would be like saying that if I understand the physics of a brick, I immediately knew why a cathedral stands, or why it is splendid.

References

Preface, and terminology and notation

[1] C. Rovelli, Loop space representation. In *New Perspectives in Canonical Gravity*, ed. A. Ashtekar *et al.* (Napoli: Bibliopolis, 1988).

[2] C. Rovelli, Ashtekar formulation of general relativity and loop space nonperturbative Quantum Gravity: a report. *Class. and Quantum Grav.* **8** (1991) 1613–1675.

[3] A. Ashtekar, *Non-perturbative Canonical Gravity* (Singapore: World Scientific, 1991).

[4] L. Smolin, Time measurement and information loss in quantum cosmology. In *Brill Feschrift Proceedings*, ed. B. Hu and T. Jacobson (Cambridge: Cambridge University Press, 1993); Recent developments in nonperturbative quantum gravity. In *Quantum Gravity and Cosmology*, ed. J. Perez-Mercader, J. Sola and E. Verdaguer (Singapore: World Scientific, 1993).

[5] J. Baez, *Knots and Quantum Gravity* (Oxford: Oxford University Press, 1994).

[6] B. Brügmann, Loop representations. In *Canonical Gravity: from Classical to Quantum.* ed. J. Ehlers and H. Friedrich (Berlin: Springer-Verlag, 1994).

[7] R. Gambini and J. Pullin, *Loops, Knots, Gauge Theories and Quantum Gravity* (Cambridge: Cambridge University Press, 1996).

[8] J. Kowalski-Glikman, Towards quantum gravity, *Lecture Notes in Physics* **541** (2000) (Berlin: Springer).

[9] A. Ashtekar, Background independent quantum gravity: A Status report. *Class. Quant. Grav.* **21** (2004) R53.

[10] L. Smolin, An invitation to loop quantum gravity, hep-th/0408048.

[11] T. Thiemann, Lectures on loop quantum gravity, *Lecture Notes in Physics* **631** (2003) 41–135, gr-qc/0210094.

[12] C. Rovelli, Loop quantum gravity, *Living Reviews in Relativity*, electronic journal, http:// www.livingreviews.org/Articles/Volume1/1998-1rovelli.

[13] A. Ashtekar, Quantum geometry and gravity: recent advances, to appear in the *Proc. 16th Int. Conf. on General Relativity and Gravitation, Durban, S Africa, July 2001*, gr-qc/9901023.

[14] M. Gaul and C. Rovelli, Loop quantum gravity and the meaning of diffeomorphism invariance, *Lecture Notes in Physics* **541** (2000) 277–324 (Berlin: Springer), gr-qc/9910079.

[15] C. Rovelli and P. Upadhya, Loop quantum gravity and quanta of space: a primer, gr-qc/9806079.

[16] J. Baez and J. Muniain, *Gauge Fields, Knots and Gravity* (Singapore: World Scientific, 1994).

[17] J.C. Baez, An introduction to spin foam models of BF theory and quantum gravity. In *Geometry and Quantum Physics*, ed. H. Gausterer and H. Grosse, *Lecture Notes in Physics* **543** (1999) 25–94 (Berlin: Springer-Verlag), gr-qc/9905087.

[18] D. Oriti, Spacetime geometry from algebra: spin foam models for non-perturbative quantum gravity, *Rept. Prog. Phys.* **64** (2001) 1489, gr-qc/0106091.

[19] A. Perez, Spin foam models for quantum gravity, *Class. and Quantum Grav.* **20** (2002), gr-qc/0301113.

[20] T. Thiemann, *Modern Canonical Quantum General Relativity* (Cambridge: Cambridge University Press, 2004, in press), a preliminary version is in gr-qc/0110034.

[21] J. Ambjorn, B. Durhuus and T. Jonsson, *Quantum Geometry* (Cambridge: Cambridge University Press, 1997).

Chapter 1: General ideas and heuristic picture

[22] M. Gell-Mann, *Strange Beauty* (London: Vintage, 2000), pp. 303–304.

[23] L. Smolin, Towards a background independent approach to M theory, hep-th/9808192; The cubic matrix model and duality between strings and loops, hep-th/0006137; A candidate for a background independent formulation of M theory, *Phys. Rev.* **D62** (2000) 086001, hep-th/9903166.

[24] L. Smolin, Strings as perturbations of evolving spin networks, *Nucl. Phys. Proc. Suppl.* **88** (2000) 103–113, hep-th/9801022.

[25] D. Amati, M. Ciafaloni and G. Veneziano, Can spacetime be probed below the string size?, *Phys. Lett.* **B216** (1989) 41.

[26] J. Hartle, Spacetime quantum mechanics and the quantum mechanics of spacetime. In *Proceedings 1992 Les Houches School, Gravitation and Quantisation*, ed. B. Julia and J. Zinn-Justin (Paris: Elsevier Science, 1995), p. 285.

[27] S.A. Fulling, *Aspects of Quantum Field Theory in Curved Spacetime* (Cambridge: Cambridge University Press, 1989).

[28] R.M. Wald, *Quantum Field Theory on Curved Spacetime and Black Hole Thermodynamics* (Chicago: University of Chicago Press, 1994).

[29] M.H. Goroff and A. Sagnotti, Quantum gravity at two loops, *Phys. Lett.* **B160** (1985) 81; The ultraviolet behaviour of Einstein gravity, *Nucl. Phys.* **B266** (1986) 709.

[30] G. Horowitz, Quantum gravity at the turn of the millenium, plenary talk at the *Marcell Grossmann Conf., Rome, 2000*, gr-qc/0011089.

[31] S. Carlip, Quantum gravity: a progress report, *Rept. Prog. Phys.* **64** (2001) 885, gr-qc/0108040.

[32] C.J. Isham, Conceptual and geometrical problems in quantum gravity. In *Recent Aspects of Quantum Fields*, ed. H. Mitter and H. Gausterer (Berlin: Springer-Verlag, 1991), p. 123.

[33] C. Rovelli, Strings, loops and the others: a critical survey on the present approaches to quantum gravity. In *Gravitation and Relativity: At the turn of the Millenium*, ed. N. Dadhich and J. Narlikar (Pune: Inter-University Centre for Astronomy and Astrophysics, 1998), pp. 281–331, gr-qc/9803024.

[34] C. Callender and N. Hugget, eds., *Physics Meets Philosophy at the Planck Scale* (Cambridge: Cambridge University Press, 2001).

[35] C. Rovelli, Halfway through the woods. In *The Cosmos of Science*, ed. J. Earman and J.D. Norton (University of Pittsburgh Press and Universitäts Verlag-Konstanz, 1997).

[36] C. Rovelli, Quantum spacetime: what do we know? In *Physics Meets Philosophy at the Planck Length*, ed. C. Callender and N. Hugget (Cambridge: Cambridge University Press, 1999), gr-qc/9903045.

[37] C. Rovelli, The century of the incomplete revolution: searching for general relativistic quantum field theory, *J. Math. Phys., Special Issue 2000* **41** (2000) 3776; hep-th/9910131.

[38] J.A. Wheeler, Superspace and the nature of quantum geometrodynamics. In *Batelle Rencontres, 1967*, ed. C. DeWitt and J.W. Wheeler, *Lectures in Mathematics and Physics*, **242** (New York: Benjamin, 1968).

[39] S. Weinberg, Ultraviolet divergences in quantum theories of gravitation. In *General Relativity: An Einstein Centenary Survey*, ed. S.W. Hawking and W. Israel (Cambridge: Cambridge University Press, 1979).

[40] S.W. Hawking, The path-integral approach to quantum gravity. In *General Relativity: An Einstein Centenary Survey*, ed. S.W. Hawking and W. Israel (Cambridge: Cambridge University Press, 1979).

[41] S.W. Hawking, Quantum cosmology. In *Relativity, Groups and Topology, Les Houches Session XL*, ed. B. DeWitt and R. Stora (Amsterdam: North Holland, 1984).

[42] K. Kuchar, Canonical methods of quantization. In *Oxford 1980, Proceedings, Quantum Gravity 2* (Oxford: Oxford University Press, 1984).

[43] C.J. Isham, Topological and global aspects of quantum theory. In *Relativity Groups and Topology. Les Houches 1983*, ed. B.S. DeWitt and R. Stora (Amsterdam: North Holland, 1984).

[44] C.J. Isham, Quantum gravity: an overview. In *Oxford 1980, Proceedings, Quantum Gravity 2* (Oxford: Oxford University Press, 1984).

[45] C. J. Isham, Structural problems facing quantum gravity theory. In *Proc. 14th Int. Conf. on General Relativity and Gravitation*, ed. M. Francaviglia, G. Longhi, L. Lusanna and E. Sorace (Singapore: World Scientific, 1997), pp. 167–209.

[46] M.B. Green, J. Schwarz and E. Witten, *Superstring Theory* (Cambridge: Cambridge University Press, 1987); J. Polchinski, *String Theory* (Cambridge: Cambridge University Press, 1998).

[47] C. Rovelli, A dialog on quantum gravity, *Int. J. Mod. Phys.* **12** (2003) 1, hep-th/0310077.

[48] L. Smolin, How far are we from the quantum theory of gravity?, hep-th/0303185.

[49] A. Connes, *Non Commutative Geometry* (New York: Academic Press, 1994).

[50] L. Smolin, *Three Roads to Quantum Gravity* (Oxford: Oxford University Press, 2000).

[51] K.S. Robinson, *Blue Mars* (New York: Bantam, 1996).

[52] G. Egan, *Schild Ladder* (London: Gollancz, 2001).

[53] E. Palandri, *Anna prende il volo* (Milano: Feltrinelli, 2000).

Chapter 2: General relativity

[54] S. Holst, Barbero's Hamiltonian derived from a generalized Hilbert-Palatini action, *Phys. Rev.* **D53** (1996) 5966–5969.

[55] L. Russo, *La rivoluzione dimenticata* (Milano: Feltrinelli, 1997).

[56] J.P. Bourguignon and P. Gauduchon, Spineurs, operateurs de Dirac et variations de metriques, *Comm. Math. Phys.* **144** (1992) 581.

[57] T. Schücker, Forces from Connes' geometry, hep-th/0111236. Lectures at the Autumn School *Topology and Geometry in Physics, Rot an der Rot, 2001*, ed. E. Bick and F. Steffen (*Lecture Notes in Physics*, Springer, 2004).

[58] L. Russo, *Flussi e riflussi* (Feltrinelli, Milano, 2003).

[59] M. Faraday, *Experimental Researches in Electricity* (London: Bernard Quaritch, 1855), pp. 436–437.

[60] R. Descartes, *Principia Philosophiae* (1644), Translated by V.R. Miller and R.P. Miller (Dordrecht: Reidel, 1983).

[61] I. Newton, *De Gravitatione et Aequipondio Fluidorum*, translation in *Unpublished Papers of Isaac Newton*, ed. A.R. Hall and M.B. Hall (Cambridge: Cambridge University Press, 1962).

[62] I. Newton, *Principia Mathematica Philosophia Naturalis*, 1687, English translation *The Principia: Mathematical Principles of Natural Philosophy* (City? University of California Press, 1999).

[63] A. Einstein and M. Grossmann, Entwurf einer verallgemeinerten Relativitätstheorie und einer Theorie der Gravitation, *Z. für Mathematik und Physik* **62** (1914) 225.

[64] A. Einstein, Grundlage der allgemeinen Relativitätstheorie, *Ann. der Phys.* **49** (1916) 769–822.

[65] M. Pauri and M. Vallisneri, Ephemeral point-events: is there a last remnant of physical objectivity?, *DIALOGOS*, **79** (2002), 263–303; L. Lusanna and M. Pauri, General covariance and the objectivity of space-time point-events: the physical role of gravitational and gauge degrees of freedom, http://philsci-archive.pitt.edu/archive/00000959/ (2002).

[66] P. Hajicek, *Lecture Notes in Quantum Cosmology* (Bern: University of Bern, 1990).

[67] A.S. Eddington, *The Nature of the Physical World* (New York: MacMillan, 1930), pp. 99–102.

[68] S.J. Earman, *A Primer on Determinism* (Dordrecht: D. Reidel, 1986).

[69] B. DeWitt, in *Gravitation: An Introduction to Current Research*, ed. L. Witten (New York: Wiley, 1962).

[70] J.D. Brown and D. Marolf, On relativistic material reference systems, *Phys. Rev.* **D53** (1996) 1835.

[71] C. Rovelli, What is observable in classical and quantum gravity?, *Class. and Quantum Grav.* **8** (1991) 297; Quantum reference systems, *Class. and Quantum Grav.* **8** (1991) 317.

[72] P.G. Bergmann, *Phys. Rev.* **112** (1958) 287; Observables in general covariant theories, *Rev. Mod. Phys.* **33** (1961) 510.

[73] B.W. Parkinson and J.J. Spilker, eds., *Global Positioning System: Theory and Applications, Prog. in Astronautics and Aeronautics, Nos. 163–164* (Amer. Inst. Aero. Astro., Washington: 1996); E.D. Kaplan, *Understanding GPS: Principles and Applications*, Mobile Communications Series (Boston: Artech House, 1996); B. Hofmann-Wellenhof, H. Lichtenegger and J. Collins, *Global Positioning System Theory and Practice* (New York: Springer-Verlag, 1993).

[74] B. Guinot, Application of general relativity to metrology, *Metrologia* **34** (1997) 261; F. de Felice, M.G. Lattanzi, A. Vecchiato and P.L. Bernacca, General relativistic satellite astrometry: I. A non-perturbative approach to data reduction, *Astron. Astrophy.* **332** (1998) 1133; T.B. Bahder, *Fermi Coordinates of an Observer Moving in a Circle in Minkowski Space: Apparent Behavior of Clocks*, Army Research Laboratory, Adelphi, Maryland, USA, Technical Report ARL-TR-2211, May 2000; A.R. Thompson, J.M. Moran and G.W. Swenson, *Interferometry and Synthesis in Radio Astronomy*, (Malabar, Florida: Krieger Pub. Co., 1994), pp. 138–139; P.N.A.M. Visser, Gravity field determination with GOCE and GRACE, *Adv. Space Res.* **23** (1999) 771.

[75] S. Weinberg, *Gravitation and Cosmology* (New York: Wiley, 1972).

[76] R.M. Wald, *General Relativity* (Chicago: The University of Chicago Press, 1989).

[77] Y. Choquet-Bruhat, C. DeWitt-Morette and M. Dillard-Bleick, *Analysis, Manifolds and Physics* (Amsterdam: North Holland, 1982).

[78] I. Ciufolini and J. Wheeler, *Gravitation and Inertia* (Princeton: Princeton University Press, 1996).

[79] H. Weyl, Electron and gravitation *Z. Physik* **56**, (1929) 330.

[80] J. Schwinger, Quantized gravitational field *Phys. Rev.*, **130**, (1963) 1253.

[81] J.F. Plebanski, On the separation of Einsteinian substructures, *J. Math. Phys.* **18** (1977) 2511.

[82] A. Sen, Gravity as a spin system, *Phys. Lett.* **119B** (1982) 89.

[83] J. Samuel, A lagrangian basis for Ashtekar's reformulation of canonical gravity, *Pramana J. Phys.* **28** (1987) L429; T. Jacobson and L. Smolin, Covariant action for Ashtekar's form of canonical gravity, *Class. and Quantum Grav.* **5** (1988) 583.

[84] R. Capovilla, J. Dell and T. Jacobson, General relativity without the metric, *Phys. Rev. Lett.* **63** (1991) 2325; R. Capovilla, J. Dell, T. Jacobson and L. Mason, Self–dual 2–forms and gravity, *Class. and Quantum Grav.* **8** (1991) 41.

[85] J.D. Norton, How Einstein found his field equations: 1912–1915, *Historical Studies in the Physical Sciences*, **14** (1984) 253–315. Reprinted in *Einstein and the History of General Relativity: Einstein Studies*, ed. D. Howard and J. Stachel, Vol. I (Boston: Birkhäuser, 1989), pp. 101–159.

[86] J. Stachel, Einstein's search for general covariance 1912–1915. In *Einstein and the History of General Relativity: Einstein Studies*, ed. D. Howard and J. Stachel, Vol. 1 (Boston: Birkhäuser, 1989), pp. 63–100.

[87] E. Kretschmann, Über den physikalischen Sinn der Relativitätpostulate, *Ann. Phys. Leipzig* **53** (1917) 575.

[88] J.L. Anderson, *Principles of Relativity Physics* (New York: Academic Press, 1967).

[89] J. Barbour, *Absolute or Relative Motion?* (Cambridge: Cambridge University Press, 1989).

[90] J. Earman and J. Norton, What price spacetime substantivalism? The hole story, *Brit. J. Phil. Sci.* **38** (1987), 515–525.

[91] J. Earman, *World Enough and Space-time: Absolute Versus Relational Theories of Spacetime* (Cambridge: MIT Press, 1989).

[92] G. Belot, Why general relativity does need an interpretation, *Phil. Sci.* **63** (1998) S80–S88.

[93] J. Earman and G. Belot, Pre-Socratic quantum gravity. In *Physics Meets Philosphy at the Planck Scale*, ed. C. Callander (Cambridge: Cambridge University Press, 2001).

[94] C. Rovelli, Analysis of the different meaning of the concept of time in different physical theories, *Il Nuovo Cimento* **110B** (1995) 81.

[95] J.T. Fraser, *Of Time, Passion, and Knowledge* (Princeton: Princeton University Press, 1990).

[96] H. Reichenbach, *The Direction of Time* (Berkeley: University of California Press, 1956); P.C.W. Davies, *The Physics of Time Asymmetry* (England:

Surrey University Press, 1974); R. Penrose, in *General Relativity: An Einstein Centenary Survey*, ed. S.W. Hawking and W. Israel (Cambridge: Cambridge University Press, 1979); H.D. Zee, *The Physical Basis of the Direction of Time* (Berlin: Springer, 1989); J. Halliwel and J.A. Perez-Mercader, eds., *Proceedings of the International Workshop: Physical Origins of Time Asymmetry, Huelva Spain, September 1991* (Cambridge: Cambridge University Press, 1992).

[97] C.J. Isham, Canonical quantum gravity and the problem of time, Lectures presented at the NATO Advanced Institute *Recent Problems in Mathematical Physics, Salamanca, June 15, 1992*; K. Kuchar, Time and interpretations of quantum gravity. In *Proc. 4th Canadian Conference on General Relativity and Relativistic Astrophysics*, ed. G. Kunstatter, D. Vincent and J. Williams (Singapore: World Scientific, 1992); A. Ashtekar and J. Stachel, eds., *Proc. Osgood Hill Conference: Conceptual Problems in Quantum Gravity, Boston 1988* (Boston: Birkhäuser, 1993).

[98] C. Rovelli, Time in quantum gravity: an hypothesis, *Phys. Rev.* **D43** (1991) 442.

[99] J. Hartle, Classical physics and hamiltonian quantum mechanics as relics of the big bang, *Physica Scripta* **T36** (1991) 228.

[100] A. Grunbaum, *Philosophical Problems of Space and Time* (New York: Knopf, 1963); T. Gold and D.L. Shumacher, eds., *The Nature of Time* (Ithaca: Cornell University Press, 1967); P. Kroes, *Time: its Structure and Role in Physical Theories* (Dordrecht: D. Reidel, 1985).

[101] C. Rovelli, GPS observables in general relativity, *Phys. Rev.* **D65** (2002) 044017, gr-qc/0110003.

[102] T.B. Bahder, Navigation in curved space-time, *Amer. J. Phys.* **69** (2001) 315–321.

[103] M. Blagojevic, J. Garecki, F.W. Hehl and Yu. N. Obukhov, Real null coframes in general relativity and GPS type coordinates, gr-qc/0110078.

Chapter 3: Mechanics

[104] V.I. Arnold, *Matematičeskie Metody Klassičeskoj Mechaniki* (Moskow: Mir, 1979). See in particular Chapter IX, Section C.

[105] J.M. Souriau, *Structure des Systemes Dynamiques* (Paris: Dunod, 1969).

[106] J.L. Lagrange, *Mémoires de la Première Classe des Sciences Mathematiques et Physiques* (Paris: Institute de France, 1808).

[107] W.R. Hamilton, On the application to dynamics of a general mathematical method previously applied to optics, *British Association Report* (1834), 513–518.

[108] Č. Crnković and E. Witten, Covariant description of canonical formalism in geometrical theories, In *Newton's Tercentenary Volume*, ed. S.W. Hawking and W. Israel (Cambridge: Cambridge University Press, 1987);

A. Ashtekar, L. Bombelli and O. Reula. In *Mechanics, Analysis and Geometry: 200 Years after Lagrange*, ed. M. Francaviglia (Amsterdam: Elsevier, 1991).

[109] M.J. Gotay, J. Isenberg and J.E. Marsden (with the collaboration of R. Montgomery, J. Sniatycki and P.B. Yasskin), Momentum maps and classical relativistic fields. Part 1: covariant field theory, physics/9801019.

[110] T. DeDonder, *Théorie Invariantive du Calcul des Variationes* (Paris: Gauthier-Villars, 1935).

[111] Hesiod, *Theogony*, translated by H.G. Evelyn-White (London: Harvard University Press, 1914), pp. 125–130. [Instigated by mother Γαῖα, Κρόνος then slaughters and castrates father, Οὐρανός.]

[112] C. Rovelli, The statistical state of the universe, *Class. and Quantum Grav.* **10** (1993) 1567.

[113] P.A.M. Dirac, Generalized Hamiltonian dynamics, *Can. J. Math. Phys.* **2** (1950) 129–148.

[114] P.A.M. Dirac, *Lectures on Quantum Mechanics* (New York: Belfer Graduate School of Science, Yeshiva University, 1964).

[115] A. Hanson, T. Regge and C. Teitelboim, *Constrained Hamiltonian Systems* (Roma: Accademia nazionale dei Lincei, 1976); M. Henneaux and C. Teitelboim, *Quantization of Gauge Systems* (Princeton: Princeton University Press, 1972).

[116] C. Rovelli, Partial observables, *Phy. Rev.* **D65** (2002) 124013, gr-qc/0110035.

[117] C. Rovelli, A note on the foundation of relativistic mechanics. I: Relativistic observables and relativistic states. In *Proc. 15th SIGRAV Conference on General Relativity and Gravitational Physics, 2002* (Bristol: IOP Publishing, 2004) in press, gr-qc/0111037.

[118] C. Rovelli, Covariant hamiltonian formalism for field theory: symplectic structure and Hamilton–Jacobi equation on the space G. In *Decoherence and Entropy in Complex Systems, Selected Lectures from DICE 2002. Lecture Notes in Physics* **633**, ed. H.T. Elze (Berlin: Springer–Verlag, 2003), gr-qc/0207043.

[119] M. Montesinos, C. Rovelli and T. Thiemann, $SL(2,R)$ model with two Hamiltonian constraints, *Phys. Rev.* **D60** (1999) 044009.

[120] H. Weil, Geodesic fields in the calculus of variations, *Ann. Math.* **36** (1935) 607–629.

[121] J. Kijowski, A finite dimensional canonical formalism in the classical field theory, *Comm. Math. Phys.* **30** (1973) 99–128; M. Ferraris and M. Francaviglia, The Lagrangian approach to conserved quantities in general relativity. In *Mechanics, Analysis and Geometry: 200 Years after Lagrange*, ed. M. Francaviglia (Amsterdam: Elsevier Sci. Publ., 1991), pp. 451–488; I.V. Kanatchikov, Canonical structure of classical field theory in the polymomentum phase space, *Rep. Math. Phys.* **41** (1998) 49; F. Hélein and J. Kouneiher, Finite dimensional Hamiltonian formalism for gauge and

field theories, math-ph/0010036; H. Rund, *The Hamilton–Jacobi Theory in the Calculus of Variations* (New York: Krieger, 1973); H. Kastrup, Canonical theories of Lagrangian dynamical systems in physics, *Phys. Rep.* **101** (1983) 1.

[122] C. Rovelli, Statistical mechanics of gravity and thermodynamical origin of time, *Class. and Quantum Grav.* **10** (1993) 1549.

[123] A. Connes and C. Rovelli, Von Neumann algebra automorphisms and time versus thermodynamics relation in general covariant quantum theories, *Class. and Quantum Grav.* **11** (1994) 2899.

[124] P. Martinetti and C. Rovelli, Diamonds' temperature: Unruh effect for bounded trajectories and thermal time hypothesis, *Class. and Quantum Grav.* **20** (2003) 4919–4932, gr-qc/0212074.

[125] M. Montesinos and C. Rovelli, Statistical mechanics of generally covariant quantum theories: a Boltzmann-like approach, *Class. and Quantum Grav.* **18** (2001) 555–569.

Chapter 4: Hamiltonian general relativity

[126] D. Giulini, Ashtekar variables in Classical General Relativity. In *Canonical Gravity: From Classical to Quantum*, ed. J. Ehlers and H. Friedrich (Berlin: Springer-Verlag, 1994), p. 81.

[127] S. Alexandrov, E. Buffenoir, P. Roche, Plebanski theory and covariant canonical formulation, gr-qc/0612071.

[128] A. Perez, C. Rovelli, Physical effects of the Immirzi parameter. *Phys. Rev.* **D73** (2006) 044013.

[129] P. Bergmann, *Phys. Rev.* **112** (1958) 287; *Rev. Mod. Phys.* **33** (1961); P. Bergmann and A. Komar, The phase space formulation of general relativity and approaches towards quantization, *Gen. Rel. Grav.*, **1** (1981), pp. 227–254; In *General Relativity and Gravitation*, ed. A. Held (1981), pp. 227–254; A. Komar, General relativistic observables via Hamilton–Jacobi functionals, *Phys. Rev.* **D4** (1971) 923–927.

[130] P.A.M. Dirac, The theory of gravitation in Hamiltonian form, *Proc. Royal Soc. London* **A246** (1958) 333; *Phys. Rev.* **114** (1959) 924.

[131] R. Arnowitt, S. Deser and C.W. Misner, The dynamics of general relativity. In *Gravitation: An Introduction to Current Research*, ed. L. Witten (New York: Wiley, 1962), p. 227.

[132] A. Ashtekar, New variables for classical and quantum gravity, *Phys. Rev. Lett.* **57** (1986) 2244; New Hamiltonian formulation of general relativity, *Phys. Rev.* **D36** (1987) 1587.

[133] T. Thiemann, Anomaly-free formulation of nonperturbative 4-dimensional Lorentzian quantum gravity, *Phys. Lett.* **B380** (1996) 257.

[134] F. Barbero, Real Ashtekar variables for Lorentzian signature spacetimes, *Phys. Rev.* **D51** (1995) 5507, gr-qc/9410014; *Phys. Rev.* **D51** (1995) 5498.

[135] G. Immirzi, Quantum gravity and Regge calculus, *Nucl. Phys. Proc. Suppl.* **57** (1997) 65; Real and complex connections for canonical gravity, *Class. and Quantum Grav.* **14** (1997) L177–L181.

[136] L. Fatibene, M. Francaviglia, C. Rovelli, On a Covariant Formulation of the Barbero-Immirzi Connection, gr-qc/0702134.

[137] C. Rovelli and T. Thiemann, The Immirzi parameter in quantum general relativity, *Phys. Rev.* **D57** (1998) 1009–1014, gr-qc/9705059.

[138] G. Esposito, G. Gionti and C. Stornaiolo, Space-time covariant form of Ashtekar's constraints, *Nuovo Cimento* **110B** (1995), 1137–1152.

[139] C. Rovelli, A note on the foundation of relativistic mechanics. II: Covariant hamiltonian general relativity, gr-qc/0202079.

[140] M. Ferraris and M. Francaviglia, The Lagrangian approach to conserved quantities in General Relativity. In *Mechanics, Analysis and Geometry: 200 Years after Lagrange*, ed. M. Francaviglia (Amsterdam: Elsevier Sci. Publ., 1991), pp. 451–488; W. Szczyrba, A symplectic structure of the set of Einstein metrics: a canonical formalism for general relativity, *Comm. Math. Phys.* **51** (1976) 163–182; J. Sniatcki, On the canonical formulation of general relativity. In *Proc. Journées Relativistes* (Caen: Faculté des Sciences, 1970); J. Novotny, On the geometric foundations of the Lagrange formulation of general relativity. In *Differential Geometry*, ed. G. Soos and J. Szenthe (Amsterdam: North-Holland, 1982).

[141] A. Peres, *Nuovo Cimento* **26** (1962) 53; U. Gerlach, *Phys. Rev.* **177** (1969) 1929. K. Kuchar, *J. Math. Phys.* **13** (1972) 758; P. Horava, On a covariant Hamilton–Jacobi framework for the Einstein–Maxwell theory, *Class. and Quantum Grav.* **8** (1991) 2069; E.T. Newman and C. Rovelli, Generalized lines of force as the gauge invariant degrees of freedom for general relativity and Yang–Mills theory *Phys. Rev. Lett.* **69** (1992) 1300; J. Kijowski and G. Magli, Unconstrained Hamiltonian formulation of General Relativity with thermo-elastic sources, *Class. and Quantum Grav.* **15** (1998) 3891–3916.

Chapter 5: Quantum mechanics

[142] E. Schrödinger, Quantisierung als Eigenwertproblem, *Ann. der Phys.* **79** (1926) 489, Part 2, English translation in *Collected Papers on Quantum Mechanics* (Chelsea Publications, 1982).

[143] E. Schrödinger, Quantisierung als Eigenwertproblem, *Ann. der Phys.* **79** (1926) 361, Part 1, English translation, op cit.

[144] M. Reisenberger and C. Rovelli, Spacetime states and covariant quantum theory, *Phys. Rev.* **D65** (2002) 124013, gr-qc/0111016; D. Marolf and C. Rovelli, Relativistic quantum measurement, *Phys. Rev.* **D66** (2002) 023510, gr-qc/0203056.

[145] F. Conrady, L. Doplicher, R. Oeckl, C. Rovelli and M. Testa, Minkowski vacuum from background independent quantum gravity, *Phys. Rev.* **D164** (2004) 064019, gr-qc/0307118.

[146] F. Conrady and C. Rovelli, Generalized Schrödinger equation in Euclidean quantum field theory, *Int. J. Mod. Phys.*, in press, hep-th/0310246.

[147] M. Montesinos, The double role of Einstein's equations: as equations of motion and as vanishing energy-momentum tensor, gr-qc/0311001.

[148] P.A.M. Dirac, *Principles of Quantum Mechanics*, 1st edition (Oxford: Oxford University Press, 1930).

[149] C. Rovelli, Is there incompatibility between the ways time is treated in general relativity and in standard quantum mechanics? In *Conceptual Problems of Quantum Gravity*, ed. A. Ashtekar and J. Stachel (New York: Birkhauser, 1991).

[150] J. Halliwell, The Wheeler–deWitt equation and the path integral in mini-superspace quantum cosmology. In *Conceptual Problems of Quantum Gravity*, A. Ashtekar and J. Stachel (New York: Birkhauser, 1991).

[151] C. Rovelli, Quantum mechanics without time: a model, *Phys. Rev.* **D42** (1991) 2638.

[152] C. Rovelli, Quantum evolving constants, *Phys. Rev.* **D44** (1991) 1339.

[153] R. Oeckl, A 'general boundary' formulation for quantum mechanics and quantum gravity, hep-th/0306025; Schroedinger's cat and the clock: lessons for quantum gravity, gr-qc/0306007.

[154] L. Doplicher, Generalized Tomonaga–Schrödinger equation, from the Hadamard formula, gr-qc/0405006

[155] S. Tomonaga, *Prog. Theor. Phys.* **1** (1946) 27; J. Schwinger, Quantum electrodynamics. I: A covariant formulation, *Phys. Rev.* **74** (1948) 1439.

[156] N.C. Tsamis and R.P. Woodard, Physical Green's functions in quantum gravity, *Annals of Phys.* **215** (1992) 96.

[157] J.B. Hartle and S.W. Hawking, Wave function of the Universe, *Phys. Rev.* **D28** (1983) 2960.

[158] D. Marolf, Group averaging and refined algebraic quantization: where are we? In *Proceedings of the IXth Marcel Grossmann Conference, Rome, Italy, July 2–9, 2000*, ed. R.T. Jantzen, G.M. Keiser and R. Ruffini (World Scientific, 1996), gr-qc/0011112.

[159] C. Rovelli, Relational quantum mechanics, *Int. J. Theor. Phys.* **35** (1996) 1637–1678.

[160] C. Rovelli, Incerto tempore, incertisque loci: Can we compute the exact time at which a quantum measurement happens?, *Foundations of Physics*, **28** (1998) 1031–1043.

[161] F. Laudisa, The EPR argument in a relational interpretation of quantum mechanics, *Foundations of Physics Letters*, **14** (2) (2001) 119–132.

[162] A. Grinbaum, Elements of information theoretic derivation of the formalism of quantum theory, *Int. J. Quant. Information* **1** (2003) 1.

[163] F. Laudisa and C. Rovelli, Relational quantum mechanics. In *The Stanford Encyclopedia of Philosophy (Spring 2002 Edition)*, ed. Edward N. Zalta, URL http://plato.stanford.edu/archives/spr2002/entries/qm-relational/.

[164] M. Bitbol, Relations et corrélations en Physique Quantique. In *Un Siècle de Quanta*, ed. M. Crozon and Y. Sacquin (Paris: EDP Sciences, 2000).

[165] J. Wheeler, Information, physics, quantum: the search for the links, *Proc. 3rd Int. Symp. Foundations of Quantum Mechanics, Tokyo 1989*, p. 354.

[166] J. Wheeler, It from Bit. In *Sakharov Memorial Lectures on Physics*, Vol. 2, ed. L. Keldysh and V. Feinberg (New York: Nova Science, 1992).

[167] C.U. Fuchs, Quantum foundations in the light of quantum information. In *Proc. NATO Advanced Research Workshop on Decoherence and its Implications in Quantum Computation and Information Transfer*, ed. A. Gonis (New York: Plenum, 2001), quant-ph/0106166,quant-ph/0205039.

[168] D. Finkelstein, *Quantum Relativity* (Berlin: Springer, 1996).

Chapter 6: Quantum space

[169] L. Freidel and E.R. Livine, Spin networks for non-compact groups, *J. Math. Phys.* **44** (2003) 1322–1356.

[170] C. Rovelli, Loop representation in quantum gravity. In *Conceptual Problems of Quantum Gravity*, ed. A. Ashtekar and J. Stachel (New York: Birkhäuser, 1991).

[171] C. Rovelli and L. Smolin, Spin networks and quantum gravity, *Phys. Rev.* **D52** (1995) 5743–5759, gr-qc/9505006.

[172] J. Lewandowski, E.T. Newman and C. Rovelli, Variations of the parallel propagator and holonomy operator and the Gauss law constraint, *J. Math. Phys.* **34** (1993) 4646.

[173] A. Ashtekar and J. Lewandowski, Quantum theory of geometry. I: Area operators, *Class. and Quantum Grav.* **14** (1997) A55; II: Volume operators, *Adv. Theor. Math. Phys.* **1** (1997) 388–429.

[174] L.H. Kauffman and S.L. Lins, *Temperley-Lieb Recoupling Theory and Invariant of 3-Manifolds* (Princeton: Princeton University Press, 1994).

[175] R. De Pietri and C. Rovelli, Geometry eigenvalues and scalar product from recoupling theory in loop quantum gravity, *Phys. Rev.* **D54** (1996) 2664, gr-qc/9602023; T. Thiemann, Closed formula for the matrix elements of the volume operator in canonical quantum gravity, *J. Math. Phys.* **39** (1998) 3347–3371, gr-qc/9606091.

[176] C. Rovelli and L. Smolin, Knot theory and quantum gravity, *Phys. Rev. Lett.* **61** (1988) 1155.

[177] C. Rovelli and L. Smolin, Loop space representation for quantum general relativity, *Nucl. Phys.*, **B331** (1990) 80.

[178] T. Jacobson and L. Smolin, Nonperturbative quantum geometries, *Nucl. Phys.* **B299** (1988) 295.

[179] R. Gambini and A. Trias, *Phys. Rev.* **D22** (1980) 1380; On the geometrical origin of gauge theories, *Phys. Rev.* **D23** (1981) 553; *Nucl. Phys.* **B278** (1986) 436; C. di Bartolo, F. Nori, R. Gambini and A. Trias, Loop space formulation of free electromagnetism, *Nuovo Cimento Lett.* **38** (1983) 497.

[180] B. Brügmann, R. Gambini and J. Pullin, Knot invariants as nondegenerate quantum geometries, *Phys. Rev. Lett.* **68** (1992) 431; Jones polynomials for intersecting knots as physical states of quantum gravity, *Nucl. Phys.* **B385** (1992) 587; *Gen. Rel. Grav.* **25** (1993) 1; J. Pullin, in *Proc. 5th Mexican School of Particles and Fields*, ed. J. Lucio (Singapore: World Scientific, 1993).

[181] R. Penrose, Theory of quantized directions, unpublished manuscript; Angular momentum: an approach to combinatorial spacetime. In *Quantum Theory and Beyond*, ed. T. Bastin (Cambridge: Cambridge University Press, 1971), pp. 151–180.

[182] L. Smolin, The future of spin networks, gr-qc/9702030.

[183] J.C. Baez, Spin networks in gauge theory, *Adv. Math.* **117** (1996) 253; J.C. Baez, Spin networks in nonperturbative quantum gravity. In *Interface of Knot Theory and Physics*, ed. L. Kauffman (Providence, Rhode Island: American Mathematical Society, 1996), gr-qc/9504036.

[184] N. Grott and C. Rovelli, Moduli spaces structure of knots with intersections, *J. Math. Phys.* **37** (1996) 3014.

[185] W. Fairbairn and C. Rovelli, Separable Hilbert space in loop quantum gravity, *J. Math. Phys.*, to appear, gr-qc/0403047.

[186] J. Zapata, A combinatorial approach to diffeomorphism invariant quantum gauge theories, *J. Math. Phys.* **38** (1997) 5663–5681; A combinatorial space for loop quantum gravity, *Gen. Rel. Grav.* **30** (1998) 1229.

[187] J. Lewandowski, A. Okolow, H. Sahlmann, T. Thiemann, *Comm. Math. Phys.* **267** (2006) 703–733.

[188] C. Fleischhack Representations of the Weyl Algebra in Quantum Geometry, math-ph/0407006.

[189] A. Ashtekar, C. Rovelli and L. Smolin, Weaving a classical metric with quantum threads, *Phys. Rev. Lett.* **69** (1992) 237, hep-th/9203079.

[190] C. Rovelli, A generally covariant quantum field theory and a prediction on quantum measurements of geometry, *Nucl. Phys.* **B405** (1993) 797.

[191] C. Rovelli and L. Smolin, Discreteness of area and volume in quantum gravity, *Nucl. Phys.* **B442** (1995) 593; Erratum, *Nucl. Phys.* **B456** (1995) 734.

[192] S. Frittelli, L. Lehner and C. Rovelli, The complete spectrum of the area from recoupling theory in loop quantum gravity, *Class. and Quantum Grav.* **13** (1996) 2921.

[193] R. Loll, The volume operator in discretized quantum gravity, *Phys. Rev. Lett.* **75** (1995) 3048; Spectrum of the volume operator in quantum gravity, *Nucl. Phys.* **B460** (1996) 143–154, gr-qc/9511030.

[194] J. Lewandowski, Volume and quantizations, *Class. and Quantum Grav.* **14** (1997) 71–76.

[195] T. Tsushima, The expectation value of the Gaussian weave state in loop quantum gravity, gr-qc/0212117.

[196] T. Thiemann, A length operator for canonical quantum gravity, *J. Math. Phys.* **39** (1998) 3372–3392, gr-qc/9606092.

[197] A. Ashtekar, A. Corichi and J. Zapata, Quantum theory of geometry. III: Noncommutativity of Riemannian structures, *Class. and Quantum Grav.* **15** (1998) 2955.

[198] S. Major, Operators for quantized directions, *Class. Quant. Grav.* **16** (1999) 3859–3877.

[199] S. Major, A Spin Network Primer, *Am. J. Phys.* **67** (1999) 972–980.

[200] A. Ashtekar and C.J. Isham, Representations of the holonomy algebra of gravity and non-abelian gauge theories, *Class. and Quantum Grav.* **9** (1992) 1433, hep-th/9202053.

[201] A. Ashtekar and J. Lewandowski, Representation theory of analytic holonomy C^*-algebras. In *Knots and Quantum Gravity*, ed. J. Baez (Oxford: Oxford University Press, 1994); Differential geometry on the space of connections via graphs and projective limits, *J. Geom. and Phys.* **17** (1995) 191.

[202] A. Ashtekar, Mathematical problems of non-perturbative quantum general relativity. In *Les Houches Summer School on Gravitation and Quantizations, Les Houches, France, Jul 5–Aug 1, 1992*, ed. J. Zinn-Justin and B. Julia (Amsterdam: North-Holland, 1995), gr-qc/9302024.

[203] J. Baez, Generalized measures in gauge theory, *Lett. Math. Phys.* **31** (1994) 213–223.

[204] A. Ashtekar, J. Lewandowski, D. Marolf, J. Mourao and T. Thiemann, Quantization of diffeomorphism invariant theories of connections with local degrees of freedom, *J. Math. Phys.* **36** (1995) 6456, gr-qc/9504018.

[205] R. De Pietri, On the relation between the connection and the loop representation of quantum gravity, *Class. and Quantum Grav.* **14** (1997) 53, gr-qc/9605064.

[206] A. Ashtekar, C. Rovelli and L. Smolin, Gravitons and loops, *Phys. Rev.* **D44** (1991) 1740–1755, hep-th/9202054; J. Iwasaki and C. Rovelli, Gravitons as embroidery on the weave, *Int. J. Mod. Phys.* **D1** (1993) 533; Gravitons from loops: non-perturbative loop-space quantum gravity contains the graviton-physics approximation, *Class. and Quantum Grav.* **11** (1994) 1653; M. Varadarajan, Gravitons from a loop representation of linearized gravity, *Phys. Rev.* **D66** (2002) 024017, gr-qc/0204067.

[207] A. Corichi and J.M. Reyes, A Gaussian weave for kinematical loop quantum gravity, *Int. J. Mod. Phys.* **D10**, (2001) 325, gr-qc/0006067.

Chapter 7: Dynamics and matter

[208] R. Borissov, R. De Pietri and C. Rovelli, Matrix elements of Thiemann's hamiltonian constraint in loop quantum gravity, *Class. and Quantum Grav.* **14** (1997) 2793, gr-qc/9703090.

[209] T. Thiemann, The phoenix project: master constraint programme for loop quantum gravity, gr-qc/0305080.

[210] M. Gaul and C. Rovelli, A generalized hamiltonian constraint operator in loop quantum gravity and its simplest euclidean matrix elements, *Class. and Quantum Grav.* **18** (2001) 1593–1624, gr-qc/0011106.

[211] T. Thiemann, QSD V: Quantum gravity as the natural regulator of the hamiltonian constraint of matter quantum field theories, *Class. and Quantum Grav.* **15** (1998) 1281–1314, gr-qc/9705019.

[212] S. Alexandrov, $SO(4;C)$-covariant Ashtekar–Barbero gravity and the Immirzi parameter, *Class. and Quantum Grav.* **17** (2000) 4255–4268; S. Alexandrov and E.R. Livine, $SU(2)$ loop quantum gravity seen from covariant theory, *Phys. Rev.* **D67** (2003) 044009, gr-qc/0209105.

[213] S. Alexandrov and D.V. Vassilevich, Area spectrum in Lorentz covariant loop gravity, gr-qc/0103105.

[214] B.S. DeWitt, Quantum theory of gravity. I: the canonical theory, *Phys. Rev.* **160** (1967) 1113.

[215] B. Brügmann, R. Gambini and J. Pullin, Jones polynomials for intersecting knots as physical states of quantum gravity, *Nucl. Phys.* **B385** (1992) 587; Knot invariants as nondegenerate quantum geometries, *Phys. Rev. Lett.* **68** (1992) 431; C. Di Bartolo, R. Gambini, J. Griego and J. Pullin, Consistent canonical quantization of general relativity in the space of Vassiliev knot invariants, *Phys. Rev. Lett.* **84** (2000) 2314.

[216] K. Ezawa, Nonperturbative solutions for canonical quantum gravity: an overview, *Phys. Repts.* **286** (1997) 271–348, gr-qc/9601050.

[217] C. Rovelli and L. Smolin, The physical hamiltonian in nonperturbative quantum gravity, *Phys. Rev. Lett.* **72** (1994) 44.

[218] C. Rovelli, Outline of a general covariant quantum field theory and a quantum theory of gravity, *J. Math. Phys.* **36** (1995) 6529.

[219] T. Thiemann, Quantum spin dynamics (QSD), *Class. and Quantum Grav.* **15** (1998) 839–73, gr-qc/9606089; QSD II: The kernel of the Wheeler–DeWitt constraint operator, *Class. and Quantum Grav.* **15** (1998) 875–905, gr-qc/9606090; QSD III: Quantum constraint algebra and physical scalar product in quantum general relativity, *Class. and Quantum Grav.* **15** (1998) 1207–1247, gr-qc/9705017; QSD IV: 2 + 1 euclidean quantum gravity as a model to test 3 + 1 lorentzian quantum gravity, *Class. and Quantum Grav.* **15** (1998) 1249–1280, gr-qc/9705018; QSD VI: Quantum Poincaré algebra and a quantum positivity of energy theorem for canonical quantum gravity, *Class. and Quantum Grav.* **15** (1998) 1463–1485, gr-qc/9705020; QSD VII: Symplectic structures and continuum lattice formulations of gauge field theories, *Class. and Quantum Grav.* **18** (2001) 3293–3338, hep-th/0005232.

[220] L. Smolin, Quantum gravity with a positive cosmological constant, hep-th/0209079; L. Freidel and L. Smolin, The linearization of the Kodama state, hep-th/0310224.

[221] T Thiemann Loop Quantum Gravity: An Inside View, hep-th/0608210.

[222] C. Rovelli and H. Morales-Tecotl, Fermions in quantum gravity, *Phys. Rev. Lett.* **72** (1995) 3642; Loop space representation of quantum fermions and gravity, *Nucl. Phys.* **B451** (1995) 325.

[223] J. Baez and K. Krasnov, Quantization of diffeomorphism-invariant theories with fermions, *J. Math. Phys.* **39** (1998) 1251–1271, hep-th/9703112.

[224] T. Thiemann, Kinematical Hilbert spaces for fermionic and Higgs quantum field theories, *Class. and Quantum Grav.* **15** (1998) 1487–1512, gr-qc/9705021.

[225] M. Montesinos and C. Rovelli, The fermionic contribution to the spectrum of the area operator in nonperturbative quantum gravity, *Class. and Quantum Grav.* **15** (1998) 3795–3801.

[226] S. O. Bilson-Thompson, F. Markopoulou, L. Smolin, hep-th/0603022.

[227] A. Alekseev, A.P. Polychronakos and M. Smedback, On area and entropy of a black hole, *Phys. Lett.* **B574** (2003) 296; A.P. Polychronakos, Area spectrum and quasinormal modes of black holes, hep-th/0304135.

[228] S. Major and L. Smolin, Quantum deformations of quantum gravity, *Nucl. Phys.* **B473** (1996) 267–290, gr-qc/9512020; R. Borissov, S. Major and L. Smolin, The geometry of quantum spin networks, *Class. and Quantum Grav.* **13** (1996) 3183–3196.

[229] L.H. Kauffman, Map coloring, q-deformed spin networks and Turaev–Viro invariants for three manifolds, *Int. J. Mod. Phys.* **B6** (1992) 1765–1794; Erratum, **B6** (1992) 3249.

[230] N. Reshetikhin and V. Turaev, Ribbon graphs and their invariants derived from quantum groups, *Comm. Math. Phys.* **127** (1990) 1–26.

[231] K. Noui and Ph. Roche, Cosmological deformation of Lorentzian spin foam models, *Class. and Quantum Grav.* **20** (2003) 3175–3214, gr-qc/0211109.

Chapter 8: Applications

[232] M. Bojowald and H.A. Morales-Tecotl, Cosmological applications of loop quantum gravity, to appear in *Proc. 5th Mexican School (DGFM), The Early Universe and Observational Cosmology*, gr-qc/0306008.

[233] M. Domagala, L. Lewandowski, Black-hole entropy from quantum geometry, *Class. Quant. Grav.* **21** (2004) 52335243.

[234] K. A. Meissner, Black-hole entropy in loop quantum gravity, *Class. Quant. Grav.* **21** (2004) 52455252.

[235] A. Ashtekar, An Introduction to Loop Quantum Gravity Through Cosmology. gr-qc/0702030.

[236] L. Modesto, Disappearance of Black Hole Singularity in Quantum Gravity, *Phys. Rev.* **D70** (2004) 124009.

[237] A. Ashtekar and M. Bojowald, Quantum geometry and Schwarzschild singularity, *Class. Quant. Grav.* **23** (2006) 391–411.

[238] A. Ashtekar, J. Baez, A. Corichi and K. Krasnov, Quantum geometry and black hole entropy, *Phys. Rev. Lett.* **80** (1998) 904, gr-qc/9710007; A. Ashtekar, J.C. Baez and K. Krasnov, Quantum geometry of isolated horizons and black hole entropy, *Adv. Theor. Math. Phys.* **4** (2001) 1–94, gr-qc/0005126.

[239] C. Rovelli, Black hole entropy from loop quantum gravity, *Phys. Rev. Lett.* **14** (1996) 3288; Loop quantum gravity and black hole physics, *Helv. Phys. Acta.* **69** (1996) 583.

[240] G.H. Hardy and S. Ramanujan, *Proc. London Math. Soc.* **2** (1918) 75.

[241] G. Amelino-Camelia, Are we at dawn with quantum gravity phenomenology? *Lectures given at 35th Winter School of Theoretical Physics: From Cosmology to Quantum Gravity, Polanica, Poland, 2–12 Feb, 1999, Lecture Notes in Physics* **541** (2000) 1–49, gr-qc/9910089.

[242] C. Rovelli and S. Speziale, Reconcile Planck-scale discreteness and the Lorentz–Fitzgerald contraction, *Phys. Rev.* **D67** (2003) 064019.

[243] M. Bojowald, Absence of singularity in loop quantum cosmology, *Phys. Rev. Lett.* **86** (2001) 5227–5230, gr-qc/0102069.

[244] M. Bojowald, Inflation from quantum geometry, *Phys. Rev. Lett.* **89** (2002) 261301, gr-qc/0206054.

[245] S.W. Hawking, Black holes in general relativity, *Comm. Math. Phys.* **25** (1972) 152.

[246] J.M. Bardeen, B. Carter and S.W. Hawking, The four laws of black hole mechanics, *Comm. Math. Phys.* **31** (1973) 161.

[247] J.D. Bekenstein, Black holes and the second law, *Nuovo Cimento Lett.* **4** (1972) 737–740; Black holes and entropy, *Phys. Rev.* **D7** (1973) 2333–2346; Generalized second law for thermodynamics in black hole physics, *Phys. Rev.* **D9** (1974) 3292–3300.

[248] S.W. Hawking, Black hole explosions, *Nature* **248** (1974) 30; Particle creation by black holes, *Comm. Math. Phys.* **43** (1975) 199.

[249] K. Krasnov, Geometrical entropy from loop quantum gravity, *Phys. Rev.* **D55** (1997) 3505; On statistical mechanics of gravitational systems, *Gen. Rel. Grav.* **30** (1998) 53–68, gr-qc/9605047; On statistical mechanics of a Schwarzschild black hole, *Gen. Rel. Grav.* **30** (1998) 53.

[250] J.W. York, Dynamical origin of black hole radiance, *Phys. Rev.* **D28** (1983) 2929.

[251] G. 't Hooft, Horizon operator approach to black hole quantization, gr-qc/9402037; L. Susskind, Some speculations about black hole entropy in string theory, hep-th/9309145; L. Susskind, L. Thorlacius and J. Uglum, *Phys. Rev.* **D48** (1993) 3743; C. Teitelboim, Statistical thermodynamics of a black hole in terms of surface fields, *Phys. Rev.* **D53** (1996) 2870–2873; A. Buonanno, M. Gattobigio, M. Maggiore, L. Pilo and C. Ungarelli, Effective Lagrangian for quantum black holes, *Nucl. Phys.* **B451** (1995) 677.

[252] A. Ashtekar, C. Beetle, O. Dreyer *et al.*, Isolated horizons and their applications, *Phys. Rev. Lett.* **85** (2000) 3564–3567, gr-qc/0006006; A. Ashtekar,

Classical and quantum physics of isolated horizons, *Lect. Notes Phys.* **541** (2000) 50–70.

[253] L. Smolin, Linking topological quantum field theory and nonperturbative quantum gravity, *J. Math. Phys.* **36** (1995) 6417–6455.

[254] A.P. Balachandran, L. Chandar and A. Momen, Edge states in gravity and black hole physics, *Nucl. Phys.* **B461** (1996) 581–596; A. Momen, Edge dynamics for BF theories and gravity, *Phys. Lett.* **394** (1997) 269–274; S. Carlip, Black hole entropy from conformal field theory in any dimension, *Phys. Rev. Lett.* **82** (1999) 2828–2831.

[255] S. Hod, Bohr's correspondence principle and the area spectrum of quantum black holes, *Phys. Rev. Lett.* **81** (1998) 4293; *Gen. Rel. Grav.* **31** (1999) 1639; Kerr black hole quasinormal frequencies, *Phys. Rev.* **D67** (2003) 081501.

[256] O. Dreyer, Quasinormal modes, the area spectrum, and black hole entropy, *Phys. Rev. Lett.* **90** (2003) 081301.

[257] A. Corichi, On quasinormal modes, black hole entropy, and quantum geometry, *Phys. Rev.* **D67** (2003) 087502, gr-qc/0212126.

[258] J.D. Bekenstein and V.F. Mukhanov, Spectroscopy of the quantum black hole, *Phys. Lett.* **B360** (1995) 7–12.

[259] L. Smolin, Macroscopic deviations from Hawking radiation? In *Matters of Gravity* 7, gr-qc/9602001.

[260] M. Barreira, M. Carfora and C. Rovelli, Physics with loop quantum gravity: radiation from quantum black hole, *Gen. Rel. Grav.* **28** (1996) 1293.

[261] R. Gambini and J. Pullin, Nonstandard optics from quantum spacetime, *Phys. Rev.* **D59** (1999) 124021, gr-qc/9809038; Quantum gravity experimental physics?, *Gen. Rel. Grav.* **31** (1999) 1631–1637.

[262] J. Alfaro, H.A. Morales-Tecotl and L.F. Urrutia, Quantum gravity corrections to neutrino propagation, *Phys. Rev. Lett.* **84** (2000) 2318–2321, gr-qc/9909079; Loop quantum gravity and light propagation, *Phys. Rev.* **D65** (2002) 103509, hep-th/0108061.

[263] C. Kozameh and F. Parisi, Lorentz invariance and the semiclassical approximation of loop quantum gravity, gr-qc/0310014.

[264] J. Collins, A. Perez, D. Sudarsky, L. Urrutia, H. Vucetich, Lorentz invariance: An Additional fine tuning problem. *Phys. Rev. Lett.* **93** (2004) 191301.

[265] G. Amelino-Camelia, J. Ellis, N.E. Mavromatos, D.V. Nanopoulos and S. Sarkar, Potential sensitivity of gamma-ray buster observations to wave disperion in vacuo, *Nature* **393** (1998) 763, astro-ph/9712103; G. Amelino-Camelia, An interferometric gravitational wave detector as a quantum gravity apparatus, *Nature* **398** (1999) 216.

[266] J. Ellis, J. Hagelin, D. Nanopoulos and M. Srednicki, Search for violations of quantum mechanics, *Nucl. Phys.* **B241** (1984) 381; J. Ellis, N.E. Mavromatos and D.V. Nanopoulos, Testing quantum mechanics in the neutral kaon system, *Phys. Lett.* **B293** (1992) 142; I.C. Percival

and W.T. Strunz, Detection of space-time fluctuations by a model matter interferometer, quant-ph/9607011; G. Amelino-Camelia, J. Ellis, N.E. Mavromatos and D.V. Nanopoulos, Distance measurement and wave dispersion in a Liouville string approach to quantum gravity, *Int. J. Mod. Phys.* **A12** (1997) 607.

Chapter 9: Spinfoams

[267] J. Barrett, Quantum gravity as topological quantum field theory, *J. Math. Phys.* **36** (1995) 6161–6179, gr-qc/9506070.

[268] M.E. Peskin and D.V. Schroeder, *An Introduction to Quantum Field Theory* (Reading: Addison Wesley, 1995).

[269] J.W. Barrett and L. Crane, Relativistic spin networks and quantum gravity, *J. Math. Phys.* **39** (1998) 3296–3302.

[270] M. Bojowald and A. Perez, Spin foam quantization and anomalies, gr-qc/0303026.

[271] J.C. Baez, J.D. Christensen, T.R. Halford and D.C. Tsang, Spin foam models of riemannian quantum gravity, *Class. and Quantum Grav.* **19** (2002) 4627–4648, gr-qc/0202017.

[272] G. Roepstorff, *Path Integral Approach to Quantum Physics: An Introduction* (Berlin: Springer-Verlag, 1994).

[273] A.S. Wightman, Quantum field theory in terms of vacuum expectation values, *Phys. Rev.* **101** (1956) 860; R.F. Streater and A.S. Wightman, *PCT, Spin and Statistics, and All That*, Mathematical Physics Monograph Series (Reading MA: Benjamin-Cummings, 1964); K. Osterwalder and R. Schrader, Axioms for euclidean Green's functions, *Comm. Math. Phys.* **31** (1973) 83; Axioms for euclidean Green's functions: 2, **42** (1975) 281.

[274] A. Perez and C. Rovelli, Observables in quantum gravity, gr-qc/0104034.

[275] A. Ashtekar, D. Marolf, J. Mourao and T. Thiemann, Constructing Hamiltonian quantum theories from path integrals in a diffeomorphism invariant context, *Class. and Quantum Grav.* **17** (2000) 4919–4940, quant-ph/9904094.

[276] J. Baez, Strings, loops, knots and gauge fields. In *Knots and Quantum Gravity*, ed. J. Baez (Oxford: Oxford University Press, 1994).

[277] J. Iwasaki, A reformulation of the Ponzano–Regge quantum gravity model in terms of surfaces, gr-qc/9410010; A definition of the Ponzano–Regge quantum gravity model in terms of surfaces, *J. Math. Phys.* **36** (1995) 6288.

[278] M. Reisenberger, Worldsheet formulations of gauge theories and gravity, talk given at the *7th Marcel Grossmann Meeting Stanford, July 1994*, gr-qc/9412035; A lattice worldsheet sum for 4-d Euclidean general relativity, gr-qc/9711052.

[279] M. Reisenberger and C. Rovelli, Sum over surfaces form of loop quantum gravity, *Phys. Rev.* **D56** (1997) 3490–3508, gr-qc/9612035.

[280] C. Rovelli, Quantum gravity as a 'sum over surfaces', *Nucl. Phys. (Proc. Suppl.)* **B57** (1997) 28–43.

[281] C. Rovelli, The projector on physical states in loop quantum gravity, *Phys. Rev.* **D59** (1999) 104015, gr-qc/9806121.

[282] F. Markopoulou, Dual formulation of spin network evolution, gr-qc/9704013.

[283] G. Ponzano and T. Regge, Semiclassical limit of Racah coefficients. In *Spectroscopy and Group Theoretical Methods in Physics*, ed. F. Bloch (Amsterdam: North-Holland, 1968).

[284] L. Crane and D. Yetter, On the classical limit of the balanced state sum, gr-qc/9712087; J.W. Barrett, The classical evaluation of relativistic spin networks, *Adv. Theor. Math. Phys.* **2** (1998) 593–600; J.W. Barrett and R.M. Williams, The asymptotics of an amplitude for the 4-simplex, *Adv. Theor. Math. Phys.* **3** (1999) 209–214; L. Freidel and K. Krasnov, Simple spin networks as Feynman graphs, *J. Math. Phys.* **41** (2000) 1681–1690.

[285] T. Regge, General relativity without coordinates, *Nuovo Cimento* **19** (1961) 558–571.

[286] V.G. Turaev and O.Y. Viro, State sum invariants of 3-manifolds and quantum 6j symbols, *Topology* **31** (1992) 865; V.G. Turaev, *Quantum Invariants of Knots and 3-Manifolds* (New York: de Gruyter, 1994).

[287] D. V. Boulatov, A model of three-dimensional lattice gravity, *Mod. Phys. Lett.* **A7** (1992) 1629–1648.

[288] E. Brézin, C. Itzykson, G. Parisi and J. Zuber, *Comm. Math. Phys.* **59** (1978) 35; F. David, *Nucl. Phys.* **B257** (1985) 45; J. Ambjorn, B. Durhuus and J. Frölich, *Nucl. Phys.* **B257** (1985) 433; V.A. Kazakov, I.K. Kostov and A.A. Migdal, *Phys. Lett.* **157** (1985) 295; D.V. Boulatov, V.A. Kazakov, I.K. Kostov and A.A. Migdal, *Nucl. Phys.* **B275** (1986) 641; M. Douglas and S. Shenker, *Nucl. Phys.* **B335** (1990) 635; D. Gross and A.A. Migdal, *Phys. Rev. Lett.* **64** (1990) 635; E. Brezin and V.A. Kazakov, *Phys. Lett.* **B236** (1990) 144; O. Alvarez, E. Marinari and P. Windey, *Random Surfaces and Quantum Gravity* (New York: Plenum Press, 1991).

[289] H. Ooguri, Topological lattice models in four dimensions, *Mod. Phys. Lett.* **A7** (1992) 2799.

[290] V.G. Turaev, *Quantum Invariants of Knots and 3-manifolds* (New York: de Gruyter, 1994).

[291] A.S. Schwartz, The partition function of degenerate quadratic functionals and Ray–Singer invariants, *Lett. Math. Phys.* **2** (1978) 247–252; G. Horowitz, Exactly soluble diffeomorphism-invariant theories, *Comm. Math. Phys.* **125** (1989) 417–437; D. Birmingham, M. Blau, M. Rakowski and G. Thompson, Topological field theories, *Phys. Rep.* **209** (1991) 129–340.

[292] L. Crane and D. Yetter, A categorical construction of 4-D topological quantum field theories. In *Quantum Topology*, ed. L. Kauffman and R. Baadhio (Singapore: World Scientific, 1993), pp. 120–130, hep-th/9301062; L.

Crane, L. Kauffman and D. Yetter, State-sum invariants of 4-manifolds I, *J. Knot. Theor. Ramifications.* **6** (1997) 177–234, hep-th/9409167; J.D. Roberts, Skein theory and the Turaev–Viro invariants, *Topology* **34** (1995) 771–787.

[293] C. Rovelli, Basis of the Ponzano–Regge–Turaev–Viro–Ooguri quantum gravity model is the loop representation basis, *Phys. Rev.* **D48** (1993) 2702.

[294] R. De Pietri and L. Freidel, $SO(4)$ Plebanski action and relativistic state sum models, *Class. and Quantum Grav.* **16** (1999) 2187–2196, gr-qc/9804071.

[295] M. Reisenberger, Classical Euclidean GR from left-handed area = right-handed area, *Class. and Quantum Grav.* **16** (1999) 1357–1371, gr-qc/9804061; On relativistic spin network vertices, *J. Math. Phys.* **40** (1999) 2046–2054, gr-qc/9711052.

[296] A. Perez, Spin foam quantization of Plebanski's action, *Adv. Theor. Math. Phys.* **5** (2002) 947–968, gr-qc/0203058.

[297] A. Barbieri, Quantum tetrahedron and spin networks, *Nucl. Phys.* **B518** (1998) 714–728.

[298] J. Baez and J. Barrett, The quantum tetrahedron in 3 and 4d, *Adv. Theor. Math. Phys.* **3** (1999) 815, gr-qc/9903060.

[299] J. Baez, Spin foam models, *Class. and Quantum Grav.* **15** (1998) 1827–1858, gr-qc/9709052.

[300] R. De Pietri, L. Freidel, K. Krasnov and C. Rovelli, Barrett–Crane model from a Boulatov–Ooguri field theory over a homogeneous space, *Nucl. Phys.* **B574** (2000) 785–806, hep-th/9907154.

[301] A. Perez and C. Rovelli, A spinfoam model without bubble divergences, *Nucl. Phys.* **B599** (2001) 255–282.

[302] D. Oriti and R.M. Williams, Gluing 4-simplices: a derivation of the Barrett–Crane spinfoam model for Euclidean quantum gravity, *Phys. Rev.* **D63** (2001) 024022.

[303] M. Reisenberger and C. Rovelli, Spinfoam models as Feynman diagrams, gr-qc/0002083; Spacetime as a Feynman diagram: the connection formulation, *Class. and Quantum Grav.* **18** (2001) 121–140, gr-qc/0002095.

[304] A. Mikovic, Quantum field theory of spin networks, gr-qc/0102110.

[305] A. Perez, Finiteness of a spinfoam model for euclidean GR, *Nucl. Phys.* **B599** (2001) 427–434.

[306] L. Freidel, Group field theory: An Overview. *Int. J. Theor. Phys.* **44** (2005) 1769–1783.

[307] D. Oriti, The group field theory approach to quantum gravity, gr-qc/0607032.

[308] L. Freidel, E.R. Levine, 3d Quantum Gravity and Effective Non-Commutative Quantum Field Theory, *Phys. Rev. Lett.* **96** (2006) 221301.

[309] L. Freidel, and D. Louapre, Nonperturbative summation over 3d discrete topologies, hep-th/0211026.

[310] L. Freidel, A Ponzano–Regge model of lorentzian 3-dimensional gravity, *Nucl. Phys. Proc. Suppl.* **88** (2000) 237–240, gr-qc/0102098.

[311] J.W. Barrett and L. Crane, A lorentzian signature model for quantum general relativity, *Class. and Quantum Grav.* **17** (2000) 3101–3118, gr-qc/9904025.

[312] A. Perez and C. Rovelli, Spin foam model for lorentzian general relativity, *Phys. Rev.* **D63** (2001) 041501, gr-qc/0009021.

[313] L. Crane, A. Perez and C. Rovelli, Perturbative finiteness in spin foam quantum gravity, *Phys. Rev. Lett.* **87** (2001) 181301; A finiteness proof for the lorentzian state sum spin foam model for quantum general relativity, gr-qc/0104057.

[314] A. Perez and C. Rovelli, 3 + 1 spin foam model of quantum gravity with spacelike and timelike components, *Phys. Rev.* **D64** (2001) 064002, gr-qc/0011037.

[315] L. Freidel, E.R. Livine and C. Rovelli, Spectra of length and area in 2 + 1 lorentzian loop quantum gravity, *Class. and Quantum Grav.* **20** (2003) 1463–1478, gr-qc/0212077.

[316] L. Freidel and K. Krasnov, Spin foam models and the classical action principle, *Adv. Theor. Phys.* **2** (1998), 1221–1285, hep-th/9807092.

[317] J. Ambjorn, J. Jurkiewicz and R. Loll, Lorentzian and euclidean quantum gravity: analytical and numerical results, hep-th/0001124; A nonperturbative lorentzian path integral for gravity, *Phys. Rev. Lett.* **85** (2000) 924, hep-th/0002050; J. Ambjorn, A. Dasgupta, J. Jurkiewicz and R. Loll, A lorentzian cure for euclidean troubles, *Nucl. Phys. Proc. Suppl.* **106** (2002) 977–979, hep-th/0201104; R. Loll and A. Dasgupta, A proper time cure for the conformal sickness in quantum gravity, *Nucl. Phys.* **B606** (2001) 357–379, hep-th/0103186.

[318] E.R. Livine and D. Oriti, Implementing causality in the spin foam quantum geometry, *Nucl. Phys.* **B663** (2003) 231–279, gr-qc/0210064.

[319] M. Arnsdorf, Relating covariant and canonical approaches to triangulated models of quantum gravity, *Class. and Quantum Grav.* **19** (2002) 1065–1092, gr-qc/0110026.

[320] E.R. Livine, Projected spin networks for lorentz connection: linking spin foams and loop gravity, *Class. and Quantum Grav.* **19** (2002) 5525–5542, gr-qc/0207084.

[321] K. Noui, A. Perez, Three-dimensional loop quantum gravity: Physical scalar product and spin foam models. *Class. Quant. Grav.* **22** (2005) 1739–1762.

[322] R. Capovilla, M. Montesinos, V.A. Prieto and E. Rojas, BF gravity and the Immirzi parameter, *Class. and Quantum Grav.* **18** (2001) L49–L52.

[323] T. Thiemann and O. Winkler, Coherent states for canonical quantum general relativity and the infinite tensor product extension, *Nucl. Phys.* **B606** (2001) 401–440, gr-qc/0102038.

[324] F. Markopoulou, An insider's guide to quantum causal histories, *Nucl. Phys. Proc. Suppl.* **88** (2000) 308–313, hep-th/9912137.

[325] A. Mikovic, Spin foam models of matter coupled to gravity, *Class. and Quantum Grav.* **19** (2002) 2335; Spinfoam models of Yang–Mills theory coupled to gravity, *Class. and Quantum Grav.* **20** (2003) 239–246.

[326] D. Oriti and H. Pfeiffer, A spin foam model for pure gauge theory coupled to quantum gravity, *Phys. Rev.* **D66** (2002) 124010.

[327] N.J. Vilenkin, *Special Functions and the Theory of Group Representations* (Providence, Rhode Island: American Mathematical Society, 1968).

[328] W. Ruhl, *The Lorentz Group and Harmonic ' Analysis* (New York: WA Benjamin Inc., 1970).

[329] L. Modesto, C. Rovelli, Particle scattering in loop quantum gravity, *Phys. Rev. Lett.* **95** (2005) 191301.

[330] C. Rovelli, Graviton propagator from background-independent quantum gravity, *Phys. Rev. Lett.* **97**, 151301 (2006) E. Bianchi, L. Modesto, C. Rovelli and S. Speziale, Graviton propagator in loop quantum gravity, *Class. Quant. Grav.* **23**, 6989 (2006).

[331] S. Speziale, Towards the graviton from spinfoams: The 3-D toy model. *JHEP* (2006) 0605:039. E.R. Livine, S. Speziale, J.L. Willis, Towards the graviton from spinfoams: Higher order corrections in the 3-D toy model. *Phys. Rev.* **D75** (2007) 024038. E.R. Livine, S. Speziale, Group Integral Techniques for the Spinfoam Graviton Propagator, gr-qc/0608131.

Chapter 10: Conclusion, and Appendices

[332] G. Galilei, *Dialogo dei massimi system* (Firenze, 1632).

[333] D.M. Brink and G.R. Satchler, *Angular Momentum* (Oxford: Clarendon Press, 1968).

[334] R. Penrose, In *Combinatorial Mathematics and its Application*, ed. D. Welsh (New York: Academic Press, 1971).

[335] L.H. Kauffman, *Int. J. Mod. Phys.* **A5** (1990) 93.

[336] L.H. Kauffman, In *Knots, Topology and Quantum Field Theories*, ed. L. Lusanna (Singapore: World Scientific, 1991).

[337] J.P. Moussoris, in *Advances in Twistor Theory, Research Notes in Mathematics*, ed. J.P. Huston and R.S. Ward (Boston: Pitman, 1979), pp. 308–312.

[338] I. Levinson, *Liet. TSR Mokslu. Acad. Darbai B Ser.* **2** (1956) 17.

[339] A.P. Yutsin, J.B. Levinson and V.V. Vanagas, *Mathematical Apparatus of the Theory of Angular Momentum* (Jerusalem: Israel Program for Scientific Translation, 1962).

[340] A. Einstein, Naeherungsweise Integration der Feldgleichungen der Gravitation, *Preussische Akademie der Wissenschaften (Berlin) Sitzungsberichte* (1916), p. 688.

[341] O. Klein, Zur Fünfdimensionalen Darstellung der Relativitaetstheorie, *Z. für Physik* **46** (1927) 188.

[342] L. Rosenfeld, Zur Quantelung der Wellenfelder, *Ann. der Physik* **5** (1930) 113; Über die Gravitationswirkungen des Lichtes, *Z. für Physik* **65** (1930) 589.

[343] M. Fierz, *Hel. Physica Acta.* **12** (1939) 3; W. Pauli and M. Fierz, On relativistic field equations of particles with arbitrary spin in an electromagnetic field, *Hel. Physica Acta* **12** (1939) 297.

[344] D.I. Blokhintsev and F.M. Gal'perin, *Pod Znamenem Marxisma* **6** (1934) 147.

[345] W. Heisenberg, *Z. für Physik* **110** (1938) 251.

[346] J. Stachel, Early history of quantum gravity (1916–1940), Presented at the *HGR5, Notre Dame, July 1999*; Early history of quantum gravity. In *Black Holes, Gravitational Radiation and the Universe*, ed. B.R. Iyer and B. Bhawal (Netherlands: Kluwer Academic Publishers, 1999).

[347] M.P. Bronstein, Quantentheories schwacher Gravitationsfelder, *Physikalische Z. der Sowietunion* **9** (1936) 140.

[348] G.E. Gorelik, First steps of quantum gravity and the planck values. In *Studies in the History of General Relativity. [Einstein Studies, Vol. 3]*, ed. J. Eisenstaedt and A.J. Kox (Boston: Birkhäuser, 1992), pp. 364–379; G.E. Gorelik and V.Y. Frenkel, *Matvei Petrovic Bronstein and the Soviet Theoretical Physics in the Thirties* (Boston: Birkhäuser-Verlag, 1994).

[349] P.G. Bergmann, Non-linear field theories, *Phys. Rev.* **75** (1949) 680; Nonlinear field theories II: canonical equations and quantization, *Rev. Mod. Phys.* **21** (1949).

[350] P.G. Bergmann, *Nuovo Cimento* **3** (1956) 1177.

[351] E.T. Newman and P.G. Bergmann, Observables in singular theories by systematic approximation, *Rev. Mod. Phys.* **29** (1957) 443.

[352] S. Gupta, *Proc. Phys. Soc.* **A65** (1952) 608.

[353] C. Misner, Feynman quantization of general relativity, *Rev. Mod. Phys.* **29** (1957) 497.

[354] P. Bergmann, The canonical formulation of general relativistic theories: the early years, 1930–1959. In *Einstein and the History of General Relativity*, ed. D. Howard and J. Stachel (Boston: Birkhäuser, 1989).

[355] R. Feynman, Quantum theory of gravitation, *Acta Physica Polonica* **24** (1963) 697.

[356] B. DeWitt, In *Conférence Internationale sur les Théories Relativistes de la Gravitation*, ed. Gauthier-Villars (Warsaw: Editions Scientifiques de Pologne, 1964).

[357] P.G. Bergmann and A. Komar, The coordinate group symmetries of general relativity, *Int. J. Theor. Phys.* **5** (1972) 15.

[358] A. Peres, *Nuovo Cimento* **26** (1962) 53.

[359] J.A. Wheeler, Geometrodynamics and the issue of the final state. In *Relativity, Groups and Topology*, ed. C. DeWitt and B.S. DeWitt (New York and London: Gordon and Breach, 1964), p. 316.

[360] B.S. DeWitt, *Phys. Rev. Lett.* **12** (1964) 742; In *Dynamical Theory of Groups and Fields* (New York: Wiley, 1965).

[361] L.D. Faddeev and V.N. Popov, Feynman diagrams for the Yang–Mills field, *Phys. Lett.* **25B** (1967) 30.

[362] M. Veltman, in *Proc. 6th Int. Symp. Electron and Photon Interactions at High Energies*, ed. H. Rollnik and W. Pfeil (Amsterdam: North Holland, 1975).

[363] L.D. Faddeev and V.N. Popov, Perturbation theory for gauge invariant fields, *Kiev Inst. Theor. Phys. Acad. Sci.* 67-036 (Fermilab Publication 72-057-T).

[364] B.S. DeWitt, Quantum theory of gravity. II: The manifestly covariant theory, *Phys. Rev.* **162** (1967) 1195; Quantum theory of gravity. III: Applications of the covariant theory, *Phys. Rev.* **162** (1967) 1239.

[365] R. Penrose, Twistor theory, *J. Math. Phys.* **8** (1967) 345.

[366] C. Misner, Quantum cosmology, *Phys. Rev.* **186** (1969) 1319.

[367] B. Zumino, Effective lagrangians and broken symmetries. In *Brandeis University Lectures On Elementary Particles And Quantum Field Theory, Vol 2*, ed. S. Deser (MIT Press, Cambridge MA, 1971), pp. 437–500.

[368] G. 't Hooft, Renormalizable lagrangians for massive Yang–Mills fields, *Nucl. Phys.* **B35** (1971) 167; G. 't Hooft and M. Veltman, Regularization and renormalization of gauge fields, *Nucl. Phys.* **B44** (1972) 189.

[369] D. Finkelstein, Space-time code, *Phys. Rev.* **184** (1969) 1261–1279.

[370] G. t'Hooft, An algorithm for the poles at dimension four in the dimensional regularization, *Nucl. Phys.* **B62** (1973) 444; G. t'Hooft and M. Veltman, One-loop divergencies in the theory of gravitation, *Ann. Inst. Poincaré* **20** (1974) 69; S. Deser and P. Van Nieuwenhuizen, One loop divergences of the quantized Einstein–Maxwell fields, *Phys. Rev.* **D10** (1974) 401; Non-renormalizability of the quantized Dirac–Einstein system, *Phys. Rev.* **D10** (1974) 411.

[371] W.G. Unruh, Notes on black hole evaporation, *Phys. Rev.* **D14** (1976) 870.

[372] G. Parisi, The theory of non-renormalizable interactions. 1: the large-N expansion, *Nucl. Phys.* **B100** (1975) 368.

[373] S. Ferrara, P. van Nieuwenhuizen and D.Z. Freedman, Progress toward a theory of supergravity, *Phys. Rev.* **D13** (1976) 3214; S. Deser and P. Zumino, Consistent supergravity, *Phys. Lett.* **B62** (1976) 335. For a review, see P. van Nieuwenhuizen, Supergravity, *Physics Reports* **68** (1981) 189.

[374] L. Brink, P. Di Vecchia and P. Howe, A locally supersymmetric and reparameterization-invariant action for the spinning string, *Phys. Lett.* **B65** (1976) 471–474; S. Deser and B. Zumino, A complete action for the spinning string, *Phys. Lett.* **B65** (1976) 369.

[375] K.S. Stelle, Renormalization of higher derivatives quantum gravity, *Phys. Rev.* **D16** (1977) 953.

[376] J.B. Hartle and S.W. Hawking, Path integral derivation of the black hole radiance, *Phys. Rev.* **D13** (1976) 2188.

[377] A.M. Polyakov, Quantum geometry of the bosonic string, *Phys. Lett.* **103B** (1981) 207; Quantum geometry of the fermionic string, *Phys. Lett.* **103B** (1981) 211.

[378] G. Horowitz and A. Strominger, in *10th Int. Conf. on General Relativity and Gravitation – Contributed Papers*, ed. B. Bertotti, F. de Felice and A. Pascolini (Padova: Università di Padova, 1983).

[379] C.J. Isham, Quantum logic and the histories approach to quantum theory, *J. Math. Phys.* **35** (1994) 2157, gr-qc/9308006.

[380] R.D. Sorkin, Posets as lattice topologies. In *General Relativity and Gravitation: Proceedings of the GR10 Conference. Volume I*, ed. B. Bertotti, F. de Felice and A. Pascolini, (Rome: Consiglio Nazionale Delle Ricerche, 1983), p. 635.

[381] M.B. Green and J.H. Schwarz, Anomaly cancellation in supersymmetric $d = 10$ gauge theory requires $SO(32)$, *Phys. Lett.* **149B** (1984) 117.

[382] D.J. Gross, J.A. Harvey, E. Martinec and R. Rohm, The heterotic string, *Phys. Rev. Lett.* **54** (1985) 502–505.

[383] P. Candelas, G.T. Horowitz, A. Strominger and E. Witten, Vacuum configurations for superstrings, *Nucl. Phys.* **B258** (1985) 46.

[384] A.A. Belavin, A.M. Polyakov and A.B. Zamolodchikov, Infinite conformal symmetry in two-dimensional quantum field theory, *Nucl. Phys.* **B241** (1984) 333.

[385] R. Penrose, Gravity and state vector reduction. In *Quantum Concepts in Space and Time*, ed. R. Penrose and C.J. Isham (Oxford: Clarendon Press, 1986), p. 129.

[386] G.T. Horowitz, J. Lykken, R. Rohm and A. Strominger, A purely cubic action for string field theory, *Phys. Rev. Lett.* **57** (1986) 283.

[387] K. Fredenhagen and R. Haag, Generally covariant quantum field theory and scaling limits, *Comm. Math. Phys.* **108** (1987) 91.

[388] H. Sato and T. Nakamura, eds., *Marcel Grossmann Meeting on General Relativity* (Singapore: World Scientific, 1992).

[389] E. Witten, Topological quantum field theory, *Comm. Math. Phys.* **117** (1988) 353.

[390] E. Witten, Quantum field theory and the Jones polynomial, *Comm. Math. Phys.* **121** (1989) 351.

[391] M.F. Atiyah, Topological quantum field theories, *Publ. Math. Inst. Hautes Etudes Sci. Paris* **68** (1989) 175; *The Geometry and Physics of Knots*, ed. Accademia Nazionale dei Lincei (Cambridge: Cambridge University Press, 1990).

[392] G.T. Horowitz, Exactly soluble diffeomorphism invariant theories, *Comm. Math. Phys.* **125** (1989) 417.

[393] E. Witten, $(2 + 1)$-dimensional gravity as an exactly soluble system, *Nucl. Phys.* **B311** (1988) 46.

[394] S. Carlip, Lectures on $(2 + 1)$-dimensional gravity, (lecture given at the *First Seoul Workshop on Gravity and Cosmology, February 24–25, 1995*).

[395] S. Deser and R. Jackiw, Three-dimensional cosmological gravity: dynamics of constant curvature, *Ann. Phys.* **153** (1984) 405; S. Deser, R. Jackiw and G. 't Hooft, Three-dimensional Einstein gravity: dynamics of flat space, *Ann. Phys.* **152** (1984) 220; A. Achucarro and P.K. Townsend, A Chern–Simon action for three-dimensional antidesitter supergravity theories, *Phys. Lett.* **B180** (1986) 89.

[396] D. Amati, M. Ciafaloni and G. Veneziano, Can spacetime be probed below the string size?, *Phys. Lett.* **B216** (1989) 41.

[397] D. Gross and A. Migdal, Nonperturbative two-dimensional quantum gravity, *Phys. Rev. Lett.* **64** (1990) 635; M. Douglas and S. Shenker, *Nucl. Phys.* **B335** (1990) 635; E. Brezin and V.A. Kazakov, *Phys. Lett.* **B236** (1990) 144; *Random Surfaces and Quantum Gravity*, ed. O. Alvarez, E. Marinari and P. Windey (New York: Plenum Press, 1991).

[398] C.G. Callan, B.S. Giddings, J.A. Harvey and A. Strominger, Evanescent black holes, *Phys. Rev.* **D45** (1992) 1005.

[399] C.W. Misner, K.S. Thorne and J.A. Wheeler, *Gravitation* (San Francisco: Freeman, 1973).

[400] G. 'tHooft, Dimensional reduction in quantum gravity, Utrecht Preprint THU-93/26, gr-qc/9310026; L. Susskind, The world as a hologram, *J. Math. Phys.* **36** (1995) 6377.

[401] A.H. Chamseddine and A. Connes, Universal formula for noncommutative geometry actions: unification of gravity and the standard model, *Phys. Rev. Lett.* **24** (1996) 4868; The spectral action principle, *Comm. Math. Phys.* **186** (1997) 731.

[402] J. Polchinski, Dirichlet branes and Ramon–Ramon charges, *Phys. Rev. Lett.* **75** (1995) 4724.

[403] C.M. Hull and P.K. Townsend, Unity of superstring dualities, *Nucl. Phys.* **B438** (1995) 109.

[404] T. Banks, W. Fischler, S.H. Shenker and L. Susskind, M-theory as a matrix model: a conjecture, *Phys. Rev.* **D55** (1997) 5112.

[405] M.J. Duff, M-Theory (the theory formerly known as strings), *Int. J. Mod. Phys.* **A11** (1996) 5623.

[406] S. Frittelli, C. Kozameh and E.T. Newman, GR via characteristic surfaces, *J. Math. Phys.* **5** (1995) 4984, 5005, 6397; T. Newman, in *On Einstein's Path*, ed. A. Harvey (New York, Berlin, Heidelberg: Springer-Verlag, 1999).

[407] A. Strominger and G. Vafa, Microscopic origin of the Bekenstein–Hawking entropy, *Phys. Lett.* **B379** (1996) 99; G. Horowitz and A. Strominger, Black strings and p-branes, *Nucl. Phys.* **B360** (1991) 197; J. Maldacena and A. Strominger, Black hole grey body factor and D-brane spectroscopy, *Phys. Rev.* **D55** (1997) 861; G. Horowitz, Quantum states of black holes. In *Proc. Symp. Black Holes and Relativistic Stars, in Honor of S. Chandrasekhar, December 1996*; gr-qc/9704072.

[408] A. Connes, M.R. Douglas and A. Schwarz, Noncommutative geometry and matrix theory: compactification on tori, *JHEP* **9802** (1998) 003.

[409] J. Ambjorn, M. Carfora and A. Marzuoli, *The Geometry of Dynamical Triangulations, Lecture Notes in Physics*, (Berlin: Springer-Verlag, 1997).

[410] J.M. Maldacena, The large-N limit of superconformal field theories and supergravity, *Adv. Theor. Math. Phys.* **2** (1998) 231; *Int. J. Theor. Phys.* **38** (1999) 1113; E. Witten, Anti-deSitter space and holography, *Adv. Theor. Math. Phys.* **2** (1998) 253.

[411] M. Gasperini and G. Veneziano, Pre-Big Bang in string cosmology, *Astropart. Phys.* **1** (1993) 317.

[412] P. Bergmann, in *Conférence Internationale sur les Théories Relativistes de la Gravitation*, ed. Gauthier-Villars (Warsaw: Scientifiques de Pologne, 1964).

[413] P. van Nieuwenhuizen, in *Proc. First Marcel Grossmann Meeting on General Relativity*, ed. R. Ruffini (Amsterdam: North Holland, 1977).

[414] P. Bergmann, in *Cosmology and Gravitation*, ed. P. Bergmann and V. De Sabbata (New York: Plenum Press, 1980).

Index